Disruptive Technologies and Digital Transformations for Society 5.0

Disruptive technologies and digital transformations for Society 5.0 aims to report innovations to enable a futuristic society in which new values and services are created continuously, making people's lives more conformable and sustainable. It aims to present how problems can be solved in different areas, including mobility, health, agriculture, food, manufacturing, disaster prevention, and energy to name a few. Society 5.0 framework is based on data captured by real-world sensors and sent to the virtual cloud world for Artificial intelligence (AI)-based analysis, which in turn will return to the real world in physical form through robots, machines, and motor vehicles. People, objects, and systems are all connected in Society 5.0 and converge in cyber and physical space to collect a large amount of data from a variety of sources using sensors and devices. In Society 5.0, new values created by social innovation eliminate regional, age, gender, and language disparities and enable the delivery of personalized products and services that meet many individuals and potential needs. Digital transformation marks a radical rethinking of how an organization uses technology, people, and processes to fundamentally change business performance. Disruptive technologies including AI, affective computing, Blockchain, biological computing, cloud computing, emotion theory, human-computer interaction, Internet of Things (IoT) predictive analysis, probabilistic methods, swarm intelligence, socio-cognitive neuroscience, quantum computing, web intelligence have monumental roles to play in digital reality and Society 5.0. These technologies are shifting the economic landscape and the time has come to imbibe these technologies and empower organizations to exploit them now and in the future. The Series accepts research monographs, introductory and advanced textbooks, professional books, and reference works.

Aim and Scope

- The series is focused to explore how disruptive technologies are helping in digital transformation and how organizations are changing the way they do business, concerning innovation processes and business model transformations.
- This series is focused on how various disruptive technologies are creating opportunities across the business landscape.
- This series provides a comprehensive guide to Industry 4.0 applications, not only introducing implementation aspects but also presenting conceptual frameworks to the design principles of Society 5.0. Besides, it discusses such effects in new business models and workforce transformation.
- Changing dynamics of global production, its complexities, high end automated processes, high-level competitiveness, and emerging technologies for new generation goods, products, and services.
- Special focus on AI, affective computing, Blockchain, biological computing, cloud computing, cognitive intelligence, digital business transformation, decision sciences, e-health services, emotion theory, Futuristic digital society, habitat Innovation, human-computer interaction, Internet of Things (IoT), Internet of Humans (IoH), IoT-oriented Infrastructure, mobile computing, neural computing, predictive analysis, probabilistic methods, resilience in cyber-physical systems, robotics and automation for futuristic applications, swarm intelligence, synergies, and tradeoffs of food, energy, and water (F-E-W) nexus, socio-cognitive neuroscience, smart homes, and smart buildings, smart mobility and transportation, smart factories, embedded devices, quantum computing, and web intelligence to name a few.

More information about this series at https://link.springer.com/bookseries/16676

Utku Kose · Junzo Watada · Omer Deperlioglu ·
Jose Antonio Marmolejo Saucedo
Editors

Computational Intelligence for COVID-19 and Future Pandemics

Emerging Applications and Strategies

Springer

Editors
Utku Kose
Department of Computer Engineering
Suleyman Demirel University
Isparta, Turkey

Omer Deperlioglu
Department of Computer Technology
Afyon Kocatepe University
Afyonkarahisar, Turkey

Junzo Watada
Graduate School of Information
Waseda Univeristy
Shinjuku, Tokyo, Japan

Jose Antonio Marmolejo Saucedo
Facultad de Ingeniería
Universidad Panamericana
Mexico City, Mexico

ISSN 2730-9061 ISSN 2730-907X (electronic)
Disruptive Technologies and Digital Transformations for Society 5.0
ISBN 978-981-16-3782-7 ISBN 978-981-16-3783-4 (eBook)
https://doi.org/10.1007/978-981-16-3783-4

This Springer imprint is published by the registered company Springer Nature Singapore Pte Ltd.
The registered company address is: 152 Beach Road, #21-01/04 Gateway East, Singapore 189721, Singapore

Foreword

COVID-19 showed us that even the most advanced technological developments cannot stand against the power of nature. As humankind, we have still too many things to understand and learn about the earth and the biological synergy among all living organisms here. At this point, technological tools have always been vital factors to deal with issues we are facing in the real life. But we still need more technological advancements to build up a better future for us. Since the middle of the 20th century, many advanced technological developments have provided opportunities to take over the routines of humans, by running electrical problem-solving methods. It is remarkable that electronics, computer-oriented systems, and even Internet have been long-time technological tools to make mysteries visible and make both the present and future world better. For around the last two years, we already understood the importance of massive health and the use of instant problem-solving tools for possible pandemics in the future. That situation enabled us to design immediate solutions against big, massive data and develop intelligent systems based on the foundations of Computational Intelligence. We can indicate that future developments will always be directed by advancements in Computational Intelligence methods and the health sector is among the most critical fields needing immediate support by these methods.

This book: 'Computational Intelligence for COVID-19 and Future Pandemics - Emerging Applications and Strategies' provides a wide reference work as including the most recent solutions of Computational Intelligence for fighting with COVID-19 and similar pandemics. As the 21st century already brought the winds of intelligent network systems such as the Internet of Things (IoT), it has been critical to examine both software and hardware-oriented advancements, by considering medical-biomedical background. Furthermore, research efforts regarding Data Science solutions have become more and more important as communication devices are providing great momentum for an instant, massive data grow from many sources. As we already know, Machine / Deep Learning has been improving the literature of Computational Intelligence, with solution systems, which are capable of learning from past samples of data to provide findings for new states. This book covers valuable chapters including touches with all of these important solution factors and contributes to the international hard work for eliminating COVID-19 and deriving new solutions for similar massive health issues. Nowadays, the literature of Artificial Intelligence and Biomedical are both very busy with reporting the most recent research works but COVID-19 is also a societal problem as it is associated with massive actions by human groups. By meeting with that, the book also gives remarkable information about not only technical findings but also societal effects, by having a good diversity of chapter topics. It is important to have a massive immediate technological action plan against COVID-19 and that book aims to trigger scientists, academicians, students, and anyone interested in that massive action plan to take their place in especially Computational Intelligence as a starting point. I liked the language and direct organization planning in each chapter so I suggest that the book will be useful in courses as well as reference readings to track the literature and have an open mind about further developments.

By covering a good diversity of chapter topics, I would like to send my also congratulations and best wishes to great scientist-editors: Drs. Kose, Watada, Deperlioglu, and Marmolejo Saucedo. Without their great efforts, that book would not be a timely and widespread work. With also their speedy work, I am very grateful to see that book in a short time, as humankind is still dealing with the COVID-19, by focusing more on vaccine works (during January 2021, while I am writing these words).

As my final words, I would like to thank all editors, and authors for their great contributions to creating such a valuable edited book work, by taking place among super-heroes fighting with COVID-19 and believing in the positive use of Computational Intelligence for a better future.

Dr. Jafar Alzubi
Al-Balqa Applied University
Jordan
j.zubi@bau.edu.jo

Preface

As a result of the latest technological developments in the last decade, it has been clearer that Artificial Intelligence is the strongest component for shaping the future of humankind. Starting from advancements in electronics, computer systems, and communication methods, intense use of intelligent automation has become a more common thing in time, by covering all fields of modern life. At this point, advantages of Artificial Intelligence-based solutions are seen in especially human-based tasks requiring iterative computational operations. On the other hand, an increased amount of data (as a result of an increased number of different data processing tools) has already made use of intelligent algorithmic techniques vital for better predictive and descriptive findings for real-world problems. Eventually, Artificial Intelligence has become not only a type of digital, data-oriented solution but also a component for developing new, smart technologies.

While Artificial Intelligence triggers the formation of the 'new brave world', its essentials associated with mathematical and logical background and data-sensitive functions that are capable of optimizing some parameters to 'learn' about target problems have been interesting research topics to be considered in research works. Here, the solution mechanism provided in the context of Computational Intelligence has allowed the field of Artificial Intelligence to be examined as a computational tool, which is used by humans, who have analog solution capabilities and seeking computerized solutions against problems. It has been seen that Computational Intelligence applications have provided cutting-edge advancements for the present and future state of the world. As associated with that, such applications have been widely realized in vital fields such as healthcare, heavy industry, and even education. Among these, healthcare has been improving its importance day-by-day as more use of technology and malicious use of earth resources are causing massive health problems in almost every half-century. As the most recent existential threat, Coronavirus Disease 2019 (COVID-19) has been the latest massive problem for us and that virus has caused already millions of deaths worldwide, in the last 2 years. Because of the rapid spread of that virus type (via especially human-caused interactions-actions), immediate technological developments using the massive data and covering the whole world have been vital research tasks to be done multidisciplinary. In the context of all these research efforts, Artificial Intelligence-Computational Intelligence has been a key element to design automated solutions analyzing the data deeply to reach some solutions such as diagnosis of COVID-19, building IoT systems, discovering drugs, organizing tracking-treatment methods, and many more alternative solutions to have strong weapons against not only COVID-19 but also for future pandemics.

As titled 'Computational Intelligence for COVID-19 and Future Pandemics - Emerging Applications and Strategies', we would like to introduce our edited book work, which is including the most recent internationally contributed chapters for the efforts on fighting with COVID-19. As covering a total of 19 chapters, we have carefully gathered the most specific research works looking at different perspectives of solutions to provide essential solutions for not the only present state but also future potential pandemics. In detail, a total of 82 authors have taken part in that book, to support the associated literature of Computational Intelligence against the COVID-19 issue. Because speedy contributions to the literature in such critical periods are essential, we run a fast-track (but careful) evaluation for the submitted works and come up with the included chapters about critical research topics. As it is also important to have an international view, we did our best to get a good diversity in terms of countries. By including us -as the editors- and all authors including respectful Dr. Jafar Alzubi, who has provided a great Foreword, this book is based on contributions from 29 countries including Austria, Brazil, Canada, China, France, India, Iran, Italy, Japan, Jordan, Mexico, Nigeria, Norway, Pakistan, Senegal, Sri Lanka, Sweden, Turkey, and United Kingdom. That situation caused the following chapters to rise with their scientific evaluations in terms of problem definitions as well as regional views:

- Chapter 1 employs the use of Machine Learning to give answers to the question of 'Can Different Impacts of COVID-19 on Different Countries Be Explained?'.
- Chapter 2 focuses on IoT and provides a good view on IoHT (Internet of Health Things) in terms of fighting with COVID-19 and future pandemics.
- Chapter 3 introduces the use of Machine Learning to predict the transmission rate of COVID-19.
- Chapter 4 considers the Human-in-the-Loop concept and designs a solution on COVID-19 detection with Transfer Learning-based CNN (Convolutional Neural Networks) models.
- Chapter 5 provides a wide survey about epidemiology forecasting of COVID-19, by thinking about Computational Intelligence-based methods.
- Chapter 6 designs a Geographical Weighted Regression approach to figure out the spatial relationship between the confirmed cases, by considering India as the case environment.
- Chapter 7 focuses on IoT and Fog Computing for designing a COVID-19 cluster tracking approach to be used in urban cities.
- Chapter 8 is based on the Reinforcement Learning strategy to think some solutions for precautions to be taken against pandemics.
- Chapter 9 follows trendy research on X-Ray image-based diagnosis and comes with a Transfer Learning-based economical, rapid COVID-19 detection system.

- Chapter 10 uses RNN (Recurrent Neural Networks) model to ensure a predictive approach for COVID-19 reoccurrence.
- Chapter 11 provides an up-to-date, great view on the usage of Computational Intelligence for different research tasks associated with the COVID-19 (and similar pandemics of course).
- Chapter 12 focuses on the use of limited data to ensure short-term COVID-19 forecasting.
- Chapter 13 follows the most recent developments by IoT-based efforts, especially COVID-19 and similar massive health issues.
- Chapter 14 focuses on Computational Intelligence and drug discovery perspective, as it is a very critical research topic in the context of the COVID-19 issue.
- Chapter 15 considers a recent research topic: XAI (Explainable Artificial Intelligence) to be used in the context of COVID-19 oriented research works.
- Chapter 16 comes with wide research on global modeling, forecasting, and evaluation of the COVID-19.
- Chapter 17 explains the societal and political effects of COVID-19 (in the context of the Turkey case) and follows the concepts of smart city and IoT to propose the model of CORONAPOLIS, an effective managerial solution for future cities.
- Chapter 18 employs Multicriteria Analysis to determine the best neonatal ventilator for COVID-19 patients.
- Chapter 19 considers a data-sensitive, secure Blockchain-oriented solution and proposes a Data-as-a-Service model for IoHT enabled epidemic management.

We believe that 'massive healthcare' will be the most important research topic of the future. At this point, advanced automation and smart tools requiring less human contribution but more computational power (in terms of time and data storage) will be common objects to deal with health issues. So, a Computational Intelligence background will be the most important factor in running true Artificial Intelligence approaches and ensuring a better world for humankind. In this sense, this book is among the most recent source works with unique chapters introducing scientific discussions for the present and future state. As the readers, you're all welcome to read further to learn about the most recent advancements and think about alternative solutions to support the associated literature. As the editors, we would like to thank all authors for their great contributions and would like to receive all readers' valuable feedback including future work suggestions for further research of COVID-19 and future pandemics. Thanks for your attention. Stay safe, and healthy!

Editors

Isparta, Turkey — Dr. Utku Kose
utkukose@sdu.edu.tr
http://www.utkukose.com/

Shinjuku, Japan — Dr. Junzo Watada
watada@waseda.jp

Afyonkarahisar, Turkey — Dr. Omer Deperlioglu
deperlioglu@aku.edu.tr

Mexico City, Mexico — Dr. Jose Antonio Marmolejo Saucedo
jmarmolejo@up.edu.mx

Acknowledgements As the editors, we would like to thank Kamiya Khatter, Silky Abhay Sinha, and the Springer Team for their valuable efforts and great support in pre-organizing the content and publishing of the book.

Disclaimer

The presentation of material and details in maps used in this chapter does not imply the expression of any opinion whatsoever on the part of the Publisher or Author concerning the legal status of any country, area or territory or of its authorities, or concerning the delimitation of its borders. The depiction and use of boundaries, geographic names and related data shown on maps and included in lists, tables, documents, and databases in this chapter are not warranted to be error free nor do they necessarily imply official endorsement or acceptance by the Publisher or Author.

Short Description

COVID-19 is currently the most dangerous variation of the coronavirus as threating the humankind. With thousands of reported deaths and the cases more than that, the world is in an active war to eliminate that virus. In the past, such situation had already been experienced with important viruses such as Ebola, H1N1, and Influenza. However, the most devastating results within last two centuries may will be associated with the active COVID-19. While the sub-fields medical as well as natural sciences have been working instantly on deriving solutions and trying to protect the humankind against such virus types, there is also a great focus on technological developments for improving the mechanism-momentum of science for effective and efficient solutions. At this point, Computational Intelligence is the most powerful tools for researchers to fight against COVID-19. In this context, the main goal of this book is to provide the latest findings-results regarding a wide variety of Computational Intelligence applications for fighting COVID-19 and future pandemics. The book consists of a total of 19 chapters and the target audience covers scientists, professionals-experts, M.Sc. and Ph.D. students, post-docs, and anyone interested in the covered subjects. The book is suitable to be used as a reference work in the courses associated with especially Artificial Intelligence, Medical, and Biomedical.

Dr. Utku Kose
Junzo Watada
Omer Deperlioglu
Jose Antonio Marmolejo Saucedo

Contents

1 **Can Different Impacts of Covid-19 on Different Countries Be Explained? a Preliminary Machine Learning Experiment** 1
Ivanoe De Falco, Antonio Della Cioppa, Umberto Scafuri, and Ernesto Tarantino

2 **Internet of Health Things (IoHT): The Significance of Virtual Tools Aiding to Overcome Novel Coronavirus (COVID-19) Pandemic** 23
Lakshmi Narasimha Gunturu and Girirajasekhar Dornadula

3 **Predicting Transmission Rate of Coronavirus (COVID-19) Pandemic Using Machine Learning Techniques** 45
David Opeoluwa Oyewola, Emmanuel Gbenga Dada, K. A. Al-Mustapha, Ayotunde Alaba Fadele, Stephen Bassi Joseph, and Asabe Ibrahim

4 **Human-in-the-Loop Enhanced COVID-19 Detection in Transfer Learning-Based CNN Models** 71
Ferhat Ozgur Catak and Kevser Şahinbaş

5 **Epidemiology Forecasting of COVID-19 Using AI—A Survey** 89
Igor Gadelha Pereira, Andouglas Gonçalves da Silva Junior, Dunfrey Pires Aragão, Emerson Vilar de Oliveira, Arthur Andrade Bezerra, Francinaldo de Almeida Pereira, Julio Gustavo F. S. Costa, Jordan Salas Cuno, Davi Henrique dos Santos, Joris Michel Guerin, Aura Conci, Esteban Walter Gonzalez Clua, Cosimo Distante, and Luiz Marcos Garcia Gonçalves

6 **Geographical Weighted Regression Approach: A Case Study on Covid-19 in India** 121
Prisilla Jayanthi, Muralikrishna Iyyanki, Michael Carlberg, and Mansour M. Ndiath

7 Integration of IoT and Fog Computing for the Development of COVID-19 Cluster Tracking System in Urban Cities 145
H. M. K. K. M. B. Herath, G. M. K. B. Karunasena, H. M. W. T. Herath, H. D. N. S. Priyankara, B. G. D. A. Madushanka, and W. R. De Mel

8 Reinforcement Learning Model for Pandemic Precautions in Healthcare Environment 171
Ramesh Sekaran, Arun Prasath Raveendran, Ashok kumar Munnangi, and Manikandan Ramachandran

9 Transfer Learning-Based Economical and Rapid COVID-19 Detection Using X-Rays Images 185
Shatakshi Singh, Keshav Jangid, Chandraprakash Sharma, Nakul Joshi, Jyoti Tomar, Manish Kumar, Dinesh Goyal, Madhushi Verma, and Sunil Kumar Jangir

10 A Predictive Modelling of Covid-19 Reoccurrence Using Recurrent Neural Network 209
Olufunke Rebecca Vincent and Jesufunbi Damilola Bolariwa

11 Computational Intelligence-Based Diagnosis of COVID-19 229
Muhammad Shoaib, Amjad Islam Aqib, Zeeshan Ahmad Bhutta, Wanxia Pu, Iqra Muzammil, and Muhammad Aamir Naseer

12 Short-term Forecasting of COVID-19 257
Hosam H. A. Mukhairez and Ahmed J. I. Alaff

13 Internet of Health Things (IoHT) Against COVID-19: A Review of Recent Development 267
Geetha Velliyangiri, Anbumani Venkatachalam, Manikandan Ramachandran, Ambeshwar Kumar, and Murugan Subramanian

14 Drug Discovery with Computational Intelligence Against COVID-19 281
Godwin M. Ubi, Joseph K. Ebigwai, Udensi O. Ugorji, and Imaobong S. Essien

15 Explainable AI for Fighting COVID-19 Pandemic: Opportunities, Challenges, and Future Prospects 315
Kazeem Moses Abiodun, Joseph Bamidele Awotunde, Dayo Reuben Aremu, and Emmanuel Abidemi Adeniyi

16 Investigation on Some Aspects of Modeling, Forecasting, and Evaluating the Impact of Global Coronavirus Disease 2019 333
Abhijit Bora and Tulshi Bezboruah

17 "Horizontal World Treatment" and a Smart City Oriented Project for Total Recovery from COVID Outbreak: CORONAPOLISES 351
Mustafa Hulki Cevizoğlu

18 Selection of the Best Neonatal Ventilator in Patients with COVID-19 Using Multicriteria Analysis 383
Alberto Ochoa-Ortiz, Oliverio Cruz-Mejía, and José Mejia

19 Blockchain-Based Secure Biomedical Data-as-a-Service for Effective Internet of Health Things Enabled Epidemic Management 405
Amirhossein Peyvandi, Babak Majidi, and Soodeh Peyvandi

Editors and Contributors

About the Editors

Utku Kose is an Associate Professor at Suleyman Demirel University, Turkey. He received a B.S. degree in 2008 from the computer education of Gazi University, Turkey. He received an M.S. degree in 2010 from Afyon Kocatepe University, Turkey, and D.S./Ph.D. degree in 2017 from Selcuk University, Turkey. He has over 100 publications including journal articles, authored and edited books, proceedings, and reports. He is also one of the series editors of the Biomedical and Robotics Healthcare Series, CRC Press. His research interest areas are artificial intelligence, machine ethics, artificial intelligence safety, optimization, chaos theory, distance education, e-learning, computer education, and computer science.

Junzo Watada received his B.Sc. and M.Sc. degrees in electrical engineering from Osaka City University, Japan, and his Ph.D. degree from Osaka Prefecture University, Japan. He is currently a Professor at the Department of Computer and Information Sciences, Universiti Teknologi PETRONAS, and Professor Emeritus at Waseda University. He received the Henri Coanda Medal Award from Inventico in Romania in 2002. He is a Life Fellow of the Japan Society for Fuzzy Theory and intelligent informatics (SOFT). Professor Watada is an IEEE senior member, Executive Chair of ISME, WCICME, a vice-president and life member, Forum of Interdisciplinary Mathematics. He contributes to editing various international journals as an editorial board member of more than 30 journals. His professional interests include artificial neural networks, human-centric data mining, soft computing, tracking systems, knowledge engineering, financial engineering, and management engineering.

Omer Deperlioglu received his B.Sc. in Electric and Electronic (1988) from Gazi University, M.Sc. in Computer Science (1996) from Afyon Kocatepe University, Ph.D. in Computer Science (2001) from Gazi University. He is an Associate Professor of Computer Programming in the Department of Science, Vocational School of

Afyon, Afyon Kocatepe University of Afyon, Turkey. His current research interests include different aspects of Artificial Intelligence applied in Power Electronics, Biomedical, and Signal Processing. He has edited a book and (co-) authored three books and over 50 papers.

Jose Antonio Marmolejo Saucedo is a Professor at Panamerican University, Mexico. He received his Doctorate in Operations Research (Hons) from the National Autonomous University of Mexico. At present, Prof. Marmolejo has the second-highest country-wide distinction granted by the Mexican National System of Research Scientists for scientific merit (SNI Fellow, Level 2). He is a member of the Network for Decision Support and Intelligent Optimization of Complex and Large Scale Systems, Mexican Society for Operations Research, and System Dynamics Society. He has authored over thirty research articles in journals, books, and conference proceedings. His research areas are operations research, largescale optimization techniques, computational techniques, analytical methods for planning, operations, and control of electric energy and logistic systems, sustainable supply chain design, and digital twins in supply chains.

Contributors

Kazeem Moses Abiodun Department of Computer Science, Landmark University, Omu-Aran, Nigeria

Emmanuel Abidemi Adeniyi Department of Computer Science, Landmark University, Omu-Aran, Nigeria

K. A. Al-Mustapha Department of Mathematics, Baze University, Abuja, FCT, Nigeria

Ahmed J. I. Alaff Institute of Information Technology, Gazi University, Ankara, Turkey

Amjad Islam Aqib Department of Medicine, Cholistan University of Veterinary and Animal Sciences, Bahawalpur, Pakistan

Dunfrey Pires Aragão Federal University of Rio Grande do Norte, Natal, Brazil

Dayo Reuben Aremu Department of Computer Science, University of Ilorin, Ilorin, Nigeria

Joseph Bamidele Awotunde Department of Computer Science, University of Ilorin, Ilorin, Nigeria

Tulshi Bezboruah Department of Electronic & Communication Technology, Gauhati University, Guwahati, India

Arthur Andrade Bezerra Federal University of Rio Grande do Norte, Natal, Brazil

Zeeshan Ahmad Bhutta The Royal (Dick) School of Veterinary Studies, University of Edinburgh, Midlothian, Scotland, UK

Jesufunbi Damilola Bolariwa Department of Computer Science, Federal University of Agriculture Abeokuta, Abeokuta, Nigeria

Abhijit Bora Department of Electronic & Communication Technology, Gauhati University, Guwahati, India

Michael Carlberg Faculty of Medicine and Health, Örebro University, Örebro, Sweden

Ferhat Ozgur Catak Simula Research Laboratory, Oslo, Norway

Mustafa Hulki Cevizoğlu Nisantasi University, Istanbul, Turkey

Esteban Walter Gonzalez Clua Fluminense Federal University, Niteroi, Brazil

Aura Conci Fluminense Federal University, Niteroi, Brazil

Julio Gustavo F. S. Costa Federal University of Rio Grande do Norte, Natal, Brazil

Oliverio Cruz-Mejía Universidad Autónoma del Estado de México, Nezahualcóyotl, México

Jordan Salas Cuno Fluminense Federal University, Niteroi, Brazil

Emmanuel Gbenga Dada Department of Mathematical Sciences, University of Maiduguri, Maiduguri, Nigeria

Ivanoe De Falco ICAR - CNR, Via P. Castellino 111, Naples, Italy

W. R. De Mel Faculty of Technology, University of Sri Jayewardenepura, Nugegoda, Sri Lanka

Emerson Vilar de Oliveira Federal University of Rio Grande do Norte, Natal, Brazil

Antonio Della Cioppa Natural Computation Lab, DIEM, Università degli Studi di Salerno, Fisciano (SA), Italy

Cosimo Distante Institute of Applied Sciences and Intelligent Systems, Lecce, Italy

Girirajasekhar Dornadula Department of Pharmacy Practice, Annamacharya College of Pharmacy, Rajampeta, India

Davi Henrique dos Santos Federal University of Rio Grande do Norte, Natal, Brazil

Joseph K. Ebigwai Department of Plant and Ecological Studies, Calabar, Nigeria

Imaobong S. Essien Department of Zoology and Environmental Biology, Faculty of Biological Sciences, University of Calabar, Calabar, Nigeria

Ayotunde Alaba Fadele Department of Computer Science, Federal College of Education, Zaria, Nigeria

Luiz Marcos Garcia Gonçalves Federal University of Rio Grande do Norte, Natal, Brazil

Dinesh Goyal Department of Computer Science Engineering, Poornima Institute of Engineering and Technology, Jaipur, Rajasthan, India

Joris Michel Guerin Laboratoire d'analyse et d'architecture des systemes, Toulouse, France

Lakshmi Narasimha Gunturu Department of Pharmacy Practice, Annamacharya College of Pharmacy, Rajampeta, India

H. M. K. K. M. B. Herath Faculty of Engineering Technology, The Open University of Sri Lanka, Colombo, Sri Lanka

H. M. W. T. Herath Faculty of Engineering, University of Moratuwa, Moratuwa, Sri Lanka

Asabe Ibrahim Department of Mathematics and Computer Science, Federal University Kashere, Gombe, Nigeria

Muralikrishna Iyyanki Former Professor and Director R&D, JNT University, Hyderabad, India

Keshav Jangid EDApply Pvt. Ltd, Jaipur, Rajasthan, India

Sunil Kumar Jangir Department of Computer Science & Engineering, Mody University of Science and Technology, Sikar, Rajasthan, India

Prisilla Jayanthi MLR Institute of Technology, Dundigal, Hyderabad, India

Stephen Bassi Joseph Department of Computer Engineering, University of Maiduguri, Maiduguri, Nigeria

Nakul Joshi Celebaltech Private Limited, Jaipur, India

Andouglas Gonçalves da Silva Junior Federal Institute of Rio Grande do Norte, Mossoró, Brazil

G. M. K. B. Karunasena Faculty of Engineering Technology, The Open University of Sri Lanka, Colombo, Sri Lanka

Ambeshwar Kumar School of Computing, SASTRA Deemed University, Thanjavur, India

Manish Kumar Department of Biomedical Engineering, Mody University of Science and Technology, Sikar, Rajasthan, India

B. G. D. A. Madushanka Faculty of Engineering Technology, The Open University of Sri Lanka, Colombo, Sri Lanka

Babak Majidi Department of Computer Engineering, Khatam University, Tehran, Iran;
Emergency and Rapid Response Simulation (ADERSIM) Artificial Intelligence Group, Faculty of Liberal Arts & Professional Studies, York University, Toronto, Canada

José Mejia Universidad Autónoma de Ciudad Juarez, Ciudad Juárez, México

Hosam H. A. Mukhairez Institute of Information Technology, Gazi University, Ankara, Turkey

Ashok kumar Munnangi Department of Information Technology, Velagapudi Ramakrishna Siddhartha Engineering College (Autonomous), Vijayawada, Andhra Pradesh, India

Iqra Muzammil Department of Veterinary Medicine, University of Veterinary and Animal Sciences, Lahore, Pakistan

Muhammad Aamir Naseer Department of Clinical Medicine and Surgery, University of Agriculture, Faisalabad-38000, Pakistan

Mansour M. Ndiath University Cheikh Anta Diop, Dakar, Senegal

Alberto Ochoa-Ortiz Universidad Autónoma de Ciudad Juarez, Ciudad Juárez, México

David Opeoluwa Oyewola Department of Mathematics and Computer Science, Federal University Kashere, Gombe, Nigeria

Francinaldo de Almeida Pereira Federal University of Rio Grande do Norte, Natal, Brazil

Igor Gadelha Pereira Federal University of Rio Grande do Norte, Natal, Brazil;
Center for Research and Development (CPqD), Campinas, Brazil

Amirhossein Peyvandi Department of Computer Engineering, Khatam University, Tehran, Iran

Soodeh Peyvandi Business Intelligence, University of Applied Sciences Upper Austria, Steyr, Austria

H. D. N. S. Priyankara Faculty of Engineering Technology, The Open University of Sri Lanka, Colombo, Sri Lanka

Wanxia Pu Key Laboratory of New Animal Drug Project, Gansu Province/Key Laboratory of Veterinary Pharmaceutical Development, Ministry of Agriculture and Rural Affairs, Lanzhou Institute of Husbandry and Pharmaceutical Sciences of Chinese Academy of Agriculture Sciences, Lanzhou, PR China

Manikandan Ramachandran School of Computing, SASTRA Deemed University, Thanjavur, India

Arun Prasath Raveendran Department of Information Technology, Velagapudi Ramakrishna Siddhartha Engineering College (Autonomous), Vijayawada, Andhra Pradesh, India

Kevser Şahinbaş Istanbul Medipol University, Istanbul, Turkey

Umberto Scafuri ICAR - CNR, Via P. Castellino 111, Naples, Italy

Ramesh Sekaran Department of Information Technology, Velagapudi Ramakrishna Siddhartha Engineering College (Autonomous), Vijayawada, Andhra Pradesh, India

Chandraprakash Sharma Wisflux Private Limited, Jaipur, Rajasthan, India

Muhammad Shoaib Institute of Microbiology, University of Agriculture, Faisalabad, Pakistan;
Key Laboratory of New Animal Drug Project, Gansu Province/Key Laboratory of Veterinary Pharmaceutical Development, Ministry of Agriculture and Rural Affairs, Lanzhou Institute of Husbandry and Pharmaceutical Sciences of Chinese Academy of Agriculture Sciences, Lanzhou, PR China

Shatakshi Singh Department of Computer Science & Engineering, Mody University of Science and Technology, Sikar, Rajasthan, India

Murugan Subramanian Department of Computer Science, Sri Aravindar Engineering College, Sedarapet, TN, India

Ernesto Tarantino ICAR - CNR, Via P. Castellino 111, Naples, Italy

Jyoti Tomar Department of Microbiology, Pacific Institute of Medical Science, Udaipur, Rajasthan, India

Godwin M. Ubi Department of Genetics and Biotechnology, Calabar, Nigeria

Udensi O. Ugorji Department of Genetics and Biotechnology, Calabar, Nigeria

Geetha Velliyangiri Department of Electronics and Communication Engineering, Kongu Engineering College, Perundurai, India

Anbumani Venkatachalam Department of Electronics and Communication Engineering, Kongu Engineering College, Perundurai, India

Madhushi Verma Department of Computer Science Engineering, Bennet University, Greater Noida, Uttar Pradesh, India

Olufunke Rebecca Vincent Department of Computer Science, Federal University of Agriculture Abeokuta, Abeokuta, Nigeria

Chapter 1
Can Different Impacts of Covid-19 on Different Countries Be Explained? a Preliminary Machine Learning Experiment

Ivanoe De Falco, Antonio Della Cioppa, Umberto Scafuri, and Ernesto Tarantino

Abstract The hypothesis that we intend to investigate here is that the extent of the impact of Covid-19 on a given country can be explained starting from a set of indicators and by using machine learning methodologies. The purpose of this chapter is not to find a way to solve the problem in an optimal way. Rather, we aim at performing a preliminary study to verify whether the aforementioned hypothesis is viable. Should it turn out so, we wish to get awareness both of which are the problems that must be solved in order to arrive at a (sub-)optimal solution, and of what are the possible limitations of the method. We firstly create a suitable data set of indicators starting from different sources available on the internet. Then, we apply onto it an evolutionary algorithm that is able to extract a set of IF–THEN decision rules allowing us to relate the values of the parameters for the different countries to the different levels of impact of Covid-19 on them.

I. De Falco (✉) · U. Scafuri · E. Tarantino
ICAR - CNR, Via P. Castellino 111, Naples, Italy
e-mail: ivanoe.defalco@icar.cnr.it

U. Scafuri
e-mail: umberto.scafuri@icar.cnr.it

E. Tarantino
e-mail: ernesto.tarantino@icar.cnr.it

A. Della Cioppa
Natural Computation Lab, DIEM, Università degli Studi di Salerno,
Via Giovanni Paolo II, 132, 84084 Fisciano (SA), Italy
e-mail: adellacioppa@unisa.it

U. Kose et al. (eds.), *Computational Intelligence for COVID-19 and Future Pandemics*,
Disruptive Technologies and Digital Transformations for Society 5.0,
https://doi.org/10.1007/978-981-16-3783-4_1

1 Introduction

After some months in which Covid-19 has hit practically every country in the world, although to different extents, epidemiologists and virologists have now wide availability of data to reason upon. Based on this empirical evidence, many researchers are using Artificial Intelligence (AI) and Machine Learning (ML) techniques to face a set of issues connected to this disease. In [1], a good review is given of some Covid-19-related aspects for which the help from ML is either already taking place or is highly welcome. From among these aspects, we can recall here identification of subjects who are most at risk [2], patients diagnosis [3], faster drug development [4, 5], prediction of the disease spread [6–8], better understanding of viruses through proteins [9], mapping of viruses provenience [10], and prediction of the next pandemic [11].

Another important issue related to Covid-19 is that of trying to understand the reasons why this coronavirus has very severely struck some countries while always not hitting some others. This becomes even more puzzling when neighboring countries, in many cases with similar life quality levels, mores, and environmental features, are considered.

Several researchers are working on this, so that different viewpoints have been proposed that lead to different possible explanations being proposed. We shortly list them here.

A first hypothesis is based on different demographic distribution of the population in different countries: some, as for example, Cincotta [12], put forward that this virus strikes more, and more severely, elderly people, while it does so less, and less severely, young subjects.

A second hypothesis takes into account the different cultural social distancing between strangers: in some countries, people never shake hands nor do they hug, whereas in others staying very close, one another is not a matter of concern. Also, in some countries a mask is used as soon as a cold is caught, which does not hold true in other places. This hypothesis is accounted for, as an example, in [13], where Jake Miller reports the view from Megan Murray, the Ronda Stryker and William Johnston Professor of Global Health and director of the Global Health and Social Medicine research core.

A third hypothesis relies on weather-related considerations: several authors, as, e.g., Triplett [14], write that higher temperatures and higher amounts of sun hours per day may make the virus weaker. An important attempt is being made by researchers at Lawrence Berkeley National Laboratory aiming at identifying which are the environmental variables Covid-19 dynamics are most sensitive to [15].

A fourth hypothesis considers the rigorousness of the lockdown measures taken by different countries. In fact, several countries have practically locked the citizens in their houses, whereas others have decided for much less strict actions. This is investigated by, e.g., Bergamin [16].

For each of these hypotheses some examples are brought that seem to prove it. Yet, for each of them, counterexamples can be easily found. Moreover, the question arises why researchers should limit to taking to account just the aspects mentioned above: maybe some others should also be investigated that could contain a correct hint to the explanation.

As a result of the above, our feeling is that each of those views is too limiting, and may consider just a small part of a much more complicated picture. Hence, in this chapter, we have decided to take into account a wider set of hypotheses, each of them represented by the values of some numerical *input* parameters, and to represent each country with their values. Namely, we have made reference to the four hypotheses listed above, and to two more, involving pollution and healthcare efficiency, respectively. Also, we represent each country with a numerical *output* parameter denoting the strength with which Covid-19 is affecting it.

Once done this, we appeal to ML, asking it to provide us with an explicit, user-friendly explanation of the connections between the values of the input parameters and those of the Covid-19 impact. In fact, within ML, and more generally in AI, a paradigm is becoming more and more important, called Interpretable Machine Learning [17]: not only should the application of an ML algorithm provide users with correct answers to their questions but should also provide on its own clear explanations, that could be easily understood by users, on the reasons for those answers. The key-point of that paper is that the use of black-box ML models added with methods for a posteriori explanation, as it is the case for deep neural networks, should be avoided, as it often leads to problems that can lead to costly consequences. Rather, the way forward is to design ML models that are inherently interpretable.

Given this goal related to directly obtaining self-explaining solutions, we appeal here to an evolutionary tool called DEREx [18], based on Differential Evolution (DE) [19, 20]. In fact, this tool is able to evolve solutions under the form of a set of easy-to-understand IF–THEN decision rules. This allows us to achieve explicit knowledge on the problem.

It should be remarked here that we consider this study as a feasibility study on the use of this approach to foresee the impact of Covid-19 on countries, rather than a conclusive paper describing a solid methodology. In fact, as it will be clearer in the following of this chapter, our research activity described here is based on some simplifications and on some assumptions that should be removed in a deeper investigation phase. Should the outcome of this study be encouraging, we would then begin a phase of more thorough research. Our aim here is to investigate the viability of a new path for the international scientific community.

However strange it may seem, to the best of our knowledge, this is the first study aiming at the above stated goal. Just another slightly similar paper exists [21] in which Onovo et al. use supervised ML and empirical Bayesian Kriging techniques to reveal correlations and patterns of Covid-19 disease outbreak in sub-Saharan Africa. Covid-19 data was merged by them with additional data on socio-demographic and health indicator survey data coming from 39 countries in that area. Their predictive model ascertained seven variables significantly associated with the risk of coronavirus infection (i.e., new HIV infections among pediatric, adolescent, and middle-aged adult PLHIV, time(days), pneumococcal conjugate-based vaccine, incidence of malaria and diarrhea treatment). Yet, theirs is a feature-selection activity, rather than one aiming at finding an explicit explanation.

Section 2 describes the way in which we have created our data set. The way in which we apply ML to it is reported in Sect. 3. The description of the experiments is given in Sect. 4, where the results obtained by DEREx are presented, and are also contrasted against those provided by a set of other ML classification algorithms. The discussion of the results and of the weaknesses of the current proposal is effected in Sect. 5. Finally, Sect. 6 contains our conclusions, and a description of our future works on this topic.

2 The Data Set

As a first step in the creation of the data set, starting from the hypotheses mentioned in Sect. 1, we have performed a brainstorming aiming to identify as many possible causes, and input parameters, that could somehow influence Covid-19 impact on the different countries as possible. As a consequence, beyond considering those four hypotheses, we have added three more hypotheses, as better specified here below.

A first further hypothesis formulated by us is that the impact could be influenced by the quality level of the health system: poorer countries, with worse health systems, could experience more problems in facing the spread of the virus.

Our second hypothesis is that this difference in impact could be related to differences in the population density of the countries: in populations in which many people live in proximity one another, the virus might spread more rapidly than in countries where people are well distanced.

Our third hypothesis consists in the pollution level in a country possibly having influence on the strength of the impact: this virus attacks lungs and the respiratory system, so people living in highly polluted countries could already have problems with lungs, which could ease further problems caused by the virus.

Based on that, a second step has been undertaken aiming at finding on the internet possible sources for the values of parameters that could well represent both the four original hypotheses and our three further ones. Of course, these parameters should be available for as-wide-as-possible sets of countries.

As a consequence, we have identified eight parameters, that are listed in the first column of Table 1. All the values for these parameters have been taken on June 10, 2020.

For some of these parameters, this phase has led us to slightly modify the original definition we had conceived for them. This is due to the problems originating in some cases to find the values of interest in either a downloadable form or at least in a tabular form. Just to give an example, for the *Average Temperature* in a country, we started with wishing to take into account the temperatures for the period February–early June, and we ended with considering the average yearly temperatures. It should be noticed that in this example the actual difference between the two approaches is not greatly significant: as the months under investigation make reference to an intermediate time span between Winter and Summer, the average temperature in this period is not very different from the average over the whole year. Had Covid-19 taken place in either Winter or Summer, this difference would have been much larger, and more significant. At the end of this phase, we have been able to select the sources where to get the values from. They are listed in the last column of Table 1.

Another issue is related to the *Stringency Index*, that is available on a daily basis since January 1, 2020. Of course, for the vast majority of countries, these daily values are equal to zero throughout January and, for many of them, also in February. Hence, we have decided to take the values of Stringency Index into consideration starting from March 1.

Moreover, the second column of the table reports a very short description for each of the parameters, the third the allowable range, whereas the fourth states the number of countries for which the values are available from the selected source.

As concerns the output parameter, this must represent the impact of Covid-19 over any given country. There are several possible parameters we could take into account. We have considered here that any parameter related to the number of deaths would not be adequate. In fact, different countries count casualties in different ways. For example, Germany counts people died *because of* Covid-19, whereas Italy counts those dead *with* Covid-19, independently of whether or not this virus has been the main cause for the deaths. Rather, we have decided to consider as the output parameter the number of positive cases per one million people (*Positive Rate*). Information about this output parameter is given in the last row of Table 1.

It should be emphasized here that all the data has been gathered on June 10, 2020. Hence, we have set this date as deadline for the data. Data related to dates after this deadline has not been considered in the present study.

The problem with the parameters in the table is their very different cardinality. For example, for *Cultural Social Distancing*, we have the values for just 42 countries, whereas for *Population Density*, the values for 225 countries are available. Quantitatively, just for 25 countries are all the eight parameters available.

This means that, if we considered all these eight parameters, we would have to work with a table containing lots of missing data. This is not a good situation for ML techniques: some methods to limit this problem exist, as, e.g., imputation [31], yet when there are too many missing data these methods may increase noise and lead to unreliable situations.

As a consequence, we have to make a decision about which of the above parameters we should use to create a reduced data set with no missing data. This decision should come from a trade-off between contrasting objectives. On the one hand, we wish to consider as many countries as possible, which would imply to use few parameters. On the other hand, from a ML viewpoint, we would like to consider as many parameters as possible, because the correct explanation we are looking for could come from one or more parameters, some of which we could cancel out from the reduced data set. This latter would yield that we could never obtain the correct explanation. Table 2 shows, for any number of parameters that can be considered, the corresponding number of countries having all of them.

After careful consideration, we have decided to consider in this study five parameters, namely those with the higher cardinality. They are: Stringency Index, Median Age, Population Density, Average Temperature, and Pollution Index. With this

Table 1 The input and output parameters used here

Parameter	Description	Range	Countries	References
Input parameters				
Stringency index	Average over March 1–June 10, the higher the value the more strict the lockdown	11.08–87.15	174	University of Oxford [22]
Mobility	Average (March 1–June 10) over three parameters (private transport, public transport, walking) , the higher the value the lower the mobility)	31.40–97.19	63	Apple [23]
Median Age	The higher the value, the older the population	15.40–53.10	212	World Population Review [24]
Population Density	The higher the value, the higher the density	0.14–26337.00	225	Statistics Times [25]
Average Temperature	Average of the yearly values until 1990, the higher the hotter	−5.10–28.25	189	Mitchell et al. [26]
Pollution Index	Data of 2020, the higher the value, the higher the pollution	11.55–93.45	110	Numbeo [27]
Healthcare Efficiency	The higher the value, the higher the efficiency of the Healthcare system	29.40–87.30	56	Bloomberg [28]
Cultural Social Distancing	Average distance in cm from a stranger, the higher the value, the farther two strangers typically stay	77.00–135.00	42	Sorokowska et al. [29]
Output parameter				
Positive Rate	Positive cases per one million people, the higher the value, the more aggressive the virus (as of June 10)	2.00–26211.00	211	Worldometers [30]

Table 2 The possible numbers of parameters, and the number of countries having all of them

Number of parameters	1	2	3	4	5	6	7	8
Number of countries	233	217	196	167	99	78	52	25

choice, the number of countries for which we have all the five values is equal to 99, which becomes the number of countries on which we can carry out our experiments. We feel that, given the general picture, this is a reasonable, although not ideal, value.

An important remark here is that the values of the output variable are real values. If we keep on considering them as real values, then, to relate the input values to the output ones, we could only perform a regression task. Of course, this can be done, yet what we could get in a case like this would be a function as, e.g.: $y = (cos(x_1) \cdot ((sin(2.57 \cdot x_3 \cdot x_5) * sqrt(x_4 + x_2))/(log(1.93 \cdot x_5)))$. Although interesting from a numerical viewpoint, this kind of result would not let us have an insight into the causes of the different strengths shown by the virus in the different countries.

What is much more interesting, instead, is to perform a classification so that we can obtain solutions under the form of decision rules linking conditions on the values of the input variables to a class representing the virus strength. To do this, we must decide the number of classes into which divide the countries. After a preliminary investigation phase on this, we have decided to consider here three classes, namely *low impact*, *medium impact*, and *high impact*. This decision leads to the following one about the ranges for the values of the output parameters for these three classes. Again, after carefully looking at the available data, we have chosen the ranges as it is shown in Table 3. Basically, we have set two threshold values, equal to 300.00 and 2000.00, respectively. This table also reports on the number of items that has been achieved for each class. As it can be seen, with the above values for the two thresholds, the data set is well balanced in terms of items per class.

With the above decisions, the problem becomes a classification task.

In the Appendix, Table 10 contains in each row the name of the country, the values for the five parameters, and the class.

3 DEREx Tool and Knowledge Extraction from the Data Set

To accomplish the task of finding a good explanatory model linking the values of the input parameters to those of the classes described above, we have decided to avail ourselves of the Differential Evolution-based Rule Extractor (DEREx) tool [18].

Table 3 The ranges of the Positive Rate for the three classes

Class	Minimum value	Maximum value	Class items
Low impact	2.00	300.00	33
Medium impact	300.01	2000.00	34
High impact	2000.01	26211.00	32

A thorough description of DEREx is beyond the scope of this study. Interested readers may refer to the seminal paper [18]. Suffice it to say here that DEREx is an evolutionary tool based on DE [19, 20] for the automatic classification of items. It automatically extracts explicit knowledge from a database under the form of a set of a number N_R of IF–THEN rules containing AND-connected literals on the database variables. Each DE individual codes for a set of N_R rules. For each of the N_C data set classes, more than one rule can be contained in the individual, and these rules can be seen as logically connected in OR. Furthermore, all the classifying rules for all the classes are found all at once in one step.

DEREx has revealed itself as a useful support to decision-making whenever explanations on why an item is assigned to a given class should be provided, as in, e.g., fall detection [32], obstructive sleep apnea [33], and multiple sclerosis lesions [34].

For any set of rules proposed during its execution, DEREx, for any data set item, checks if one and only one of the rules covers it, i.e., if the item can be assigned to one and only one class. If this holds true, the item is assigned to the class corresponding to the rule covering it. Of course, if more rules corresponding to a same class cover that item, it is assigned to that class as well. If, instead, an item is either covered by no rule or by two or more rules related to different classes, then it is an *indeterminate* item, respectively of the type *no–no* or *yes–yes*. As we wish that any item is assigned to a class, on each indeterminate item DEREx perfors a recovery action, so as to assign it to a class. Details on how this is made can be found in our seminal paper [35].

DEREx performs supervised learning, so it requires to split the data set into two subsets: during its evolution, it tries to generalize from the items assigned to the *train set*, while at the end of the run the best found individual, in terms of best fitness function value over the train set, is evaluated over another set of never-previously-seen items that make up the *test set*. The result of this latter evaluation is the value we are interested in, as it tells us about the generalization ability of the set of rules found.

As for any evolutionary algorithm, DEREx too requires the definition of a suitable fitness function with which evaluate the quality of any solution proposed. We have decided to make use here of the classification accuracy A_{cc}, defined as the percentage of the ratio between the number of correctly classified items and the total number of items. This fitness function has values in $[0.0 - 100.0]$. Given this definition, the problem becomes a maximization problem, for which the higher the fitness the better the solution.

Once decided the typology of the problem to face, i.e., classification, we have had to choose which are the operators that are allowed to participate in the rules, together with the symbols representing the problem variables. They can be found in Table 4.

With our choices, the basic literal is of the type:

$$< x_i >< rel_{op} >< c >$$

where x_i is the generic ith problem variable, rel_{op} is one of the first five relational operators in Table 4, and c is a constant value in a suitable range for x_i. The relational

Table 4 The operators used in this study

Operator class	Items
Logical operators	AND, OR, NOT
Relational operators	<, <=, =, >=, >, IN, OUT
Problem variables	x[1], x[2], x[3], x[4], x[5]
Constant values	

Table 5 The Correspondence between the Parameters Used in this Study and their Positions in the Array x

x	x[1]	x[2]	x[3]	x[4]	x[5]
Parameter	*Stringency Index*	*Median Age*	*Population Density*	*Average Temperature*	*Pollution Index*

operators IN and OUT, instead, need two constant values c_1 and c_2 ($c_1 < c_2$) to represent a range: IN means the problem variable is within that range, whereas OUT means that it is outside.

Our input parameters are represented as a real-valued array x. The correspondence of the x parameters with our input parameters is shown in Table 5.

4 The Experiments and the Results

A very important step in the use of an evolutionary algorithm lies in the choice of the values for its parameters. Many such parameters exist in DE, and, consequently, in DEREx. This latter has some more parameters specific of its own.

From among them, we can cite here the Rule_Threshold RT and the Literal_Threshold LT. They act as penalty functions and allow modulating the number of active rules and the number of active literals in rules, respectively. Their ranges lie within 0.0 and 1.0, the higher their values the higher the penalty for large sets of rules and large number of literals in the rules, and hence, the more compact the rule sets proposed. More precisely, small (respectively, large) values for them determine on the one hand a larger (respectively, lower) numerical quality of the solution obtained, in terms of a larger (respectively, lower) accuracy value, and, on the other hand, a higher (respectively, lower) complexity of the rules, which affects understandability of the solution for the user.

In this study, we have tried to balance both needs, aiming in any case at preferring readability. Hence, we have set their values at 0.50 and 0.60, respectively, which means that we expect to obtain quite compact, explanatory sets of rules, even to the detriment of higher accuracy values that could be achieved through larger sets and larger rules. In fact, given the small size of the data set, we fear that DEREx could perform a kind of an overfitting over the items shown to it in the training set. In this way, the resulting rules would depend too much on the specific data set items in the train set, which would cause that generalization would be very poor. With short rules, instead, better generalization should be achieved.

As regards the other settings required by DE, typically a tuning phase both for the choice of the most suitable operators and for the parameter values is carried out aiming at improving the quality of the achieved solution. Being this a preliminary study, we have skipped this phase. Basically, we have kept both parameter values and operator types as they typically are in DE. Table 6 shows the values used for all the DEREx parameters.

To perform supervised learning over the data set, we have divided it into train and test sets. For the items, we have considered an alphabetical ordering on the data set in terms of country names. Consequently, we have assigned the first 66 items to the train set and the last 33 to the test set.

We have run the DEREx algorithm, set with the above parameter values, 25 times, each time with a different random seed for the random number generator. Then we have considered the final best fitness values obtained in the 25 runs over both train and test sets, and we have computed their average, standard deviation, minimum and maximum values. These are reported in Table 7.

We have considered as the best run the one providing the highest accuracy value over the test set, i.e., that showing the highest generalization ability over previously unseen items. The corresponding rules obtained are:

Table 6 The DEREx parameter values used

Parameter	Value
Population size	200
Number of generations	1000
Fitness function	Accuracy
Search strategy	Rand/1/bin
Crossover probability	0.30
Mutation probability	0.70
Replacement	Generational
Elitism	No
Maximum number of rules N_R	6
Rule threshold RT	0.50
Literal threshold LT	0.60

Table 7 The statistics over the 25 runs

Set	*Average*	*Std. Dev.*	*Minimum*	*Maximum*
Train	63.39	1.06	60.61	65.15
Test	39.64	2.81	27.27	60.61

IF ((*Population Density*) $\leq$ 766.11) AND (*Average Temperature* $\geq$ 1.47)) THEN class = *low impact*

IF ((*Stringency Index* $\leq$ 27.47) AND (*Population Density* $\geq$ 3993.84) AND (*Average Temperature* $\leq$ 4.97)) THEN class = *medium impact*

IF (*Average Temperature* IN (9.20–21.28)) THEN class = *medium impact*

IF (*Stringency Index* $\geq$ 57.68) THEN class = *medium impact*

IF (*Stringency Index* < 19.92) THEN class = *high impact*

IF ((*Stringency Index* < 73.85) AND (*Pollution Index* < 71.44)) THEN class = *high impact*

which have accuracy values equal to 63.64% and 60.61% over train and test sets, respectively. The accuracy over the whole data set is, hence, equal to 62.63%.

Let's take a closer look at the rules we have found.

For the low impact class, the only rule found says that, if the density of the population is not high and the average temperature is not too low, the impact is low. The second part of this rule fits very well the third hypothesis of Sect. 1, while a first check is added on the population density, which is related to our own second hypothesis. This latter seems very sensible, as in countries where the density is not high the virus may have more difficulties in spreading.

As concerns the three rules found for the medium impact class, the first rule is a good summary of several hypotheses: if the stringency is not that strong, the population density is high, and the average temperature is relatively low, then the virus can impact, although not at its worst. The second rule, instead, refers to the average temperature: countries with mild climate will be mildly impacted. This latter is in good accordance with the third hypothesis presented in Sect. 1. Finally, the third such rule is straightforward with respect to medium-to-high values of the stringency index, and well agrees with the first hypothesis of the Introduction.

Finally, let's take into account the two rules found for the high impact class. The first of them says that if countries have a very low stringency (lock-down) value (lower than 19.92), then this value on its own suffices to suggest that they are likely to be strongly impacted by Covid-19. This is in excellent accordance with the fourth hypothesis presented in Sect. 1. The second such rule, instead, says that even if this value is higher (lower than 73.85), the country can be highly hit as well, provided that the pollution index is below 71.44.

A first remark over these rules is that they contain four of the five parameters, namely Stringency Index, Population Density, Average Temperature, and Pollution Index. Median Age, instead, is not considered.

Another important remark here is that none of the hypotheses presented earlier is sufficient on its own to explain the different impact of Covid-19 on the different countries. Rather, a mix of many of them should be considered.

It is to note that, by using smaller values for RT and LT, higher accuracy values have been obtained over the train set. Yet, this has had two drawbacks, the former being the attainment of very large sets of very large rules, and the second being the achievement of low accuracy values over the test set. Both issues suggest that overfitting has occurred when training, and that such parameter configurations should be avoided.

Table 8 reports the three confusion matrices obtained by using the above set of decision rules over the train set (left), the test set (middle), and over the whole data set (right). In the table, for the sake of conciseness, the low-impact class is represented with a 1, the medium-impact with a 2, and the high-impact one with a 3.

As a general comment to these tables, the set of rules causes that the low-impact class is very stable, as only six of its 33 items are incorrectly classified as medium-impact items, while none of them is seen as a high-impact country. As regards the medium-impact items, instead, many of them are seen as high-impact class. Finally, as regards the high-impact countries, many of them are erroneously seen as low-impact countries.

4.1 Comparison

To evaluate the quality of the solution shown in this study, we have performed a comparison against a set of other well-known and well-performing classification algorithms. This has been effected within the Waikato Environment for Knowledge Analysis (WEKA) environment [36], version 3.8.4. This latter contains a wide set of ML algorithms, based on different ideas and working mechanisms. From among them, following the results and the suggestions from the scientific literature, we have chosen here Random Forest (RF) [37], Support Vector Machine (SVM) [38], Multi-Layer Perceptron (MLP) [39], Bagging (Bag) [40], as these typically perform very well for classification tasks. Moreover, we have also run algorithms based on concepts as trees or rules, namely Repeated Incremental Pruning (JRip) [41], Reduced Error Pruning Tree (REPT) [42], and Partial Decision Tree (PART) [43].

As regards their parameter values, we have not performed any preliminary tuning, rather we have used those contained as default in WEKA. We have executed each

Table 8 The confusion matrix over the train set for the decision rule set shown in this study. 1 = low impact, 2 = medium impact, 3 = high impact

True class	Train			Test			Train + Test		
	Assigned class			Assigned class			Assigned class		
	1	2	3	1	2	3	1	2	3
1	17	3	0	10	3	0	27	6	0
2	0	15	13	0	4	2	0	19	15
3	8	0	10	8	0	6	16	0	16

Table 9 The statistics for accuracy over the test set for the algorithms investigated

	DEREx	RF	SVM	MLP	Bag	JRip	REPT	PART
Average	39.64	46.13	43.63	**46.26**	41.13	38.88	35.50	40.63
Std. dev.	2.81	2.26	0.62	3.13	6.36	2.23	9.79	**0.00**
Highest	**60.61**	50.00	43.75	50.00	56.25	40.63	50.00	40.63
Lowest	27.27	**40.63**	**40.63**	37.50	28.13	31.25	25.00	**40.63**

of these algorithms for the same number of runs as we have done for DEREx, i.e., 25, by using different values for the initial random seed. As for DEREx, we have considered as the result of each run the final accuracy value over the testing set.

Table 9 shows the related statistics in terms of average final accuracy value, standard deviation, highest and lowest values. For each of these indicators, the best value is reported in bold in the table.

As it can be seen, DEREx is the algorithm that has been able to yield the highest accuracy value over the test set. Other algorithms have provided higher average values, or lower standard deviation values, or higher lowest values. Yet, for the problem at hand, what is important is to provide the highest accuracy value, which is where DEREx performs better.

Moreover, it is important to note that many of these algorithms, although performing well, do not provide any explicit knowledge automatically extracted from the data set, which is the goal of this study. This is the case, for example, of SVM, MLP, and Bagging.

5 Discussion

Until here, we have seen how the methodology we have proposed seems to provide interesting results. Yet, we must keep in mind that we have made several assumptions and simplifications. Hence, a careful discussion about them should be reported about these delicate issues. In the following, we shortly comment on some of them.

5.1 Further Hypotheses

In this study, we have taken into account four popular hypotheses about possible causes for the different impact of Covid-19 on different countries. Moreover, we have advanced three further hypotheses. Of course, other hypotheses can be put forward as well. These latter, together with the seven reported here, could lead to a more variegated picture in which a wider set of parameters is considered. Possibly some hypothesis not investigated here could lead to a better understanding.

5.2 Time Period of the Data

The data refers to different periods. Especially that related to average temperature has two problems: (a) it refers to years until 1990; (b) it is the yearly average value.

The first issue would probably require an increase of 1–3 degrees in those values to keep track of global warming taking place in these last years, however, the differences between countries would remain about the same.

The second point is conceptually more serious, but fortunately in this case it does not impact much. This is because throughout the world the period February–early June is related to spring (boreal hemisphere) or autumn (austral hemisphere), therefore the average yearly temperatures coincide roughly with the temperatures of the period under investigation. If the pandemic had occurred in Summer or in Winter, this data would not have been valid.

Obviously, the best solution is to search for databases that offer for each country, in a simple downloadable form, the average monthly values of temperatures and, if possible, the index of humidity, the amount of rainfall, and other meteorological indicators.

5.3 Daily Parameters Versus Non-daily Parameters

The majority of the parameters considered in this study are represented by yearly average values. Instead some data, specifically Stringency Index, Mobility, and the Positive Rate, are daily.

In this feasibility study, we proceeded with a simplification for the Stringency Index, calculating the average over the period investigated.

In addition, mobility actually consists of three distinct parameters. For it, firstly the daily average of the three parameters provided by Apple was calculated, and then the average of these average values was computed over the period under investigation.

It should be noted that for the mobility, in addition to Apple, Google also provides six parameters for each country, and for many countries, including Italy, the data are even divided by regions [44]. The problem with this data is that it is not provided in a tabular form, nor in a simply downloadable form, rather for each individual country the data of each single day must be downloaded one at a time. Therefore, in this feasibility study, it was decided not to consider them for reasons of time.

For mobility, perhaps all the parameters could be considered separately (the three from Apple and/or the six from Google) instead of mediating them. This would yield a higher number of input parameters.

Furthermore, for each parameter, the average over a given time range could perhaps be considered (a week? two weeks? a month?) and each of these averages could be considered as an additional parameter. This would then lead to the input parameters becoming a larger number.

For the Positive Rate, the value of June 10 was used, so as of today its values for all countries are certainly greater than, or at least equal to, the values used here.

5.4 Number of Classes

It is not trivial to say what is the optimal number of classes in which to divide countries. Maybe three classes are a good choice, or maybe a finer division in, e.g., five classes (very low, low, medium, high, very high) could be better. Vice versa, a coarser subdivision into just two classes could be more fruitful. An experimentation phase should be carried out by varying the number of classes.

5.5 Value Range for Each Class

Similarly, the range of values for the output parameter Positive Rate to be associated to each class should be carefully studied. An investigation phase as these intervals vary appears important.

5.6 DEREx

Obviously, to generate a set of decision rules one can think of other settings for DEREx, different from the one used here. Important decisions are whether or not consider in the grammar the OR operator, the NOT operator, the IN and OUT operators, and so on.

Also, parameter values may influence the results.

Furthermore, the /rand/1/bin search strategy has been used here, but DE, and DEREx as well, may run many such different strategies. This could impact on the quality of the solution.

Each of these choices can influence the final result, therefore it would be important to take a thorough examination in a preliminary set-up phase. Here, instead, a rather rapid one was made, which allowed us to obtain a sufficient fine-tuning for the algorithm, but obviously we can, and we should, carry out this phase in a much deeper manner.

5.7 Readability of the Solution

The problem of how to obtain solutions that are easily readable and informative for the user, and at the same time of good quality (in terms of high accuracy), must be addressed. Obviously, decreasing the values for the parameters related to the size of the solutions can give better accuracy values, but the obtained sets of decision rules could be too large, thus not very well understandable: On the other hand, increasing the values for these parameters could yield easier-to-read solutions, but their quality in terms of accuracy values could be lower.

Differently from using the two above parameters, we can think of using a multi-objective algorithm having as two conflicting objectives the accuracy of a solution and its complexity in terms of solution size. This feature is not currently covered by DEREx, hence we could decide to use other evolution-based tools instead. With reference to this, PonyGE2 [45], a freely downloadable Grammatical Evolution [46, 47] tool, offers this possibility and we could avail ourselves of it in a future investigation phase.

6 Conclusions and Future Work

The preliminary study presented in this chapter investigates the use of a ML methodology to acquire explicit knowledge on the reasons for which Covid-19 virus is impacting with different strength levels the different countries.

The experiments suggest that the proposed approach is actually feasible, and can yield interesting results.

Our approach allows obtaining a set of easy-to-understand rules, at least one per class, that permit us to classify the countries with a good degree of correctness with respect to the different levels of impact.

The proposed explicit knowledge achieved in our experiments represents a mixture of several hypotheses put forward, none of which can on its own completely explain the different impact of Covid-19 on different countries.

In order to continue working on this path, we have clearly evidenced in Sect. 5 a set of weaknesses of the current approach, as well as a list of to-do issues. All of these shall be faced with care and solved. Conceptually, these are problems with low difficulty degrees, yet effectively solving them may require a far-from-trivial amount of time.

Appendix

See Table 10.

Table 10 The countries, their parameter values, and their class. For the classes: 1 = low impact, 2 = medium impact, 3 = high impact

Country	String. index	Median age	Populat. density	Average temper.	Pollut. index	Class
Afghanistan	56.12	18.80	59.63	12.60	93.45	2
Albania	66.60	32.90	105.00	11.40	77.77	2
Algeria	63.83	28.10	18.41	22.50	65.75	1
Argentina	69.04	31.70	16.52	14.80	50.67	2
Australia	58.06	38.70	3.32	21.65	23.46	1
Austria	63.72	44.00	109.30	6.35	22.19	2
Azerbaijan	74.18	31.30	122.70	11.95	78.12	2
Bahrain	17.51	32.30	2239.00	27.15	71.43	3
Bangladesh	68.95	26.70	1265.00	25.00	86.21	2
Belarus	61.58	40.00	46.57	6.15	43.63	3
Belgium	75.35	41.40	382.70	9.55	52.94	3
Bolivia	58.50	24.30	10.78	21.55	74.6	2
Bosnia a. H.	65.74	42.10	64.33	9.85	62.33	2
Brazil	67.19	32.00	25.43	24.95	54.98	3
Bulgaria	50.42	42.70	64.01	10.55	65.33	2
Cambodia	50.57	25.30	94.71	26.80	79.5	1
Canada	64.83	42.20	4.15	-5.35	27.83	3
Chile	52.79	34.40	25.71	8.45	65.78	3
China	66.06	37.40	153.30	6.95	80.77	1
Colombia	68.82	30.00	45.86	24.50	62.83	2
Costa Rica	63.02	31.30	99.77	24.80	46.97	1
Croatia	69.62	43.00	73.36	10.90	30.46	2
Cyprus	57.27	36.80	130.70	18.45	53.55	2
Czech Republic	59.35	42.10	138.60	7.55	40.23	2
Denmark	58.58	42.20	136.50	7.50	21.33	3
Dominican Rep.	68.83	28.10	224.50	24.55	77.9	2
Ecuador	78.56	27.70	71.04	21.85	57	3
Egypt	66.69	23.90	102.80	22.10	85.65	2
Estonia	56.18	42.70	31.29	5.10	19.81	2
Ethiopia	50.39	17.90	115.00	22.20	79.19	1
Finland	77.46	42.50	18.23	1.70	11.55	2
France	68.14	41.40	119.20	10.70	43.56	3

(continued)

Table 10 (continued)

Country	String. index	Median age	Populat. density	Average temper.	Pollut. index	Class
Georgia	56.05	38.10	57.41	5.80	71.09	1
Germany	61.44	47.10	240.40	8.50	29.03	3
Ghana	64.79	21.10	136.60	27.20	89.2	2
Greece	69.12	44.50	80.86	15.40	52.55	1
Guatemala	60.94	22.10	167.20	23.45	70.65	2
Hungary	63.18	42.30	106.70	9.75	48.29	2
Iceland	58.51	36.50	3.40	1.75	16.21	3
India	78.03	27.90	464.10	23.65	78.87	1
Indonesia	71.66	30.20	151.00	25.85	66.56	1
Iran	51.50	30.30	51.58	17.25	77.45	3
Iraq	85.20	20.00	92.61	21.40	78.83	2
Ireland	45.48	36.80	71.68	9.30	33.99	3
Israel	74.91	29.90	400.00	19.20	57.25	2
Italy	80.45	45.50	205.60	13.45	55.63	3
Jamaica	73.45	26.00	273.40	24.95	68.92	1
Japan	75.28	47.30	346.90	11.15	39.59	1
Jordan	42.23	22.50	114.90	18.30	77.78	1
Kazakhstan	71.83	30.60	6.96	6.40	75.15	2
Kenya	76.52	19.70	94.48	24.75	76.6	1
Kuwait	49.07	29.30	239.70	25.35	68.69	3
Lebanon	60.39	30.50	667.20	16.40	88.37	1
Lithuania	74.75	43.70	43.44	6.20	28.8	2
Luxembourg	65.32	39.30	241.70	8.65	24.28	3
Malaysia	79.19	28.50	98.51	25.40	63.18	1
Mauritius	61.35	35.30	626.50	22.40	55.88	1
Mexico	50.99	28.30	66.33	21.00	66.1	2
Morocco	45.36	29.30	82.70	17.10	70.64	1
Nepal	64.20	24.10	203.30	8.10	84.77	1
Netherlands	55.33	42.60	508.20	9.25	27.41	3
New Zealand	43.87	37.90	18.31	10.55	23.4	2
Nigeria	65.48	18.40	226.30	26.80	87.63	1
Norway	59.06	39.20	14.84	1.50	20.35	2
Oman	80.32	25.60	16.50	25.60	37.74	3
Pakistan	64.48	23.80	286.50	20.20	74.25	2
Panama	77.89	29.20	58.04	25.40	63.09	3
Peru	86.21	28.00	25.76	19.60	84.13	3
Philippines	50.98	23.50	367.50	25.85	74.28	1
Poland	68.42	40.70	123.60	7.85	54.46	2
Portugal	82.29	42.20	111.30	15.15	30.89	3
Qatar	80.03	33.20	248.20	27.15	61.06	3
Romania	87.15	41.10	83.58	8.80	58.42	2

(continued)

Table 10 (continued)

Country	String. index	Median age	Populat. density	Average temper.	Pollut. index	Class
Russia	72.48	39.60	8.91	−5.10	62.79	3
Saudi Arabia	70.81	27.50	16.20	24.65	65.09	3
Serbia	77.11	42.60	99.90	10.55	60.32	2
Singapore	63.35	34.60	8358.00	26.45	33.48	3
Slovakia	37.09	40.50	113.50	6.80	39.66	1
Slovenia	57.39	44.50	103.20	8.90	24.06	2
South Africa	46.55	27.10	48.89	17.75	57.3	2
South Korea	72.65	41.80	527.30	11.50	62.48	1
Spain	63.46	42.70	93.74	13.30	39.99	3
Sri Lanka	67.08	32.80	341.50	26.95	59.14	1
Sweden	68.03	41.20	24.61	2.10	18.09	3
Switzerland	60.56	42.40	219.00	5.50	22.39	3
Syria	67.44	24.30	95.30	17.75	68.42	1
Tanzania	32.14	17.70	67.44	22.35	62.93	1
Thailand	31.53	37.70	136.60	26.30	75.07	1
Trinidad a. T.	71.03	36.00	272.80	25.75	68.31	1
Tunisia	74.58	31.60	76.07	19.20	69.53	1
Turkey	67.06	30.90	109.60	11.10	67.35	3
Ukraine	72.24	40.60	75.49	8.30	65.08	2
Utd. Arab Em.	78.42	30.30	118.30	27.00	51.15	3
Utd. Kingdom	53.68	40.50	280.60	8.45	40.56	3
United States	64.00	38.10	36.19	8.55	36.88	3
Uruguay	75.88	35.00	19.85	17.55	44.84	1
Venezuela	68.98	28.30	32.24	25.35	73.76	1
Vietnam	40.11	30.50	313.90	24.45	86.47	1
Zimbabwe	68.69	20.00	38.42	21.00	74.61	1

References

1. Schmitt M (2020) How to fight covid-19 with machine learning, https://towardsdatascience.com/fight-covid-19-with-machine-learning-1d1106192d84. Accessed 17 Dec 2020
2. DeCaprio D, Gartner J, Burgess T, Kothari S, Sayed S, Building a covid-19 vulnerability index, arXiv:2003.07347
3. Balasubramanian M (2020) Covid 19-The new age pandemic. Notion Press
4. Richardson P, Griffin I, Tucker C, Smith D, Oechsle O, Phelan A, Stebbing J (2020) Baricitinib as potential treatment for 2019-ncov acute respiratory disease. Lancet (London, England) 395(10223):e30
5. Zhang H, Saravanan KM, Yang Y, Hossain MT, Li J, Ren X, Pan Y, Wei Y (2020) Deep learning based drug screening for novel coronavirus 2019-ncov. Computational Life Sciences, Interdisciplinary Sciences, p 1

6. Li L, Yang Z, Dang Z, Meng C, Huang J, Meng H, Wang D, Chen G, Zhang J, Peng H et al (2020) Propagation analysis and prediction of the covid-19. Infect Dis Modell 5:282–292
7. De Falco I, Della Cioppa A, Scafuri U, Tarantino E, Coronavirus covid-19 spreading in italy: optimizing an epidemiological model with dynamic social distancing through differential evolution. arXiv:2004.00553
8. Zhao Z, Li X, Liu F, Zhu G, Ma C, Wang L (2020) Prediction of the covid-19 spread in african countries and implications for prevention and controls: a case study in south africa, egypt, algeria, nigeria, senegal and kenya. Science of the Total Environment 138959
9. Senior AW, Evans R, Jumper J, Kirkpatrick J, Sifre L, Green T, Qin C, Žídek A, Nelson AW, Bridgland A et al (2020) Improved protein structure prediction using potentials from deep learning. Nature 1–5
10. Babayan SA, Orton RJ, Streicker DG (2018) Predicting reservoir hosts and arthropod vectors from evolutionary signatures in rna virus genomes. Science 362(6414):577–580
11. Moore KA, Lipsitch M, Barry JM, Osterholm M, Covid-19: the cidrap viewpoint: Part 1: the future of the covid-19 pandemic: Lessons learned from pandemic influenza, CIDRAP: University of Minnesota. April 30th
12. Cincotta R (2020) Population age structure: the hidden factor in covid-19 mortality, https://www.newsecuritybeat.org/2020/05/population-age-structure-hidden-factor-covid-19-mortality/. Accessed 07 Dec 2020
13. Miller J (2020) Covid-19 context clues: Social factors shape the curve, https://medicalxpress.com/news/2020-04-covid-context-clues-social-factors.html. Accessed 07 Dec 2020
14. Triplett M (2020) Evidence that higher temperatures are associated with lower incidence of covid-19 in pandemic state, cumulative cases reported up to march 27. medRxiv. https://www.medrxiv.org/content/early/2020/04/12/2020.04.02.20051524
15. Berkeley Lab, Using machine learning to estimate covid–19's seasonal cycle, https://newscenter.lbl.gov/2020/05/20/using-machine-learning-to-estimate-covid-19s-seasonal-cycle/. Accessed 10 Jun 2020
16. Bergamin F (2020) Lockdown measures are having an impact in spread of covid-19, https://medicalxpress.com/news/2020-04-lockdown-impact-covid-.html. Accessed 10 Jun 2020
17. Rudin C (2019) Stop explaining black box machine learning models for high stakes decisions and use interpretable models instead. Nature Mach Intell 1(5):206–215
18. De Falco I (2013) Differential evolution for automatic rule extraction from medical databases. Appl Soft Comput 13(2):1265–1283
19. Storn R, Price K (1997) Differential evolution-a simple and efficient heuristic for global optimization over continuous spaces. J Global Optim 11(4):341–359
20. Price K, Storn RM, Lampinen JA (2006) Differential evolution: a practical approach to global optimization. Springer Science & Business Media, Berlin
21. Onovo A, Atobatele A, Kalaiwo A, Obanubi C, James E, Gado P, Odezugo G, Ogundehin D, Magaji D, Russell M, Using supervised machine learning and empirical bayesian kriging to reveal correlates and patterns of covid-19 disease outbreak in sub-saharan africa: Exploratory data analysis. Available at SSRN 3580721
22. Hale T, Webster S, Petherick A, Phillips T, Kira B, Oxford covid-19 government response tracker, Blavatnik School of Government. Data use policy: Creative Commons Attribution CC BY standard 25
23. Apple, Covid–19 report on mobility trends, https://www.apple.com/covid19/mobility. Accessed 10 June 2020
24. World Population Review, Countries by median age 2018, https://worldpopulationreview.com/countries/median-age/. Accessed 07 Dec 2020
25. Statistics Times, List of countries by population density at april 20, 2020, http://statisticstimes.com/demographics/countries-by-population-density.php. Accessed 07 Dec 2020
26. Mitchell TD, Carter TR, Jones PD, Hulme M, New M, et al (2004) A comprehensive set of high-resolution grids of monthly climate for europe and the globe: the observed record (1901–2000) and 16 scenarios (2001–2100), Tyndall centre for climate change research working paper 55(0), p 25

27. Numbeo, Pollution index by country 2020, https://www.numbeo.com/pollution/. Accessed 07 Dec 2020
28. Bloomberg Quint, These are the economies with the most (and least) efficient health care, https://www.bloombergquint.com/global-economics/u-s-near-bottom-of-health-index-hong-kong-and-singapore-at-top. Accessed 07 Dec 2020
29. Sorokowska A, Sorokowski P, Hilpert P, Cantarero K, Frackowiak T, Ahmadi K, Alghraibeh AM, Aryeetey R, Bertoni A, Bettache K et al (2017) Preferred interpersonal distances: a global comparison. J Cross-Cultural Psychol 48(4):577–592
30. Worldometers, Covid-19 coronavirus pandemic, https://www.worldometers.info/coronavirus/. Accessed 07 Dec 2020
31. Rubin DB (2004) Multiple imputation for nonresponse in surveys, vol 81. Wiley, New York
32. Sannino G, De Falco I, De Pietro G (2014) Effective supervised knowledge extraction for an mhealth system for fall detection, pp 1378–1381
33. Sannino G, De Falco I, De Pietro G (2014) An automatic rules extraction approach to support osa events detection in an mhealth system. IEEE J Biomed Health Inf 18(5):1518–1524
34. De Falco I, Scafuri U, Tarantino E (2016) A differential evolution approach for classification of multiple sclerosis lesions. In: IEEE symposium on computers and communication (ISCC). IEEE, pp 141–146
35. De Falco I, Della Cioppa A, Tarantino E (2002) Discovering interesting classification rules with genetic programming. Appl Soft Comput 1(4):257–269
36. Garner SR, et al (1995) Weka: the waikato environment for knowledge analysis. In: Proceedings of the New Zealand computer science research students conference, pp 57–64
37. Breiman L (2001) Random forests. Mach Learn 45(1):5–32
38. Zeng Z-Q, Yu H-B, Xu H-R, Xie Y-Q, Gao J (2008) Fast training support vector machines using parallel sequential minimal optimization. In: 3rd International conference on intelligent system and knowledge engineering, vol 1. IEEE, pp 997–1001
39. Rumelhart DE, Hinton GE, Williams RJ, et al, Learning representations by back-propagating errors. Cognit Mod 5(3)
40. Breiman L (1996) Bagging predictors. Mach Lear 24(2):123–140
41. Cohen WW (1995) Fast effective rule induction. In: Machine learning proceedings 1995, Elsevier, Amsterdam, pp 115–123
42. Quinlan JR (1987) Simplifying decision trees. Int J Man-Mach Studies 27(3):221–234
43. Frank E, Witten IH, Generating accurate rule sets without global optimization
44. Google, Covid–19 report on movements within the community, https://www.google.com/covid19/mobility/. Accessed 07 Dec 2020
45. Fenton M, McDermott J, Fagan D, Forstenlechner S, Hemberg E, O'Neill M (2017) Ponyge2: grammatical evolution in python, pp 1194–1201
46. Ryan C, Collins JJ, Neill MO (1998) Grammatical evolution: evolving programs for an arbitrary language. In: Banzhaf W, Poli R, Schoenauer M, Fogarty TC (eds), European conference on genetic programming, vol 1391 of LNCS. Springer, Berlin, pp 83–96
47. Brabazon A, O'Neill M, McGarraghy S (2015) Grammatical evolution. In: Natural computing algorithms, natural computing series. Springer, Berlin (Chap 19), pp 357–373

Chapter 2
Internet of Health Things (IoHT): The Significance of Virtual Tools Aiding to Overcome Novel Coronavirus (COVID-19) Pandemic

Lakshmi Narasimha Gunturu and Girirajasekhar Dornadula

Abstract Severe Acute Respiratory Syndrome virus 2 (SARS-CoV-2), a viral infection of novel coronavirus that originated in Wuhan, China, led to panic among the public and health care professionals due to its rapid spread around the world. Health care professionals and researchers at present days are targeting various approaches for the detection and prevention of viral growth. One of the best approaches to tackle this pandemic is the implementation of the Internet of Health things (IoHT) based on Artificial Intelligence (AI) approaches in healthcare. Internet of Health Things (IoHT) is a smart technology-based approach where a person can take self-care by the use of existing technology approaches. The first objective of this chapter is to provide a comprehensive review of IoHT-based approaches used in the prevention, detection, and control of SARS CoV-2 transmission, and the second objective is to provide knowledge on the future pandemics and their control by IoHT. We conducted an electronic search strategy in the Pub Med database, Google Scholar, and other technology-related websites that utilized the IoHT against the novel coronavirus. All the suitable free full-text articles are extracted and included in the study. In this chapter, by gathering various recent literature evidence, we clearly explained the implementation of IoHT technologies at various stages of corona pandemic for both health care settings and the common public to effectively cope up with the COVID-19 outbreak and also for future pandemics.

Keywords Internet of health things (IoHT) · Artificial intelligence (AI) · Virtual tools · COVID-19 · Future pandemics · Prevention and control strategies

---

L. N. Gunturu (✉) · G. Dornadula
Department of Pharmacy Practice, Annamacharya College of Pharmacy, Andhra Pradesh, Rajampeta, India

U. Kose et al. (eds.), *Computational Intelligence for COVID-19 and Future Pandemics*, Disruptive Technologies and Digital Transformations for Society 5.0,
https://doi.org/10.1007/978-981-16-3783-4_2

1 Introduction

The pandemic of novel corona virus is caused by Severe Acute Respiratory Syndrome 2 (SARS) and originated in Wuhan, China in December 2019. It belongs to the genera of the Beta coronavirus, and its genetic material is similar to viral genetic material of SARS and Middle East Respiratory Syndrome virus (MERS) [1]. Since its outbreak, the viral transmission is at a rapid pace among the public. As of December 19, 2020, World Health Organization has confirmed 74,299,042 cases of pandemic, and also, it took the lives of 1,669,982 people globally [2]. Therefore, the outburst of the virus is declared as Pandemic by the World Health Organization (WHO). Researchers find that the basic reproduction number of the novel coronavirus is 2–3 which is high when compared to the other viruses of the same genera [3]. The basic reproduction number is defined as an average number of secondary cases generated by the infected patients in a group of susceptible populations. Probably due to this reason, there is a rapid viral spread [4]. Drugs used to treat this pandemic are mostly Antivirals like Favipiravir, Lopinavir/Ritonavir, Remdesivir, Ribavirin, and Antibody therapies like Convalescent Plasma, Imdevimab, and Bamlanivimab. However, most of these experimental therapies are resulting in drug interactions in the affected Covid-19 patients with other Co-morbidities [5]. Therefore, it has become a difficult task for the researchers to identify a potent active molecule that can benefit the public without producing any toxic results. Other options in identifying the potential drug therapy are the use of the existing or shelved drugs incorporating technology like Artificial Intelligence (AI) to decrease the viral transmission [6–8]. Hence, there is no specific drug therapy for the cure of Covid-19. Existing approaches like pharmacological or experimental treatments only aim at reducing the complications.

Other aspects like preventive measures are also important for reducing the spread of the virus. Hence, to reduce the transmission of the virus and to detect the cases, researchers are looking at alternative approaches based on technology. Technology, nowadays, has rapidly changed the living world. Most of the public use technology-related applications every day. They can also be used in health care settings as Artificial Intelligence and its subgroup as the Internet of Things or Internet of Health Things (IoHT) [9].

2 Artificial Intelligence and Internet of Health Things

2.1 Artificial Intelligence

In the contemporary world, human thinking has investigated the several aspects of science concerned with predictive modeling or mimicking. Each day, studies confined to this area come up with ultra-modern thoughts to benefit the public. This thinking can be imbibed by performing several aspects like basic studies, research, and literature evidence that would benefit society. This thinking or Intelligence emulated by mechanical devices and sensors that are combined with software technology is termed as Artificial Intelligence. It includes various mathematical models like neuronal networks, decision trees, and Machine learning algorithms. As we all are aware of the fact that a human brain can easily perform mathematical functioning within a short stipulated time. However, the finishing of complex operations by human thinking needs time. So, AI-based computer applications or programming are utilized in robust ways to analyze the data in the given short period.

These applications of AI in the health field mainly focus on, firstly, providing the specific interventional parameters to the right group of people in a given time [10]. Secondly, it helps the researchers and public to utilize the available literature evidences in the data sets and other repositories or arithmetic (digital) information regarding the epidemiological disease outbreaks, like pandemics, and drug surveillance process to assess the treatment modalities that would help the physicians to heal the diseases [11–13]. Thirdly, AI-based approaches can also be implemented for its use in the development of Precision or Personalized medicine to treat the various disorders. Despite these merits, the question that arises here is how often the public utilizes or adapts AI applications during emergency needs [14].

2.2 *Internet of Health Things*

IoHT is a network of devices that consists of biosensors, wearables, healthcare devices, and other technology-related systems that would analyze the data provided. Simply, it had interconnection of physical devices, internet use, artificial intelligence, and data analyzing sensors or hardware [15–17]. It is considered a wireless means of exchanging information. The basis for the IoHT is IoT that was coined by Kevin Ashton in 1999. It was also called machine technology [18]. In Fig. 1 below, we presented the various elements that are part of IoHT, like mobiles, apps, sensors, devices, and other healthcare workers.

IoHT is used in various ways like smart transportation, smart farming, smart city, analysis of logistic regressions, and also in health care. Due to the present situation of the Covid-19 pandemic, many active cases and death have been reported worldwide. As a result, the public is very conscious regarding the safety of their health. Because of all these reasons, IoHT-based applications have become an effective research area where the scientists and researchers are having their focus [19].

3 Scope of IoHT in Healthcare

IoHT is used for various purposes in healthcare settings as:

1. Vehicle monitoring.
2. Surveillance.
3. Telemedicine healthcare.

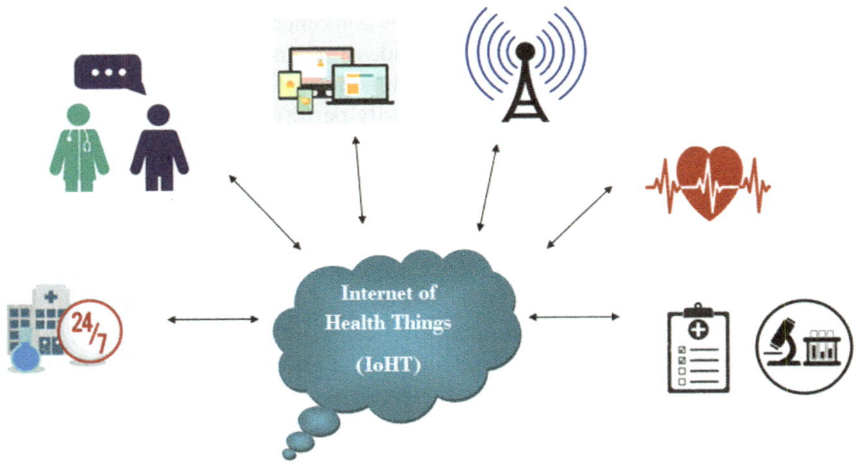

Fig. 1 Schematic representation of components of the Internet of Health Things

3.1 Vehicle Monitoring

IoHT integrated with AI is used in the vehicles for determining the destinations, to trace the route of vehicles, and particularly in the ambulances, this technology saves the lives of many people, as it provides the first aid measures for the emergencies of which all the equipment is embedded with this feature. Apart from these, vehicles integrated with IoHT devices can send the signals to the respective interconnected networks as a result of the vehicle had been met with an accident, then an alerting signal is sent to the confined emergency authorities that save the life of the person injured. It also provides information about the blocked areas on road with heavy traffic [20].

3.2 Surveillance

It is also used in the surveillance of health care settings, people, and also the database of the patient records to assess the huge amount of information. This mechanism of surveillance is also applicable to monitor the air pollution levels in the environment. This will help the environment agencies or health officials to predict the amounts of toxic exposure of the public to various poisonous gases [21].

3.3 Telemedicine and Healthcare

Here, this technology of IoHT is used as active trackers of patient condition, and biomedical sensors that provide the physiological activity and record the vital parameters. IoHT-based health care devices collect the data, analyze the information recorded, and monitor the patient's condition regularly. They can minimize the stress levels and also record the amount of physical activity done by the person each day. The best example for this is Blood pressure monitoring devices, Pulse recording devices, Pacemakers, Hearing aids, smartwatches. Such an application is Biostrap, a wearable device to monitor heartbeat [22]. They can help the patient to administer the appropriate drug or medication for existing disease and act as alarm-giving tools. So, that without the need of caregiver, a patient can take his medication. For all these reasons, IoHT is also known as Smart Health or Smart Technology [23–25].

4 Objectives

The first part of this chapter aims to provide a comprehensive review of the impact of the Internet of Health things against the Covid-19, and to assess all the possible ways to prevent the Covid-19 transmission and its detection by the use of IoHT. In the second part of this study, we discussed the various IoHT applications in use for the control of COVID-19 pandemic at different phases, and in the last section of the study, we discussed the use of IoHT for future pandemics.

5 Materials and Methods

a. Constructing an appropriate question.
b. Using the specific key terms and the appropriate database.
c. Advanced searching in PubMed was carried out with MeSH terms (Table 1).

Table 1 PICO (Patients, Intervention, Comparator, Outcome) for designing our question

Parameters	Included	Excluded
Patients	Covid-19	Other diseases
Intervention	Internet of health things, artificial intelligence	5G, big data
Comparator	Public, healthcare professionals	—
Outcome	Predicting and detection of increased cases and ways to control Covid-19	Mortality reports

PubMed, Google scholar, and other technology websites were searched with the keywords Internet of Health Things, Covid-19, Healthcare, and Future pandemics till December 20, 2020. Title and abstract screening were done manually. Full texts of these received articles were further screened for possible inclusion and exclusion criteria. We excluded Case studies, Un-authored proofs, and Fuzzy systems.

6 Results and Discussion

Our search results concluded that various ways of IoHT are used in preventing the spread of Covid-19 and are mentioned in this section that starts with below in Table 2.

6.1 IoHT-Driven Cluster Identification and Technology Advancements to Combat Covid-19

As mentioned earlier, IoHT used the technology in medical services, thus it easily exchanges the data and acquires the information. It helps the public, Healthcare workers, and other Government civilians to keep in touch with each other. As a result of the implementation of this technology, during the crisis of Covid-19 Pandemic, effective patient tracing is done, and also, suspected patients can be easily identified. But, one must take control that symptoms reported should be proper and accurate diagnosis is needed, because as most of the symptoms coincides with the normal seasonal diseases. So, therefore, it is the responsibility of the doctors, Health policy-makers, Nurses, and Patient caretakers to report the symptoms and isolate the patients [31, 32].

Table 2 Studies that targeted ways to counteract the Covid-19 spread by IoHT

Study	Proposed ways to counteract the virus
Singh et al. [26]	a. Contact tracing and cluster identification b. Technology advancement apps
Swayamsiddha [27]	a. Real-time tracking of patients b. Remote monitoring of patients c. Rapid Diagnosis d. Screening and Surveillance e. Workload reduction
Rehm et al. [28]	a. Assisting in Patient-ventilator Asynchrony (PVA) b. Developing mobile-based applications for ventilator waves
Moro Visconti et al. [29]	Digital platforms improve patient care, satisfaction, and treatment outcomes
Ben Hassen et al. [30]	Development of more sophisticated hospitals and facilities

To make public the potential symptoms and the care that has to be followed during the corona pandemic, the Governments of various countries had used the IoHT-based other AI technologies that combined to provide the public mobile-based Smart apps. The best example for this is the Government of India launched a new mobile-based application named AarogyaSetu. This app consists of various instructions along with how to self-isolate the infected persons. Similarly, the country China released the app called Close contact that enables the users to identify the place of residence and their mobility. Taiwan, which consisted of most infected corona cases after China, also instructed specific strategies for the suppression of the pandemic. It integrated the national health data based on the cloud-based and IoHT applications to analyze the patient's disease symptoms. It also issued the appropriate warnings for the public and also generated a QR-based technology of IoT that enables the users to report their travel history [33].

6.2 IoHT as a Tracing Tool

Daily real-time tracing of Covid-19 severities is very important for the Government. So that the number of cases registered, people who have been on active treatment, the number of mortality cases that occurred, and people who have been in isolation due to the infection of Corona can be easily traced with the IoHT-based AI algorithms and other smart applications. This would help the physicians and health policymakers to shortlist the defined pathological concepts of pandemic and also can aid in the drug discovery process. Treatment measures and precautions can also be updated daily by the use of the Internet of Health Things to assure safety to the public [34–36].

6.3 IoHT-Based Technology for Remote Screening

As we all know that pandemic is contagious and easily spreads, therefore, it is also difficult for the healthcare workers and doctors to provide treatment to the infected patients. But by the integration of IoHT in smart devices as sensors, they can easily monitor the health vitals of the patients like blood pressure, temperature, electrocardiogram, heart rate, and breathing activity. Another advantage of this is doctors can easily access the patients remotely without direct contact with the person. Therefore, IoHT is also used for the providence of patient treatment [37].

6.4 IoHT as a Useful Technique in the Diagnostic Procedure

Diagnosis of corona patients is an important step in the identification of the disease, because there might be the chances of false positives and also false negatives. That

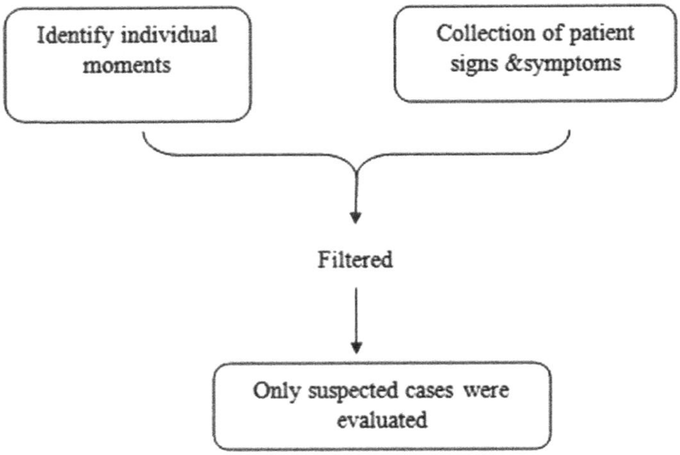

Fig. 2 Use of IoHT-based simplified diagnostic procedure

is the reason immigrant persons and even suspected patients are quarantined. IoHT-integrated AI-based devices can easily diagnose the disease from other forms of conditions like pneumonia. Most of the studies confirmed that AI-based CT is more effective in identifying the different forms of pneumonia that are caused by Covid-19 [38]. A study of Ko et al. developed a fast track Covid-19 network to diagnose Covid-19 pneumonia with an accuracy of 96.97% [39]. Another study by Li et al. confirmed the detection of Covid-19 from community-acquired pneumonia and other lung diseases [40]. All these are possible by the integration of sensors and the development of the Internet of Health Things.

Screening of the public had become mandatory at all public places like airports, hotels, malls, etc. This screening is done by the thermal-based imaging techniques that easily identify even the minor changes in the body regarding the variations in the temperature which is an important clinical sign for the Covid-19 patients. All these thermal-based imaging technologies work with sensors of IoT backed with Cognitive internet of medical things (CloMT) [41]. A simplified procedure for use of IoHT in use is described in Fig. 2.

6.5 IoHT for the Work Load Reduction During Corona Pandemic

Another important way in which IoHT is used against the Corona is the reduction of the workload. A normal conventional procedure takes a long time for a disease to get identified. They also require skilled professionals to analyze and interpret the data. All these manual handling techniques require some time to provide the report,

Table 3 Comparison of IoHT and conventional approaches for the detection of Covid-19

Parameter	IoHT-integrated applications	Conventional
Specificity	More and it also provides the best confirmatory methods	Less when compared to IoHT because there might be handling errors, interpretation by lab technicians who do not have sufficient skills
Sensitivity	More as it is integrated with highly advanced technology approaches	Less sensitivity is observed with the conventional methods; this might be due to the overlapping of the other disease conditions and comorbid conditions
Cost	High initially but there is a decrease in cost when it is widely used among the public	Sometimes high and sometimes low depending upon the type of the service providers
Duration	Within a short time report is provided	It takes nearly 2–3 days for the actual report

whereas IoHT-integrated AI-based devices can easily minimize these time taken, and provide the report very quickly for a huge amount of population. As the public can easily be traced with the use of smart devices and sensors, so it is not therefore needed for the healthcare professionals like doctors and lab technicians to visit the nearby houses for the monitoring of patient vital symptoms [42].

Below is Table 3 that compares the IoHT-based applications with conventional results.

During the corona pandemic, the use of the ventilators had become the most prominent. Particularly, the older age people, who are with existing co-morbidities, along with this infection are more vulnerable, and their respiratory parameters are worsened. Therefore, ventilator requirement had become a must strategy. However, due to lack of ventilator therapies, many deaths had also been recorded. Most of the present-day ventilators in the health care settings do not have automated systems or devices to record the variations in patient ventilator asynchrony. As a result, clinicians had to be near the patients to record the vitals of respiration. In the study of the Rehm et al., it is reported that they developed a system that can improve the accuracy of ventilator-based respiratory changes. It is integrated with the IoHT and other AI-based algorithms that can detect the different types of breath sounds. Therefore, the internet of things can also be used in the ventilator-assisted devices to monitor the respiration of patients.

Through the use of the Internet of Health things, various technologies of medical applications with proper regulation from health care professionals are available. The utility of such applications-based services will surely benefit the public as well as also implement the internet of health things against the corona pandemic in a wider way. This also addresses the emerging problems in health care. It also acts as a tool for Patient health records maintaining, Clinical decision making, medical training, and medical education. Numerous technologies along with the IoHT that are involved in the medical software applications are due to the rapid growth of the use of smart phones and mobile devices.

7 Scope of Various IoHT Applications Used in Covid-19 Tackling

Internet of Health things (IoHT) on the Covid-19 is one of the potential preventive measure approaches. IoHT that works with AI is a better way to prevent the spread of the virus and to create awareness among the public regarding the potential symptoms of the pandemic outbreak. There are three main phases of the Covid-19 pandemic. These include:

1. Early diagnosis.
2. Self-isolation or Quarantine time.
3. Post-recovery stage.

Various IoHT technologies utilized efficiently in all these phases to prevent viral transmission are discussed in this section.

7.1 Early Diagnosis

This is the key phase of the disease. As evidence suggests, because of high reproductive number of the virus, there is a rapid pace of viral transmission. Even the patients who do not show the symptoms (asymptomatic) are also capable of spreading the virus at this stage. Hence, if the patient is diagnosed quickly during this stage, then we can control the transmission of the virus. The diagnosis symptoms include a high fever of 100 F, nausea, vomiting, sore throat, breathlessness, and loss of taste [43]. All these alterations can be easily identified by the IoHT devices by the use of the sensors and can rapidly identify infected persons. Various IoHT applications are used to record these alterations as discussed below.

7.1.1 Wearables

These are considered efficient applications to identify the disease at an early stage. An IoHT wearable can easily detect the respiratory alterations in a patient and can conclude whether the patient is infected with the virus. This will provide better diagnostic accuracy and also helps to fight the pandemic efficiently. The best examples of wearables are Smart thermometers, Smart helmets, and Smart Glasses.

Smart thermometers like iFever and Tempdrop can record and identify even the minute changes in the body temperature. They are cost-effective, easy to wear, and provide accurate results. In the conventional approach, we can use infrared thermometers for the detection of temperature variations. However, due to the proximity between the patient and the healthcare provider, there might be a chance of viral transmission. So, always a smart thermometer is recommended [44].

Fig. 3 Detection of suspected cases by use of Smart helmet

Smart helmets like KC N901 that are mostly used in the Chinese countries are provided with a thermal camera which can be considered as efficient ways of recording temperature rise when compared to the infrared thermometer. Whenever there is a rise in the body temperature, these sensors in the thermal camera recognizes it and send the signals to respective health care authorities and agencies in such a way that the face of the person can easily appear on the monitor or system. They also provide the location history of the patient [45]. Therefore, it becomes easy for the health authorities to identify the asymptomatic persons infected with a virus. Figure 3 briefs out the technology associated with Smart helmet.

Another application is the Smart Glasses that are similar to the smart helmets in the technology features. These consist of a thermal camera, face recognition features, and a location identifying or tracking system, so that it can easily identify suspicious persons with temperature changes in a group of people. As it also consists of a location tracker, therefore, it also gives information regarding the places the suspected patient visited. The best example of these smart glasses is Rokid that had the capability of monitoring 200 persons in public or crowd [46].

7.1.2 Drones

These drones are used to identify infected people in a place or a crowd. These are technically termed as Unmanned Aerial Vehicles (UAV) used in the diagnosis and control of Covid-19. Drones fixed with thermal imaging features are capable of sensing the rise in temperature of a person and can send an alarming signal to the health agencies. Such drones are combined with wearable virtual reality features to identify the infected person location. Advantages include less human involvement and also a thermally virtual zone that can reach any place to monitor the pandemic where humans cannot. Canadian company developed a drone application that can

Table 4 Social apps and their uses

Apps	Associated Uses
NCapp [49]	Provide long-term health benefits and health counseling parameters
Stop Corona [50]	Updates viral symptoms and identify the high-risk areas
Social distance monitoring apps [51]	Helps to identify the patient infected with Covid-19 and also provides this information to Government officials

monitor and detect human temperatures and also identify respiratory changes like heartbeat and coughing [47, 48].

7.1.3 IoHT Buttons and Smart Phone Applications

IoHT-developed buttons act as alerting system for a health agency to identify and diagnose the infected cases. An IoHT button called Wanda quick touch developed by Vision state is implemented in hospital settings as an alerting security system.

In the present world, each person has his/her mobile phones; hence, these are also designed in such a manner where they can perform the limited tasks within the mobile screens. The best examples for this smartphone developments use in pandemic control are namely nCapp, Stop Corona, and Social distance monitoring apps [49–51]. Below is Table 4 describing the uses of these apps.

7.2 Self-isolation or Quarantine Period

Once the active cases or the infected patients are diagnosed and identified, it is important to isolate them to prevent the transmission of the virus. This isolation may be in the home or in the hospital. The use of IoHT-based applications discussed in this section can reduce the transmission of the virus and also monitor the vital signs of patients like heart rate, blood pressure, and respiratory signs.

7.2.1 Wearables

During the self-isolation period, there is a huge chance of viral spread from the infected patients. So, to prevent this and to monitor the patients, the effective method is the use of wearable bands. These are cost-effective and provide accurate results. They are connected to a patient smartphone application with Bluetooth fixed with sensors. Whenever the patient tries to escape from the quarantine place, it alerts the health officials through the web interface. The best example for this is IoT-Q-Band, a

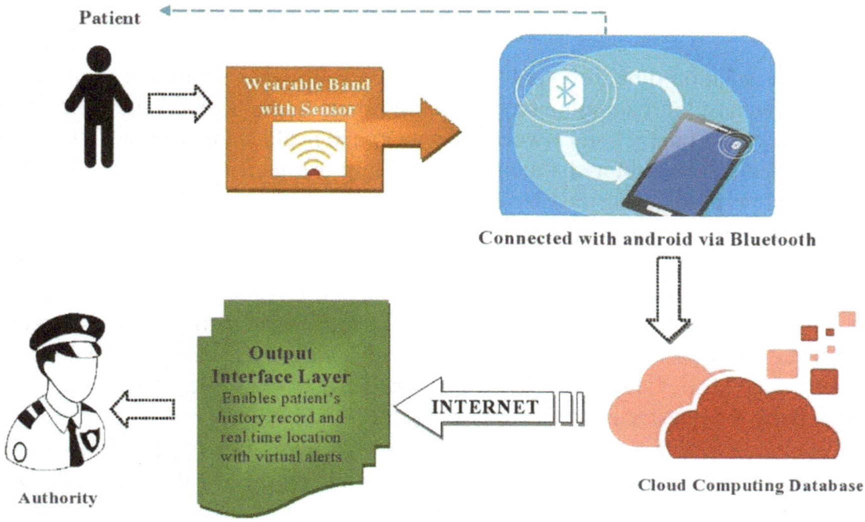

Fig. 4 Detection of suspected patient by using wearable band devices

type of wearable device that can detect the person's location [52]. Figure 4 explains the process of working of this device.

7.2.2 Drones

Drones during this phase play a vital role in preventing disease transmission. By the use of these drones, there can be no or minimal interaction between the infected or suspected cases and also from the infected areas. The most popular drones used in the quarantine period are the Disinfectant drone and Medicine delivery drone [53].

Disinfectant drones sanitize the place of the infected areas because sanitization plays a key role in the killing of virus and this can be achieved by the Disinfectant drone. Another advantage here is healthcare professionals and other sanitation authorities are not infected with the virus. The first disinfectant drone developed by the DJI company can disinfect approximately 100 m in 1 hour [54].

On the other side, the Medication delivery zone transfers the samples, kits, and medications between the patients and labs. This reduces human interactions. They also reduce hospital visits among the patients. This type of medicine drone delivery process used in China had increased the chance of diagnosis rate [55].

7.2.3 Robots

In the self-isolation time, robotic tools play a crucial tool in inhibiting viral transmission. Because of the employment of these robots in the hospital staff, there are fewer chances of interaction between the patient and health care workers. They also provide medications and assist the patients to improve their health condition [56]. Various robots used during the Covid-19 pandemic are Telerobots, Social robots, and Collaborative robots.

Telerobots are operated from a distance by humans and are involved in activities like disease identification, performing remote operations, and treatment of disease. Because of these special advantages, there are no human interactions that ultimately reduce the disease spread of the pandemic and also infections among the health care workers. The best example for this is Da Vinci, a type of surgical robot, operated remotely by the physicians to perform the operations [57].

Collaborative robots are used whenever there are any urgent operations to be performed. These are not as beneficial as telerobots during the corona pandemic; however, these types of robots are useful during the self-isolation time to prepare food and provide medications to the infected patients which would lessen the human interactions, health care expenses, and also decrease the use of manual power during the emergency measures. The best examples for this type are Asimov robots and eXtreme Disinfection Robot [58].

During the lockdown or self-isolation period, there can be chances of mental distraction, as a result, possibility of occurrence of psychological disturbances is more. So, to reduce this mental fatigue, social robots are employed in quarantine areas to interact with the people. This interaction between the patient and the robot can help to ease the mental stress. The best example for this is the Paro robot that reduces stress by having human interaction [59].

7.2.4 Mobile Applications

During the quarantine period, important aspect to consider is the monitoring of the patient in the isolated area to prevent viral transmission. Hence, for this purpose, smartphones are widely used. An app called Social Monitoring is made mandatory by the Russian Government for its netizens to install it on the mobile devices. Once the app is installed, patients are asked to scan for the quick response code, whenever they are leaving the quarantine place [51]. As a result, government officials have control over the infected patients and can track the patients. Similarly, Civitas app can help the physicians to monitor the health status of the infected patients remotely, it also provides information to the health authorities regarding the location of suspected patients who left the quarantine center without permission [60].

Table 5 Summary of wearable types with uses

Type of wearable	Use
1. Easyband [62]	These are integrated with sensors, Internet of Medical Things. These wearable devices provide a beep sound when there is a violation of the social distancing norm (>6 m distance) between the two persons
2. Proximity tracer [63]	This is used mainly for industrial workers. As there can be close contact within the working persons, these devices alert them whenever this happens. This helps the workers to do their work freely without any thought of viral contamination

7.3 Post-recovery Phase

It is the phase the patient is completely cured of the virus. Because of the lockdown restrictions, there has been a big impact on the business and market areas, since they are shut down due to the corona transmission. At present, lockdown restrictions had been lifted, so extra care is needed, and the onus lies on public and health officials to prevent the reinfection with the virus during the post-recovery stage. In this regard, we had discussed some of the IoHT applications that play a crucial role in combating viral transmission after the lockdown.

7.3.1 Wearables

After the lockdown, all the activities were restored like schools, businesses, market areas, malls, and other crowded places. So many people visit these areas frequently and daily. Hence, in this context, the two main things that should be focused on are contact tracing and maintaining social distance. These goals can be achieved by wearable devices. These devices can alert the public when they are violated by social distance norm [61]. Table 5 explains the types of wearables.

7.3.2 Drones

These drones are used to monitor crowded places after the recovery phase. They are also used to provide awareness among the public regarding the safety measures to be followed. This can be well explained by drones, namely, Surveillance drones and Announcement drones. Surveillance drones like Micro Multi copter [64] developed by China are being in use to provide safety measures among the public in crowded areas. They are installed with speakers and help to communicate the important information from the health care officials to the public. Other drones like Announcement drones are confined to remote areas where there is a lack of internet facility. In these areas, these drones provide important announcements to the public regarding the safety measures like social distance and covid-19 guidelines.

Table 6 Types of smartphone applications with their advantages

App	Uses
1. Trace Together [65]	This app records information about close-contact individuals, their duration of contact. Data is stored for nearly 20 days for further investigation
2. eRouska [66]	This can alert the public when a healthy person comes into contact with the virus-infected person
3. Hamagen [67]	This is very similar to the eRouska app and identifies the Covid-19 positive patients in a group
4. Coalition [68]	This is combined with IoHT and block chain technology and act as a tracing tool

7.3.3 Smartphone Applications

These smartphones are effective in giving measures to the public and also because of the wider reach of mobiles in the present world. Smart phone apps play an important role in the post-recovery stage. Such smartphone applications are discussed in Table 6 below.

8 Future Pandemics and IoHT-based Emerging Applications and Strategies

Pandemics are considered as the biggest threats to society and affect the socio-marketing conditions of an economy. Among the existing technologies in the present world, Artificial intelligence integrated with IoHT had gained the special spotlight for Covid-19 and future pandemics. Internet of Health Things (IoHT) using AI technology has shown convincing results right from early detection to recovery from COVID-19. The main goal of IoHT and Internet of Medical Things IoMT is to minimize the direct interactions between the patient and healthcare workers. This is the key principle for future pandemics also, and to achieve this, use of touchless technology should be developed by using other inputs like voice and chatbots.

8.1 SWOT Analysis of IoHT in Future Pandemics

1 **Strengths**

a. Early detection and diagnosis of patients through IoT using remote monitoring strategies [69].
b. IoT wearable devices for temperature and vitals monitoring of suspected and infected patients.

c. Disinfection with the help of IoT-based devices like robots and drones.
d. Delivery of essential goods and medicines through drones during the quarantine.
e. Telemedicine to treat infected patients.
f. Smartphone applications to alert and track infected patients.

2. **Weakness**

a. Breach privacy of data
b. Storage and retrieval of data
c. Expensive
d. Patient satisfaction may not be met

3. **Opportunities**

a. Challenges to tackle future pandemics and try to find out new solutions.
b Incorporation of AI technology in the healthcare system will be precise and accurate.

4. **Threats**

a. Vulnerability to the hacking of applications.
b. Misuse of data among individuals.

8.2 Strategies to Be Adapted for Future Pandemics

a. To develop new algorithms for unanswered problems using IoHT.
b. .Robust computational models for differential diagnosis of suspected and infected patients [38].
c. Automatic alert systems and software programs to minimize overcrowding and social distancing.
d. Sensor-based technologies for disinfection and sanitization of hospitals and infected areas.
e. Incorporation of AI to predict about the viral propagation models.
f. Development of high accuracy and precision models based on AI in early detection, diagnosis, and tracing the infected persons [38].
g. Development of therapeutic molecules based on docking simulation models.
h. Protect integrity, and privacy of data shared should be maintained confidentially during the use of technology.

Apart from these, various other uses are IoHT-based technologies like Cloud-based drug quality checking in the Quality Assurance and Quality control stage of drug discovery, IoHT-based patient education platforms to determine the Drug metabolism and Organic Chemistry, and BOT platforms for determining the pharmacokinetics and Pharmacodynamics of the drugs. All these improvements surely benefit researchers and scientists to develop a lead molecule in a short time and cut off the lengthy process of drug discovery.

9 Conclusions

Various technologies are in use to tackle the covid-19 outbreak. Of these technologies, Internet of Health Things had gained special limelight because of its varied applications in the health care sector and industry fields. In this scenario of pandemic outbreak, IoHT had shown significant results to prevent the viral spread and transmission. As still there is no promising therapy for the Covid-19, hence, we need additional precaution measures and should use all the existing resources to protect ourselves from the virus. Other technologies like AI integrated with IoHT is also used as one of the potential measure, not only as a preventive measure but also used in screening the lead molecules for the drug discovery that will save the time. By the application of these technologies in daily life, patients can check the health status at regular time intervals, thus leading to better achievement of results.

References

1. Sanders JM, Monogue ML, Jodlowski TZ, Cutrell JB (2020) Pharmacologic Treatments for Coronavirus Disease 2019 (COVID-19): A Review. JAMA 323(18):1824–1836. https://doi.org/10.1001/jama.2020.6019
2. WHO Coronavirus disease (COVID-19) Dashboard. https://covid19.who.int. Accessed 20 Dec 2020
3. Liu Y, Gayle AA, Wilder-Smith A, Rocklov J (2020) The reproductive number of COVID-19 is higher compared to SARS coronavirus. J Travel Med 27(2):1–4. https://doi.org/10.1093/jtm/taaa021
4. Heffernan JM, Smith RJ, Wahl LM (2005) Perspectives on the basic reproductive ratio. J R Soc Interface 2(4):281–283. https://doi.org/10.1098/rsif.2005.0042
5. Liverpool COVID-19 Interactions. https:// www.covid19-druginteractions.org/prescribing-resources. Accessed 20 Dec 2020.
6. Gao K, Nguyen DucDuy, Chen J, Wang R, Wei G-W (2020) Repositioning of 8565 existing drugs for COVID-19. J Phys Chem Lett 11(13):5373–5382. https://doi.org/10.1021/acs.jpclett.0c01579
7. Zhou Y, Wang F, Tang J, Nussinov R, Cheng F (2020) Artificial intelligence in COVID-19 drug repurposing. Lancet Digit Health 2(12):e667–e676. https://doi.org/10.1016/S2589-7500(20)30192-8
8. Ke YY, Peng TT, Yeh TK, Huang WZ, Chang SE, Wu SH et al (2020) Artificial intelligence approach fighting COVID-19 with repurposing drugs. Biomed J 43(4):355–362. https://doi.org/10.1016/j.bj.2020.05.001
9. Stanfill MH, Marc DT (2019) Health information management: implications of artificial intelligence on healthcare data and information management. Yearb Med Inform 28(1):56–64. https://doi.org/10.1055/s-0039-1677913
10. Flahault A, Geissbuhler A, Guessous I, Guérin PJ, Bolon I, Salathé M, Escher G (2017) Precision global health in the digital age. Swiss Med Wkly 147:w14423. https://doi.org/10.4414/smw.2017.14423
11. Salathé M (2018) Digital epidemiology: what is it, and where is it going? Life SciSoc Policy 14(1):1–5. https://doi.org/10.1186/s40504-017-0065-7
12. Nguyen P, Tran T, Wickramasinghe N, Venkatesh S (2017) A convolutional net for medical records. IEEE J Biomed Health Inform 21(1):22–30. https://doi.org/10.1109/JBHI.2016.2633963

13. Dernoncourt F, Lee JY, Uzuner O, Szolovits P (2017) De-identification of patient notes with recurrent neural networks. J Am Med Inform Assoc 24(3):596–606. https://doi.org/10.1093/jamia/ocw156
14. Lecun Y, Bengio Y, Hinton G (2015) Deep learning. Nature 521(7553):436–444. https://doi.org/10.1038/nature14539
15. Atzori L, Iera A, Morabito G (2010) The internet of things: a survey. ComputNetw 54(15):2787–2805. https://doi.org/10.1016/j.comnet.2010.05.010
16. Stankovic JA (2014) Research directions for the internet of things. IEEE Internet Things J 1(1):3–9. https://doi.org/10.1109/JIOT.2014.2312291
17. Basatneh R, Najafi B, Armstrong DG (2018) Health sensors, smart home devices, and the internet of medical things: an opportunity for dramatic improvement in care for the lower extremity complications of diabetes. J Diabetes SciTechnol 12(3):577–586. https://doi.org/10.1177/1932296818768618
18. Ashton K (2009) That 'Internet of Things' thing. Accessed 20 Dec 2020. https://www.scrip.org/ReferenceID=1578164
19. Zhang G, Li K, Gu D, Wang X, Yang X, Zhu K et al (2019) Visualizing knowledge evolution and hotspots of rural environment and health: a systematic review and research direction. IEEE Access 7:72538–72550. https://doi.org/10.1109/ACCESS.2019.2919549
20. Thomas MO, Rad BB (2017) Reliability evaluation metrics for internet of things, car tracking system: a review. Int J Inf Technol Comput Sci 9(2):1–10. https://doi.org/10.5815/ijitcs.2017.02.01
21. Zanella A, Bui N, Castellani A, Vangelista L, Zorzi M (2014) Internet of things for smart cities. IEEE Internet Things J 1(1):22–32. https://doi.org/10.1109/JIOT.2014.2306328
22. Biostrap (2020) Health Tracker. https://biostrap.com. Accessed 20 Dec 2020
23. da Costa CA, Pasluosta CF, Eskofier B, da Silva DB, da Rosa RR (2018) Internet of health things: toward intelligent vital signs monitoring in hospital wards. Artif Intell Med 89:61–69. https://doi.org/10.1016/j.artmed.2018.05.005
24. Gatouillat A, Badr Y, Massot B, Sejdic E (2018) Internet of medical things: a review of recent contributions dealing with cyber-physical systems in medicine. IEEE Internet Things J 5(5):3810–3822. https://doi.org/10.1109/jiot.2018.2849014
25. Abawajy JH, Hassan MM (2017) Federated internet of things and cloud computing pervasive patient health monitoring system. IEEE Commun Mag 55(1):48–53. https://doi.org/10.1109/MCOM.2017.1600374CM
26. Singh RP, Javaid M, Haleem A, Suman R (2020) Internet of things (IoT) applications to fight against COVID-19 pandemic. Diabetes MetabSyndr 14(4):521–524. https://doi.org/10.1016/j.dsx.2020.04.041
27. Swayamsiddha S, Mohanty C (2020) Application of cognitive Internet of medical things for COVID-19 pandemic. Diabetes MetabSyndr 14(5):911–915. https://doi.org/10.1016/j.dsx.2020.06.014
28. Rehm GB, Woo SH, Chen XL, Kuhn BT, Cortes-Puch I, Anderson NR, Adams JY, Chuah CN (2020) Leveraging IoTs and Machine Learning for Patient Diagnosis and Ventilation Management in the Intensive Care Unit. IEEE Pervasive Comput 19(3):68–78. https://doi.org/10.1109/mprv.2020.2986767
29. Moro Visconti R, Morea D (2020) Healthcare digitalization and pay-for-performance incentives in smart hospital project financing. Int J Environ Res Public Health 17(7):2318. https://doi.org/10.3390/ijerph17072318
30. Ben Hassen H, Ayari N, Hamdi B (2020) A home hospitalization system based on the Internet of things, Fog computing and cloud computing. Inform Med Unlocked 20:100368. https://doi.org/10.1016/j.imu.2020.100368
31. Hanna TP, Evans GA, Booth CM (2020) Cancer, COVID-19 and the precautionary principle: prioritizing treatment during a global pandemic. Nat Rev ClinOncol 17(5):268–270. https://doi.org/10.1038/s41571-020-0362-6
32. Roy D, Tripathy S, Kar SK, Sharma N, Verma SK, Kaushal V (2020) Study of knowledge, attitude, anxiety & perceived mental healthcare need in Indian population during COVID-19 pandemic. Asian J Psychiatr 51:102083. https://doi.org/10.1016/j.ajp.2020.102083

33. Wang CJ, Ng CY, Brook RH (2020) Response to COVID-19 in Taiwan: big data analytics, new technology, and proactive testing. J Am Med Assoc 323(14):1341–1342. https://doi.org/10.1001/jama.2020.3151
34. Vaishya R, Javaid M, Khan IH, Haleem A (2020) Artificial intelligence (AI) applications for COVID-19 pandemic. Diabetes MetabSyndr 14(4):337–339. https://doi.org/10.1016/j.dsx.2020.04.012
35. Ting DSW, Carin L, Dzau V, Wong TY (2020) Digital technology and COVID-19. Nat Med 26(4):459–461. https://doi.org/10.1038/s41591-020-0824-5
36. Yang T, Gentile M, Shen CF, Cheng CM (2020) Combining point-of-care diagnostics and internet of medical things (IoMT) to combat the COVID-19 pandemic. Diagnostics 10:224. https://doi.org/10.3390/diagnostics10040224
37. Sood SK, Mahajan I (2017) Wearable IoT sensor-based healthcare system for identifying and controlling chikungunya virus. Comput Ind 91:33–44. https://doi.org/10.1016/j.compind.2017.05.006
38. Shi F, Wang J, Shi J, Wu Z, Wang Q, Tang Z et al (2020) Review of artificial intelligence techniques in imaging data acquisition, segmentation and diagnosis for COVID-19. IEEE Rev Biomed Eng 2(1):220–235. https://doi.org/10.1109/RBME.2020.2987975
39. Ko H, Chung H, Kang WS, Kim KW, Shin Y, Kang SJ et al (2020) COVID-19 pneumonia diagnosis using a simple 2D deep learning framework with a single chest CT image: model development and validation. J Med Internet Res 22(6):e19569. https://doi.org/10.2196/19569
40. Li L, Qin L, Xu Z, Yin Y, Wang X, Kong B et al (2020) Using artificial intelligence to detect COVID-19 and community-acquired pneumonia based on pulmonary CT: evaluation of the diagnostic accuracy. Radiology 296(2):E65–E71. https://doi.org/10.1148/radiol.2020200905
41. SrinivasaRao ASR, Vazquez JA (2020) Identification of COVID-19 can be quicker through artificial intelligence framework using a mobile phone-based survey when cities and towns are under quarantine. Infect Control Hosp Epidemiol 41(7):826–830. https://doi.org/10.1017/ice.2020.61
42. Madurai Elavarasan R, Pugazhendhi R (2020) Restructured society and environment: a review on potential technological strategies to control the COVID-19 pandemic. Sci Total Environ 725:138858. https://doi.org/10.1016/j.scitotenv.2020.138858
43. Definitions of Symptoms for Reportable Illnesses (2020) https://www.cdc.gov/quarantine/air/reportingdeathillness/definitionssymptoms-reportableillnesses.html. Accessed 20 Dec 2020
44. Tamura T, Huang M, Togawa T (2018) Current developments in wearable thermometers. Adv Biomed Eng 7:88–99
45. Ghosh S (2020) Police in China, Dubai, and Italy are using these surveillance helmets to scan people for COVID-19 fever as they walk past and it may be our future normal. https://www.businessinsider.com/coronavirus-italy-holland-china-temperature-scanning-helmets. Accessed 20 Dec 2020
46. Bright J, Liao R (2020) Chinese startup rokid pitches COVID-19 detection glasses in the us. Available at. https://defence.pk/pdf/threads/chinese-startup-rokid-pitches-covid-19-detection-glasses-in-us.661845/. Accessed on 20 December 2020
47. Draganfly - Driving the future of drones since the pre-drone era. Available at. https://draganfly.com/. Accessed 20 Dec 2020
48. Cozzens T (2020) Pandemic drones to monitor, detect those with COVID19. Available at. https://www.sage.unsw.edu.au/news/pandemic-drones-to-monitor-detect-those-with-covid-19. Accessed on 20 December 2020
49. Bai L, Yang D, Wang X, Tong L, Zhu X, Zhong N, Bai C, Powell CA, Chen R, Zhou J, Song Y (2020) Chinese experts' consensus on the internet of things-aided diagnosis and treatment of coronavirus disease 2019 (COVID-19). Clinical eHealth 3:7–15. https://doi.org/10.1016/j.ceh.2020.03.001
50. United against coronavirus! Stop corona App (2020) https://stopcorona.app/. Accessed 20 Dec 2020
51. Kelion L (2020) Coronavirus: Moscow rolls out patient-tracking app. https://www.bbc.com/news/technology-52121264. Accessed 20 Dec 2020

52. Singh VK, Chandna H, Kumar A, Kumar S, Upadhyay N, Utkarsh K (2020) IoT-Q-Band: a low cost internet of things based wearable band to detect and track absconding COVID-19 quarantine subjects. EAI Endorsed Trans Internet Things 20(21):163997. https://doi.org/10.4108/eai.13-7-2018.163997
53. Lancet T (2020) COVID-19: protecting health-care workers. Lancet 395(10228):922. https://doi.org/10.1016/S0140-6736(20)30644-9
54. Sharma M (2020). How drones are being used to combat COVID-19. https://www.geospatialworld.net/blogs/how-drones-are-being-used-to-combat-covid-19/. Accessed 20 Dec 2020
55. Yang J (2020) 3 ways china is using drones to fight coronavirus. https://www.weforum.org/agenda/2020/03/three-ways-china-is-using-drones-to-fight-coronavirus/. Accessed 20 Dec 2020.
56. O'Meara S (2020) Hospital ward run by robots to spare staff from catching virus. New Sci 245(3273):11. https://doi.org/10.1016/S0262-4079(20)30526-1
57. Yang G, Lv H, Zhang Z, Yang L, Deng J, You S et al (2020) Keep healthcare workers safe: application of tele operated robot in isolation ward for COVID-19 prevention and control. Chin J Mech Eng 33(47):1–4. https://doi.org/10.1186/s10033-020-00464-0
58. Tao M (2020) Cobots v COVID: How universal robots and others are helping in the fight against coronavirus. https://bit.ly/2ATJ5vH. Accessed 21 December 2020
59. Knibbs K (2020). There's no cure for COVID-19 loneliness, but robots can help. https://www.wired.com/story/covid-19-robot-companions/. Accessed December 21, 2020
60. Wright T (2020) Block chain app used to track COVID-19 cases in Latin America. https://cointelegraph.com/news/blockchain-app-used-to-track-covid-19-cases-in-latin-america. Accessed 21 December 2020
61. Zielinski D (2020). Wearables enable distancing, contact tracing at work. https://www.shrm.org/resourcesandtools/hr-topics/technology/pages/wearables-enable-distancing-contact-tracing-work.aspx. Accessed 21 December 2020
62. Tripathy AK, Mohapatra AG, Mohanty SP, Kougianos E, Joshi AM, Das G (2020) Easy band: a wearable for safety-aware mobility during pandemic outbreak. IEEE Consumer Electron Mag 9(5):57–61. https://doi.org/10.1109/MCE.2020.2992034
63. Contact tracing IoT solution (2020). https://bit.ly/2B60rF2. Accessed 21 December 2020
64. Marr B (2020) Robots and drones are now used to fight COVID-19 (forbes.com). Accessed on 20 December 2020
65. Trace Together, safer together (2020). https://www.tracetogether.gov.sg/. Accessed 21 December 2020
66. I protect both you and me (2020). https://erouska.cz/en. Accessed 21 Dec 2020
67. Stub ST (2020). Israeli phone apps aim to track coronavirus, guard privacy. https://www.usnews.com/news/best-countries/articles/2020-04-20/new-tech-apps-in-israel-aim-to-track-coronavirus-guard-privacy. Accessed 21 Dec 2020
68. Token Post (2020) IoT block chain platform launches a COVID-19 contact tracing app. https://bit.ly/3eS2VGt. Accessed 21 Dec 2020
69. Alwashmi MF (2020) The use of digital health in the detection and management of COVID-19. Int J Environ Res Public Health 17(8):2906. https://doi.org/10.3390/ijerph17082906

Chapter 3
Predicting Transmission Rate of Coronavirus (COVID-19) Pandemic Using Machine Learning Techniques

David Opeoluwa Oyewola, Emmanuel Gbenga Dada, K. A. Al-Mustapha, Ayotunde Alaba Fadele, Stephen Bassi Joseph, and Asabe Ibrahim

Abstract COVID-19 is a single-stranded enclosed virus that was first discovered in Wuhan in December 2019. The 2019 novel coronavirus also known as COVID-19 has quickly transmitted to virtually all the nations of the world. This novel virus has posed a serious health challenge, and many countries are finding it difficult to preventing the spread of the deadly virus. In spite of all the money, time and energy invested in research toward finding a lasting solution to this problem, no remedy or vaccine has been recognized to effectively cure this disease. Under such circumstances, comprehending the rate of transmission of this virus is very crucial to curb its promulgation. Conventional techniques to decide the circulation rate of this infection are time wasting, and there is acute shortage of such technology in some places as well. Therefore, the need for a really quick and efficient machine learning (ML) model to tackle this menace has defied all known solutions. This paper used four high-performing ML methods to predict the rate of spread of COVID-19. The performance of these ML models was evaluated using the data obtained from patients that have been infected with COVID-19 illness between January 22, 2020 and August 7, 2020. Results of our experiments indicate that logistic regression (LR) outperforms every other model considered in this work. The implication of this is that LR model is a promising classifier that can be used to effectively predict the rate of transmission of COVID-19. We also discovered from our results that there is growing tendency of COVID-19 transmission in every other continent of the world with the

D. O. Oyewola · A. Ibrahim
Department of Mathematics and Computer Science, Federal University Kashere, Gombe, Nigeria

E. G. Dada (✉)
Department of Mathematical Sciences, University of Maiduguri, Maiduguri, Nigeria
e-mail: gbengadada@unimaid.edu.ng

K. A. Al-Mustapha
Department of Mathematics, Baze University, Abuja, FCT, Nigeria

A. A. Fadele
Department of Computer Science, Federal College of Education, Zaria, Nigeria

S. B. Joseph
Department of Computer Engineering, University of Maiduguri, Maiduguri, Nigeria

U. Kose et al. (eds.), *Computational Intelligence for COVID-19 and Future Pandemics*, Disruptive Technologies and Digital Transformations for Society 5.0,
https://doi.org/10.1007/978-981-16-3783-4_3

exception of some parts of Africa and Oceania where COVID-19 transmission has been decelerated and curve flattened.

Keywords COVID-19 · Machine learning (ML) · Logistic regression (LR) · Support vector machine (SVM) · Recurrent neural network (RNN)

1 Introduction

Wuhan Jinyintan Hospital was the first to report coronavirus (COVID-19) case on 30 December 2019 from a patient with chronic inflammation of the lungs of strange etiology. Based on the result, the virus had characteristics of the coronavirus family named Betacoronavirus 2B. Bat SARS-like coronavirus showed a closed relationship with COVID-19 virus. World Health Organization (WHO) on 30 January 2020 declared coronavirus as global public health emergency due to an outbreak of respiratory disorder. The new coronavirus was identified by WHO as severe acute respiratory syndrome coronavirus 2 (SARS-COV-2) and was called the coronavirus disease 2019 (COVID-19). Some symptoms are associated with COVID-19 infections as explained in Ref. [1]. In extreme cases of coronavirus pandemic disease, it may cause kidney failure, death and severe acute respiratory syndrome. At this stage, national and global health systems are currently under severe pressure by this pandemic (COVID-19). Brazil has, as at the period of writing this manuscript has been one of the countries the experienced the greatest difficulties in tackling the surge in 2019-nCoV, in public health departments, emergency medical services and hospitals [2]. Unfortunately, in many countries, infection rates, irrespective of their system of health care, are projected to increase, exponentially. Therefore, it is necessary in every health-care system and hospital to execute urgent actions to provide high-quality medical equipment and high-quality information in Nigeria.

In this regard, translational science can provide stakeholders and clinicians with appropriate evidence-based medicine concepts. This pandemic was reported in 199 countries as early as March 29, 2020 with 724,759 cases confirmed, 34,020 deaths and 152,087 cases recovered. Dry cough, shortness of breath, fatigue, pneumonia, sore throat, diarrhea and dyspnea have been reported as signs of the disease. Several methods for minimizing transmission have been investigated and presented such as using masks, hygienically supervise hands practices, avoiding public touches, case detection, touch monitoring and quarantines. Infected people of COVID-19 rely on symptomatic treatment because there is no proven effective antiviral therapy [3].

Since the beginning of the pandemic, a wide range of studies have been conducted to provide a better understanding about the cause, avoidance and diagnosis of COVID-19. Nemati et al. [4] proposed the combination of statistical methods, Support Vector Machine (SVM), and ensemble techniques that use COVID-19 data of patients to predict the date they are likely to be discharged from isolation center. It also appraises the clinical information to determine the duration. The downside of this work is that it is just a framework and there is no practical implementation of any machine learning or statistical algorithm. Their technique was also not assessed. Later on, Lalmuanawma et al. [5] presented a study on the importance of artificial intelligence (AI) and ML for investigating and predicting the transmission rate of COVID-19. They also examine how these techniques can be used for recognizing, evaluating and handling individuals who have been exposed to COVID-19 to avert further transmission. Moreover, the authors examined how AI and ML can help in the process of bringing a new pharmaceutical drug to clinical practice for SARS-CoV-2 and its associated endemic. Findings from this study indicated that AI and ML have appreciably enhanced the management, treatment, testing, prediction and cure/immunization steps needed to take COVID-19 drugs from concept to market availability.

More recently, Tuli et al. [6] applied Robust Weibull ML technique to predict the escalation of COVID-19 in various nations, globally. They demonstrated that their approach has the potential to produce superior results to create a prediction framework. Their approach was implemented on a cloud platform to annex its advantages to extract valuable information from COVID-19 datasets in real time. Malki et al. [7] applied several machine learning models to obtain the correlation between various features and the rate of transmission of COVID-19. The ML models were used to evaluate the effect of atmospheric condition factors on the spread of COVID-19 by mining the connection amid the number of reported confirmed cases and the atmospheric condition variables in some counties. The authors opined that atmospheric variables have great significance in forecasting the death toll as a result of COVID-19 compared with other factors mentioned in the paper.

At the same time, Kavadi et al. [8] proposed a new technique for predicting COVID-19. Their approach uses the partial derivative regression (PDR) to explore the dataset for optimal parameters with little computer resource usage. Thereafter, the ML model was used to normalize attributes that are employed to make better predictions. The drawback of their work is that the proposed method was not compared with other approaches to determine its effectiveness. Also, only prediction time was used as performance metric, which is not enough to evaluate the performance of a new model. In a more specific study, Amar et al. [9] applied different ML and statistical techniques to predict the transmission of COVID-19 in Egypt. The purpose of their study was to use their approach to assist the Egyptian government in managing the pandemic in the following months. Experimental results showed that the exponential model outperforms other models compared in the paper. The authors deduced from their results that COVID-19 pandemic in Egypt is not likely to end soon.

In a different study, Adhikari et al. [10] presented a scoping review using the approach proposed by Arksey and O'Malley [11]. About 65 articles were analyzed and examined in order to know how to prevent coronavirus. The study covered publication dates, journal terminology, affiliations of authors and methodological features. The results indicated that the majority of the reports were written in the English language (89.2%). Findings also indicated that Chinese scholars have published 67.7% of the articles related to this issue. They also showed that early research papers concentrated on causes, but over a period of time there have been growing numbers of papers that give attention to prevention and control.

The works cited above are indications that the use of machine learning algorithms is growing exponentially in the medical domain to tackle issues related to this pandemic by analysis and interpretation of large quantities of data to identify important patterns [12]. Many researchers have used Machine Learning (ML) algorithms to overcome other medical problems before the advent of COVID-19. Oyewola et al. [13] used five ML techniques to predict breast cancer. The authors classified the patients infected with breast cancer into two categories (malignant and benign). The authors achieved 95.8% accuracy with SVM learning classification in diagnosing breast cancer.

So Ref. [14] carried out a study on the diagnosis of dementia by means of ML techniques. The authors considered six different algorithms in which the patients were classified as regular, moderate intellectual deficiency and senility. The overall accuracy of the diagnosis system was 97% for support vector machine. Dengue disease patients are increasing rapidly in a paper by Ref. [15] compared with the performance of seven ML models. Findings indicated that the LR model achieved 92%, 90% and 94% in terms of sensitivity, specificity and accuracy, respectively. A paper by Ref. [16] examines a method for predicting heart disease known as an ensemble classification. In this research, a comparative study was done to decide how ensemble methods can improve the prediction of heart disease by weak classifiers. Findings show that bagging and boosting increase poor classifiers accuracy.

In this study, ML algorithm was applied to predict the transmission rate of COVID-19 endemic. ML has proven over time to be a very efficient and robust algorithm that can cope with huge data effectively. It can, therefore, be used to judiciously predict the transmission of the COVID-19 pandemic. The key contributions of this work include:

- i A survey of machine learning and statistical algorithms that have been applied to the prevention, diagnosis and control of COVID-19 was presented.
- ii Determine the most ideal predictive model for day-to-day confirmed cases in different continents of the world that have become the hotspot COVID-19 cases in the world.
- iii Predict the spread of COVID-19, number of confirmed and death cases using LR, SVM, RNN, LSTM models
- iv To evaluate the proposed model.

The rest of this paper is organized as follows: Sect. 2 discusses the machine learning models used in this work as well as our employed performance measurements. The results and the discussion are presented in Sect. 3 and Sect. 4 is the conclusion.

2 Materials and Method

2.1 Dataset

For this experiment, the Coronavirus (COVID-19) dataset was obtained from the Kaggle database [17]. This dataset contains information obtained from 214 countries. The information in this dataset contains the number of the confirmed cases, death cases and recovery cases of COVID-19 incidents from patients who have contacted coronavirus disease from 31 December 2019 to 7 August 2020. The description of the dataset is shown in Figs. 1, 2, 3, 4, 5, 6. The dataset was divided into continents, which are Africa, Asia, South America, North America, Europe and Oceania for confirmed cases of coronavirus. Based on the dataset in Fig. 1, Djibouti has the highest confirmed cases followed by South Africa and Sao Tome and Principe while

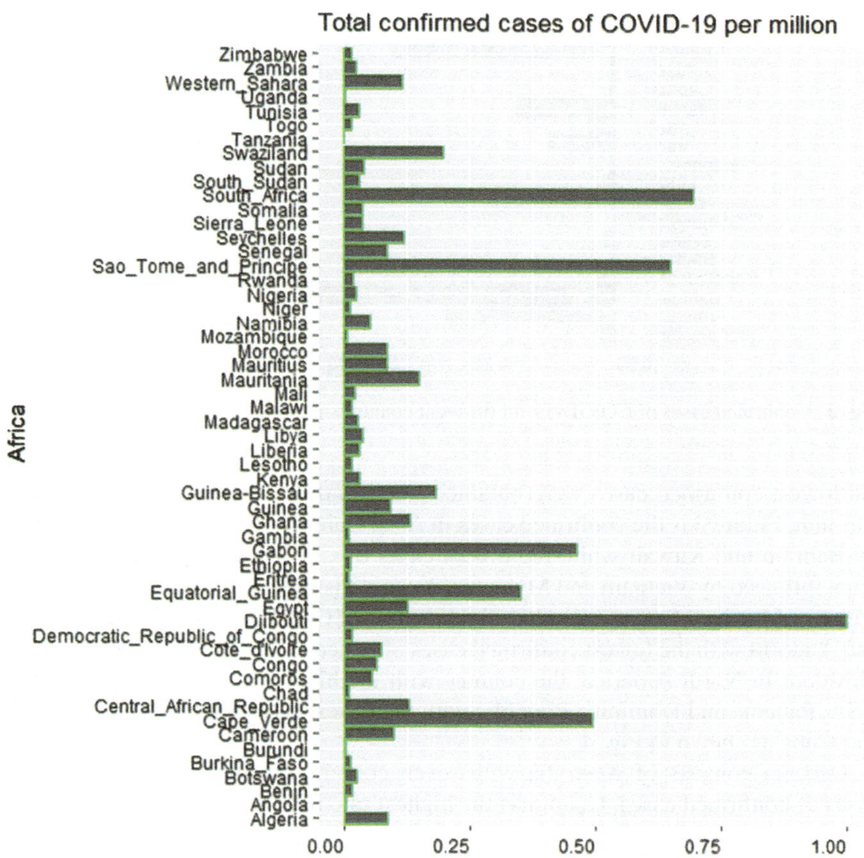

Fig. 1 Confirmed cases of COVID-19 for different countries in Africa

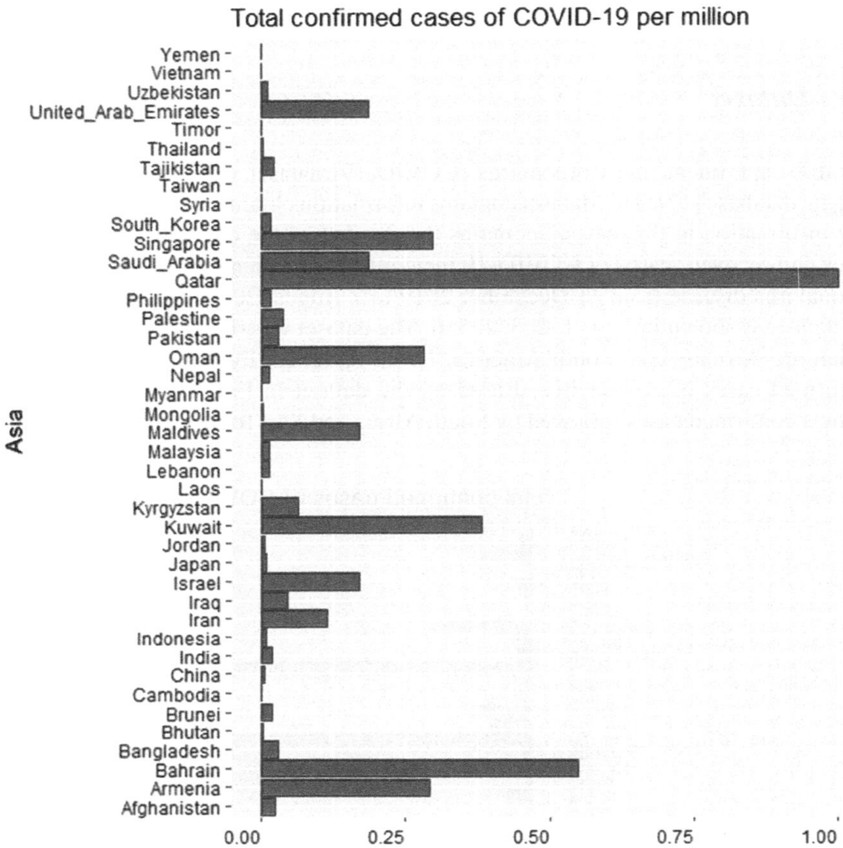

Fig. 2 Confirmed cases of COVID-19 for different countries in Asia

the lowest confirmed cases were obtained from Tanzania followed by Angola and Burundi. Figure 2 is the confirmed cases in Asia. Qatar has the highest rate followed by Bahrain and Kuwait while the lowest cases of COVID-19 were obtained from Laos followed by Vietnam and Myanmar. Figure 3 is the confirmed cases in Europe with San Martino has the highest confirmed cases of COVID-19 followed by Vatican and Andorra with the lowest confirmed cases obtained from Georgia, Greenland and Slovakia. In North America, the country with the highest confirmed cases are the USA, Panama and Dominica Republic with the lowest cases are Belize, Saint Lucia and Cuba as shown in Fig. 4.

Oceania consists of Australia with seven countries inclusive with COVID-19 cases of confirmed cases. Guam has the highest rate followed by Northern Mariana Islands and Australia while Papau New Guinea, Fiji and New Caledonia have the lowest (Fig. 5). As shown in Fig. 6, in South America countries with the highest confirmed cases of COVID-19 are Chile, Peru and Brazil while countries with

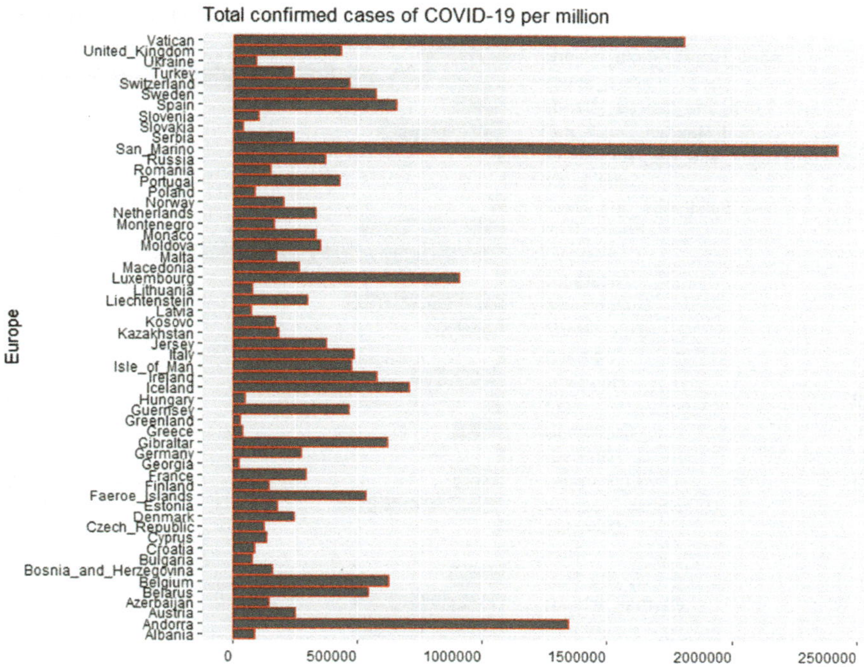

Fig. 3 Confirmed cases of COVID-19 for different countries in Europe

the lowest confirmed cases of COVID-19 are Trinidad and Tobago, Curacao and Venezuela (Fig. 6). The bar chart presented in Fig. 1 was plotted using R Studio from the confirmed cases of COVID-19 dataset for various African nations obtained from Kaggle website. The chart depicted in Fig. 2 was plotted using R Studio from the confirmed cases of COVID-19 dataset for various nations in Asia obtained from Kaggle website. The chart represented in Fig. 3 was plotted using R Studio from the confirmed cases of COVID-19 dataset for the European countries obtained from Kaggle website. The chart presented in Fig. 4 was plotted using R Studio from the confirmed cases of COVID-19 dataset for different North American countries obtained from Kaggle website. Figure 5 is a pictorial representation of the confirmed cases of COVID-19 dataset for various Oceania countries obtained from Kaggle website which we plotted using R Studio. Finally, Fig. 6 is a graphical representation of the confirmed cases of COVID-19 dataset for various South American countries retrieved from Kaggle website which we plotted using R Studio.

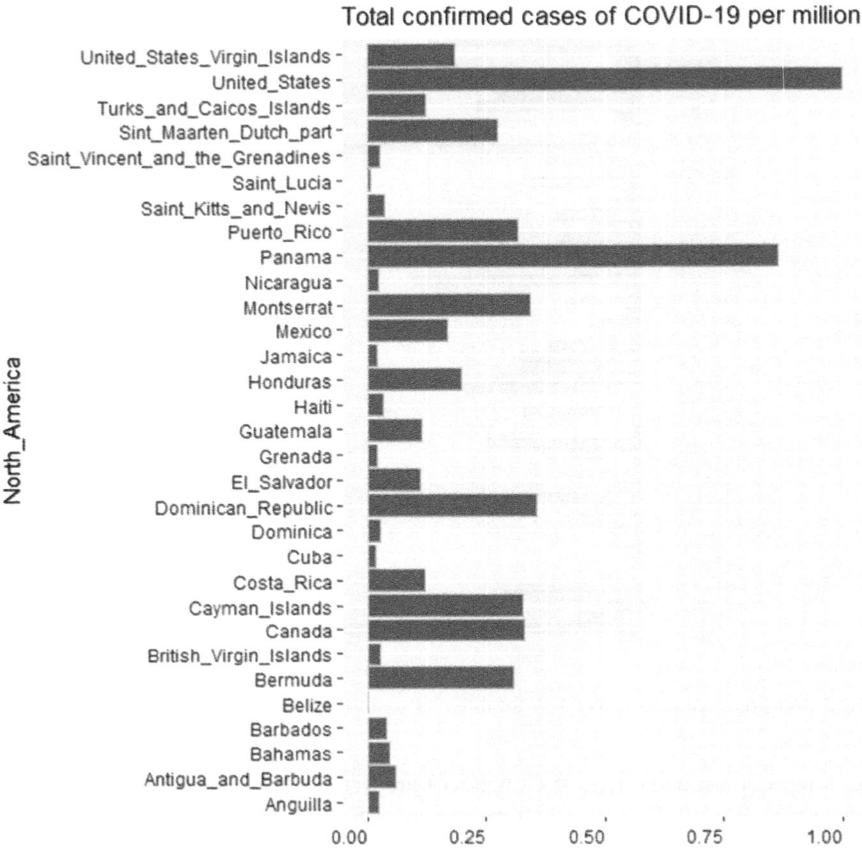

Fig. 4 Confirmed cases of COVID-19 for different countries in North America

2.2 *Machine Learning Algorithms*

2.2.1 Logistic Regression (LR)

The LR is the appropriate regression technique for analysis when the dependent variable is binary. When there are only two possible outcomes (so-called binary classification task) for prediction logistic regression is very effective in predicting the outcomes. LR is a very simple ML model. They are easy to understand, interpret and have the capacity to generate very good prediction or classification results. LR can be binomial for predicting binary variable, ordinal for predicting ordinal variable and multinomial for predicting multinomial variable. The results in the multinomial variable can have more than two possible types [13]. The mathematical equation of logistic regression is given as:

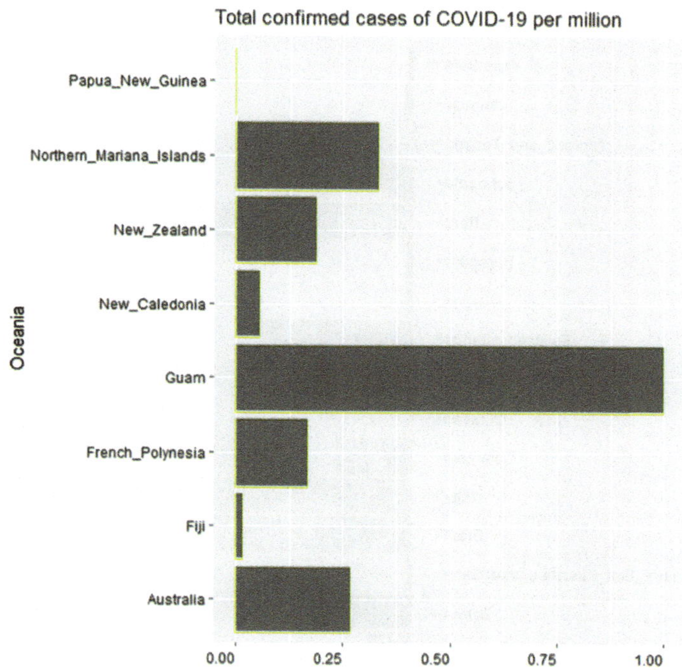

Fig. 5 Confirmed cases of COVID-19 for different countries in Oceania

$$p = \frac{e^{a+bX}}{1 + e^{a+bX}} \tag{1}$$

where p is the probability, e is the base of the natural logarithm, while a and b are the parameters of the model. A LR will model the probability of an output depending on specific characteristics. This is because probability is a ratio, and it is the logarithm of the probability that will essentially be modeled and can be expressed:

$$\log\left(\frac{\pi}{1-\pi}\right) = \beta_0 + \beta_1 x_1 + \beta_2 x_2 + \beta_3 x_3 + +\beta_n x_n \tag{2}$$

assuming π specifies the likelihood of an event, and β_i are the regression coefficients related to the reference class and x_i represents the descriptive variables. The reference class, signified by β_O, is composed of those entities exhibiting the degree of reference of every one of the variable $x_{1\ldots m}$.

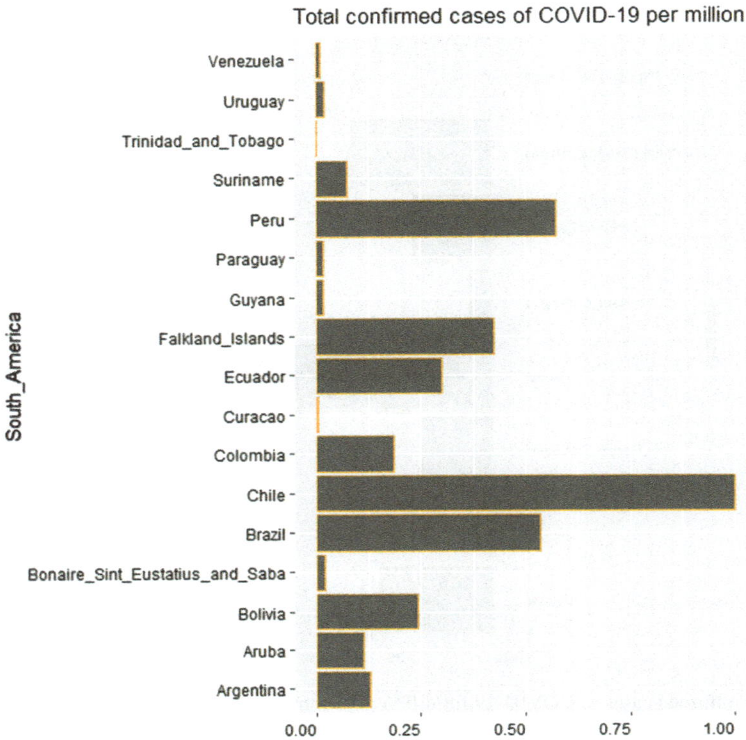

Fig. 6 Confirmed cases of COVID-19 for different countries in South America

2.2.2 Support Vector Machine (SVM)

The SVM is a machine learning technique for solving both regression and classification problems. It was developed by Vapnik [18]. SVM is based on the idea of finding the maximum margin hyperplane to reduce the prediction error. For non-linear and linear data, the technique can be utilized. SVM classifies data vectors using an hyperplane or a group hyperplanes in a high-dimensional space [19]. The approach creates a hyperplane set, which can be employed for both classification and regression [18]. When taking hyperplane to the next point of the two classes, a maximum distance is reached such hyperplane is considered an optimal hyperplane. It can use many hyperplanes for classification by division. However, the most ideal hyperplane generates the greatest margin amid two groups of data. In several situations, the data are divided nonlinearly for all possible inputs. Therefore, they require increases or decreases of linear relationships between variables into a geometric setting and the linear classifier that gives the highest associated distance from the decision boundary for each sample can be used. This is what the Kernel functions are used for to achieve. During the training phase, they are used to choose the support vectors near the function plane. Subsequently, SVM is used to classify the data through the support vectors that form the hyperplane in the n-dimensions where your variables reside. Choosing a suitable kernel for a particular classification problem affect the effectiveness of SVM as various kernel function builds diverse SVMs and influences the capacity of SVM to generalize and learn [20]. There is no existence of any known hypothetical method for choosing kernel function and its parameters. Currently, the Gaussian kernel is the only known kernel function that is typically employed for its suitable qualities [21]. The equation is expressed in mathematical form as:

$$H = E.T + a = 0 \tag{3}$$

where H is the hyperplane, E is the weight, T is the input and a is the bias. The hypothesis function h of the SVM classifier can be expressed as:

$$h(x_i) = \begin{cases} +1 if e.t + a \geq 0 \\ -1 if e.t + a < 0 \end{cases} \tag{4}$$

The point that is greater than or on the same level as the hyperplane is categorized as class + 1, and the point that is less than the hyperplane is categorized as class −1. The soft margin of the SVM classifier can be minimized by using the mathematical equation below:

$$\left[\frac{1}{n}\sum_{n+1}^{n} \max(0, 1 - y_i(w.x_i - b))\right] + \lambda\|w\|^2 \tag{5}$$

2.2.3 Recurrent Neural Network (RNN)

RNN is unique class of neural network because of its additional set of feedbacks from the generated output of layer that is hidden in between input and output layers. It can be defined as a network that receives feedback signals from neurons across all layers. In RNN, the result of processing in a preceding phase can be carried over and used in the current period phase. This important attribute of the RNN is very beneficial for real-time applications. RNN can have an unrestricted memory level and can, therefore, learn connections through time in addition to learning via the all current possible inputs [22].

Deep learning is a subdivision of machine learning in artificial intelligence with networks that can learn from data that is unstructured or unlabeled without any human supervision. Figure 7 illustrates the RNN unit, which has an internal state that is fed back to the input state and shares corresponding weights and bias between them. The cell feedback loop allows the network to communicate with the hidden state d_t to the future time t. X_t denotes the cell input at time t and Y_t denotes the cell output at time t. Input to the hidden layer equation is given as:

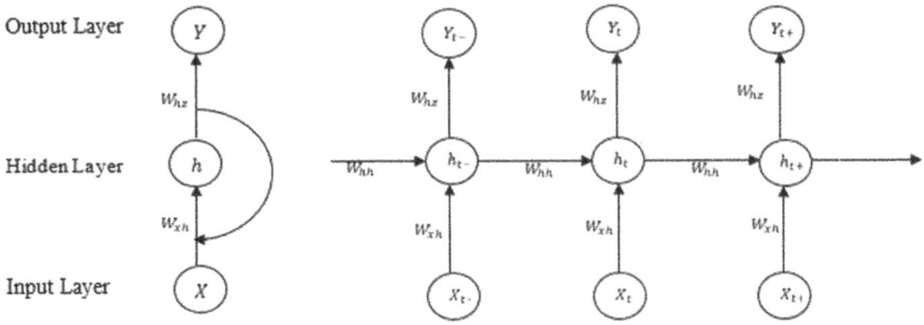

Fig. 7 Schematic diagram of Recurrent Neural Network

$$d_t = g_n(W_{xh}X_t + W_{hh}h_{t-1} + b_h) \tag{6}$$

where h_t is the hidden layer at t^{th} instant, g_n is the function, W_{xh} is the input to the hidden layer of the weight matrix, X_t is the input at t^{th} instant, h_{t-1} is the hidden layer at $t - 1$ instant, b_h is the bias or threshold value.

The hidden to output layer equation is given as:

$$Z_t = g_n(W_{hz}h_t + b_z) \tag{7}$$

whereas Z_t is the output vector, W_{hz} is the hidden to output layer weight matrix, b_z is the bias or threshold.

2.2.4 Long Short Term Memory (LSTM)

LSTM deep neural network cell is different from the recurrent neural network from two viewpoints [23]. To start with, long-term state c_t and the short-term state h_t are divided into two components. The sigmoid function is an activation function ρ of the gates, which outputs values within the range [0, 1]. The three control gates on the state path were included to control the cell states and these are the forget gate, the input gate as well as the output gate. The forget gate f_t indicates that nothing from the previous long-term state c_{t-1} needs to be carried forward. Output gate o_t checks the development of current short-term state h_t utilizing the long-term state c_t information [23]. The activation function for the candidate cell state is a hyperbolic tangent function, which produces values in the range [−1, 1]. The LSTM cell differs considerably from the simple RNN cell because the output o_t is equivalent to the hidden state h_t. We can describe it mathematically using the following equations:

$$f_t = \rho\left(W_f.\left[h_{t-1}, x_t\right] + b_f\right) \tag{8}$$

$$i_t = \rho\left(W_i.\left[h_{t-1}, x_t\right] + b_i\right) \tag{9}$$

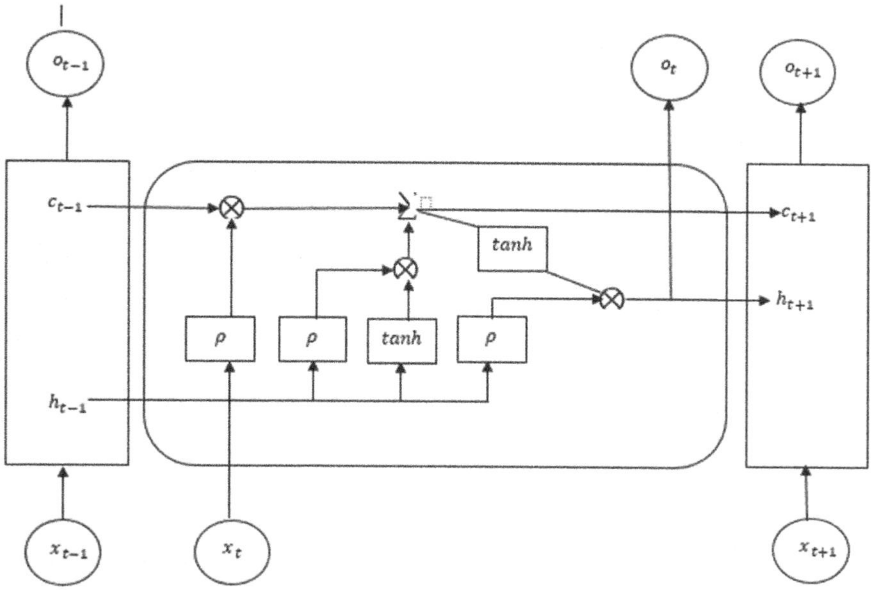

Fig. 8 Schematic diagram of LSTM

$$c_t = tanh\left(W_i.[h_{t-1}, x_t\right] + b_c) \tag{10}$$

$$o_t = \rho\left(W_o.[h_{t-1}, x_t\right] + b_o) \tag{11}$$

$$h_t = o_t \times \tanh(c_t) \tag{12}$$

where f_t forget gate vector, i_t input gate vector, c_t cell state vector, o_t output gate vector, x_t is the input vector, h_t is the output vector, ρ is the sigmoid function and W, b is the parameter matrix and vector. The schematic diagram of LSTM is depicted in Fig. 8.

2.3 Description of Variables

The dataset used in this paper provides regular details on the number of reported cases, deaths and recovery of novel coronavirus COVID-19. It can be presented in a form of time series with the total number of cases is reported on a daily basis. In this paper, we considered only the confirmed cases of those infected with the disease, which represents target vectors Y_t at time t while the input vectors are given in Eqs. (13)–(18). The target variable for each continent is given as:

$$y(t) = [y_1(t), y_2(t), \dots y_n(t)]^T \tag{13}$$

We assume the input variable is:

$$b_t = y_t - y_{t-n} \tag{14}$$

$$m_t = \left|\sum_{t=0}^{n} b_t\right| - \left(\sum_{t=0}^{2n} |b_t| - \sum_{t=0}^{n} |b_t|\right) \tag{15}$$

$$p_t = \frac{1}{t}\sum_{t=0}^{t} y_t \tag{16}$$

$$h_t(y) = \alpha y_t + (1 - \alpha)h_{t-1} \tag{17}$$

$$r_t = h_t(s) - h_t(l) \tag{18}$$

where α is the weighting parameter of the confirmed cases in the past having the values from 0 to 1, b_t is the momentum at time t, m_t is trend direction index at time t, p_t is the simple moving average at time t, h_t is the exponential moving average at time t, r_t are the moving average convergence divergence at time t and s, l is the 12 and 26 days exponential weighted moving average, respectively.

2.4 Performance Measures

We assess prediction performance utilizing three measures namely, Mean Absolute Error (MAE), Root Mean Square Error (RMSE) and Mean Absolute Scaled Error (MASE).

Mean Absolute Error (MAE).

MAE is defined as the average of the difference between predicted and actual values in the test [24]. Consider real or actual returns y_t^n and their estimated values $\hat{y}_t^n$. MAE is defined as follows:

$$\frac{1}{n}\sum_{n=1}^{n} |y_t^n - \hat{y}_t^n| \tag{19}$$

Root Mean Square Error (RMSE).

RMSE is defined as the standard deviation of prediction errors in a test [24]. RMSE is defined as:

$$\sqrt{\frac{1}{n}\sum_{n=1}^{n}\left(y_t^n - \hat{y}_t^n\right)^2} \tag{20}$$

Mean Absolute Scaled Error (MASE).

MASE is defined as is a measure of the accuracy of predictions. It is a scale-free error metric that gives each error as a percentage in comparison to a standard mean error [24]. MASE is defined as follows:

$$\frac{1}{n}\sum_{n=1}^{n}\left(\frac{\left|y_t^n - \hat{y}_t^n\right|}{\frac{1}{n-m}\sum_{n=m+1}^{n}\left|y_t^n - y_{t-m}^n\right|}\right) \tag{21}$$

where m is the seasonal period of return r_t^n.

3 Results and Discussion

This section describes the results of predicting COVID-19 transmission based on the findings of ML techniques investigated in this study, which are: LR, SVM, RNN and LSTM. We then analyze it to predict the transmission of coronavirus more accurately. The COVID-19 dataset used for this work was obtained from Kaggle database. Data were divided into five parts to statstically compare and test learning algorithms. Figure 9 demonstrates the time series for each of the continents used in this. As shown in this figure, COVID-19 in Europe is at its peak compared with other continents. During January and February, most continents have stable patterns. In the mid-February, there are exponential trends of confirmed cases of pandemic diseases, which show that the diseases spread faster in other continents. The accuracy of these algorithms can be demonstrated using mean absolute error, root means square error and mean absolute scaled error to predict the spread of COVID-19. Each of these accuracies has been obtained from Eqs. (16) to (18). As explained in previous page, Fig. 9 is a pictorial representation of each continent and their rate of spread of COVID-19.

From the times series of cases of spread of COVID-19 presented in Fig. 9, Europe has the highest number of cases. The figure depicted the actual values of the confirmed cases of COVID-19 for each of the continents.

Four algorithms are considered in this study, which are LR, SV, RNN and LSTM. Using machine learning technique, the best algorithms among them were determined. COVID-19 spreads in each continent are effectively predicted, as shown in Figures 8, 9, 10, 11, 12, 13. The machine learning technique was used to determine the best algorithms among the four algorithms, which can effectively predict COVID-19 spreads in each continent as shown in Figures 8, 9, 10, 11, 12, 13. The overall performance of every one of the confirmed cases of COVID-19 is compared with other algorithms like LR, SVM, RNN and LSTM.

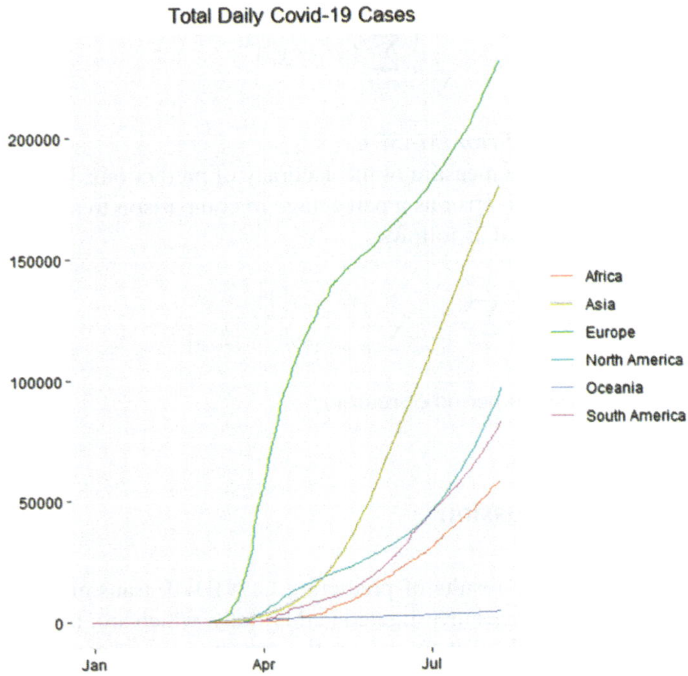

Fig. 9 Time series combined cases of spread of COVID-19

Figures 10, 11, 12, 13, 14, 15 are the real continents and estimated value of each of the algorithms. In this study, trend-spread decision for COVID-19 based on the Uptrend and Downtrends from all the four algorithms was also considered. The findings show in relation to the actual values of each of the continents.

Table 1 reports the predicted values using three criteria: MAE, RMSE and MASE. Logistic regression outperforms the other three algorithms, which means the approaches are more reliable than any other prediction. The results show that logistic regression can predict the spread of diseases. The transmission of COVID-19 in Europe, Asia, North America and South America is very high while trend of transmission of the disease is stable in Africa and Oceania.

Figure 10 represents the comparison of the actual and predicted values of confirmed cases of COVID-19 for Africa using LR, SVM, RNN and LSTM. From the figure, LR has the best predictive result.

Figure 11 depicts the comparison of the actual and predicted values of confirmed cases of COVID-19 for Asia using LR, SVM, RNN and LSTM. From the figure, the performance of LR to that of other predictive models compared.

Figure 12 shows the comparison of the actual and predicted values of confirmed cases of COVID-19 for Europe using LR, SVM, RNN and LSTM. It can be seen from the figure that LR outperforms other predictive models compared.

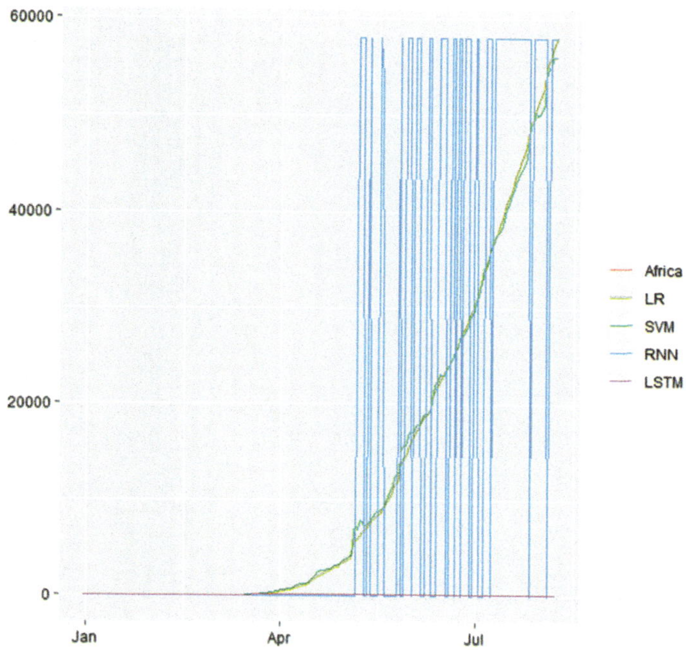

Fig. 10 Output prediction obtained from LR, SVM, RNN, LSTM model

Figure 13 depicts the comparison of the actual and predicted values of confirmed cases of COVID-19 for North America using LR, SVM, RNN and LSTM. From the figure it is clear that LR has the best performance among all the other predictive models compared.

Figure 15 indicates the comparison of the actual and predicted values of confirmed cases of COVID-19 for Oceania using LR, SVM, RNN and LSTM. It is evident from the figure that LR outperforms all the other predictive models compared.

Figure 15 presents the comparison of the actual and predicted values of confirmed cases of COVID-19 for South America using LR, SVM, RNN and LSTM. The performance of LR outclassed that of other predictive models compared.

Figures 8, 9, 10, 11, 12, 13 are the output prediction of each of the total confirmed cases of COVID-19 in each continent. The values generated for MAE, MASE and RMSE are shown in Table 1 are very promising due to the fact that the SVM, RNN and LSTM fail to accurately predict COVID1-19. This is likely to be due to the nature of the dataset used for the experiment. As the study involved huge data size of the daily COVID-19 confirmed and death cases. It is only logistic regression that was able to effectively handle the large data involved and accurately predict COVID-19 transmission rate. This also shows in the MAE, MASE and RMSE low values it generated compared with that of other models.

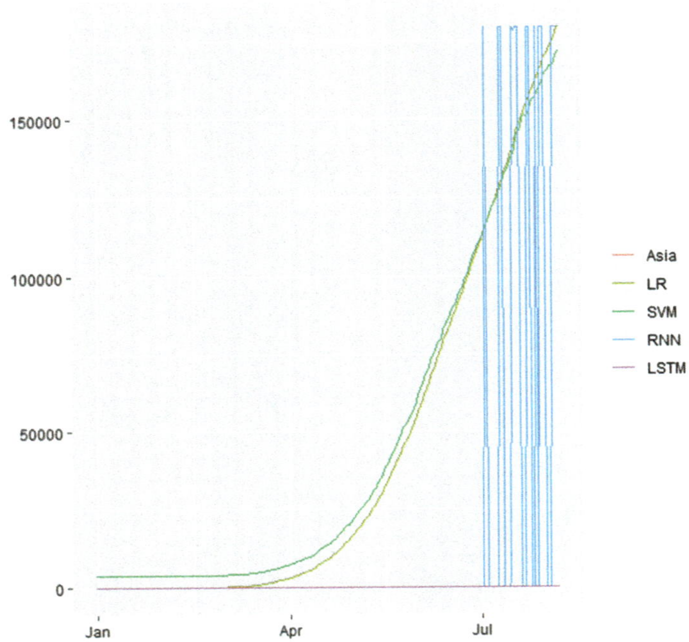

Fig. 11 Output prediction obtained from LR, SVM, RNN, LSTM model

The AUC or ROC curve cannot be used in this work as the COVID-19 data used is not a classification dataset. The dataset used for this study is sequential dataset and therefore neither AUC nor ROC curve is presented in this paper.

3.1 Application of Machine Learning to Future Pandemic

After World War II, COVID-19 is the worst global crisis and the biggest global endemic since the 1918–19 Spanish Flu. People all over the world used a great part of their day monitoring the death toll of the speedy and geometric growth in the number of confirmed cases of this new strain of the COVID-19 virus as it rises on daily basis. Software apps that use machine learning algorithms like the ones presented in this paper can be applied to predict how contagious a COVID-19 patient or anyone suffering from a future pandemic is likely to be. This way, machine learning techniques in combination with other modern-day and go-getting instruments can be deployed to manage and assuage the effects of any future pandemic.

Machine learning (ML) will be a great tool in the fight against future endemics. ML is commonly affirmed as being the game-changing and most auspicious technological advancement of twenty-first century development and is correctly projected

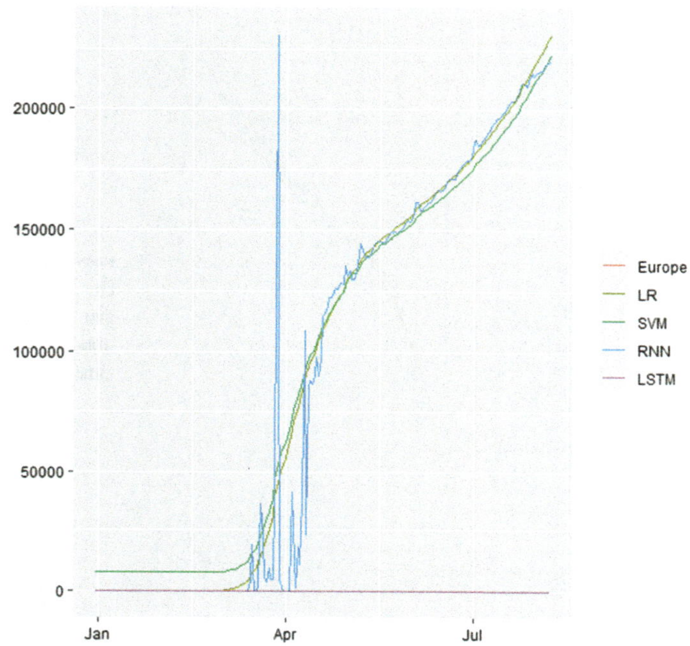

Fig. 12 Output prediction obtained from LR, SVM, RNN, LSTM model

to aid humanity combat future pandemics [5, 25]. First, this can be attained by using machine learning for adapted analysis of risks to assist in designing epidemic disease models [26]. Machine learning models can be trained using data from several sources to model and predict the risk level (or as a minimum a measure of potentiality) of a person experiencing acute consequences if infected with any future endemic like COVID-19. This helps in predicting the optimal allocation and use of critical care supplies in a particular health care service to meet the health needs of target populations.

Second, ML can be applied in the area of vaccine development to help curb future pandemic [27]. A very good example is the Human Immunomics Initiative of Harvard T.H. Chan School of Public Health and the Human Vaccines Project [28]. It is a combined effort that will apply machine learning models to expedite vaccine development. The project is expected to bring together specialists in the field of machine learning, data science, computational/systems biology, epidemiology, causal inference, immunology and statistical modeling to develop artificial intelligence energized models of human immune system and response mechanisms that will be employed to speed up the design and testing of vaccines and medications for a wide array of infections [29]. These AI energized models will provide researchers with a platform to practically test prospective vaccines and predict the vaccines and treatments that will be most suitable across a particular group of people.

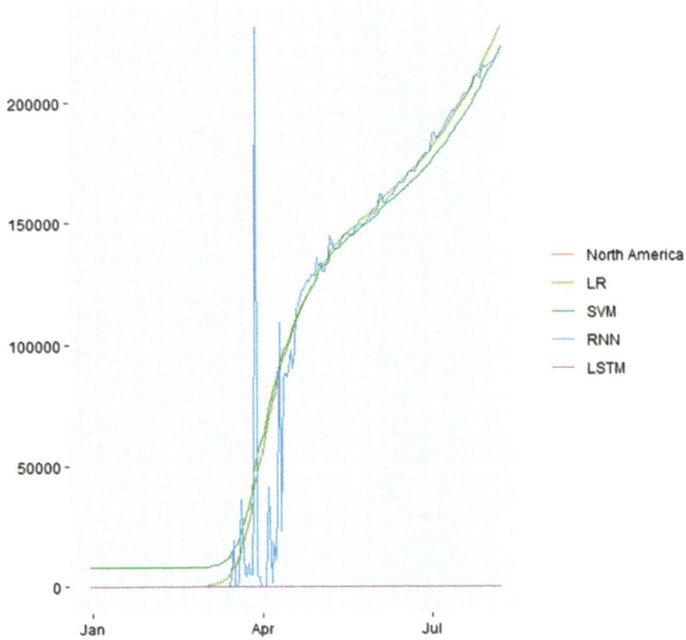

Fig. 13 Output prediction obtained from LR, SVM, RNN, LSTM model

This has the potential to immensely accelerate development of vaccines and drugs for future pandemic with reduced costs for testing and trials [30].

Third, ML can be used for predicting protein secondary structure of future pandemic like COVID-19 [31, 32]. Worldwide endemics like COVID-19 are mostly due to viruses. From simplest level, a virus comprises of one (or a small number of) strands of DNA/RNA. Ascertaining the 3D protein structure (e.g., the structure of amino acid molecules from the changes in chromosomes, genes or proteins data is fundamental to develop specific groups, subgroup and class of nucleic acids of vaccines. This job although computable would take far too many resources to actually compute irrespective of the volume of hardware resources expended on it when traditional protein-folding algorithms are used. ML and deep learning techniques will play an important role in solving this problem. With AI and ML energized inventions such as DeepMind [33] and AlphaFold [34], it has now become possible to accurately predict the secondary structure of protein using its genetic sequence [35]. This will help in combating future pandemic and minimize it effects to the barest minimum.

Fourth, ML techniques can be deployed for grouping people together according to the risks they present, and clustering for better identification of people who have recently been in contact with someone diagnosed with an infectious disease so as to quarantine them [36]. Large-scale testing and tracing of potentially infectious people have been recognized as the most feasible strategies to confront any future endemic

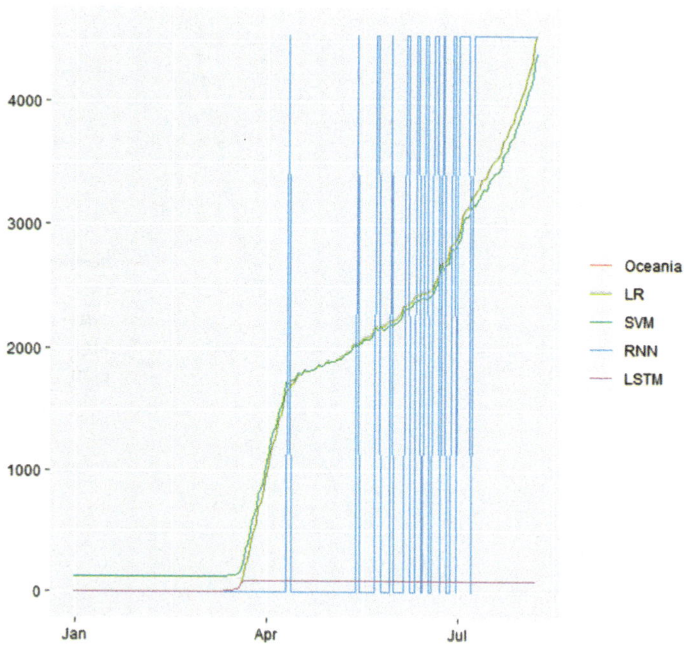

Fig. 14 Output prediction obtained from LR, SVM, RNN, LSTM model

that has passed the most critical stage so as to lessen the transmission of such virus. Conventional contact tracing methods rely on pieces of detached data collected from different testing centers and government/health authorities. By the time, several thousands of data points start flowing in continuously, traditional methods can without any form of resistance stop working. Therefore, ML techniques can be a robust tool for real-time classification and clustering of small or specialized populations, who are at higher risk of catching or transmitting the disease. This important attribute of ML techniques will be amazingly useful for quarantine and tracing of contacts, even with inadequate resources [5].

Finally, ML techniques can be used for digital surveillance of future pandemics [37]. With the ever-increasing volumes of data available on social networking websites, weblogs, chatrooms, and local newspapers that provide us with daily evidences of disease outbreaks occurrence [38]. This development is on the increase since more people in countries that are worst hit by the pandemic are now going online to inform the online community about their worry and symptoms, search for drugs and medical experts, government policies and others. ML techniques can be used to track these discussions, data packets, systematic way in which search for such information was done, and the related digital demographic data that can be used to model, predict and caution public health systems and political leaders about novel endemics all over the globe [39]. All the above potential application of ML to future

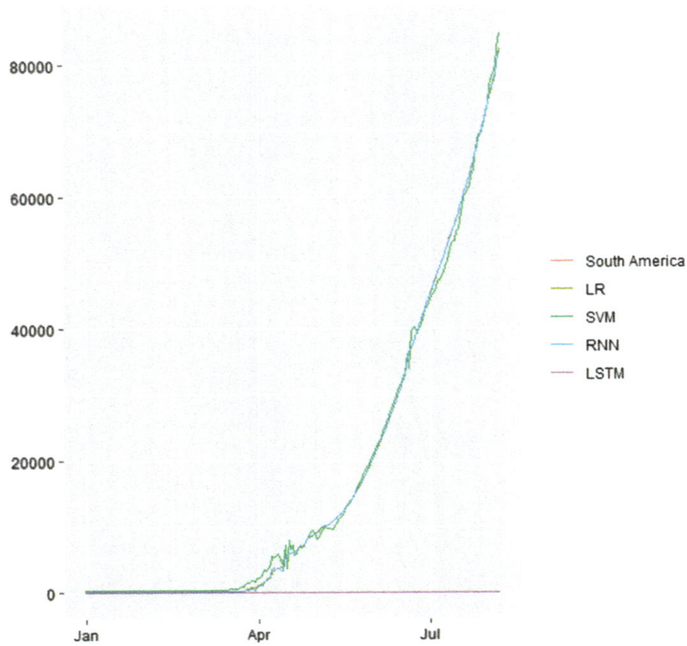

Fig. 15 Output prediction obtained from LR, SVM, RNN, LSTM model

pandemic is proven that ML is a promising technique for effectively handling any endemic that might emerge in the nearest future.

COVID-19 has shown that the world is not prepared to effectively handle any outbreak of pandemic. We need to prepare now in order to mitigate human and financial losses arising from potential global pandemics. With the latest report on declining and spike of COVID-19 in some countries, we have to focus on how this latest world pandemic has been mismanaged and device ways that can be used to curtail any outbreak of the next global pandemic. The rising occurrence of pandemics in recent decades is a matter of great concern. With the aid of political and financial commitments, we can encourage and strengthen public health institutions to minimize the catastrophic effects of any future pandemics and safeguard our future generations.

Table 1 Comparison of COVID-19 models

Continent	Algorithm	MAE	RMSE	MASE
Africa	LR	4.80E-01	9.80E-01	0.00E + 00
	SVM	2.98E + 02	5.09E + 02	2.80E-01
	RNN	1.01E + 04	1.77E + 04	9.62E + 00
	LSTM	1.21E + 04	2.08E + 04	1.16E + 01
Asia	LR	3.60E-01	6.80E-01	0.00E + 00
	SVM	3.97E + 03	4.22E + 03	1.21E + 00
	RNN	3.25E + 04	5.61E + 04	9.96E + 00
	LSTM	4.24E + 04	7.03E + 04	1.30E + 01
Europe	LR	1.11E + 00	2.34E + 00	0.00E + 00
	SVM	5.66E + 03	6.20E + 03	1.34E + 00
	RNN	6.49E + 03	1.91E + 04	1.54E + 00
	LSTM	9.34E + 04	1.24E + 05	2.22E + 01
North America	LR	3.20E-01	5.50E-01	0.00E + 00
	SVM	1.33E + 03	1.48E + 03	7.70E-01
	RNN	1.08E + 02	1.74E + 02	6.00E-02
	LSTM	2.14E + 04	3.32E + 04	1.25E + 01
Oceania	LR	5.00E-02	9.00E-02	0.00E + 00
	SVM	8.25E + 01	9.64E + 01	1.01E + 00
	RNN	1.02E + 03	1.40E + 03	1.26E + 01
	LSTM	1.44E + 03	1.97E + 03	1.78E + 01
South America	LR	3.50E-01	8.30E-01	0.00E + 00
	SVM	7.54E + 02	9.65E + 02	5.10E-01
	RNN	9.61E + 01	1.25E + 02	6.00E-02
	LSTM	1.75E + 04	2.91E + 04	2.91E + 04

4 Conclusion and Future Work

This paper presents different ML models based on information obtained from Kaggle database for the spread of COVID-19 that can be used to effectively predict the transmission of COVID-19. The study also discussed the applications of ML techniques for handling future pandemic. Our findings reveal that there are exponential drifts of COVID-19 in Africa, Europe, Oceania, North America and South America whereas the drift of spread of COVID-19 has been stabilized in some part of Africa and Oceania. As COVID-19 cases keep growing in Africa, Europe, Oceania, North America and South America, there are a to curtail transmission of the diseases. The main strategies employed in Asia and Diamond Cruise Ship can also be employed, which include: the first method is by preventing the exportation of diseases to other places to avoid further disease spreading. The second method is to reduce pandemic severity and delay the rise in the affected patients. The third method is to reduce the clusters of confirmed pandemics and self-quarantine cases of COVID-19. In the future, we will develop different deep learning and machine learning models, which can accurately predict COVID-19 transmission by using patients' information and epidemiological data, perform contact tracing, and give useful advice on suggested drugs that the infected person can use.

References

1. National Center for Immunization and Respiratory Diseases (NCIRD) DoVD (2020) Coronavirus disease 2019 (COVID-19) situation summary: centers for disease control and prevention. https://www.cdc.gov/coronavirus/2019-nCoV/summary.html. Accessed 30 Mar 2020
2. Gardner L (2020) Update January 31: modeling the spreading risk of 2019-nCoV. Johns Hopkins University Center for Systems Science and Engineering. https://systems.jhu.edu/research/public-health/ncov/. Accessed 30 Mar 2020
3. Huang C, Wang Y, Li X, Ren L, Zhao J, Hu Y, Cheng Z (2020) Clinical features of patients infected with 2019 novel coronavirus in Wuhan, China. The Lancet 395(10223):497–506. https://doi.org/10.1016/S0140-6736(20)30183-5
4. Nemati M, Ansary J, Nemati N (2020) Machine-learning apprgoaches in COVID-19 survival analysis and discharge-time likelihood prediction using clinical data. Patterns 1:100074. https://doi.org/10.1016/j.patter.2020.100074
5. Lalmuanawma S, Hussain J, Chhakchhuak L (2020) Applications of machine learning and artificial intelligence for Covid-19 (SARS-CoV-2) pandemic: a review. Chaos, Solitons Fractals 139:110059. https://doi.org/10.1016/j.chaos.2020.110059
6. Tuli S, Tuli R, Gill SS (2020) Predicting the growth and trend of COVID-19 pandemic using machine learning and cloud computing. Internet Things 11:100222. https://doi.org/10.1016/j.iot.2020.100222
7. Malki Z, Atlam E, Hassanien AE, Dagnew G, Elhosseini MA, Gad I (2020) Association between weather data and COVID-19 pandemic predicting mortality rate: Machine learning approaches. Chaos, Solitons Fractals 138:110137. https://doi.org/10.1016/j.chaos.2020.110137
8. Kavadi DP, Patan R, Ramachandran M, Gandomi A (2020) Partial derivative Nonlinear Global Pandemic Machine Learning prediction of COVID 19. Chaos, Solitons Fractals 139:110056. https://doi.org/10.1016/j.chaos.2020.110056
9. Amar LA, Taha AA, Mohamed MY (2020) Prediction of the final size for COVID-19 epidemic using machine learning: A case study of Egypt. Infectious Disease Modelling 5:622–634. https://doi.org/10.1016/j.idm.2020.08.008
10. Adhikari SP, Meng S, Wu YJ, Mao YP, Ye RX, Wang QZ, Sun C, Sylvia S, Rozelle S, Raat H, Zhou H (2020) Epidemiology, causes, clinical manifestation and diagnosis, prevention and control of coronavirus disease (COVID-19) during the early outbreak period: a scoping review. Infect Dis Poverty 9(1):1–12
11. Arksey H, O'Malley L (2005) Scoping studies: towards a methodological framework. Int J Soc Res Methodol 8:19–32
12. Sarwar A, Sharma V (2014) Comparative analysis of machine learning techniques in prognosis of type II diabetes. AI & Soc 29(1):123–129
13. Oyewola D, Hakimi D, Adeboye K, Shehu MD (2016) Using five machine learning for breast cancer biopsy predictions based on mammographic diagnosis. Int J Eng Technol 2(4):142–145
14. So A, Hooshyar D, Park KW, Lim HS (2017) Early diagnosis of dementia from clinical data by machine learning techniques. Appl Sci 7(7):651
15. Iqbal N, Islam M (2019) Machine learning for dengue outbreak prediction: a performance evaluation of different prominent classifiers. Informatica 43(3):363–371

16. Goel S, Deep A, Srivastava S, Tripathi A (2019) Comparative analysis of various techniques for heart disease prediction. 2019 4th international conference on information systems and computer networks (ISCON). IEEE, New York, pp 88–94
17. Devakumar KP (2020) COVID-19 dataset: number of confirmed, death and recovered cases every day across the globe. https://www.kaggle.com/imdevskp/corona-virus-report [Accessed May 2020]
18. Vapnik V (1995) The nature of statistical learning theory. Springer, Berlin
19. Ahmad I, Abdulah A, Alghamdi A (2010) Towards the designing of a robust ıntrusion detection system through an optimized advancement of neural networks. In: Kim T-H, Adeli H (eds) Advances in computer science and ınformation technology, LNCS, vol 6059. Springer, Berlin, pp 597–602
20. Huang Y, Zhao L (2018) Review on landslide susceptibility mapping using support vector machines. CATENA 165:520–529
21. Broomhead DS, Lowe D (1988) Multivariable functional ınterpolation and adaptive networks. Comp Syst 2:321–355
22. Sherstinsky A (2020) Fundamentals of recurrent neural network (RNN) and long short-term memory (LSTM) network. Phys D: Nonlinear Phenomena 404:132306
23. Hochreiter S, Schmidhuber J (1997) Long short-term memory. Neural Comput 9(8):1735–1780
24. Oyewola DO, Al-Mustapha KA, Dada EG, Kennedy OA (2019) Stock market movement direction with ensemble deep learning network. J Nigerian Assoc Math Phys Nigeria 53(2019):103–116
25. Kumar A, Gupta PK, Srivastava A (2020) A review of modern technologies for tackling COVID-19 pandemic. Diabetes Metab Syndr 14(4):569–573
26. Kaundal R, Kapoor AS, Raghava GP (2006) Machine learning techniques in disease forecasting: a case study on rice blast prediction. BMC Bioinf 7(1):485
27. Lee EK, Nakaya HI, Yuan F, Querec TD, Burel G, Pietz FH, Pulendran B (2016) Machine learning for predicting vaccine immunogenicity. Interfaces 46(5):368–390
28. Koff WC, Gust ID, Plotkin SA (2014) Toward a human vaccines project. Nature İmmunol 15(7):589–592
29. Romero P, Banchereau J, Bhardwaj N, Cockett M, Disis ML, Dranoff G, Kvistborg P (2016) The human vaccines project: a roadmap for cancer vaccine development. Sci Transl Med 8(334):334ps9–334ps9
30. Ong E, Wong MU, Huffman A, He Y (2020) COVID-19 coronavirus vaccine design using reverse vaccinology and machine learning. Frontiers İmmunol 11:1581
31. Muggleton S, King RD, Stenberg MJ (1992) Protein secondary structure prediction using logic-based machine learning. Protein Eng Des Sel 5(7):647–657
32. Baldi P, Brunak S, Frasconi P, Soda G, Pollastri G (1999) Exploiting the past and the future in protein secondary structure prediction. Bioinformatics 15(11):937–946
33. AlQuraishi M (2019) AlphaFold at CASP13. Bioinformatics 35(22):4862–4865
34. Wei GW (2019) Protein structure prediction beyond AlphaFold. Nat Machine Intell 1(8):336–337
35. Heo L, Feig M (2020) Modeling of severe acute respiratory syndrome coronavirus 2 (SARS-CoV-2) proteins by machine learning and physics-based refinement. bioRxiv.
36. Prakash A, Muthya S, Arokiaswamy TP, Nair RS (2020) Using machine learning to assess covid-19 risks. medRxiv Preprint. https://doi.org/10.1101/2020.06.23.20137950
37. Calvo RA, Deterding S, Ryan RM (2020) Health surveillance during covid-19 pandemic: how to safeguard autonomy and why it matters. BMJ, 369. http://eprints.whiterose.ac.uk/159306/3/BMJ_Health_Surveillance_COVID19.pdf. [Accessed 30 Jul 2020]
38. Gupta A, Katarya R (2020) Social media based surveillance systems for healthcare using machine learning: a systematic review. J Biomed Inf 108:103500
39. Desai AN, Kraemer MU, Bhatia S, Cori A, Nouvellet P, Herringer M, Lassmann B (2019) Real-time epidemic forecasting: challenges and opportunities. Health Security 17(4):268–275

Chapter 4
Human-in-the-Loop Enhanced COVID-19 Detection in Transfer Learning-Based CNN Models

Ferhat Ozgur Catak and Kevser Şahinbaş

Abstract Recently, many countries have concentrated on protecting human health and struggle with COVID-19 outbreak. Now that the virus is showing up in many places worldwide and the numbers of infected people are increasing rapidly, the governments of some countries are going to have to act quickly. The diagnostic tools are expensive, and the diagnostic kits can be challenging to obtain in many countries which makes it more difficult to stop the epidemic. Therefore, we need a fast and straightforward diagnostic tool based on AI technology that will save time, money, and resources in the prevention and control of the virus. To achieve this problem, researchers are developing X-ray scan images-based COVID-19 detection models which are based on convolutional neural networks and transfer learning algorithms. These models are trained on visual pictures in order to predict the presence of the virus in the sample. On the other hand, predictive modeling can make some false predictions like false negative which means that the sample may not be infected with COVID-19, but it will show positive COVID-19. It is important to remark that the wrong diagnostic will lead to an incorrect approach to the treatment and prevention of the virus. For example, if a patient is infected with COVID-19, but the test is negative for the virus, the patient can lose their life. Such an incorrect diagnosis will be very costly for the government. For this reason, model performance and the reliability of the classification result should be considered together. Thus, we recommend applying to a medical doctor for low-quality estimates by using uncertainty. In this study, we propose uncertainty quantification enhanced transfer learning-based convolutional neural network models for predicting the presence of COVID-19 in X-ray scan images publicly available on a GitHub repository. During the training phase, we used 278 positive and 66 negative COVID-19 X-rays and 39 positive and 30 negative images during the testing phase. We used VGG16, VGG19, ResNet, DenseNet, and InceptionV3 pre-trained deep CNN models to cope with

F. O. Catak (✉)
Simula Research Laboratory, Oslo, Norway
e-mail: ozgur@simula.no

K. Şahinbaş
Istanbul Medipol University, Istanbul, Turkey
e-mail: ksahinbas@emedipol.edu.tr

U. Kose et al. (eds.), *Computational Intelligence for COVID-19 and Future Pandemics*,
Disruptive Technologies and Digital Transformations for Society 5.0,
https://doi.org/10.1007/978-981-16-3783-4_4

the complex nature of the images that require deep architecture. Findings from the results demonstrate that the pre-trained VGG16 model can detect COVID-19 from non-COVID-19 cases with the highest classification performance among the other proposed models with 75% accuracy and can be used as an additional tool in the radiology department.

Keywords COVID-19 · Transfer learning · CNN · X-ray · Uncertainty quantification

1 Introduction

COVID-19 is a respiratory virus in the human (pulmonary) lung with the highest infectious potential of all respiratory viruses. It can be transmitted by aerosol particles and by respiratory droplets (e.g., from coughing, sneezing from an infectious agent). The transmission rate of COVID-19 is very high, and the disease can be fatal if it is not treated in time. The clinical studies based on COVID-19 diagnosis can be time-consuming and expensive, and it is challenging to detect in the population (as in many other viral diseases). This issue makes it extremely important to identify the disease with cheap and fast diagnostic tools that can be used in hospitals. Artificial intelligence (AI) is one of the most promising tools for the diagnosis of this disease. AI can detect the presence of COVID-19 from the X-ray images (as in the case of the present study), and it can also be used to perform the diagnostic steps and to assess the severity of the disease in the population.

There are some limitations in machine learning-based approaches for the diagnosis of COVID-19, such as computational complexity. These limitations can be overcome with the use of the transfer learning models, which allow combining the information obtained from the different data sources to improve the accuracy of the machine learning models. This type of transfer learning has been applied in many fields (e.g., medicine, computer vision, speech recognition) where the data is relatively large (e.g., thousands of images) and the training data is relatively small (e.g., few hundred images). Transfer learning models are also useful for the diagnosis of COVID-19 since they can be applied to a large number of data sources (e.g., X-ray images), and they are very efficient.

On the other hand, AI-based models for the prediction should be reliable, and we need to quantify the quality of the model's predictions which is essential for making reliable recommendations to the patients. The uncertainty is a promising issue for the reliability of the machine learning models, and the future of AI-based diagnostics is still a matter of many years. For this reason, the present study is the first one to use AI for the diagnosis of this disease with uncertainty quantification. We believe that the present study can serve as a basis for future investigations in this field.

We have summarized some studies on transfer learning-based diagnosis of COVID-19 using chest X-rays. Recently, many radiology images are widely used to detect COVID-19. Wang [1] introduced an artificial intelligence that could diagnose corona virus. The developers named artificial intelligence COVID-Net which obtained 92.4% accuracy. The developers, who collected 5,941 X-ray images of thousands of patients with viral and bacterial infections, uploaded this data to the database of COVID-Net. As a result of this great devotion, COVID-Net only examined lung X-rays and was able to diagnose corona virus based on these data in the database. Ucar and Korkmaz [2] present SqueezeNet model network design that is called as COVIDDiagnosis-Net, a COVID-19 diagnosis with Bayesian optimization additive in X-rays images. In the study, a deep learning architecture-based rapid diagnosis system was proposed using thousands of internationally shared lung X-ray images by focusing on chest radiography imaging and analysis due to abnormalities in chest X-ray images of COVID-19-infected patients. The developed system can only quickly determine the diagnosis of Normal (not infected), Pneumonia (not from Covid-19), or COVID-19 Infected with lung X-rays. Ioannis et al. [3] presents Convolutional Neural Network (CNN) architectures with transfer learning. They obtain a 97.82% overall accuracy in the detection of COVID-19. Hemdan et al. [4] demonstrate deep learning models that are named as COVIDX-Net to classify COVID-19 in X-ray images. Narin et al. [5] presented pre-trained ResNet50, InceptionV3, and InceptionResNetV2 convolutional neural network-based models with 98% accuracy, which is the best of the pre-trained ResNet50 model to detect COVID-19 cases on chest X-ray. Zhang et al. [6] present a model that predicts the prognosis of COVID-19 patients and distinguish patients' SARS CoV-2 against other viral infections. Xu et al. [7] proposed ResNet model that is a pre-trained deep CNN ResNet model to detect COVID-19 patients by using CT images and has a 86.7% accuracy in performance rate.

In our previous work, we showed that the transfer learning-based CNN models can be used to predict the COVID-19 based on X-ray images, and this work is based on the same experimental data (i.e., the same set of images from which we extracted) [8].

In our best knowledge, the current literature is based on the same approach to detect the COVID-19 based on X-ray images and CNN models. On the other hand, we believe that a medical doctor should enhance the ability of the model to predict the prognosis of the COVID-19 in some cases. The current deep neural network models predict the label (positive or negative). To accurately find the COVID-19 patients, the model's predictions have to be reliable. A traditional neural network predicts the label based on softmax probability outputs. Recently, a new paradigm in deep learning has been proposed to calculate the reliability (or uncertainty) of the prediction. The next step would be to calculate the confidence of the COVID-19 model based on the predicted probability. This approach has been applied to a variety of different imaging techniques, but it will be necessary to improve the accuracy of the COVID-19 model by incorporating uncertainty quantification to assess the predictors. The most important feature of our research is that we will be able to improve the accuracy of the COVID-19 model by applying uncertainty quantification to the prediction models.

In this book chapter, we will describe the basic concepts and use of uncertainty quantification to derive reliable predictions for COVID-19. We will show the importance of the uncertainty in machine learning models for the classification performance of the diagnostic model. We will also discuss the practical application of uncertainty quantification for the COVID-19 diagnosis. Section 2 includes Convolutional Neural Network (CNN), transfer learning concept, and pre-trained algorithms. In Sect. 3, model uncertainty in Bayesian deep learning is explained. Data Collection and explanation are given in Sect. 4. Section 5 contains system model and simulation results. Lastly, we conclude the chapter in Sect. 7.

2 Convolutional Neural Network (CNN)

There is increasing demand in many areas such as medical image processing, processing of large datasets, processing digital images, and supporting the movement of autonomous cars [9]. CNN studies have been carried out since the late 70s [10]. CNN introduced by LeCun in 1989 for computer vision applications is a type of multi-layer feedforward ANN [11]. CNN is a class of deep and feedforward artificial neural network that is frequently used in image processing and computer vision problems and is considered the latest technology for image classification [12]. It is designed to mimic the structure of the animal visual cortex [13] .The Convolutional Neural Network (CNN) is a deep learning technique used to represent data using the hierarchical layers of an image (or other input data). This feature makes CNN the right choice for many tasks, such as face detection, object detection. CNN is used in a wide variety of applications and is a powerful tool for classification and image segmentation. The convolution layer represents the input image in a three-dimensional space, while the activation layer is responsible for determining the output of the network. The convolutional layer is the central part of the CNN, and the network architecture can be described as follows: a convolutional layer with two input layers, a max-pooling layer and a fully connected layer. The feature map is created by applying the high- and low-level kernels in each of the input.

In Eq. 1, the convolution process is shown.

$$S(i,j) = (I \times K)(i,j) = \sum_m \sum_n I(m,n) K(i-m, j-n), \qquad (1)$$

where m and n are the kernel size, in the above equation, the kernel is the input of the pooling layer and the feature map is the output of the pooling layer. Pooling layer is generally applied between convolution layers. The main feature of the pooling layer is to reduce the number of computation steps needed to reduce the amount of the feature map. Another reason for the use of pooling layer is to increase the effectiveness of the network by removing the dominant features from the model. There are different types of pooling layers, and maximum and average pooling layers are the most common in the CNN. In the fully connected layers, the nodes are connected to the previous layer in the network.

Pooling layer discards unnecessary and redundant attributes, which reduces network computing cost and therefore makes the network more efficient [14].

2.1 Transfer Learning

This section provides some of the well-known deep learning image classifiers that are needed to apply the clinical aim of the detection of COVID-19 framework. Transfer learning is the reuse of information learned by a network in another dataset to improve the performance of another network [15]. Classical machine learning techniques can work efficiently only when there is sufficient and appropriate training data. In real life, it is difficult to find training data that covers all situations and, in some cases, one of the most significant difficulties encountered by researchers is the limited number of the available dataset [16]. It is difficult to find training data, such as in the analysis of medical data. Besides, with the change of conditions, the training data at hand can become outdated and become unavailable. In such cases, all the approaches that try to reduce the need for training data for machine learning by allowing information exchange between similar tasks are called transfer learning that provides the training of data with a small dataset and needs fewer calculation costs [17]. It is the most significant advantage of using transfer learning. In this paper, we structured deep CNN-based VGG16, VGG19, InceptionV3, and DenseNet models for the classification of COVID-19 chest X-ray images to detect COVID-19 classes.

2.1.1 VGG16 and VGG19

One of the deep learning algorithms is the Visual Geometry Group Network (VGG) architectural structure by Simonyan and Zisserman [18]. It is a model developed by Oxford University to give the best results compared to other algorithms. In this model, the number of convolutions has been increased to 13 for better results. Three fully connected structures were used. It is a model consisting of a total of 41 layers with all layers. The size of the images in the input layer is $224 \times 224 \times 3$. As in AlexNet architecture, the classification process is performed in the result layer, and the result is produced [18]. The deep network architecture is classified into two main categories, namely, VGG16 and VGG19. These models have different depths and layers [19]. VGG19 is deeper, and the number of parameters is larger and more expensive than VGG16 to train the network. The VGG19 architecture includes approximately 138 million computational parameters [20]. Figure 1 indicates the VGG16 architectures [21].

2.1.2 InceptionV3

The input images are given as $224 \times 224 \times 3$ and consist of 22 layers of depth. GoogLeNet consists of modular structures connected in parallel instead of sequentially ordered convolution and pooling layers. GoogLeNet is a deep learning network created by Christian Szegedy et al. in 2015 using the ImageNet dataset. This model with 48 layers depth takes the image input dimensions as $299 \times 299 \times 3$. Incep-

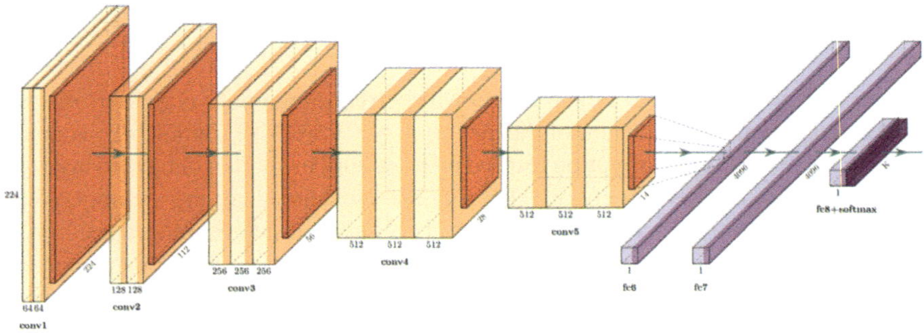

Fig. 1 VGG16 architecture [21]

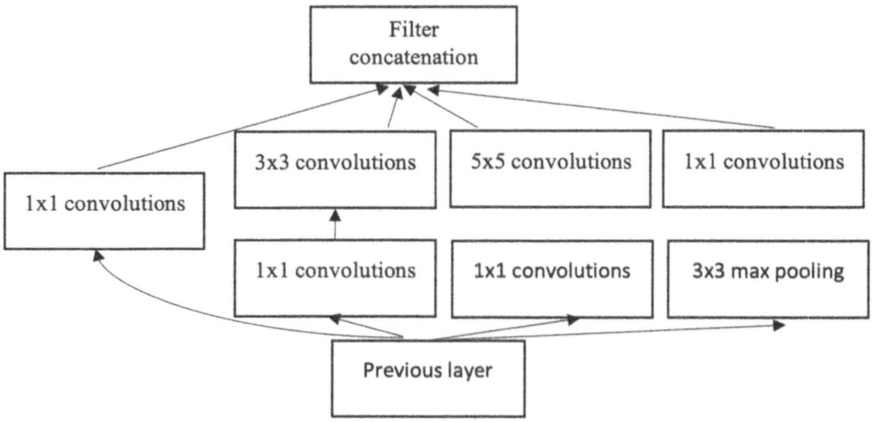

Fig. 2 InceptionV3 architecture [23]

tionV3 won 2014 Image net challenge with 93.3% top-5 accuracy [22]. Figure 2 demonstrates the compressed architecture of the InceptionV3 convolutional neural network [23].

In this network, fully connected layers are replaced by the average pooling layer. Since fully connected layers contain a large number of parameters, this modification process significantly reduces the number of parameters [24].

2.1.3 DenseNet

This network is designed to overcome the vanishing gradient problem caused by the depth of the network. In particular, the connection architectures of all layers are used to ensure the maximum flow of information between layers. In this configuration, each layer receives entries from all previous layers and passes its attribute maps to all subsequent layers. Attribute maps are combined on each layer to transmit

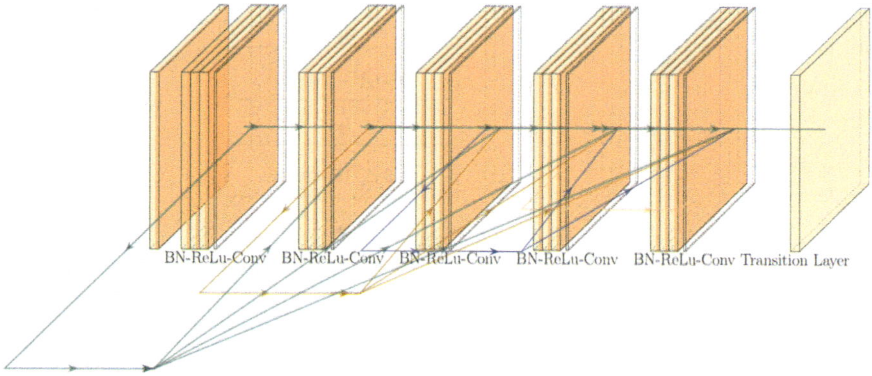

Fig. 3 DenseNet architecture [25]

information to layers after previous layers. This network architecture eliminates the need to learn redundant information and, accordingly, the number of parameters is significantly reduced. It is also effective in protecting information thanks to the linking feature of all layers. In the study by Huang et al., the network performed very well for classifications made with a small training dataset. Figure 3 shows a compressed image of DenseNet [25].

2.1.4 Resnet

In 2015, ResNet was the winner of the contest with a 3.57% error in the ImageNet dataset in the ILSVRC fight. He et al. [26] instead of teaching the layers the functions that are not distinctive in ResNet, they reformulate the remaining functions as learning by referring to layer inputs. Optimizing residual networks now ensures accuracy at increasing depths. This was revealed in tests with Resnet [23].

Under normal conditions, the part from the input to the output is expressed with the nonlinear $H(\mathbf{x})$ function. In ResNet architecture, the part from the input to the output is expressed by the nonlinear function defined as $F(\mathbf{x}) = H(\mathbf{x}) - \mathbf{x}$. The original mapping is repeated as $F(\mathbf{x}) + \mathbf{x}$. That is, the input value $\mathbf{x}$ is arithmetically added to the function $F(\mathbf{x})$. After this process, it is passed through ReLU with the function $F(\mathbf{x}) + \mathbf{x}$ as shown in Fig. 4. In this way, the values in the past layers are strongly transmitted to the next layers [26].

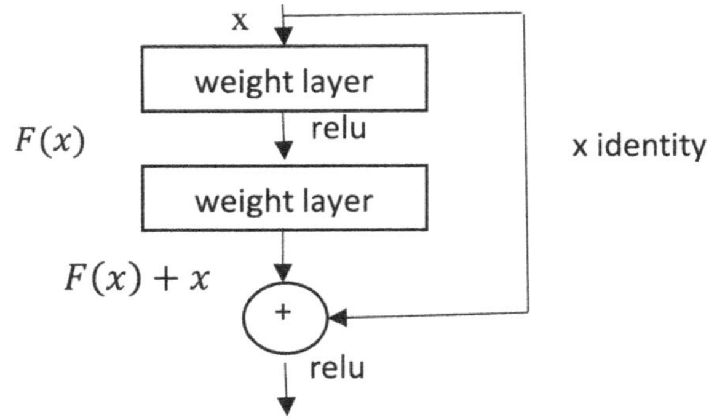

Fig. 4 Residual mapping structure [26]

3 Model Uncertainty in Bayesian Deep Learning

$\mathcal{D} = \{(\mathbf{x}_1, y_1), \ldots, (\mathbf{x}_N, y_N)\} \subset X \times Y$ is a given dataset where $X \in \mathbb{R}^m$ is a m-dimensional input vector and $Y \in \{1 \cdots C\}$, training instances $\{\mathbf{x}_i, y_i\} \forall X$ are a set of independent and identically distributed (i.i.d.) training samples from some unknown probability measure P on $X \times Y$. The main idea behind $h : X \mapsto Y$ is to get the hypothesis applying weights of neural net parameters w and a loss function $\ell : Y \times Y \mapsto \mathbb{R}$ as close as possible to the original function that has generated the labels.

Gal et al. [27] introduced a neural network model with prediction time dropout that is similar to a particular variational inference on a Bayesian neural network. The prediction hypothesis uncertainty is approximated by averaging probabilistic feedforward Monte Carlo dropout sample over the forecast period.

Monte Carlo dropout-based inference is a practical method for approximate inference in large and complex models [27]. This prediction is made by training a neural network with dropout before every weight layer. The prediction hypothesis uncertainty is approximated by averaging probabilistic feedforward Monte Carlo dropout sampling during prediction time. The predictive mean over dropout iterations, T, is used as the final inference,$\hat{y}$, on the test instance $\hat{\mathbf{x}}$.

$$p(\hat{y} = c|\hat{\mathbf{x}}, \mathcal{D}) \approx \hat{\mu}_{pred} = \frac{1}{T} \sum_{y \in T} p(\hat{y}|\theta, \mathcal{D}), \tag{2}$$

where θ is the model weights, $\mathcal{D}$ is the input dataset.

Figure 5 shows the general overview of the Monte Carlo dropout-based prediction in the inference time. In the inference time, random neurons in each layer are removed according to the p from the base neural network model to create another model. As a result, T different classification model will be created.

For each testing input sample $\mathbf{x}$, the predicted label is assigned with the highest predictive mean. We used entropy and predictive variance measures in order to quantify how much the model is uncertain about its labeling predictions. The entropy is

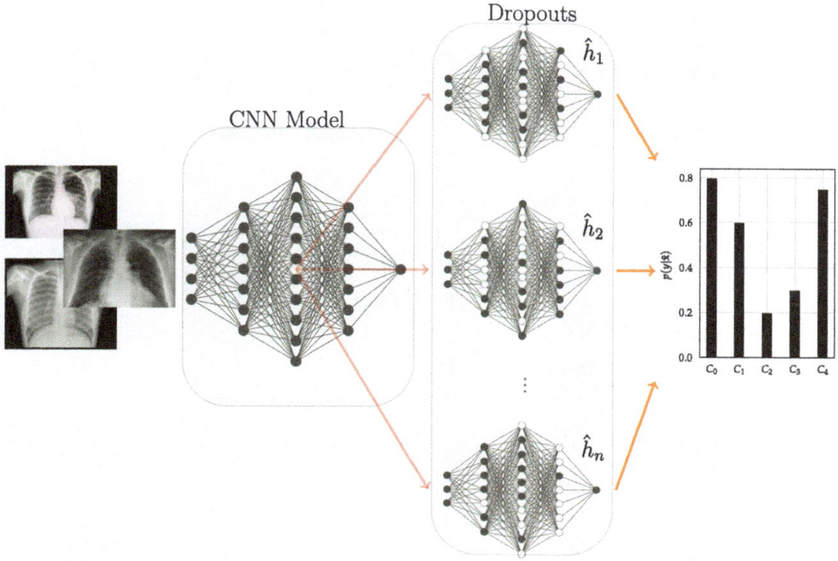

Fig. 5 Illustration of the Monte Carlo dropout-based Bayesian prediction

a measure defined as the average level of uncertainty fixed in the likely outcomes of a random variable [27].

$$H(\hat{y}|\theta, \mathcal{D}) = -\sum_{c\in C} p\left(f(\mathbf{x}) = c\right) \log p\left(f(\mathbf{x}) = c\right), \tag{3}$$

where $f(\mathbf{x})$ is the classification model prediction and p is the probability of the predicted label, which is class c.

$$VAR(\mathbf{y}) = \mathbb{E}\left[\hat{f}(\mathbf{x}|\theta, \mathcal{D}) - \mathbb{E}\left[\hat{f}(\mathbf{x}|\theta, \mathcal{D})\right]\right]^2. \tag{4}$$

4 Data Collection, Explanation

In this work, a publicly available GitHub dataset is used for the model development. The dataset contains chest X-ray and CT images of several lung diseases, including positive, suspected positive of COVID-19, or other viral and bacterial types of diseases such as MERS, SARS, and ARDS. In our previous work [x29], we had only 50 positive and 50 negative X-ray scan images for the training and 20 positive and 20 negative samples for the testing. In this work, we have increased the number of posi-

Table 1 Chest X-ray labels

Label	Number
Pneumonia/Viral/COVID-19	196
Pneumonia	38
Pneumonia/Fungal/Pneumocystis	21
Todo	12
Pneumonia/Bacterial/Streptococcus	12
No Finding	10
Tuberculosis	9
Pneumonia/Viral/SARS	8
Pneumonia/Bacterial/Klebsiella	7
Pneumonia/Bacterial/Mycoplasma	4
Pneumonia/Bacterial/Nocardia	4
Pneumonia/Bacterial/Legionella	4
Pneumonia/Lipoid	4
Pneumonia/Viral/MERS-CoV	4
Pneumonia/Viral/Varicella	4
Pneumonia/Bacterial	2
Pneumonia/Bacterial/Staphylococcus/MRSA	1
Pneumonia/Bacterial/Chlamydophila	1
Pneumonia/Viral/Influenza	1
Pneumonia/Viral/Influenza/H1N1	1
Pneumonia/Fungal/Aspergillosis	1

tive and negative samples for the training and testing sets to 278 and 66, respectively (Table 1).

5 System Model

We executed the experiments 30 times and tested the models for the following question: How accurate are the softmax output prediction's variance and prediction entropy in COVID-19 X-ray images dataset?

As the dataset is not balanced, and the data is not uniform, the traditional accuracy-based classification performance evaluation method is not suitable to assess the models. However, our experiments will provide four standard measurement metrics in information retrieval for evaluating the models such as overall prediction accuracy [28], average recall [29], average precision [30], and F_1 [31].

Precision is the ratio of the true value of the prediction to the prediction error and true value. In Eq. 5, mathematical calculation of precision is illustrated as follows:

$$Precision = \frac{Correct}{Correct + False}. \tag{5}$$

Recall is calculated by the fraction of retrieved relevant samples.

$$Precision = \frac{Correct}{Correct + Missed}. \tag{6}$$

Firstly, we calculated the precision and recall values for each class and then the mean value. In Eqs. 7 and 8, average precision and recall are presented, respectively.

$$Precision_{avg} = \frac{1}{n_{classes}} \sum_{i=0}^{n_{classes}-1} Prec_i. \tag{7}$$

$$Recall_{avg} = \frac{1}{n_{classes}} \sum_{i=0}^{n_{classes}-1} Recall_i. \tag{8}$$

The F_1 values represent the harmonic mean of precision and sensitivity values.

$$F_1 = 2 \times \frac{Prec_{avg} \times Recall_{avg}}{Prec_{avg} + Recall_{avg}}. \tag{9}$$

6 Simulation Results

6.1 Dataset

We used the publicly available dataset of chest X-rays and CT images of positive or suspected COVID-19 patients for this study. The dataset was obtained indirectly from both public sources, physicians and different hospitals. This dataset that contains images and data can be reached and published in the GitHub repository. X-ray images and CT scans are available in the original dataset. We selected X-ray scan images to construct a COVID-19 detection model for our proposed model. While building the model, we used 278 positive and 66 negative COVID-19 posteroanterior X-ray scan images for training and 39 positive and 30 negative images for testing phases.

We rescaled the X-ray scan images to 256 × 222 size. We have applied different image augmentation methods to create more advanced input instances such as flipping right/left and up/down, rotation and translation utilizing random five various angles.

This dataset is publicly accessible. But, it is important to remark that the number of datasets and the number of samples they include could increase in the future. Besides, we consider the tagging issue. The dataset comprises X-ray scan images containing the disease. There are only three records in the dataset that contain COVID-19 neg-

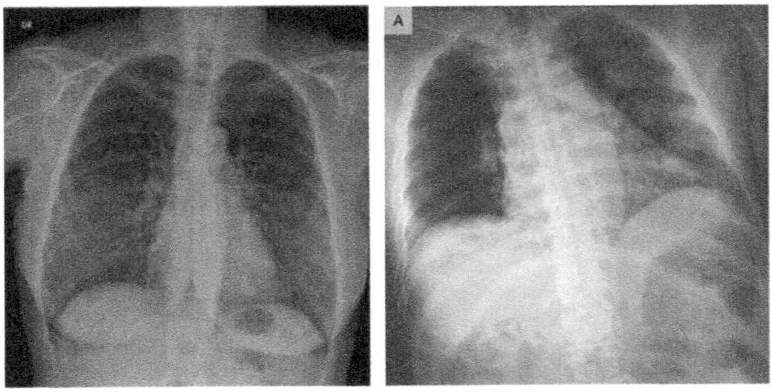

Fig. 6 An example of two X-ray scan images taken from the dataset **a** with a label of COVID-19 negative, **b** COVID-19 positive

ative samples, and the "finding" field was selected as No finding. Therefore, X-ray images with ARDS and Streptococcus results are labeled to select COVID-19 negative samples. Additional samples of COVID-19 negative patients will increase in this type of data collection for finding reliable results (Figs. 6 and 7).

In order to enhance the performance of the classification model and increase the samples in the dataset, we apply image augmentation techniques. The parameters of image augmentation are zoom range 15, rotation range 20, width shift range 0.2, height shift range 0.2, shear range 0.15, and horizontal flipping. In Fig. 8, an image augmentation technique is presented.

In Fig. 8, it is seen that the first image is the original image and the rest of the images are augmented images.

6.2 *Experiments*

In the experiments, we tested the uncertainty quantification methods and their relations with the classification performance. Simulations were performed on the following machine: 2.8 GHz Quad-Core Intel Core i7 with 16 GB of RAM. The experiments were performed using the Python scripts and machine learning libraries: Keras, Tensorflow, and Scikit-learn. Two models, with and without MC dropouts, were trained to obtain COVID-19 detection model's prediction results for the dataset. The hyperparameters such as the number of hidden layers and the number of neurons in the hidden layers, the activation function, the loss function, and the optimization method are the same for both models. However, the only difference between the two models is the prediction time dropout. In the first model, the drop out is applied at the prediction time, and the second model uses the drop out at the training time. The dataset has been divided into two parts: 33% test sets and 67% training sets. In the

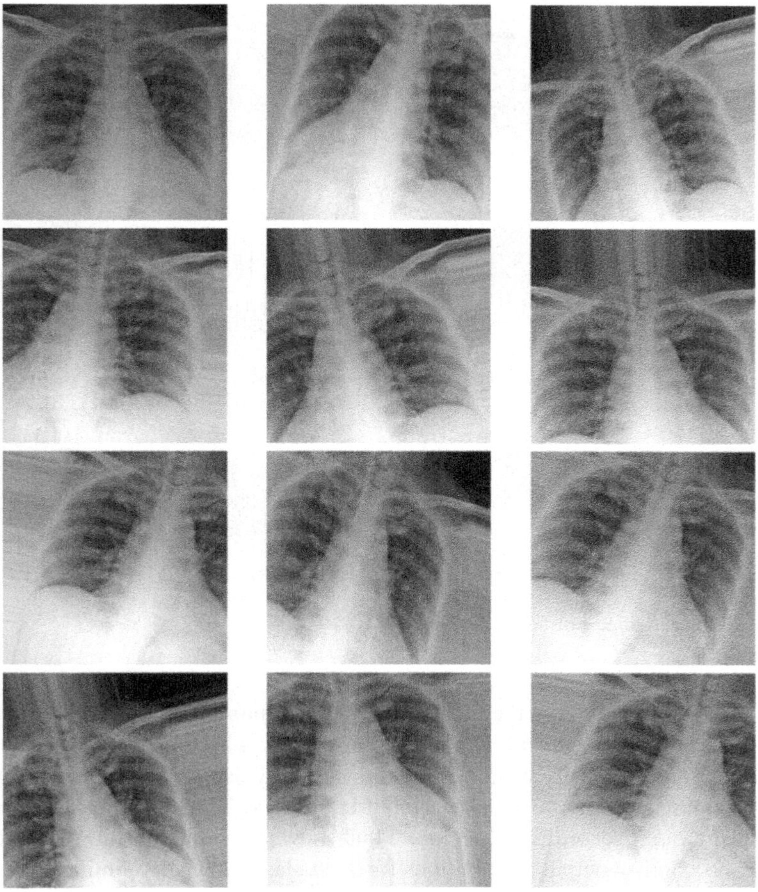

Fig. 7 Example of Data augmentation

test sets, we randomly choose from a subset of the datasets and use them to evaluate the prediction performance of the models. The number of epochs was set to 500, and the epoch size was fixed for all models. Prediction time dropout values were set to 0.0 and increased to 0.9 with an increment of 0.02. Monte Carlo dropout models were used for each test input with an $T = 100$ times for prediction of the test dataset. The predicted labels for the test set are calculated by averaging the output of the last layer's softmax function over T iterations. The prediction uncertainty for each input in the test sets is quantified using Shannon entropy and prediction variance, respectively. Table 2 shows the classification performance of the Monte Carlo dropout-based model and the standard model for each dataset.

Firstly, we used the entropy and prediction variance values of the COVID-19 detection model's softmax probability outputs to compare each uncertainty quantifi-

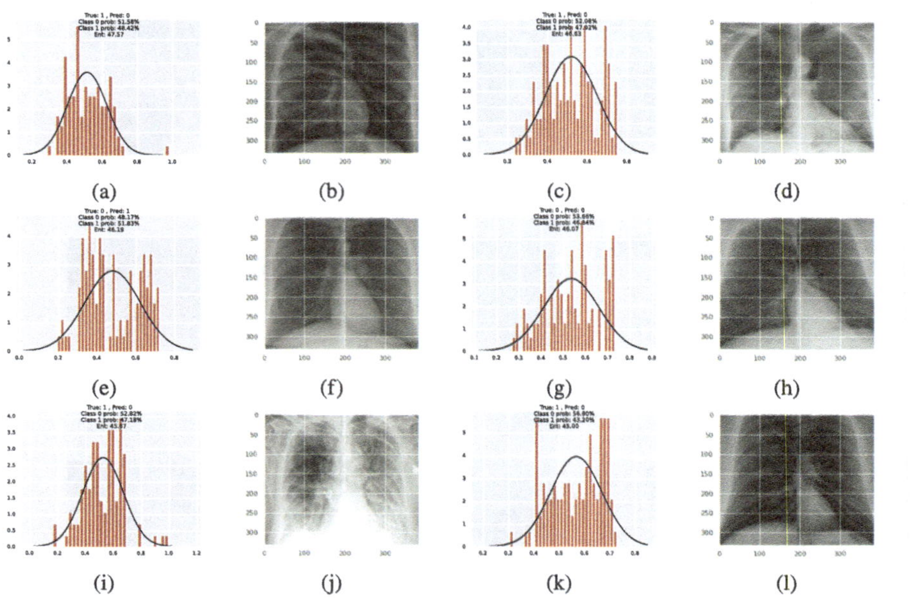

Fig. 8 Most uncertain test instances for the **COVID-19** dataset

cation method to determine the most and least uncertain predictions on the dataset. Figure 8 shows the six most uncertain test instances with entropy-based and softmax prediction probability variance-based uncertainty quantification for the dataset. Each sub-figure in the figures shows an input example with prediction probabilities (T times) histogram. According to the figure that shows the highly uncertain examples, softmax probability output distributions for each class show very similar patterns in the results. According to Fig. 8, the model's prediction is COVID-19 positive, and the true label is also COVID-19 positive for the input example. On the other hand, the uncertainty is very high for the prediction results, which has an uncertainty value of 0.98. Also, the class assignment probabilities for this input example are almost identical for each class label 49.81% and 50.19%, respectively, for all classes, and it can be difficult reliable. Another example is shown in Fig. 8. The model's class prediction is wrong for this example: it shows the highest probability for all COVID-19 negative 50.04 %, and the value for the COVID-19 positive is 49.96% with an uncertainty value of COVID-19 positive. The class softmax probabilities histogram distribution shows a very similar pattern for each class label.

Table 2 provides the pre-trained CNN algorithms used to build detection models for COVID-19. The findings from Table 2 show that pre-trained VGG16 algorithm presents the highest accuracy performance with 75% among the other well-known trained algorithms. The VGG19, ResNet50, and InceptionV3 indicate 74%, 57%, and 72%, respectively. It can also be observed that VGG16 model achieved the highest F-score value with 75%.

Table 2 Performance metrics for CNN pre-trained models for the diagnosis of COVID-19 of the proposed model

Model	Precision	Recall	F_1	Acc.
VGG16	0.76	0.75	0.75	0.75
VGG19	0.74	0.74	0.74	0.74
Resnet50	0.32	0.57	0.41	0.57
InceptionV3	0.75	0.72	0.71	0.72

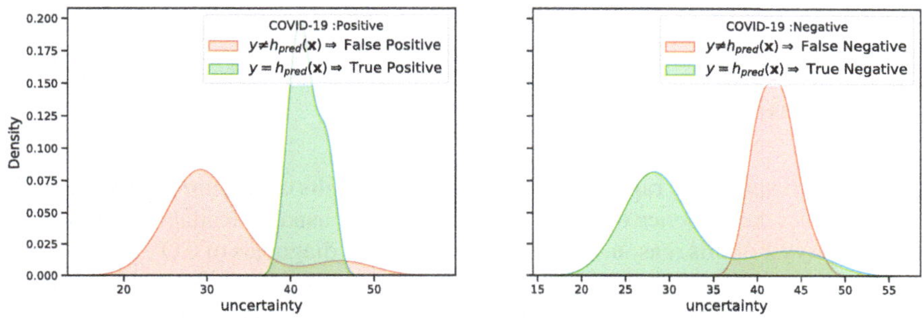

Fig. 9 Bayesian model uncertainty for the COVID-19 dataset

Figure 9 shows the COVID-19 model's entropy-based uncertainty with the $p = 0.2$ value. In the figure, the x-axis shows the model uncertainty value and the y-axis shows the density of data points. The green lines are the true-positive and true-negative values of the classification model. For instance, the sub-figure on the left shows the uncertainty of positive example (true *COVID-19* labels) predictions with true-positive and false-positive labels, respectively. The true-positive and true-negative labeled instances' uncertainty histogram values are clustered in the lower values (the left side of the figures), which means the low uncertainty in the correctly labeled samples. On the other hand, the false-positive and false-negative labeled instances' uncertainty histogram values are more scattered in the upper values (the right side of the figures), indicating the high uncertainty in the incorrectly labeled samples.

6.3 Discussion

Our results show that the COVID-19 model provides a precise understanding of the classification model. It shows how the COVID-19 model can be described in an even more comprehensive and well-defined way and the use of the classifier. It uses an extended classification framework to help it describe the classification model, as well as an explanation of the model's structure, more concisely. Figure 9 shows the uncertainty distribution for correct and wrong predictions using the uncertainty framework. According to the figure, the uncertainty expectations are low for the correctly labeled samples, and the incorrect labels are clustered in the upper values (the left side of the figures). This suggests that the model has a more consistent classification (high confidence) for the positive labels (COVID-19 positive); for the incorrectly predicted labels, the model's predictions have low confidence. It concludes that, if a deep learning-based COVID-19 detection system is applicable for the real-case scenario, it is essential to examine the possible prediction uncertainty to catch the wrong predictions.

7 Conclusion

COVID-19 is spreading rapidly around the world. Results of sputum or blood sample tests to detect the presence of this virus in humans are usually available within a few hours or days. For this reason, the importance of faster diagnosis of COVID-19 detection is increasing considerably. In this chapter, we propose uncertainty quantification enhanced transfer learning-based convolutional neural network models by exploiting the VGG16 model for predicting the presence of COVID-19 in X-ray scan images to provide a fully automatic and faster diagnosis. As shown by findings from results, the proposed VGG16 model detects whether a chest X-ray is COVID-19 or healthy with the highest classification performance as 0.75 accuracy values. The findings suggest that additional data is needed to find proper accuracy performance. Using only a deep learning model in corona detection can lead to irreversible results. The quality of the prediction made by the deep learning model must also be measured. While the studies conducted so far are concerned only with the quality of the model, in this study, the quality of each estimate is also measured.

References

1. Wang L, Wong A (2020) Covid-net: a tailored deep convolutional neural network design for detection of covid-19 cases from chest x-ray images. arXiv:2003.09871
2. Ucar F, Korkmaz D (2020) Covidiagnosis-net: deep bayes-squeezenet based diagnostic of the coronavirus disease 2019 (covid-19) from x-ray images, *Medical Hypotheses*, p 109761
3. Apostolopoulos ID, Mpesiana TA (2020) Covid-19: automatic detection from x-ray images utilizing transfer learning with convolutional neural networks, p 1
4. Hemdan EED, Shouman MA, Karar ME (2020) Covidx-net: a framework of deep learning classifiers to diagnose covid-19 in x-ray images. arXiv:2003.11055
5. Narin A, Kaya C, Pamuk Z (2020) Automatic detection of coronavirus disease (covid-19) using x-ray images and deep convolutional neural networks. arXiv:2003.10849
6. Zhang K, Liu X, Shen J, Li Z, Sang Y, Wu X, Zha Y, Liang W, Wang C, Wang K et al (2020) Clinically applicable ai system for accurate diagnosis, quantitative measurements, and prognosis of covid-19 pneumonia using computed tomography. Cell
7. Xu X, Jiang X, Ma C, Du P, Li X, Lv S, Yu L, Chen Y, Su J, Lang G (2020) Others, deep learning system to screen coronavirus disease 2019 pneumonia. arXiv:2002.09334

8. Sahinbas K, Catak FO (2020) In: Kose U, Gupta D, de Albuquerque VHC, Khanna A (eds) Transfer learning based convolutional neural network for covid-19 detection with x-ray images. In: Data Science for COVID-19, Computational Perspectives, vol 24. Elsevier: Elsevier, pp 1–24
9. Michalski P, Ruszczak B, Tomaszewski M (2018) Convolutional neural networks implementations for computer vision. Springer, Berlin, pp 98–110
10. Litjens G, Kooi T, Bejnordi BE, Setio AAA, Ciompi F, Ghafoorian M, Van Der Laak JA, Van Ginneken B, Sánchez CI (2017) A survey on deep learning in medical image analysis. Med Image Anal 42:60–88
11. LeCun Y, Boser B, Denker JS, Henderson D, Howard RE, Hubbard W, Jackel LD (1989) Backpropagation applied to handwritten zip code recognition. Neural Comput 1(4):541–551
12. Sultana F, Sufian A, Dutta P (2018) Advancements in image classification using convolutional neural network. IEEE, pp 122–129
13. Beysolow T II (2017) Introduction to deep learning using R: a step-by-step guide to learning and implementing deep learning models using R. Apress
14. Goodfellow I, Papernot N, McDaniel P, Feinman R, Faghri F, Matyasko A, Hambardzumyan K, Juang Y-L, Kurakin A, Sheatsley R, et al (2016) Cleverhans v0. 1: an adversarial machine learning library, vol 1. arXiv:1610.00768
15. Yosinski J, Clune J, Bengio Y, Lipson H (2014) How transferable are features in deep neural networks? In: Advances in neural information processing systems, pp 3320–3328
16. Zou L, Yu S, Meng T, Zhang Z, Liang X, Xie Y (2019) A technical review of convolutional neural network-based mammographic breast cancer diagnosis. Computational and mathematical methods in medicine, vol 2019
17. Pan SJ, Yang Q (2009) A survey on transfer learning. IEEE Trans Knowl Data Eng 22(10):1345–1359
18. Simonyan K, Zisserman A (2014) Very deep convolutional networks for large-scale image recognition
19. Zhen X, Chen J, Zhong Z, Hrycushko B, Zhou L, Jiang S, Albuquerque K, Gu X (2017) Deep convolutional neural network with transfer learning for rectum toxicity prediction in cervical cancer radiotherapy: a feasibility study. Phys Med & Biol 62(21):8246
20. Toğaçar M, Ergen B, Cömert Z (2020) Application of breast cancer diagnosis based on a combination of convolutional neural networks, ridge regression and linear discriminant analysis using invasive breast cancer images processed with autoencoders. Med Hypoth 135:109503
21. Fu Y, Aldrich C (2018) Using convolutional neural networks to develop state-of-the-art flotation froth image sensors. IFAC-PapersOnLine 51(21):152–157
22. Szegedy C, Liu W, Jia Y, Sermanet P, Reed S, Anguelov D, Erhan D, Vanhoucke V, Rabinovich A (2015) Going deeper with convolutions. pp 1–9
23. Mahdianpari M, Salehi B, Rezaee M, Mohammadimanesh F, Zhang Y (2018) Very deep convolutional neural networks for complex land cover mapping using multispectral remote sensing imagery. Remote Sens 10(7):1119
24. Chollet F (2016) Building powerful image classification models using very little data. Keras Blog
25. Huang G, Liu Z, Van Der Maaten L, Weinberger KQ (2017) Densely connected convolutional networks. pp 4700–4708
26. He K, Zhang X, Ren S, Sun J (2016) Deep residual learning for image recognition, pp 770–778
27. Gal Y, Ghahramani Z (2016) Dropout as a bayesian approximation: Representing model uncertainty in deep learning. pp 1050–1059
28. Witten IH, Frank E (2002) Data mining: practical machine learning tools and techniques with java implementations. Acm Sigmod Record 31(1):76–77
29. Han J, Pei J, Kamber M (2011) Data mining: concepts and techniques. Elsevier, Amsterdam
30. Turpin A, Scholer F (2006) User performance versus precision measures for simple search tasks, pp 11–18
31. Makhoul J, Kubala F, Schwartz R, Weischedel R et al (1999) Performance measures for information extraction. In: Proceedings of DARPA broadcast news workshop. Herndon, VA, pp 249–252

Chapter 5
Epidemiology Forecasting of COVID-19 Using AI—A Survey

Igor Gadelha Pereira, Andouglas Gonçalves da Silva Junior, Dunfrey Pires Aragão, Emerson Vilar de Oliveira, Arthur Andrade Bezerra, Francinaldo de Almeida Pereira, Julio Gustavo F. S. Costa, Jordan Salas Cuno, Davi Henrique dos Santos, Joris Michel Guerin, Aura Conci, Esteban Walter Gonzalez Clua, Cosimo Distante, and Luiz Marcos Garcia Gonçalves

Abstract This chapter introduces a thorough discussion about the current tools to estimate the epidemiology of COVID-19 focusing on AI strategies. We present in detail both the traditional methods for pandemics forecasting, namely, the SIR, SEIR, SEIRS, SIRD, and ARIMA and novel data-driven approaches that presented promising results, the LSTM, AE, MAE, CNN, and Random Forest. The article aims to give a comprehensive understanding of the main advantages and drawbacks of these data-driven strategies when compared to traditional strategies, in order to give a research direction for this area.

I. G. Pereira · D. P. Aragão · E. V. de Oliveira · A. A. Bezerra · F. de A. Pereira · J. G. F. S. Costa · D. H. dos Santos · L. M. G. Gonçalves (✉)
Federal University of Rio Grande do Norte, Natal, Brazil
e-mail: lmarcos@dca.ufrn.br

I. G. Pereira
e-mail: igorgad@dca.ufrn.br; igpereira@cpqd.com.br

D. P. Aragão
e-mail: dunfrey@gmail.com

E. V. de Oliveira
e-mail: evilaroliveira@gmail.com

A. A. Bezerra
e-mail: arthur.bezerra.049@ufrn.edu.br

F. de A. Pereira
e-mail: falmeida@dca.ufrn.br

J. G. F. S. Costa
e-mail: juliogustavocosta@gmail.com

D. H. dos Santos
e-mail: davihenriqueds@gmail.com

I. G. Pereira
Center for Research and Development (CPqD), Campinas, Brazil

U. Kose et al. (eds.), *Computational Intelligence for COVID-19 and Future Pandemics*, Disruptive Technologies and Digital Transformations for Society 5.0,
https://doi.org/10.1007/978-981-16-3783-4_5

1 Introduction

The coronavirus pandemics started in the end of 2019 and quickly spread throughout the world, directly affecting approximately 70 million people worldwide and having profound social and economic effects on many more. Just recently there's substantial advances in the development of a vaccine for the COVID-19 disease but that's yet not fully deployed, and the best way to reduce deaths is still to prevent the spread of the virus through simple yet effective preventive methods, such as social distancing, cleaning the hands and wearing masks. Understanding the dynamics of the spread of the virus is one way to help the planning of actions for reducing its dissemination. Prediction of a dynamic model could also be helpful to improve management strategies in hospitals in preparation for imminent second waves. This can be done by studying the epidemiology of the disease, mainly how it can generate important knowledge to help predicting the pandemic behavior and help public authorities in decision-making to reduce the spread of the virus and the number of deaths.

1.1 Classical Approaches

There are several known methods that can be applied in Epidemiology to estimate the spread of pandemics. The most popular is undoubtedly the SIR model [8, 9, 36, 37, 42] followed by SEIR [16, 31, 44], SEIRD [30], and ARIMA [14], between others. These are known to be classical methods for virus dynamics determination and in general they follow some rules and equations that are defined according to studies performed from past experienced cases of occurrence. In these models, basically, the phase transitions of the disease are modeled as instantaneous rates in

A. G. da S. Junior
Federal Institute of Rio Grande do Norte, Mossoró, Brazil
e-mail: andouglas.goncalves@ifrn.edu.br

J. S. Cuno · A. Conci · E. W. G. Clua
Fluminense Federal University, Niteroi, Brazil
e-mail: jordansalas@id.uff.br

A. Conci
e-mail: aconci@ic.uff.br

E. W. G. Clua
e-mail: esteban@ic.uff.br

J. M. Guerin
Laboratoire d'analyse et d'architecture des systemes, Toulouse, France
e-mail: jorisguerin.research@gmail.com

C. Distante
Institute of Applied Sciences and Intelligent Systems, Lecce, Italy
e-mail: cosimo.distante@cnr.it

differential equations or as probabilities of transition in discrete-time differences or matrix equations [39]. These models provide accurate estimates of the position of equilibrium points, when the rate at which individuals enter each stage is equal to the rate at which they exit. However, they do not accurately capture the distribution of the time an individual spends at each stage. Therefore, they do not accurately capture the transitory dynamics of the epidemics. However, if the matrix models are modified, they can provide a direct route to accurately model the duration of each stage and thus correctly reproduce the dynamics of the epidemic.

In resume, these methods are based on statistical and/or mathematical models that use epidemiological parameters such as susceptible population, number of infected cases, and the spread rate of the virus, among other parameters, to model the dynamics of the virus spread. Approaches, such as SIR, SEIRD, and SEITR, are helpful to understanding the dynamics of the Sars-CoV-2 in the short term. Nonetheless, determining the long-term forecasting in a most precise way is still a challenge because of the lack of ground truth data. And, despite these models representing somehow the virus dynamics, they must be improved so that they can be applied with higher precision. It has been shown, for example, for the SEIR model, in the literature [16] that it underestimate peak infection rates (by a factor of three, using published parameter estimates based on the progress of the epidemic in Wuhan) and to substantially overestimate the persistence of the epidemic after the peak has passed. So, despite being useful, these traditional models (SIR, SEIR, SEIRD) need improvements so that they can be applied to the study of the dynamics of the new virus.

1.2 Data-Driven Approaches

Several difficulties appear in this novel pandemic dynamics determination mainly due to the fact that the Sars-CoV-2 is a pretty new virus and there is not, currently, a complete modeling available for it. Actually, Sars-CoV-2 is a virus that has a spread function not so easy to be modeled with any specific classical model because of the influence of the several terms that affect its dissemination [39]. Besides, the worst factor is because of the non-linearity of data, due to the under-notifications and also the lack of efficient measures by governs for stopping its spread, which makes changing on the several parameters on the way. So in these situations it seems a good idea to try with data-driven (AI-based) methods.

To make it more complex, its spread has been very different in the several countries around the world and even regions inside a country, mostly depending on the people behavior and several other factors such as pollution, population density, poverty rate, and medical resources available. So, in order to avoid making very complex models, recent efforts are also focusing on data-driven (machine learning) methods based on AI tools, which could be an alternative to the classical models above. AI approaches work based on the available data instead of using predefined model based on parameters and variables. In fact, another motivation for using AI is that

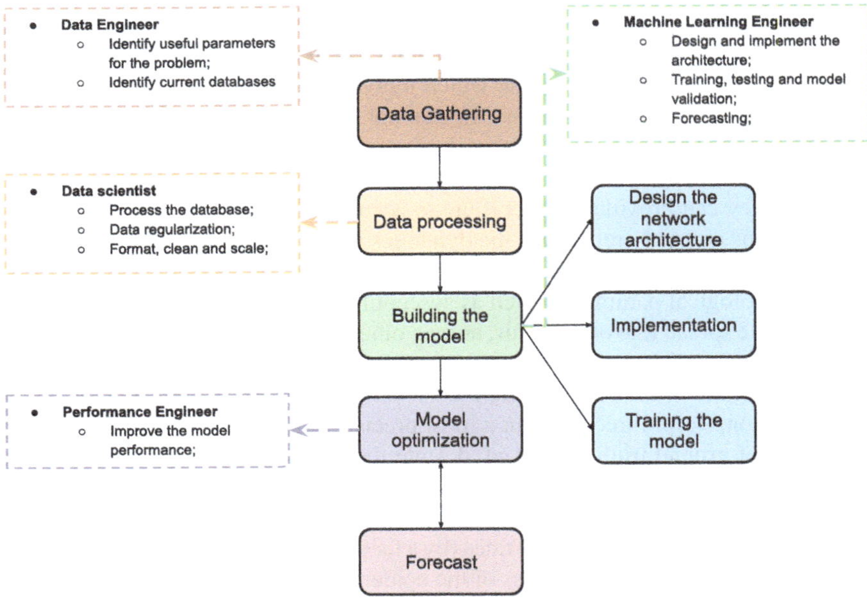

Fig. 1 Typical workflow for time series forecasting using AI. Each step can be performed by specialized personal

experienced data from more advanced pandemic regions can be used for predicting the dynamics of the pandemics in the late occurring ones.

A common methodology that we adapt here for time series forecasting using machine learning techniques is shown in Fig. 1. Each step presents its own challenges and can be performed by specialized personal. The performance of a machine learning model is dependent on the quality and quantity of data used in training, so the first step is to gather large amounts of raw data about the subject one is trying to forecast. In the case of COVID-19, there are some aggregate datasets that compile information given by many governmental agencies around the world, notably the Johns Hopkins University dataset [12]. These datasets present untreated raw data that has different levels of granularity and must be treated to be used for machine learning training. The second step is feature extraction, which is done in order to improve the quality of the raw data according to the problem one is trying to solve. This can be done by cleaning the data, selecting meaningful instances of the data (random sampling), feature tuning (scaling and normalizing), transforming the data (categorical to numeric), feature extraction, feature selection, and feature construction. Next step is to build the model. After deciding which network one wants to use, the architecture is defined, deciding the number of layers and/or the number of units in each layer and also other parameters such as bias and saturation function. The next step is to implement and train the model. The training of the model can also be done in a multitude of ways. And deciding which performance metric to use to evaluate the model.

The main contribution of this chapter is toward helping to solve the problem of dynamics modeling in a data-driven way. The chapter introduces a thorough discussion presenting both types of methods (classical and data-driven), including current tools to estimate the epidemiology of COVID-19 using AI. We aim to give a comprehensive understanding of the main advantages and drawbacks of the last, data driven, strategies when compared to traditional strategies, in order to draw a research direction to this area. We will discuss several data-driven approaches as CNN, Autoencoders, LSTM, MAE, starting with a general approach for AI based. We start next with a section showing how to get clean data, with data regularization, clustering, and error pruning. Then we describe the main classical approaches. Finally, data-driven methods are put with emphasis on MAE and LSTM that we implemented in order to test their performance.

2 Gathering Data, Data Regularization, and Clustering

In this section we summarize some techniques from the perspective of the datasets and the clustering techniques they adopted. However, before we proceed, let us present a brief impression of how we read what follows. The reading of the related articles implied, in the end, the perception of a certain degree of subjectivity that ends up resulting in the absence of guarantees regarding the degree of generalization of their results. However, we understand that there is some inherent degree of empiricism in the prediction methods evaluated, if not in the clustering techniques used, in the choice of data that reach them.

In this sense, even among the techniques discussed below, it is possible to observe that at least one result is different in relation to our expectations. Such disparity caused an interpretive effort based on subjective aspects to validate it. For another work [56], not presented here among the summaries that follow, in the same sense, it is observed that the removal or addition of a single variable (the country area, for example) in the dataset completely changes the results and therefore, the analysis. This variable, however, did not appear in any other article read to write these summaries. So only the socioeconomic, climatic, and demographic aspects are used as a background to understand the epidemiological aspects of the COVID-19 pandemic. We also found that issues related to the quality of the notifications generate serious problems in the interpretation of the results [3]. However, the relevance of aspects of data measurement is also observed [39], which may vary from country to country and even from region to region in the same country. Through the summaries that follow, even algorithms considered to be more robust as the UMAP, require initialization parameters that, although they do not define the number of clusters a priori, conform with them.

Finally, as a rule, collecting/gathering data and grouping it are relevant techniques for AI forecasting processes [15, 39]. However, the issues raised in the previous paragraph, suggest the need to pay close attention to the data, as to how they are collected and compose the datasets. In addition, regardless of the methods employed, it is possible to conclude that there is no solution that can be widely applied successfully.

2.1 Data Regularization

The COVID-19 behavior data presents some challenges that we need to consider before developing data-driven models. As such, the development of data-driven models is generally preceded by a full exploratory data analysis followed by a process of cleaning and transformation.

A few aggregation services provide data concerning COVID-19 pandemic state in a ready-to-use format, such as the Johns Hopkins University [12] that collect data from media services and several countries' health agencies. We also recognize the existence of other data providers, like *Dipartimento della Protezione Civile* [1] and an independent Brazilian data collector Wesley Cota [10].

In general, the data regarding the COVID-19 pandemic is represented as a time series with features expressing important marks, such as daily new cases, daily deaths, daily recovered, number of patients in UCI, among others. Therefore, we take into consideration the same measures adopted to deal with time series data.

First, we analyze the data searching for outliers. An outlier is a value that exceeds any practical limit for an observed variable, such as negative values of daily deaths for example. Hence, outliers should be interpreted taking into account practical considerations. After outlier removal, we search for any sort of discontinuity, given that most of the data-driven models cannot handle sparse values.

After the process of outlier removal, we must reason about the data collecting methods and how the countries count the number of cases. For example, the Hubei province in China changed the criteria to recognize a person infected by the coronavirus. Before February 12th of 2020, only positive cases confirmed by laboratory tests were accounted. After this date, any respiratory disease with clinical symptoms similar to COVID-19 was considered as a positive case. Such a change in the counting method created a huge discrepancy regarding the initial behavior of the pandemic and must be anticipated prior to the development of data-driven models.

Another problem related to the data collecting procedure regards the availability of tests on a daily basis. As an example, some countries perform more tests during the week than on the weekends, leading to an increased amount of confirmed cases on a single day of the week representing the cases that were not tested on the weekend. This way, we argue that a moving average must be applied to the time series in order to redistribute equally the accumulated cases over each day. Also, the pandemic began on different days in different countries, and the time series must be padded to start on a single day and end on the last day of available data for all countries.

Last but not least, the time series data should be normalized according to the specifications of the model's input as data-driven approaches generally use normalized data.

Table 1 Survey of methods discussed further in this Section

Refs.	Method	Title
[43]	Hierarchical Cluster	Estimation of COVID-19 dynamics in the different states of the United States using Time Series Clustering
[26]	K-Means	Time Warping Clustering for the Forecast and Analysis of COVID-19
[3]	Minimal Spanning Tree and Hierarchical Trees	Comparisons of COVID-19 dynamics in the different countries of the World using Time Series clustering
[23]	K-Means	Time series clustering for home dwell time during COVID-19: what can we learn from it
[50]	ST-DBSCAN	Risk Assessment of the Overseas Imported COVID-19 of Ocean-Going Ships Based on AISand Infection Data
[39]	UMAP	Forecasting COVID-19 Dynamics in Brazil: A Data-Driven Approach

2.2 *Clustering Methods*

Here, we evaluate the challenges of applying clustering methods, presented in the literature, in order to give us a perspective on how the use of such methods, applied to the data, and improving their results. Table 1 presents an overview of the clustering methods that we detail in this Section.

2.2.1 Simultaneous Analysis of Time Series

One of the main questions is to determine the similarity of the behavior of COVID-19 in different countries/states using time series data. When the pandemics has an irregular spread in different states, the time series has different dimensions. A way to overcome this is to carry out an analysis of the time series simultaneously from the states of the United States using dynamic time warping distance (DTW), and subsequently creating a hierarchical cluster considering the behavior of each of the states, as proposed by Rojas [43]. They use Gompertz and SAIR logistic models can be used to predict the progress of cases of infected and deceased people.

This need arises from the necessity to compare time series of different dimensions, due to the variation in the way they are presented in each of the states of the United States. To deal with this, the dynamic time warping distance measurement method (DTW) is used to measure the similarity between the time series, in the context of their variations. This is a condition for the hierarchical clustering method to be applied.

Experiments can be found with simultaneous analysis of time series using the dataset of Johns Hopkins University [43]. The time series of the number of people infected and the number of deaths was weighted with value = {0, 0.5, 1} for the calculation of a distance matrix of the states of the United States. To analyze the accuracy of the hierarchical cluster, the Cophenetic Correlation Coefficient is used together with the Calinski–Harabasz criterion to find the optimal number of clusters for grouping states.

2.2.2 Clustering with K-Means

K-Means is a clustering algorithm that aims to partition n observations among k groups where each observation belongs to the group closest to the mean [27]. Basically, given an initial but not optimal clustering, the idea is to relocate each point to its new nearest center, updating the clustering centers by calculating the mean of the member points, and repeating the relocating-and-updating process until convergence criteria (such as predefined number of iterations, difference on the value of the distortion function) gets satisfied [27]. Its result is a division of the data space in a structure which is similar to the Voronoi Diagram. It is considered to be a computationally difficult (NP-hard) problem. Nonetheless, there are efficient heuristic algorithms that are commonly used to help fast convergence to the optimum result. K-means clustering tends to result in comparable spatial extension clusters.

K-Means can be applied in several ways. In practice, the county-level grouping can be done by way of using k-means clustering algorithm. However as the method is unsupervised for pattern extraction, an unrestricted neural network should also be employed together. In an application to Covid-19 example [26], the New York Times set of cumulative daily data at the United States county level is used. A detailed analysis of the k-means clustering algorithm is applied at the fine-grained county level for the United States. Exploratory analyzes are carried out and smoothing techniques applied, with outlier filters and data normalization. In this sense, the municipalities are divided into layers, based on their cumulative cases. Time series of unequal length are pre-padded with zeros. Only then were the data made available for AI training. Here, the clustering method was K-Means (TimeSeriesKMeans, an implementation of tslearn library) and the cluster size was fixed at 25. Two types of clustering metrics are considered, Euclidean distance and dynamic Time warping (DTW). As a prediction model, the Monte-Carlo dropout Feed-forward (MC-NN) is proposed, which consists of two dense layers of 64 units with ReLU activation. The MC-NN receives data for 84 days and produces a 28-day forecast. MC-NN is trained for 200 seasons with size 64. After varying parameters of the preprocessing techniques, and combining them with two different metric forms, a total of 24 sets of clusters is found.

Another study with K-Means in order to understand how the recommendations given to the population to remain in their homes for a certain period of time were assimilated is done in the metropolitan area of Atlanta–Sandy Springs–Roswell [23]. This study carries out an analysis of the data using the K-Means data clustering method. In addition, demographic and socioeconomic aspects are also investigated. Data are collected from the company's website SafeGraph (a company that collects location data through applications, anonymously), with the observation period starting from January 1, 2020 to August 31, 2020. The information that these data express only the average time spent in the home in minutes for all devices in a given region. K-Means is used for the partitioning of data and the Euclidean Distance is used to calculate the proximity between series. This method, however, previously requires the desired number of clusters (3), so the result of the method generated 3 groupings: one that suggested a significant increase in residence time, another that implied a moderate increase and finally, a third with imperceptible changes.

2.2.3 Minimal Spanning Tree and Hierarchical Trees as a Method to Clustering

Minimal Spanning Tree [18] and Hierarchical Trees [20] can be used as non-parametric techniques for clustering. Minimum spanning tree clustering algorithm is known to be capable of detecting clusters with irregular boundaries. Variations of it can include an initial setting for the number of clusters or not [18]. The basic idea behind hierarchical clustering is to obtain a recursive partitioning of the input data in a tree-like fashion such that the leaves one-to-one represent the single items and all inner points represent clusters of various granularity degrees [20]. The use of hierarchical trees is known to be NP-hard.

In practice, these techniques have been applied in order to identify groups of countries with similar ways of spreading the pandemic COVID-19 [3]. Authors start with considerations that concern the fact that countries have taken different actions, obtaining different results in combating the pandemic, implying dynamics of accelerating or decelerating the rate of contamination and different deaths [3].

In this approach, it is necessary to choose a variable through which the technique reaches the clusters. For the authors, this variable is the number of daily infections per country during a period of at least 100 days after confirmation of the tenth case. As for the data, the study compares active cases by population in 191 countries. Although it focuses on the analysis of this variable, it also analyzed another set of indicators related to the pandemic itself—demographic and socioeconomic.

To follow this approach, the time series of new cases represent the countries in the nodes, while the mode of relations between these series corresponds to Pearson's correlation distances for each pair of nodes. From these correlation distances, the MST is constructed, a minimal graph that covers all nodes without loops.

However, the authors noted that Pearson's correlation distance matrix needs to be modified because its inputs do not meet Euclidean metric requirements. Thus, showing the transformation algorithm, it is transformed in such a way that a distance correlation matrix compatible with the Euclidean metric is arrived. It is, therefore, this last matrix that is used to determine the MST, connecting all nodes. At the end of the process, the generated MST is used to calculate the distance between each pair of nodes in order to also generate a Hierarchical Tree—whose structure is represented by a *Dendrogram*.

As a result, the method found 4 clusters. One with 104 countries, another 2 with 43 countries each, and a fourth with one country. It was also observed that partitions 1 and 3 had greater distance diversity between nodes; partition 1 had more homogeneous distances between nodes. Evaluating the average of the contagion curves of each cluster, and comparing them with each of the curves in the same cluster, it was noticed a clear similarity between them. Furthermore, the authors did not say whether they performed any data processing to perform the clustering.

In addition, the authors carry out a series of analyzes that have nothing to do with the analysis of the quality of the clustering method, but only a discussion about the curves are as they are, considering the health measures of each country, as well as socioeconomic conditions.

2.2.4 ST-DBSCAN as Method to Clustering

DBSCAN is an algorithm for scanning data to find spatial densities of points, representing these densities as clusters. However, to be employed in the current problem, it should be modified for investigating the points from both spatial and temporal perspectives. This variation is known as Spatiotemporal DBSCAN, or ST-DBSCAN. From the point of view of initialization, in addition to the DBSCAN space difference control variable between points, ST-DBSCAN adds one more variable at its initialization: the maximum distance in the temporal dimension between points. After defining these parameters, ST-DBSCAN is executed and scanned the AIS dataset in order to find clusters of ship stop patterns. The value of each parameter, when starting the ST-DBSCAN algorithm, the authors point out, can have a significant impact on the results of the clusters analysis.

The concern that guides the study [50] of these authors is determined by the need to define labels to mark ships as to the risk they represent for their destinations, in the context of COVID-19. In this sense, the authors propose an approach capable of offering information that suggests how a particular ship presents a risk to its destinations, through the automatic detection of stop regions. For this purpose, in order to recognize patterns related to these stops along the trajectory, a variation of DBSCAN, the Spatio-Temporal Density-Based Spatial Clustering of Applications with Noise (ST-DBSCAN), was used.

To perform the task, four different datasets are used: global ship AIS dataset, global port location dataset, global COVID-19 infection, and global administration boundary. The first has spatiotemporal information on ship displacement. The second has information from ports around the world. The third reports epidemiological data on the pandemic for each country in the world, and finally, the fourth contains information from border administrations.

The AIS dataset contains information on the space-time displacement of ships. From the data in this dataset, clusters that involve the recognition of port stops over time will be extracted. These data, explain the authors, are updated every minute or second and this form of updating favored the identification of stop regions, namely, point densities by time interval. In addition, it was the authors explain in the text how to detect such point densities, making them correspond to the port areas where they dock. Through these data it is also possible to extract the duration of these stops or to order them.

As a result, the authors reported a case study in which they were able to safely identify the series of stop clusters on an international cargo ship. The investigation of these points also took into account aspects of the length of stay in the stop state and the proximity of the vessel to the port (since, although stopped, the vessel could not be moored to this one). All these aspects were relevant to determine the entrance to the risk classification phase—calculated not only in relation to these variables, but also in relation to the epidemiological variables of the stopping region.

Finally, the authors conclude by recommending the solution as a way to classify the risk levels of ships as to what they may represent for their destinations in the context of COVID-19.

2.2.5 UMAP and Affinity Propagation as Method to Clustering

In our study we are interested in analyzing and predicting contagion curves of the pandemic COVID-19, motivated by offering a method capable of assisting public and private health authorities in managing the reaction to fight the spread of the COVID-19 disease. So the first step is the clustering. In order to implement this approach [39], we have chosen a method for scanning and clustering the data, which usually deals not only with epidemiological time series: the Uniform Manifold Approximation and Projection (UMAP) [32].

UMAP is an unsupervised embedding method that tends to preserve the global distances from the initial dataset. Its lower dimensional feature space makes it easier to visualize and understand data. And also to improve the results of the clustering for algorithms in which the number of clusters is not predetermined.

Although UMAP identifies clusters, it does not inform what characterizes each of the clusters. It only produces a new transformed space not generating directly the clusters assignments, which would be necessary to select the countries for training our further data-driven models. Therefore, another tool has to be used to make these characteristics known and, then, the countries/regions to be properly put in their clusters. The tool used in conjunction with UMAP is the Affinity Propagation. Only after that, the data were effectively grouped and used to carry out training and model predictions.

In our previous work result, we observe that this combined clustering strategy generates 7 distinct and well-formed clusters. We suggest that the format should reflect the coping policies adopted by government authorities in the investigated countries/regions—recognizing, however, with association's subjectivity. We also assess how much each feature contributed to the characterization of clusters, partitioning to partitioning [39].

We remark that the final result is not so good, with some issues being identified to justify it. Among them, here we highlight two because they were identified in other works, namely, problems related to the amount of available samples and problems related to under reporting that occurred in several countries/regions. We highlight these two issues because they are closely related to the result of employing data clustering methods.

3 Classical Approaches for Epidemiology Forecasting

Traditional approaches have been employed to predict the pandemic diseases behavior and help specialists and authorities to make better decisions to stop or at least slow down the progress of contamination. In these methods, several kinds of models that use a different set of features like the number of infected, removed, and susceptible people are applied rely on the pandemic available information. In this section we introduce the most commons traditional methods and that can be used as a comparative approach to data-driven methods application.

Table 2 presents an overview of the traditional models used as the basis of some studies related to COVID-19 we detailed in this Section.

3.1 SIR

The SIR is a model for disease transmission and dissemination, developed by Ronald Ross, William Hame, and other researchers [51]. The name is derived from the three sub-populations that the methodology observes, the *susceptible*, the *infected*, and the *removed*. The main idea (Fig. 2) is that susceptible people come into contact with infected ones at some probability (β) becoming part of the infection group. Then, they pass from infected to removed step spontaneously, at a fixed rate (ν), as the infected recover.

The method assumes several assumptions, as large and closed population, the outbreak is short-lived, the total population does not change during the period, recovering from infection confers immunity, etc. Because of its simplicity, this methodology was studied and varied several times over the years. Initially, we denote the sub-populations sizes at time t by $S(t)$, $I(t)$, and $R(t)$. The simplest form from the SIR is composed of the following system of quadratic ODEs [51]:

Table 2 Survey of methods discussed further in this Section

Refs.	Method	Title
[9]	SIR	A SIR model assumption for the spread of COVID-19 in different communities
[52]	SEIR	Modified SEIR and AI prediction of the epidemics trend of COVID-19 in China under public health interventions
[11]	SEIRS	SARS-CoV-2 waves in Europe: A 2-stratum SEIRS model solution
[5]	SIRD	Data-based analysis, modeling and forecasting of the COVID-19 outbreak
[4]	ARIMA	Forecasting the spread of the COVID-19 pandemic in Saudi Arabia using ARIMA prediction model under current public health interventions

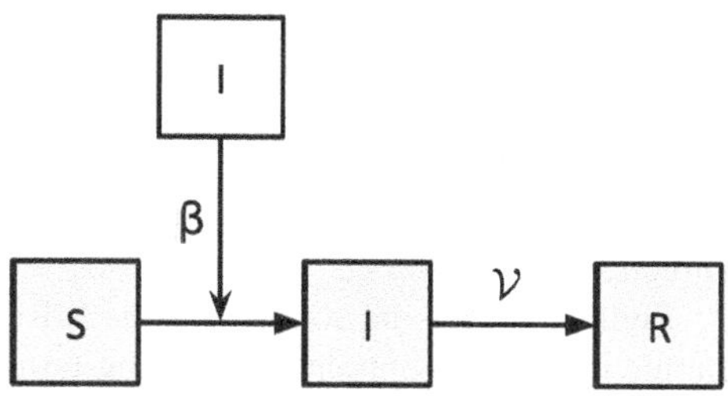

Fig. 2 The sub-populations and rates relations in the SIR model

$$
\begin{aligned}
\frac{dS}{dt} &= -\beta\ S\ I \\
\frac{dI}{dt} &= \beta\ S\ I - \nu\ I \\
\frac{dR}{dt} &= \nu\ I;
\end{aligned}
\tag{1}
$$

where $\beta > 0$ is the transmission rate and $\nu > 0$ the recovery rate. This way the infection duration can be denoted by $D = 1/\nu$. The term $\beta\ S\ I$ is the homogeneous mixing of the infected and susceptible classes. The total population size is always constant, then equations left side sum is the derivative of total population size, and the right side sum is zero. Mathematically,

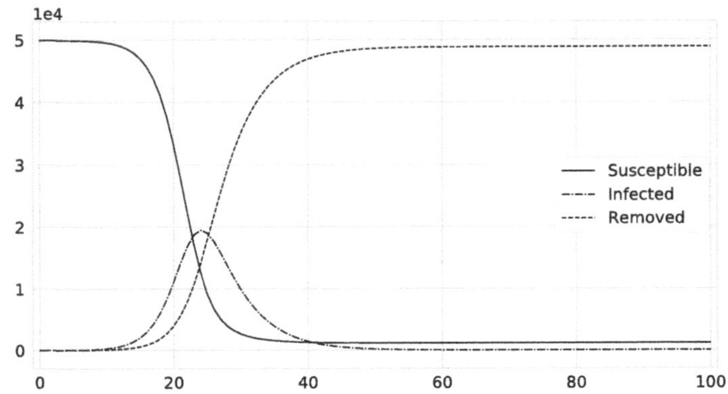

Fig. 3 SIR solution for ODEs system to population $N = 50000$, $\beta = 0.7/N$, $\nu = 1/5$, $S(0) = 49955$, $I(0) = 5$, $R(0) = 0$

$$\frac{dS}{dt} + \frac{dI}{dt} + \frac{dR}{dt} = 0. \tag{2}$$

The letter N represents the population size. Thereby, once $R(t) = N - S(t) - I(t)$, the two first ODEs can represent the system.

In a situation that each infected has sufficient contact for contamination with κ other individuals on population per time unit, and κ is independent of the population size, $\kappa\ S/N$ of these contacts are with susceptible individuals. If a number τ of these contacts result in transmission, each $\kappa\ \tau\ S/N$ infected can infect another susceptible per unit time, thus $\beta = b/N$ where $b = \kappa\ \tau$. Then, τ is the transmissibility of the infectious disease.

Figure 3 shows the solution of the SIR system simulating a highly virulent epidemic in a 50,000 population. To long-term limits, epidemic threshold theorem, and other analyses see [51].

3.2 SEIR

One major limitation of the SIR model is that it is not taken into account during the incubation period. This period is related to how long someone that was classified as a suspected case must keep in observation until be sure that will not be an infection case. To overcome this hurdle, a new step called exposed can be added to the SIR model.

The idea is to consider when a susceptible person comes into contact with someone that is infected, they get into the exposed step. The transition to the infectious step occurs at a fixed rate ϵ. At this point, the infected person is possibly in the symptoms beginning but is not yet able to infect other people. Therefore, with this change, the model becomes the susceptible-exposed-infectious-removed (SEIR) model (Fig. 4).

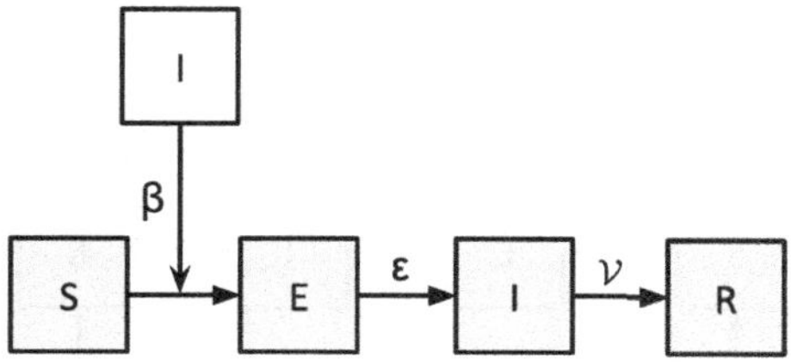

Fig. 4 Exposed group added to SIR approach resulting on SEIR model

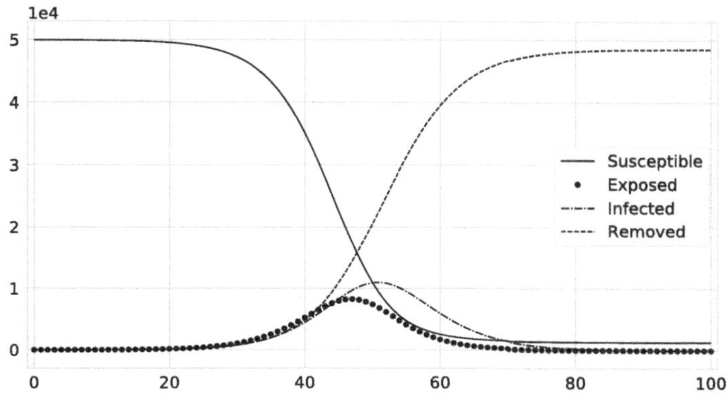

Fig. 5 SEIR model with ϵ rate equals 0.3

Taking the same parameters used in the SIR model and rate ϵ equals 0.3, the graph showing the behavior of the three previous sub-populations and the exposed population is depicted in Fig. 5.

As the exposed group was added to the model, the infected group curve decreased the spike when balanced with the SIR model. This change was expected because now only a number of the sub-population exposed related by ϵ rate became part of the infected group.

3.3 SEIRS

Another major point at epidemic analysis is the possibility of the recovered people becomes of the susceptible group over again. COVID-19 reinfection cases already have been reported and studies started to analyze these cases [7, 47]. More one variation of the SIR model that takes into account the exposed group (as the SEIR model) and the reinfection possibilities is the susceptible-exposed-infected-removed-susceptible (SEIRS) model (Fig. 6). In this case, removed people return to the susceptible group at a ρ fixed rate.

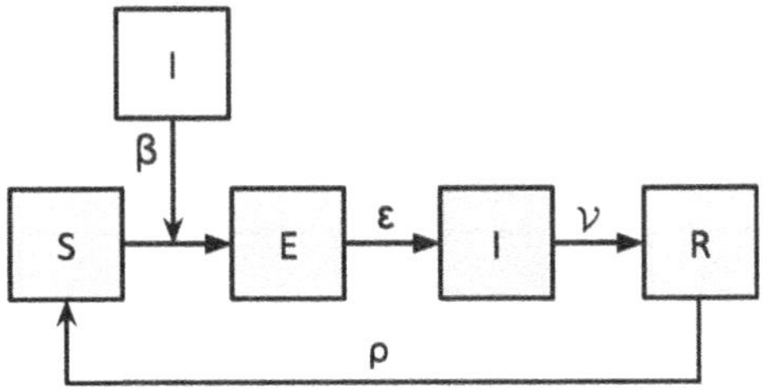

Fig. 6 SEIRS model with feedback from the removed group to susceptible one adding the reinfection possibilities

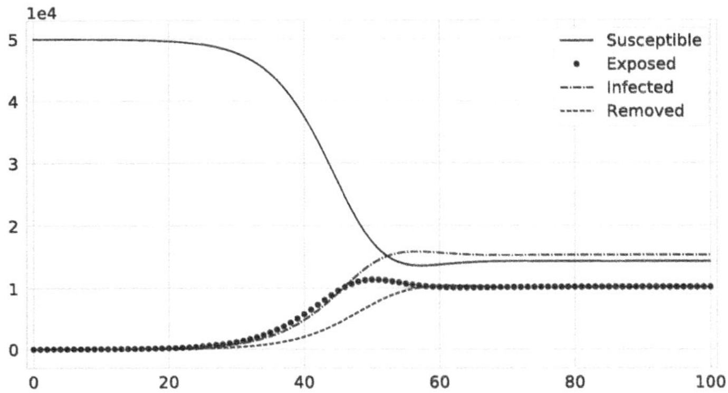

Fig. 7 SEIRS model with ρ rate equals 0.3

Over again, using the same parameters of SIR and SEIR models, and adding the feedback from removed to the susceptible group with ρ equals to 0.3, the graphs of the four sub-populations are shown in Fig. 7.

Notice that in this case, the number of people of each group makes stable in some value different from zero. This occurs due to the reinfection cases that will always keep the feedback of the closed system.

3.4 SIRD

In this variation, it is added the mortality rate to the traditional SIR model. In this case, the removed group consists of two other groups: number of recovered people and dead ones (Fig. 8).

The mortality rate *gamma* is one of the most important parameters in epidemic models because is directly related to how dangerous is the disease. Another point of this parameter is that it is a value very close to the real, due to, generally, all the people of this group are tested after the death is confirmed (Fig. 9).

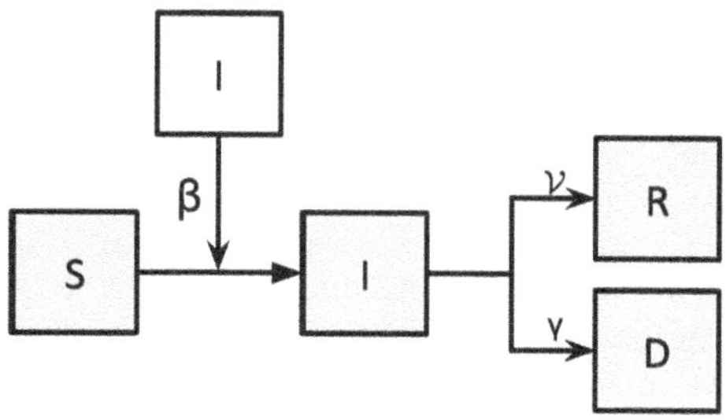

Fig. 8 SIRD model diagram with the dead population added to the SIR model

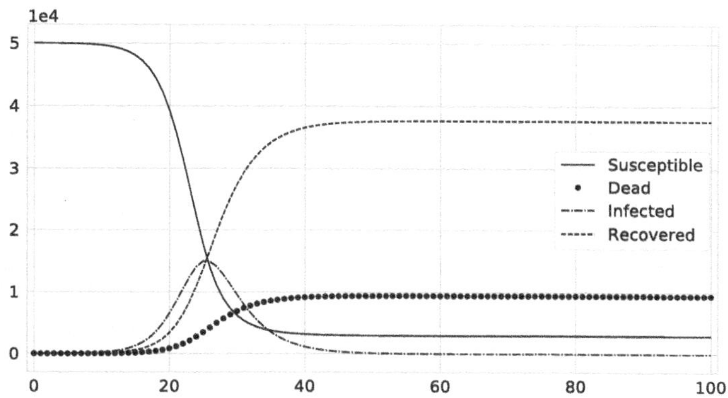

Fig. 9 SIRD model that taken into account the death group

When balanced with the SIR model, the difference of the SIRD graph is the removed curve turning out to the recovered and dead sub-populations curves. As can be seen, the sum of these two curves is equal to the removed one of the SIR traditional model.

3.5 ARIMA

ARIMA models are widely used to analyze one-variate time series. A mixture of three procedures composes the method. The auto-regressive (AR), the differentiating (I), and the moving average (MA) procedures [2].

The auto-regressive form of order o, $AR(o)$, can be described as

$$x_t = c + a_1 x_{t-1} + a_2 x_{t-2} + \cdots + a_o x_{t-o} + \varepsilon_t; \quad t = 1, 2, \ldots T, \tag{3}$$

where ε_t is the error term, indicating is a withe noise, with $E(\varepsilon_t) = 0$ and $\nu(\varepsilon_t) = \sigma^2$, there is a normal distribution where $\varepsilon_t \approx N(0, \sigma^2)$. This kind of procedure take in consideration all past values, there is, a comulative impact that configure a long-term memory model. The MA procedure has an order u, then $MA(u)$ if:

$$x_t = \varepsilon_t - \theta_1 x_{t-1} - \theta_2 x_{t-2} - \cdots - \theta_u x_{t-u}. \tag{4}$$

In this procedure, the model will accumulate only u errors, do not take into consideration higher orders error that configures a short memory model.

The auto-regressive moving average ($ARMA$) procedure has an order o and u, there is $ARMA(o, u)$ if:

$$x_t = c + a_1 x_{t-1} + \cdots + a_o x_{t-o} + \varepsilon_t - \theta_1 \varepsilon_{t-1} - \cdots - \theta_u \varepsilon_{t-u}. \tag{5}$$

Thus, this model procedure can mixture both types, AR, and MA models.

The ARMA models can be generalized to the non-stationary time series by letting the resulting series from ARIMA models to be differentiable. The general non-stationary model is known as ARIMA(o, d, u) where o, d, u are the order of auto-regressive, the degree of differencing, and the order of moving average, respectively. In an example that x_t is a non-stationary time series, the x_t first-difference is taken so that Δx_t gets stationary, thus Eq. (6) defines the ARIMA (o, 1, u) model, where $\Delta x_t = x_t - x_{t-1}$,

$$\Delta x_t = c + a_1 \Delta x_{t-1} + \cdots + a_o \Delta x_{t-o} + \varepsilon_t - \theta_1 \varepsilon_{t-1} - \cdots - \theta_u \varepsilon_{t-u}. \tag{6}$$

4 Data-Driven Approaches to COVID-19 Dynamics Prediction

Most data-driven approaches for the Epidemiology of COVID-19 follow a similar methodology, derived from the one exposed in Sect. 1: first, a large amount of data is necessary to obtain meaningful results. There are several databases that aggregate useful information for COVID-19 forecasting, such as number of confirmed cases, number of confirmed deaths, and number of recovered. Other variables, such as weather information and population movement, can also help with the forecasting. The second step consists of data regularization. This step improves the quality of the data by removing redundant information and/or reducing the dimensionality of the dataset in order to improve training. As will be shown next, many machine learning architectures can be used for time series forecasting, so the next step is to decide which one to use. Next, the training of the network is performed and a model is found. Evaluation metrics are used to indicate the performance of the training and indicate how accurately the model represents the training data and also the model ability to predict outside the training dataset.

Zeroual [57] presents a brief explanation of LSTM, RNN, BiLSTM, Gated Recurrent Units (GRUs), and Variational Autoencoders (VA). Here we complete those models with the ones shown in Table 3, which presents an overview of the data-driven methods that we detail in this section.

4.1 Long Short-Term Memory (LSTM)

The usual recurring neural networks have one or more explicit information re-feeding into the structure. This type of function, combined with the backpropagation algorithm, presents problems that lead to "blow up" or vanishing gradients. The "blow up" (or explosion) may generate an oscillation over the neural weight values. The vanish can make the network fitting takes an impracticable time for assimilating the necessary time dependence, and yet may not work as expected. To address the explosion/vanish of the gradient, [21] proposed the Long Short-Term Memory (LSTM). The LSTM neural network can retain information for a greater amount of past data entries in contrast to usual RNNs (long-term memory) while keeping relevance of not long past. In order to do this, the LSTM has two internal states, the cell state, and the hidden state. These states make the information transport forward ahead to the network neurons. Figure 10 shows an LSTM neuron (also calls cell) representation.

Figure 10 is possible to observe in a given time t that LSTM cell has as input the information source x_t, the hidden state h_{t-1} and the cell state c_{t-1}, both recurrent states of the past time instant. The cell outputs are the current time cell state c_t and hidden state h_t, and the output information y_t. If the cell belongs to the final layer, h_t is comprehended as y_t, in case of the neuron belongs to a hidden layer, h_t will be the h_{t-1} to the forward layer.

The LSTM cell is composed of five regions (or gates). The forgot gate (more left on Fig. 10) is where the neuron forget information if it is necessary. The input and cell gates (left-center and center on Fig. 10) are responsible for deciding the part of new information will be inserted in the network states. The cell update (top on Fig. 10) corresponds to the set of operations made to obtain (actualize) the new cell state value. The output gate (right on Fig. 10) is the last operation that will transform (and mix) the cell state information in the hidden state (or output) value. The set of Eqs. (7) show the operations done in the LSTM cell [17].

$$
\begin{aligned}
f_t &= \varphi(W_{if}x_t + b_{if} + W_{hf}h_{(t-1)} + b_{hf}) \\
i_t &= \varphi(W_{ii}x_t + b_{ii} + W_{hi}h_{(t-1)} + b_{hi}) \\
g_t &= tanh(W_{ig}x_t + b_{ig} + W_{hg}h_{(t-1)} + b_{hg}) \\
c_t &= f_t * c_{(t-1)} + i_t * g_t \\
o_t &= \varphi(W_{io}x_t + b_{io} + W_{ho}h_{(t-1)} + b_{ho}) \\
h_t &= o_t * tanh(c_t),
\end{aligned}
\tag{7}
$$

where W_{if}, W_{ii}, W_{ig}, and W_{io} are the matrixes weights for inputs on forgot gate, input gate, cell gate and output gate, respectively, all them $\in \mathbf{R}^{H\times X}$. Hidden weights

Table 3 Survey of methods discussed further in this Section

Refs.	Model	Title	Approach
[13]	LSTM, GRU, bi-LSTM	Predictions for COVID-19 with deep learning models of LSTM, GRU and bi-LSTM	28 days forecast of confirmed, death and recovered in 7 countries
[45]	bi-LSTM conv-LSTM	Time series forecasting of COVID-19 using deep learning models: India-usa comparative case study	1 month comparative forecast of USA and India
[53]	AE	Bifurcated Autoencoder for segmentation of COVID-19 infected regions in ct images	Image Segmentation of the Lungs
[28]	AE	COVIDomaly: A Deep Convolutional Autoencoder approach for detecting early cases of COVID-19	Early detection from chest radiography
[34]	AE	Improved Sparse Autoencoder based artificial neural network approach for prediction of heart disease	Predict heart disease following COVID-19 infection
[19]	AE	Similarity maps and pairwise predictions for transmission dynamics of COVID-19 with neural networks	Prediction of Transmission dynamics
[58]	Survey	Deep learning methods for forecasting COVID-19 time series data: A comparative study	Survey comparing RNN, LSTM, BiLSTM, VAE and GRUs
[24]	VAE LSTM sa-LSTM	Variational-LSTM Autoencoder to forecast the spread of coronavirus across the globe	Predict Value Spatial Relations
[40]	CNN	A comprehensive study on classification of COVID-19 on computed tomography with pre-trained Convolutional Neural Networks	Detection from tomography images
[22]	CNN	Multiple-Input Deep Convolutional Neural Network Model for COVID-19 Forecasting in China	Forecast of new cases in China
[49]	Random Forest	Forecasting COVID-19 cases in the Philippines using various mathematical models	Forecast of new cases in Philippines
[41]	Random Forest	Analysis, prediction and evaluation of COVID-19 datasets using machine learning algorithms	Comparison with other ML models in forecasting new cases
[25]	Random Forest	Patient Health Prediction Using Boosted Random ForestAlgorithm	Patient Health Prediction

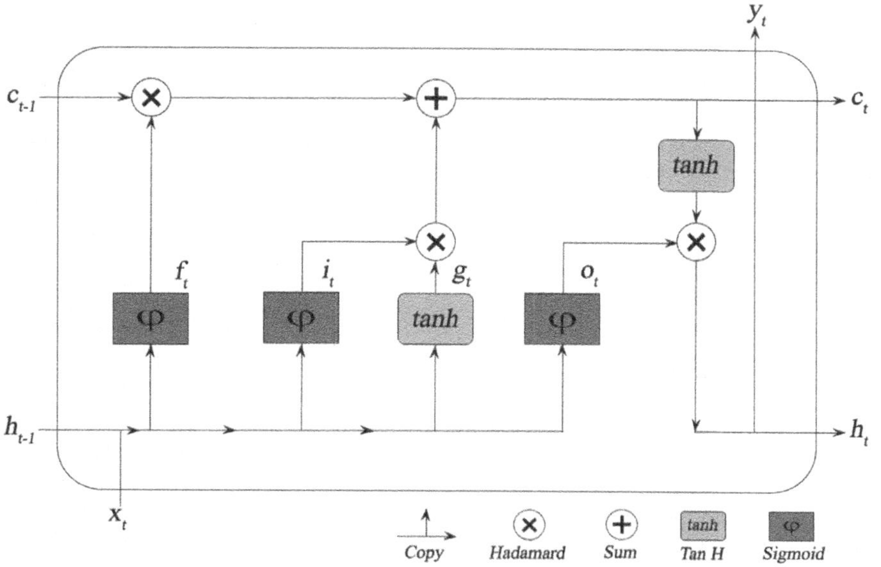

Fig. 10 Long Short-Term Memory cell representation

are represented for W_{hf}, W_{hi}, W_{hg} and W_{ho} for forgot gate, input gate, cell gate, and output gate, respectively, and them $\in \mathbf{R}^{H \times H}$. All b matrixes represent the bias weights for the respective gate, and as the input and hidden weights, they $\in \mathbf{R}^{X}$ and $\in \mathbf{R}^{H}$, respectively. The gates outputs are represented by f_t, i_t, g_t, o_t, and the hidden and cell state are h_t, c_t. The symbols φ and $*$ are the *Sigmoid* activation function and the element-wise multiplication. The *tanh* is the Tanget Hiperbolic activation function.

According to [21, 54], after getting into the equations that govern how LSTMs fit together and an analogy to quickly understand how they work, the system also has to be able to store information for an arbitrary period, be resilient to noise (i.e., fluctuations of inputs that are random or irrelevant to predict a correct output) and the system parameters are trainable in a reasonable time.

Since LSTMs are effective in capturing long-term time dependencies without suffering the optimization attempts that plague simple recurring networks (SRNs), they have been used to improve the state of the art for many difficult problems [55]. This includes handwriting recognition and generation, language modeling and translation, speech acoustic modeling, speech synthesis, secondary protein structure prediction, audio and video data analysis, among others.

Recently, [13] used LSTM networks to perform COVID-19 data prediction from seven different countries (Brazil, Germany, Italy, Spain, UK, China, India, Israel, Russia, and USA) with 110 days of network training, to predict the next 48 days. The dataset comprises three characteristics of cases, namely, confirmed, deaths, and recovered cases, with r2_score for countries like China and the United Kingdom reaching 0.9996 for recovered cases.

Not only relating to LSTM algorithms and making a comparison with similar algorithms (bi-LSTM and Convolutional-LSTM) [45] proposed models based on deep learning for confirmed cases and deaths from SaRS-CoV-2 with a month in advance with LSTM. Its variants (bi-LSTM and Convolutional-LSTM) were also used to predict future values with an average accuracy of 97.82% for India and 98% for the USA in the confirmed cases, and a average accuracy of 96.66% for India and 97.5% for USA in the death cases.

4.2 Autoencoder (AE)

Autoencoders are an unsupervised technique that imposes a bottleneck based on the minimization of the square error between observation and nonlinear reconstruction in the network, acting as an input data display compressor; and then rebuild the compressed data after [33]. Because of that, scenarios that require dimensional reduction can use features selection [48] by Euclidean cost function.

$$\underset{W,W'}{\text{argmin}} \left\| X - W'\phi(WX) \right\|^2, \tag{8}$$

where X consists in the hole training samples, W the connection weights from all input nodes, and $\phi(WX)$ the output of the hidden nodes, reconstructing the error between original data and a decoded latent space, which lies solution to non-convex problems caused by the gradient descent, such as the shown in [46].

Autoencoders, in general, are formed in two paths, one $h(\circ)$, called encoder, and $g(\circ)$, named decoder, as shown in Fig. 11. The input data x performs at $Re(x) = g(h(x))$, which can permits to obtain the reconstruction accuracy by Eq. 8, and an average error given by $loss(x, Re(x))$. Besides, the model can be used forward to limit the number of neurons in the hidden layers, and hence reducing the number of input features. This process permits us to get a fundamental idea behind Autoencoders, which is to (re)construct a knowledge from the data.

Considering the high-dimensional data as an input and compressing the information into the latent space and reconstruct the full flow field data, this type of neural network model can operate in several applications and, its adaptability may change some architecture of the model.

Recently, the use of Autoencoders has been applied focused within COVID-19 disease case, for multiple approaches such as presented in [53], which performs a shared encoder path, where at the end of the encoder is bifurcated in two separated decoder paths. The approach is used to support cardiologists in diagnosing the COVID-19 disease by the use of image segmentation of the healthy region of the lungs and infected regions.

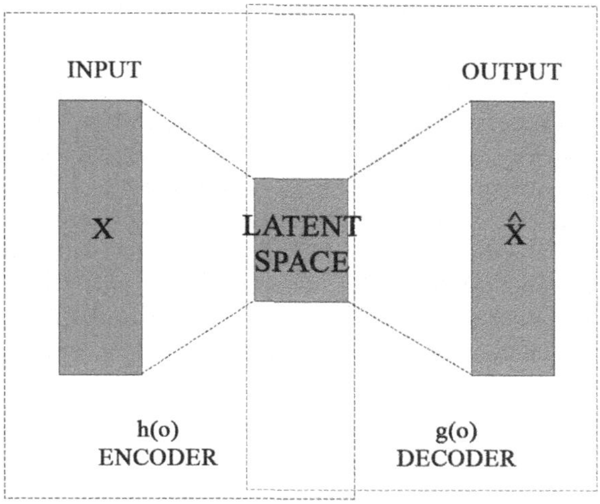

Fig. 11 Autoencoder macro-representation. An Autoencoder works to compact input data to a latent space representation, and then tries to rebuild the data representation as a result

Besides, another application aimed at detecting COVID-19 through image manipulation is the work mentioned by COVIDomaly, described in [28]. In this work, a Convolutional Autoencoder uses chest radiography datasets and helps detect COVID-19 early. Another approach is [34], which uses a Sparse Autoencoder to predict heart disease caused by the COVID-19-related issues.

Multiple efforts are being made to address COVID-19 related trends. Hartono [19] introduces an Autoencoder that reproduces the transmission dynamics of the country in the output of the decoder and that uses features extracted from the topological map generated at the end of the encoder as input to an LSTM model, which generates the similarity map and produces matches about the dynamics of the disease for different countries. The two-dimensional topological map selected here is the hidden layer based on [29].

Not limited in use only Autoencoders, [58] made a comparative study regarding multiple models, including simple Recurrent Neural Network (RNN), Long Short-term Memory (LSTM), Bidirectional LSTM (BiLSTM), Variational Autoencoder (VAE), and Gated Recurrent Units (GRUs). During the test phase 17 days were adopted for the forecast. The variational autoencoder model provides closer prediction to confirmed and new recovered cases when compared to other models considered for almost all countries scenarios.

Another work using variational autoencoder is shown in Ibrahim [24], which takes assumptions to build the model such as if it is possible to extract features regarding the historical data of spread, be able to predict values similar to the initial case, take into consideration spatial relationships between different countries, and make associations about the spread case of the virus. Additionally, LSTM and Self-Attention LSTM is connected to the network connected to the output of the encoder, to regress information on coronavirus spread, government measurements, and urban features.

4.3 Convolutional Neural Network (CNN)

Convolutional Neural Networks (CNNs) are a specialized type of neural networks for processing data grid-like topology. An example is time series, which can be modeling as a 1D grid taking samples at regular intervals. Another example is image data, which can be modeling as a 2D pixels grid. This kind of network has been widely used and successful in practical applications in different fields [15]. The name CNN indicates that the network applies a convolution in your functionality. Convolution is a specialized kind of linear operation. In its most general form, the convolution is an operation on two functions with real arguments. Equation (9) shows the mathematical convolution operation, your discrete formula, and your common denomination as a $*$, respectively [15].

$$\begin{aligned} s(t) &= \int p(a)q(t-a)da \\ s(t) &= \sum_{a=-\infty}^{\infty} p(a)q(t-a) \\ s(t) &= p(t) * q(t). \end{aligned} \tag{9}$$

In CNN terminology, the first argument p is the input and the second argument q is the kernel. The CNN output sometimes receives the name as a feature map. Discrete convolution can be viewed as multiplication by a matrix [15].

Furthermore, we often used convolutions over more than a single axis at a time. For example, in a two-dimensional image denoted as IMG as input, probably is also necessary a two-dimensional kernel, as

$$S(i,j) = (IMG * K)(i,j) = \sum_m \sum_n IMG(m,n)K(i-m, j-n). \tag{10}$$

Convolution is a commutative operation and can be equivalently written, like:

$$S(i,j) = (IMG * K)(i,j) = \sum_m \sum_n IMG(i-m, j-n)K(m,n). \tag{11}$$

Due to the less variation in m and n valid values, the formula in Eq. (11) is more implement on machine learning libraries [15]. Convolutional neural networks are simple neural networks that use convolution in place of general matrix multiplication in at least one of their layers. Figure 12 shows an illustration of a 2D convolution neuron.

In the context of the COVID-19 disease. Pham [40] performs a comprehensive study with pre-trained CNNs applied to the classification of COVID-19 from computed tomography images. They demonstrated that with only 6 epochs, the pre-trained model DenseNet-201 achieved very high performance on the classification task.

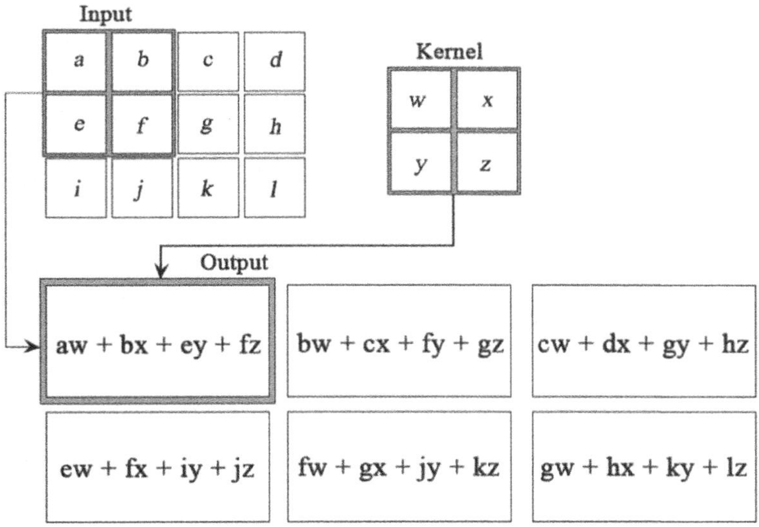

Fig. 12 An example of a 2D convolution operation in input matrix

Furthermore, CNNs have been widely used for COVID-19 detection from x-ray images [6, 38]. In [6] the authors use transfer learning with a dataset composed of confirmed common bacterial pneumonia, confirmed COVID-19 disease, and normal conditions. The study suggests that X-ray images may show significant biomarkers related to the COVID-19 disease and use it joined with CNNs can achieve satisfactory accuracy to the proposal.

Regarding the forecast of new cases, [22] uses data from new confirmed cases, deaths, and recovered, to train a multiple-input CNN model. Later, they used this model to forecast the cumulative number of confirmed cases.

Staying in that same context, [35] examines multiple CNN-based models for COVID-19 forecasting. The work uses a French dataset with classical CNNs, and TCNs (Temporal Convolutional Networks) to provide short-term confirmed cases and hospitalizations, even when few data are available.

4.4 *Random Forest*

Random forest is a supervised machine learning algorithm that can be used for classification and regression problems, mixing a technique that works well to create information through randomly added decision trees, to prevent over-fitting and misstatement. Applying the bagging process to the samples and the feature randomness, Random forest used several decision trees and merged them to stabilize the prediction (to ensure a low correlation between decision trees).

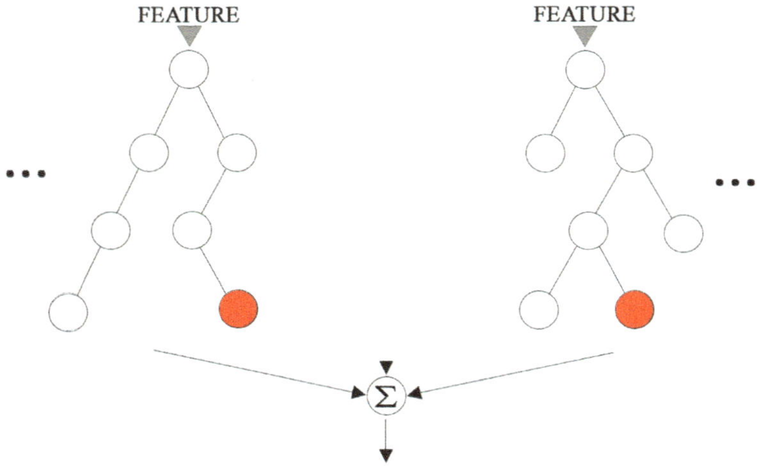

Fig. 13 Random forest example with two decision trees. The Random forest model may include many decision trees, in which all the decision trees are constructed and their votes are combined to classify an example, through majority voting

For the prediction of a regression and to classification problems it may be used decision trees mean values:

$$\hat{\mu} = \frac{1}{N}\sum_{i=1}^{N} E(x_i), \tag{12}$$

where $E(x)$ is the expected value to the event occurred at the decision trees. The misstatement can be regarding by standard deviation of the predictions, by

$$\sigma = \sqrt{\frac{\sum_{i=1}^{N} E(x_i) - \hat{\mu}}{N-1}}. \tag{13}$$

One simplest example is illustrated in Fig. 13, where are two (and maybe more) decision trees, getting a bag of features and, at the end of each, the results are summed to decides about what to do with the value obtained.

A comparison using Random forest in [49] shows that the model, along with the other five mathematical models applied in the Philippines, underestimates the tendency of COVID-19 on the set of tests, being almost a constant value. According to these results, it is supported by the amount of data to improve the machine learning algorithm. However, in [41], the comparison of multiple machine learning models demonstrated at the test stage that the Random Forest Classifier and Random Forest Regression were closer to predict the COVID-19 trend than other ML models.

To achieve a fine-tuned Random forest model, [25] proposes to enhance the model with the AdaBoost algorithm in order to improve the performance, applied to the geographical, travel, health, and demographic data of the patient, allowing an F1 score of 0.86 across the COVID-19 patient dataset, and more accurate predictions even on imbalanced datasets

5 Data-Driven Versus Traditional

As demonstrated throughout this paper, the prediction of the behavior of a pandemic state poses several challenges to traditional methods as well as to data-driven methods.

For the traditional models, we observe the complexity in approximating the parameters related to the spread of the virus, such as the transmission rate and recovery rate, making the modeling of the pandemic behavior a function of the choice of the parameters.

From the data-driven viewpoint, we acknowledge the hassle in handling data from several sources gathered in a non-standard format. Given that countries do not face the pandemic with a single strategy leading to different testability rates and political decisions, such as social distancing, the impose of lockdowns between others, we are required to perform a detailed exploratory analysis of the available data and cluster them according to their behavior.

Nonetheless, data-driven methods prove to approximate the spread parameters of the pandemic, even if it is a hidden representation, with a smaller error than the traditional methods, given its low error rate approximating the output pandemic curve on training data. Another advantage of the data-driven methods regards the adaptability to track the dynamics of the spread parameters, dealing with political decisions to avoid a greater spread of the virus, and possibly consider these choices in its hidden representation to better forecast the dynamics of the pandemic state.

We highlight, in Table 4, the primary differences between data-driven and traditional models.

In Table 4 above, we realize the ease of implementation as common strengths of the traditional models. Indeed, these models are based on differential equations and can be implemented with no domain-specific knowledge except mathematics and numerical analysis. However, due to the manual choice of parameters, the traditional models are harder to tune, and one must update its parameters according to new political decisions that might affect the spread of the virus. Additionally, traditional models inherently perform linear modeling, which might not be the case with pandemic data.

On the opposite, data-driven models handle nonlinear data and are relatively easier to tune regardless that a reasonable amount of data is collected and analyzed to evaluate and fix minor problems, such as wrong values and the presence of outliers. Nevertheless, data-driven models present drawbacks concerning the requirement of specific knowledge about Machine Learning and must be implemented by data scientists with proper frameworks using automatic differentiation for fitting the model to the existing data. Still, data-driven models might be unstable in the sense that they might not converge to a plausible pandemic state, leading to never-ending pandemics.

Table 4 Data-Driven vs. Traditional Models Comparison

Model	Type	Pros	Cons
SIR	Traditional	Easy Implementation	Tunning requires setting parameters manually
SEIR	Traditional	Easy Implementation Account Incubation Period	Difficult to Tune The incubation Period is relatively unknown for new diseases
SEIRS	Traditional	Easy Implementation Account Incubation Period Account Reinfection cases	Difficult to Tune The incubation period and reinfection cases are relatively unknown for new diseases
ARIMA	Traditional	Easy to Tune	Limited by a relative short past behavior Unable to track the dynamic state of the pandemic Linear Modeling
RNN LSTMs	Data-Driven	Easy to Tune regardless data aggregation and analysis Non-linear Modeling	Difficult Implementation and Deploy Requires Deep Learning knowledge Requires a reasonable amount of data Possibly unstable
AE VAE MAE	Data-Driven	Easy to Tune regardless data aggregation and analysis Non-linear Modeling	Difficult Implementation and Deploy Requires Deep Learning knowledge Requires a reasonable amount of data Less unstable than LSTMs
CNN TCN	Data-Driven	Easy to Tune regardless data aggregation and analysis Non-linear Modeling	Difficult Implementation and Deploy Requires Deep Learning knowledge Requires reasonable computational resources

In the following, we describe the developments made to increase the data-driven model's stability and generates realistic results to a pandemic state forecast.

5.1 *Our Implementations*

In a previous work [39] we have contributed with a new approach to train a (Modified) Autoencoder (MAE) to forecast virus spreading. We also tested with the LSTM model, however the MAE demonstrated better results than the preliminary LSTM tests approach and was thus chosen as our model, at that time, for predicting data for Brazilian states. That work has shown the applicability of data-driven approaches to predicting the time series in COVID-19. The obtained forecasts were reported in a site available at the work (www.natalnet.br/covid). As different countries and states reacted differently to the pandemic, taking different actions as explained earlier, there was no guarantee that using data from any given country for training could generalize well to any state for prediction.

In order to overcome this problem, we did an initial data clustering, based on a set of features: early mortality, days until $10\times$, and early acceleration. Data was trained with different prediction networks within each cluster, using data from the clustered countries and regions that had a more advanced stage of the pandemic at that time, as for example China, Italy, Spain, and the United States. So these networks were used to make predictions on Brazilian states that belonged to the cluster on which it was trained. In a comparison with traditional approaches as SIR and SEIR, this approach performed better, with promising results [39].

6 Conclusion

One question that we have brain stormed in this study is about what type of preprocessing is necessary in order to have data ready for the data-driven models. As far we found that the need of performing a clustering by countries and regions that have similar behaviors seems a good idea. And this worked well in the MAE approach that we implemented. The LSTM did not work as well as the MAE, besides it has promising results, that, to this time seems more acceptable than the MAE. At the beginning we could not even think about the spreading getting these high numbers, neither a second wave was thought bout because the vaccine was supposed to coming earlier to this. So while most of the works in literature focus on using new ML methods and architectures, an important aspect that is often not given great attention is the dataset prepossessing. Choosing the appropriate parameters to train the network can be greatly benefited by using specialized epidemiological knowledge. Parameters such as immunity and asymptomatic transmission are not taken into account in the data-driven models.

The number of papers and techniques exploring the use of AI methods shows that there's some initiative to use this kind of approach to COVID-19 forecasting. Still, there are relatively few works dealing with the use of AI to COVID-19 epidemiology and the performance of models generated by ML methods can be incrementally improved with the increase of the dataset used for training. With the continuum increases of the dataset size and quality, we expect that the performance of AI techniques is going to improve and that ML methods can generate important results in COVID-19 forecasting.

Despite the hardness of finding a curve that mathematically and exactly represents the COVID-19 dissemination [39], being hard to determine a relation to modeling COVID-19 using statistical distributions. Actually, the COVID-19 curve cannot be considered a Gaussian probability distribution. Accordingly, the curve is not a distribution in the sense usually meant in probability and statistics. Due to this fact the data-driven approaches can be better explored and produce better results.

References

1. Github covid-19 repository data (2020). https://github.com/pcm-dpc/COVID-19
2. Abotaleb MSA et al (2020) Predicting covid-19 cases using some statistical models: an application to the cases reported in china Italy and USA. Acad J Appl Math Sci 6(4):32–40
3. Alvarez E, Brida JG, Limas E (2020) Comparisons of covid-19 dynamics in the different countries of the world using time-series clustering. medRxiv
4. Alzahrani SI, Aljamaan IA, Al-Fakih EA (2020) Forecasting the spread of the COVID-19 pandemic in saudi arabia using ARIMA prediction model under current public health interventions. J Inf Publ Health 13(7):914–919
5. Anastassopoulou C, Russo L, Tsakris A, Siettos C (2020) Data-based analysis, modelling and forecasting of the COVID-19 outbreak. PLOS ONE 15(3):e0230405
6. Apostolopoulos ID, Mpesiana TA (2020) Covid-19: automatic detection from x-ray images utilizing transfer learning with convolutional neural networks. p 1
7. de Araujo Torres D, do Carmo Bueno Ribeiro L, de Freitas Linhares Riello AP, Horovitz DDG, Pinto LFR, Croda J (2020) Reinfection of COVID-19 after 3 months with a distinct and more aggressive clinical presentation: case report. J Med Virol. https://doi.org/10.1002/jmv.26637
8. Bastos SB, Cajueiro DO (2020) Modeling and forecasting the early evolution of the covid-19 pandemic in brazil (Second version). Accessed 10 Apr 2020
9. Cooper I, Mondal A, Antonopoulos CG (2020) A SIR model assumption for the spread of COVID-19 in different communities. Chaos, Solitons & Fractals 139:110057
10. Cota W (2020) Monitoring the number of COVID-19 cases and deaths in brazil at municipal and federative units level. SciELO Preprints, 362. https://doi.org/10.1590/scielopreprints.362
11. Djaparidze L, Lois F (2020) SARS-CoV-2 waves in europe: a 2-stratum SEIRS model solution. https://doi.org/10.1101/2020.10.09.20210146
12. Dong E, Du H, Gardner L (2020) An interactive web-based dashboard to track covid-19 in real time. The Lancet. Inf. Dis. 20(5):533–534
13. Zameer Farah Shahid Aneela MM (2020) Predictions for covid-19 with deep learning models of lstm, gru and bi-lstm. Chaos, Solitons & Fractals 140:110227
14. Francis H, Kusiak A (2017) Prediction of engine demand with a data-driven approach. Proc Comput Sci 103:28–35
15. Goodfellow I, Bengio Y, Courville A, Bengio Y (2016) Deep learning, vol 1. MIT Press, Cambridge

16. Grant A (2020) Dynamics of covid-19 epidemics: seir models underestimate peak infection rates and overestimate epidemic duration. medRxiv. https://doi.org/10.1101/2020.04.02.20050674, https://www.medrxiv.org/content/early/2020/04/12/2020.04.02.20050674
17. Greff K, Srivastava RK, Koutník J, Steunebrink BR, Schmidhuber J (2016) Lstm: a search space odyssey. IEEE Trans Neural Netw Learn Syst 28(10):2222–2232
18. Grygorash O, Zhou Y, Jorgensen Z (2006) Minimum spanning tree based clustering algorithms. In: 2006 18th IEEE international conference on tools with artificial intelligence (ICTAI'06), pp 73–81. https://doi.org/10.1109/ICTAI.2006.83
19. Hartono P (2020) Similarity maps and pairwise predictions for transmission dynamics of covid-19 with neural networks. Inf Med Unlocked 20:100386. https://doi.org/10.1016/j.imu.2020.100386
20. Hartung S, Guo J, Komusiewicz C, Niedermeier R, Uhlmann J (2010) Exact algorithms and experiments for hierarchical tree clustering
21. Hochreiter S, Schmidhuber J (1997) Long short-term memory. Neural Comput 9(8):1735–1780
22. Huang CJ, Chen YH, Ma Y, Kuo PH (2020) Multiple-input deep convolutional neural network model for COVID-19 forecasting in China. medRxiv p. 2020.03.23.20041608. https://doi.org/10.1101/2020.03.23.20041608, https://www.medrxiv.org/content/10.1101/2020.03.23.20041608v1
23. Huang X, Li Z, Lu J, Wang S, Wei H, Chen B (2020) Time-series clustering for home dwell time during covid-19: what can we learn from it? ISPRS Int J Geo Inf 9(11):675
24. Ibrahim M, Haworth J, Lipani A, Aslam N, Cheng T, Christie N (2020) Variational-LSTM Autoencoder to forecast the spread of coronavirus across the globe. medRxiv p. 2020.04.20.20070938. https://doi.org/10.1101/2020.04.20.20070938
25. Iwendi C, Bashir AK, Peshkar A, Sujatha R, Chatterjee JM, Pasupuleti S, Mishra R, Pillai S, Jo O (2020) COVID-19 patient health prediction using boosted random forest algorithm. Front Public Health 8:357. https://doi.org/10.3389/fpubh.2020.00357
26. Jin Q (2020) Time warping clustering for the forecast and analysis of covid-19
27. Jin X, Han J (2010) K-means clustering, pp 563–564. Springer US, Boston
28. Khoshbakhtian F, Ashraf AB, Khan SS (2020) Covidomaly: a deep convolutional autoencoder approach for detecting early cases of covid-19
29. Kohonen T (2013) Essentials of the self-organizing map. Neural Netw 37:52–65. https://doi.org/10.1016/j.neunet.2012.09.018. Twenty-fifth Anniversay Commemorative Issue
30. Loli Piccolomiini E, Zama F (2020) Monitoring italian covid-19 spread by an adaptive seird model. medRxiv. https://doi.org/10.1101/2020.04.03.20049734, https://www.medrxiv.org/content/early/2020/04/06/2020.04.03.20049734
31. Lyra W, do Nascimento Jr. JD, Belkhiria J, de Almeida L, Chrispim PPM, de Andrade I (2020) Covid-19 pandemics modeling with modified determinist seir, social distancing, and age stratification. the effect of vertical confinement and release in brazil. PLOS ONE 15(9):1–17. https://doi.org/10.1371/journal.pone.0237627
32. McInnes L, Healy J, Melville J (2020) Umap: uniform manifold approximation and projection for dimension reduction
33. Memisevic R, Konda KR, Krueger D (2015) Zero-bias autoencoders and the benefits of co-adapting features. arXiv:abs/1402.3337
34. Mienye ID, Sun Y, Wang Z (2020) Improved sparse autoencoder based artificial neural network approach for prediction of heart disease. Inf Med Unlocked 18:100307. https://doi.org/10.1016/j.imu.2020.100307
35. Mohimont L, Chemchem A, Alin F, Michaël K, Steffenel L (2020) Convolutional neural networks and temporal cnns for covid-19 forecasting in france. Appl Intell
36. Nesteruk I (2020) Comparison of the coronavirus epidemic dynamics in italy and mainland china. Preprint ResearchGate
37. Nesteruk I (2020) Statistics-based predictions of coronavirus epidemic spreading in mainland China (2020)
38. Ozturk T, Talo M, Yildirim EA, Baloglu UB, Yildirim O, Acharya UR (2020) Automated detection of covid-19 cases using deep neural networks with x-ray images, p 103792

39. Pereira IG, Guerin JM, Silva Jùnior AG, Garcia GS, Piscitelli P, Miani A, Distante C, Gonçalves LMG (2020) Forecasting covid-19 dynamics in brazil: a data driven approach. Int J Environ Res Public Health 17(14):5115. https://doi.org/10.3390/ijerph17145115
40. Pham TD (2020) A comprehensive study on classification of COVID-19 on computed tomography with pretrained convolutional neural networks. Sci Rep 10(1):16942
41. Prakash KB, Imambi SS, Ismail M, Pavan Kumar T, Naga Pawan YV (2020) Analysis, prediction and evaluation of covid-19 datasets using machine learning algorithms. Int J Emerg Trends Eng Res 8(5):2199–2204. https://doi.org/10.30534/ijeter/2020/117852020
42. Roda WC, Varughese MB, Han D, Li MY (2020) Why is it difficult to accurately predict the covid-19 epidemic? Infect Dis Model 5:271–281
43. Rojas I, Rojas F, Valenzuela O (2020) Estimation of covid-19 dynamics in the different states of the united states using time-series clustering. medRxiv. https://doi.org/10.1101/2020.06.29.20142364, https://www.medrxiv.org/content/early/2020/06/29/2020.06.29.20142364
44. Semenova Y, Glushkova N, Pivina L, Khismetova Z, Zhunussov Y, Sandybaev M, Ivankov A (2020) Epidemiological characteristics and forecast of covid-19 outbreak in the republic of kazakhstan. J Kor Med Sci 35(24)
45. Shastri S, Singh K, Kumar S, Kour P, Mansotra1 V (2020) Time series forecasting of covid-19 using deep learning models: India-usa comparative case study. Chaos, Solitons & Fractals 140:110212
46. Tariyal S, Majumdar A, Singh R, Vatsa M (2016) Deep dictionary learning. IEEE Access 4:10096–10109. https://doi.org/10.1109/ACCESS.2016.2611583
47. Tillett RL, Sevinsky JR, Hartley PD, Kerwin H, Crawford N, Gorzalski A, Laverdure C, Verma SC, Rossetto CC, Jackson D, Farrell MJ, Hooser SV, Pandori M (2020) Genomic evidence for reinfection with SARS-CoV-2: a case study. Lancet Infect Dis
48. Tomar D, Prasad Y, Thakur MK, Biswas KK (2018) Feature selection using autoencoders. https://doi.org/10.1109/MLDS.2017.20
49. Torres MC, Alvin Buhat CH, Paul Dela Cruz BB, Francis Felix EO, Gemida EB, Mamplata JB (2020) Forecasting COVID-19 cases in the Philippines using various mathematical models. medRxiv p. 2020.10.07.20208421. https://doi.org/10.1101/2020.10.07.20208421
50. Wang Z, Yao M, Meng C, Claramunt C (2020) Risk assessment of the overseas imported covid-19 of ocean-going ships based on ais and infection data. ISPRS Int J Geo Inf 9(6):351
51. Weiss HH (2013) The sir model and the foundations of public health. Mat Mat 0001–0017
52. Yang Z, Zeng Z, Wang K, Wong SS, Liang W, Zanin M, Liu P, Cao X, Gao Z, Mai Z, Liang J, Liu X, Li S, Li Y, Ye F, Guan W, Yang Y, Li F, Luo S, Xie Y, Liu B, Wang Z, Zhang S, Wang Y, Zhong N, He J (2020) Modified SEIR and AI prediction of the epidemics trend of COVID-19 in china under public health interventions. J Thor Dis 12(3):165–174. https://doi.org/10.21037/jtd.2020.02.64
53. Yazdekhasty P, Zindar A, Nabizadeh-ShahreBabak Z, Roshandel R, Khadivi P, Karimi N, Samavi S (2020) Bifurcated autoencoder for segmentation of covid-19 infected regions in ct images
54. Simard Yoshua Bengio Patrice PF (1994) Learning long-term dependencies with gradient descent is difficult. IEEE Trans Neural Netw 5:157–166
55. Zaremba W, Sutskever I, Vinyals O (2014) Recurrent neural network regularization. arXiv:1409.2329
56. Zarikas V, Poulopoulos SG, Gareiou Z, Zervas E (2020) Clustering analysis of countries using the covid-19 cases dataset. Data Brief 31:105787
57. Zeroual A, Harrou F, Dairi A, Sun Y (2020) Deep learning methods for forecasting covid-19 time-series data: a comparative study. Chaos, Solitons & Fractals 140:110121
58. Zeroual A, Harrou F, Dairi A, Sun Y (2020) Deep learning methods for forecasting covid-19 time-series data: A comparative study. Chaos, Solitons & Fractals 140:110121. https://doi.org/10.1016/j.chaos.2020.110121

Chapter 6
Geographical Weighted Regression Approach: A Case Study on Covid-19 in India

Prisilla Jayanthi, Muralikrishna Iyyanki , Michael Carlberg, and Mansour M. Ndiath

Abstract The pandemic disease termed Covid-19 has been increasing in every corner of the geosphere, leaving none untouched. The spread has been unimaginable with the count rising to 6,48,315 (confirmed cases) as of July 4, 2020. The study aims to find the spatial relationship between the confirmed cases of the different months and to classify it into three sets. The findings show that the Monte-Carlo simulations that generated probability distribution mapped using GIS map vary from each explanatory variable according to each case. This clearly states that GWR simulations are a way to describe uncertainty in variables of a risk analysis (Covid-19). It also provides a correlation of inputs with possible interdependent relationships between input variables in any model. The models of case-I ($R^2 = 0.9998$) and case-III ($R^2 = 0.9993$) were found to be better compared to case-II ($R^2 = 0.8903$). The GWR and OLS models were performed on the variable testing labs (response), and GWR was found to be a better model with R^2 (0.9207) and AICc (266.9961) over OLS. GWR using STATA was applied on testing samples (response) of Covid-19 and the significance test shows observed (5.2132) and the p-value (0.030) was statistically significant. The testing lab clearly implies that the testing centres in India are very few and most of the data is unavailable in few states.

Keywords Geographical weighted regression · Spatial data · Covid-19 · Monte-Carlo simulation · Testing samples · Confirmed cases · OLS

P. Jayanthi (✉)
MLR Institute of Technology, Dundigal, Hyderabad, India

M. . Iyyanki
Former Professor and Director R&D, JNT University, Hyderabad, India

M. Carlberg
Faculty of Medicine and Health, Örebro University, 701 82 Örebro, Sweden
e-mail: michaelrita@yahoo.se

M. M. Ndiath
University Cheikh Anta Diop, Dakar, Senegal

U. Kose et al. (eds.), *Computational Intelligence for COVID-19 and Future Pandemics*, Disruptive Technologies and Digital Transformations for Society 5.0,
https://doi.org/10.1007/978-981-16-3783-4_6

1 Introduction

The unexpected pandemic of Covid-19 has shaken humans from the beginning of this year. Covid-19 has affected 213 countries and territories around the world and 2 international conveyances according to the sources available. The number of cases rising daily and the unfortunate lives that closed their eyes in the very first week of March is unforgettable. The Covid-19 has put the world economy at a major risk and a major loss to a number of families. Few countries, such as China (81,848), France (22,025) and the UK (8081), have seen many Covid-19 cases as of March 25, 2020, and India (562) had very few cases. Now, India (697,069) is in third place in the world and countries such as France (166,960), China (83,553) and the UK (285,416) have shown declination in the Covid-19 cases as of July 6, 2020. These countries adopted various prevention methods to reduce the positive cases from inclinations. The spread of the virus is so fast with the touch and is highly transmitted through air particles from human to human. Respiratory droplets from any infected person will spread the virus more rapidly. Any contact with the person who has symptoms of respiratory (running nose, coughing) is at risk of being exposed to potentially infective respiratory droplets. The authors of [5, 14] discussed the virus and called Wuhan coronavirus (2019-nCov) by Chinese researchers. Later, the International Committee on Taxonomy of Viruses termed the virus as SARS-CoV-2 and the disease as Covid-19. In (with) this concern of Covid-19 disease spread all over the country, the study was made using GWR on STATA with the statistical data available from http://www.mohfw.gov.in and Statista. The data was collected from March 12, 2020 to July 4, 2020 from http://www.mohfw.gov.in. The simulation produced by the Monte-Carlo test has proposed GWR solution to investigate spatial heterogeneity in data relationships across space using ArcGIS [3].

The authors of [4] explored OLS and GWR models on the spatial factors of Covid-19 in New York City. In [10] spatial lag and spatial error models to examine spatial dependence and GWR and multiscale GWR (MGWR) models to locally examine spatial non-stationarity data were used, and the Covid-19 incidence rate in the continental USA was further explained in the study. In [11] GWR for the spatial variability of relationships between malaria incidence and persistence and few selected socio-economic and human predictors was explored. For the study, the vector dataset was considered as spatial data, and statistical data (non-spatial) on malaria confirmed cases, socio-economic status, population data and environmental factors such as temperature and rainfall were considered for exploratory analysis. The study was carried out at Keur Soce health and demographic surveillance site. The authors of [1] developed a technique coined as GWR which captures the variation by calibrating multiple regressions at various points. In [12] multiscale (GWR) with a Python-based that focuses on the multiscale analysis of spatial heterogeneity process was implemented. The efficient functionality for inference and local spatial processes with unique diagnostics for a multiscale local model was explored. The authors of [15] used the GWR model to examine the relationship between the index of frequency of extreme precipitation and other climate indices such as the frequency of warm days, cold days, warm nights and cold nights. The study was carried out on a daily basis data of the temperature and precipitation during the period 1961–2005 in China from 753 stations. The authors of [2] in their study of bovine tuberculosis used three methods for analysis—least-squares linear regression with a spatial autocorrelation term, GWR and boosted regression tree analysis to identify the factors that influence the spread of endemic (bTB) in England and Wales. The results showed some consistency in the identification of certain factors related to flooding, history of disease and the existence of multiple genotypes of bTB.

2 Methodology

2.1 Data Collection

The Covid-19 cases of India for all the states were collected from March 12, 2020 to July 4, 2020 from http://www.mohfw.gov.in, Statista, ICMR, and https://en.wikipedia.org/wiki/COVID-19_pandemic_in_India. The study involves 562 cases and aims to understand the spatial relationship of confirmed cases of Covid-19 of a month (any date) with confirmed cases of previous months. Here, the confirmed cases are used as the response variable with confirmed cases of other months as explanatory variables. The case study on Covid-19 is classified into three sets:

(1) Confirmed cases of July 2, 2020 (response) and confirmed cases of June 30, May 28, April 28 and March 15, 2020 as explanatory variables.
(2) Confirmed cases of June 30, 2020 (response) and confirmed cases of May 1, April 7 and March 15, 2020 as explanatory variables.
(3) Confirmed cases of May, 2020 (response) and confirmed cases of April 28, April 7 and March 15, 2020 as explanatory variables.

2.2 Geographical Weighted Regression

Geographically weighted regression (GWR) is a spatial regression and non-stationary [6] technique used in statistics, geography and other disciplines. It's a local model of the variable that is used to understand or predict by fitting a regression equation to every feature in the dataset. GWR builds the equations distinctly by integrating the response and explanatory variables of features approaching within the bandwidth of each target feature. The shape and size of the bandwidth are determined by user input for the kernel type, bandwidth method, distance and number of features.

The studies [1, 7, 8] have given a general form for the GWR model as:

$$y_i = \alpha_{i0} + \sum_{k-1}^{n} \alpha_{ik} x_{ik} + \varepsilon_i \tag{1}$$

where y_i is the response variable at location i; x_{ik} is the kth explanatory variable at the same location i; n is the total explanatory variables; α_{i0} is the intercept parameter at location i; α_{ik} is the local regression coefficient for the kth explanatory variable at location i; and ε_i is the random error at location i. In essence, GWR measures the inherent relationships around each regression point i, where each set of regression coefficients is estimated by weighted least squares.

3 Discussion and Results

In this study of Covid-19 cases in India, the confirmed cases of all the states were observed and grouped into three different sets. Here, it was assumed that the relationship between confirmed cases (response) in a month to be examined with confirmed cases (explanatory) in other previous months. The dates for the study were selected randomly and Table 1 describes the variables used in GWR for three sets of cases of Covid-19.

The spatial distribution of the cases per state indicates that the cases increased due to communal spread and also with air in that the state with living style of the residential location; generally, lower cases were recorded in states of Sikkim (102), Mizoram (162), Nagaland (539) and Meghalaya (62) which are located in the northeast, but Assam (9673) and Tripura (1525) are also located in the northeast with more number of confirmed cases. High levels of cases were in the northwest states, namely Maharashtra (1,92,990) and Gujarat (34,600), the central region of Delhi (94,695) and Tamil Nadu (1,02,721) located in the southeast as of July 4, 2020.

Table 1 Variables mentioned for GWR for the Covid-19 cases, 2020

Variable	Details	Dependent or independent
CC_May28	Confirmed cases as of May 28, 2020	I
CC_MAY1	Confirmed cases as of May 1, 2020	I
CC_APR28	Confirmed cases as of April 28, 2020	I
CC_APR7	Confirmed cases as of April 7, 2020	I
CC_Mar15	Confirmed cases as of March 15, 2020	I
CC_June30	Confirmed cases as of June 30, 2020	D
CC_July2	Confirmed cases as of July 2, 2020	D

Table 2 GWR regression results—case-I

Parameters	Coeff.	Std. error	t-value	p-value	CI
CC_June30	1.0836	0.0095	113.22	0.000	1.064–1.103
CC_May28	0.0154	0.0495	0.31	0.757	−0.086–0.117
CC_APR28	−0.5115068	0.2035	−2.51	0.018	−0.928–0.094
CC_Mar15	−5.874116	15.4038	−0.38	0.706	−37.427–25.679
_cons	136.542	104.6869	1.30	0.203	−77.899–350.983
$R^2 = 0.9998$	Convergence: bandwidth = 8.4203				

3.1 *Outputs from GWR: Case I*

In this case, the spatial relationship between cases after unlock week and previous months were examined. GWR regression results (Table 2) as of July 2, 2020 cases (response) for the case-I are shown, with the R^2 value for the model is 0.9998, indicating a high degree of fit to the data. The spatial distribution of simulation (probability distribution) is shown using the explanatory variables obtained by MCS in Fig. 1 with choropleth maps.

Table 3 discusses the results that are useful to assess the significance of the spatial variations in the parameter estimates as determined by computing the **si** statistics using the Monte-Carlo method. It is found as the only coefficient to vary significantly over space associated with levels of confirmed cases in other previous weeks. The standard deviation of the spatially varying parameter estimates for the variable CC_Mar15 (10.7066) is over two-thirds times greater than the standard error of the global parameter estimate (15.4038 as shown in Table 2). The other variables CC_June30, with si (0.0271) and standard error (0.0095) in Table 1, is found to vary in slight marginal quantity. Thus it ensures that the rise in cases is not affected by the regions of being cold or hot, either north or south. But GWR is useful in exploring the data and identifying underlying geographical patterns for explanatory variables incorporated into a model.

3.2 *Case-II: Outputs of GWR*

The standard error value of CC_Mar15 (394.9389) in Table 4 is 26 times higher than Table 2 (15.4038) and is higher by 1.7 times of Table 5 si of CC_Mar15 (226.7075). The model is found to be fit with the R^2 value of 0.8903. The standard deviation of CC_APR7 (si) value (2.0891) in Table 5 is 12 times greater than the standard error (394.9389). This indicates that the cases were found less compared to the unlocked period, i.e. cases as of July, 2. The spatial variation of standard deviation produced by the MCS is shown using choropleth maps in Fig. 2.

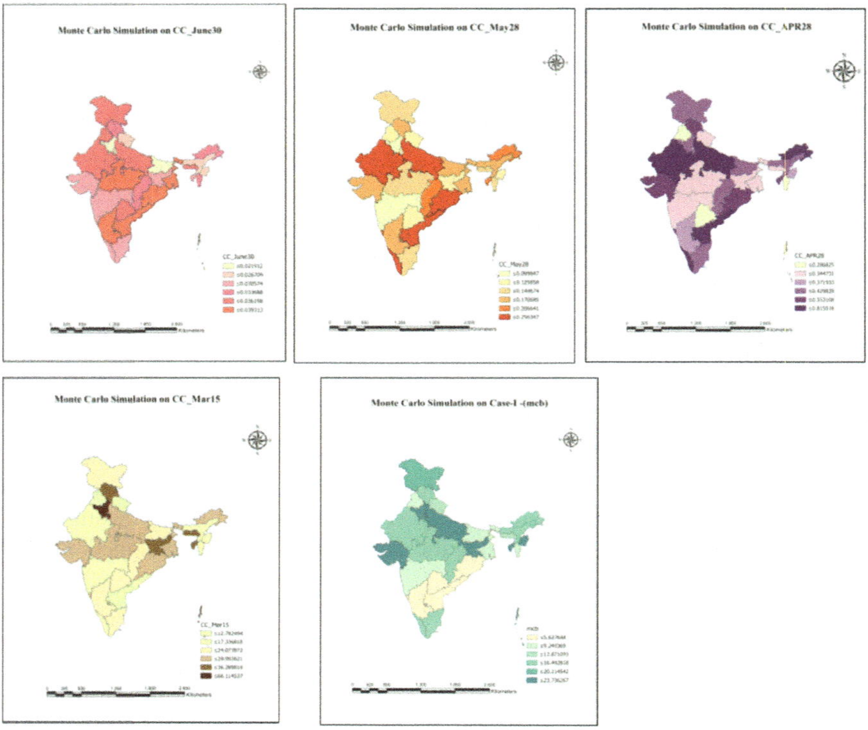

Fig. 1 The Choropleth map for each explanatory variable in case-I

Table 3 Significance tests for non-stationarity—case-1

Variable	si	p-value
Constant	173.3595	0.000
CC_June30	0.0271	0.788
CC_May28	0.1074	0.879
CC_APR28	0.2565	0.939
CC_Mar15	10.7066	0.939

Table 4 GWR regression results—case-II

Parameters	Coeff.	Std. error	t-value	p-value
CC_MAY1	10.3714	1.689375	6.14	0.000
CC_APR7	103.5724	25.88379	4.00	0.000
CC_Mar15	−186.0827	394.9389	−0.47	0.641
_cons	−2881.561	2521.004	−1.14	0.262
$R^2 = 0.8903$	Convergence: bandwidth = 23.7345			

Table 5 Significance tests for non-stationarity—case-III

Var	si	p-value
Constant	848.3460	0.000
CC_MAY1	0.8168	0.242
CC_APR7	2.0891	1.000
CC_Mar15	226.7075	0.061

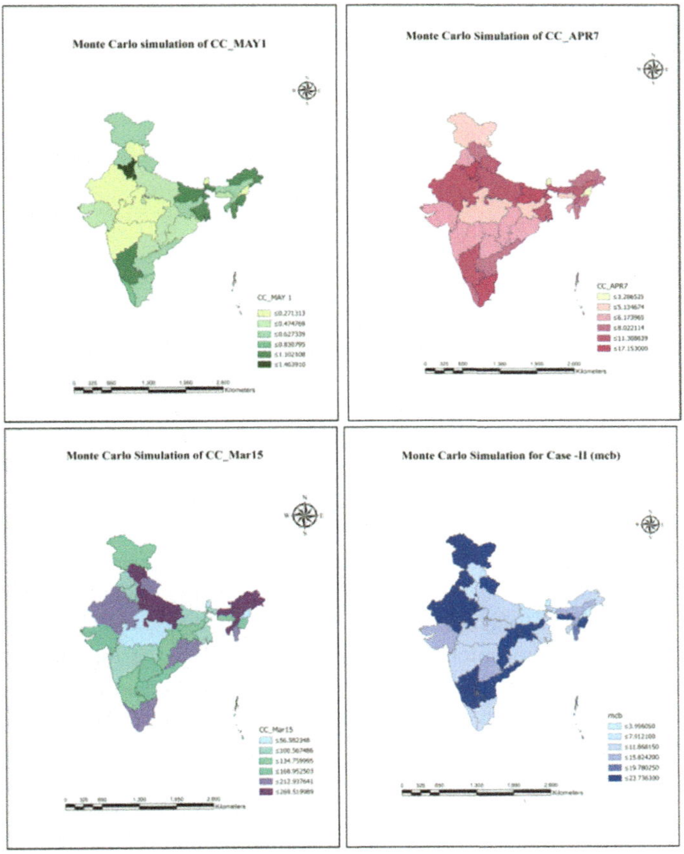

Fig. 2 The choropleth map for each explanatory variable in case-II

3.3 *Case-III: Outputs of GWR*

In this case, the third-month cases were assumed as the response (May 1, 2020) to the first and second-month cases and found that no variation in the standard error of CC_mar15 which is 11.87184 (Table 6) and si (1.6855) from Table 7. Let's consider CC_APR7 (0.2398) is almost the same as SE (0.12457). The model is a better fit with

Table 6 GWR regression results—case-III

Parameters	Coeff.	Std. error	t-value	p-value
CC_APR28	1.247108	0.009932	125.56	0.000
CC_APR7	−0.5518284	0.12457	−4.43	0.000
CC_Mar15	0.2620784	1.859631	0.14	0.889
_cons	−9.286613	11.87284	−0.78	0.440
$R^2 = 0.9993$	Convergence: bandwidth = 10.8109			

Table 7 Significance tests for non-stationarity—case-III

Variable	si	p-value
Constant	6.7080	0.152
CC_APR28	0.0117	0.545
CC_APR7	0.2398	0.545
CC_Mar15	1.6855	0.758

$R^2 = 0.9993$ which is very close to 1 and bandwidth (10.8109). The choropleth map is presented in Fig. 3 with the spatial distribution of simulations over all the states of India. Thus, from the results it infers that the cases were few as the lockdown was issued with strict discipline, but as the relaxation was given the cases increased in the case-I and case-II study.

Figure 4 shows the Monte-Carlo simulation results of all three cases which infer that the lines extend from the boxes (*whiskers*), indicating inconsistency outside the upper and lower quartiles. The results of observed values for case-I (8.4204), case-II (23.7345) and case-III (10.8109) are shown in Table 8, and from the graph depicted in Fig. 5, it is found that case-I has the least p-value (0.152), which is slightly good over other two models of the cases.

Figure 6 shows the choropleth maps for the probability distribution of each individual case study with all the explanatory variables plotted in a single map with dots. The choropleth map provides a visualization of how a variable varies across a geographic area. The choropleth maps in the figure shows the spatial probability distribution of the explanatory variables which are obtained using Monte-Carlo simulations for the confirmed cases. This simulation helps to understand the risks and decision making for the coming future.

3.4 *Tested Samples and Confirmed Cases*

In [13], a cross-sectional observational study on critical care beds in Asian countries and regions was carried out and concluded that low- and lower-middle-income countries or regions have a lesser number of hospital care beds to accommodate the Covid-19 patients. According to the sources available, India has performed the total

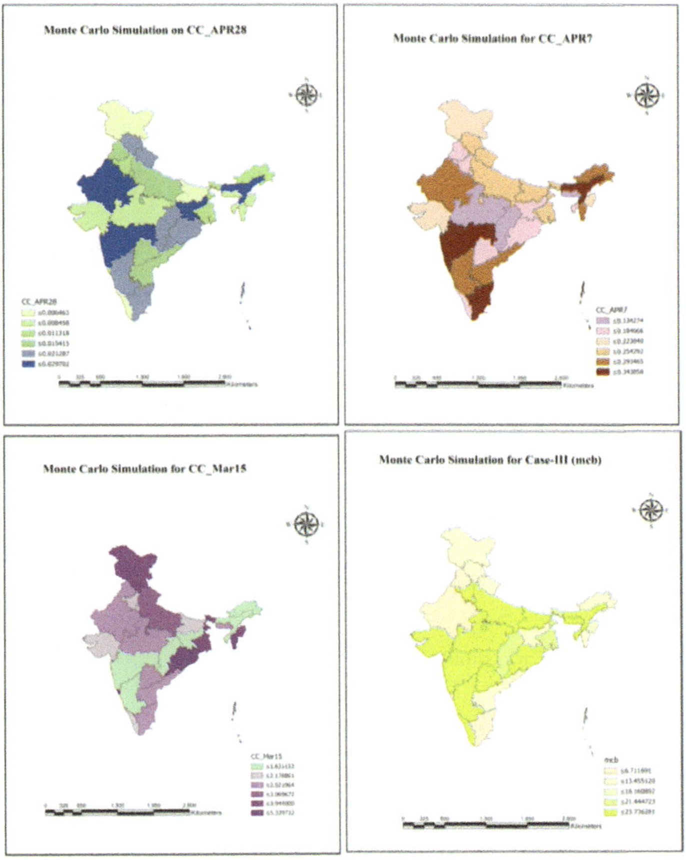

Fig. 3 The choropleth map for each explanatory variable of case-III

tests (9,969,662) on the total population (1,380,122,667) for Covid-19 whereas China performed total tests (90,410,000) on its population (1,439,323,776) (https://www.worldometers.info/coronavirus/#countries), this is because of the non-availability of testing labs in India (Appendix-I Tables 17 and 18). However, according to WHO, the death rate has declined in the world to 2.94% and there is no need to panic. Isolation of the Covid-19 patient is essential to avoid the spread of the virus. Table 9 shows the total number of Covid-19 cases of 10 leading countries in the world.

Figure 7 shows the day-wise increase of confirmed cases in India as of July 4, 2020, and the day-wise increase of confirmed cases of 12 different states is depicted in Fig. 8. Figure 9 presents the rise of confirmed cases in 12 different states.

Table 10 shows the availability of intensive care beds in the world. However, India stands tenth in position in the world with 1.5 nurses per1000 [9]. India provides a 2.3 total number of critical care beds per 100,000 inhabitants according to the source

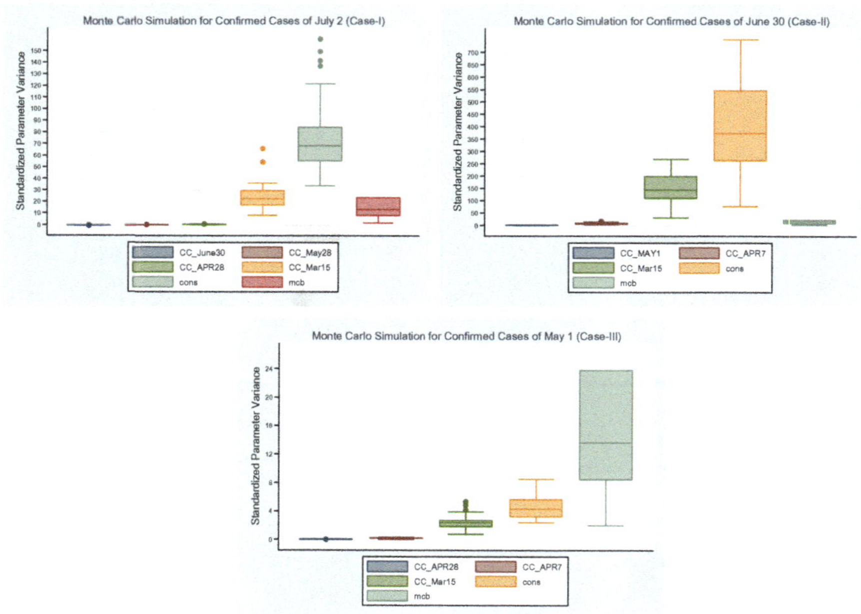

Fig. 4 Boxplot of Monte-Carlo simulation

Table 8 Significance test for bandwidth

S. no.	Observed	p-value
Case-I	8.4204	0.152
Case-II	23.7345	0.515
Case-III	10.8109	0.424

Fig. 5 All three cases significance test for bandwidth

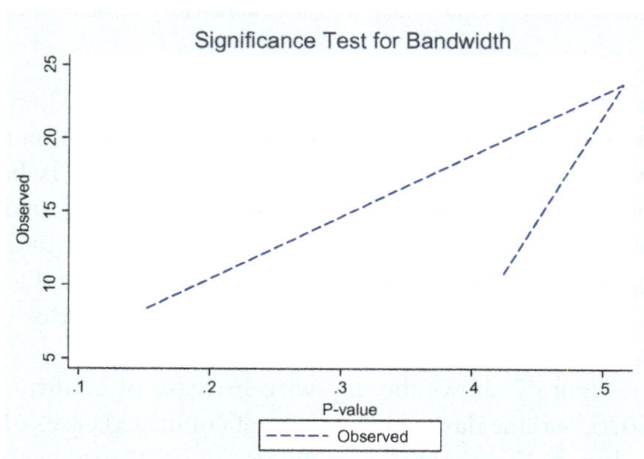

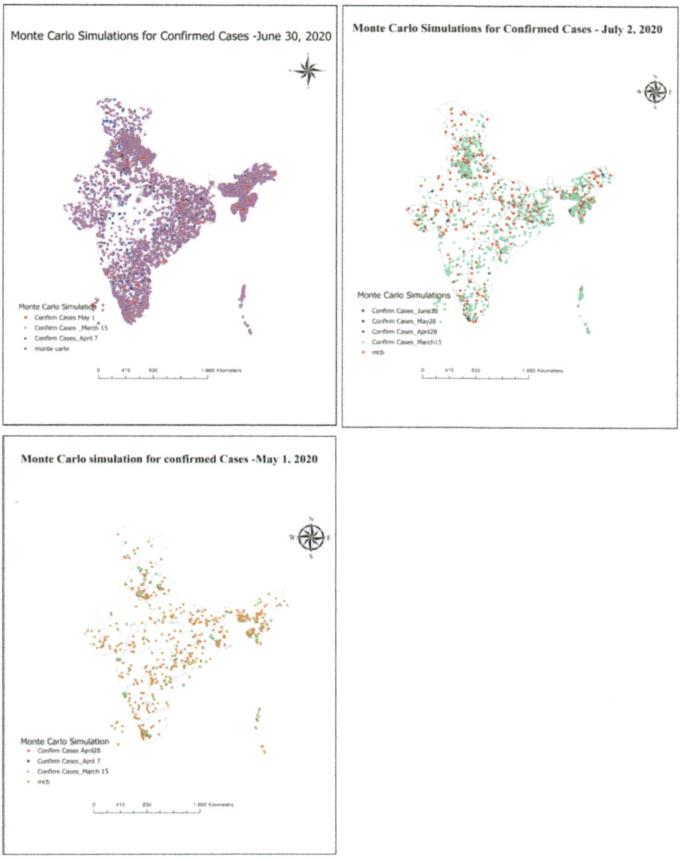

Fig. 6 Choropleth maps for the three cases

available (https://www.statista.com/chart/21163/nurses-per-1000-inhabitants/). The media stated that in India, there are around 1,065 testing labs across the country with public sector labs (768) and private labs (297). Covid-19 daily barometer helps in measuring the mood, attitudes and economic impact of the virus on citizens (https://www.statista.com/study/72001/covid-19-barometer/). Figure 10 depicts the ratio of positive cases in India (July 4, 2020) and Fig. 11 shows the graph of total people tested with the confirmed cases of Covid-19 in India as of July 4, 2020. The Covid-19 test does not involve any blood examination but it includes examining the throat swab or the nasal swab to check for infection. These samples are taken from the people to test for Covid-19; later if the tests are positive, the patient has to remain quarantined for at least 14 days, till recovery (WHO). However, according to the National Health Profile 2018 (https://cdn.downtoearth.org.in/pdf/NHP-2018.pdf), in urban areas there are 431,173 government hospital beds (no ICU beds in rural areas). It is projected that 2–4% of these beds are ICU beds, i.e. 3% = 12,935 public sector

Table 9 Worldwide Covid-19 cases

Number of Covid-19 confirm cases as of July 6, 2020 in few countries	
World	11,480,171
USA	2,964,127
Brazil	1,579,837
India	697,069
Russia	681,251
Peru	299,080
Spain	297,625
Chile	295,532
UK	285,416
Mexico	252,165
Italy	241,611

Source https://www.statista.com/

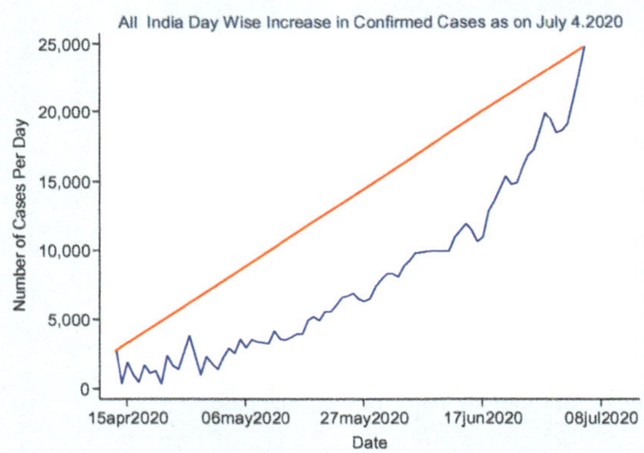

Fig. 7 All India day-wise increase in confirmed cases as of July 4, 2020

ICU beds. According to a database maintained by Cipla Inc., a global pharmaceutical company based in India, there are 90,008 private hospital beds, of which 17,062 are private sector ICU beds. These give a total of 29,997 ICU beds. But still, India lacks maintaining the correct database as of the year 2020. It was reported that there are not enough beds in the hospitals to care for the Covid-19 patients as such, so many schools, hostels and restaurants have been rehabilitated as Covid-19 centres for the patients.

Covid-19 has been a nightmare for doctors and other medical staff across the globe. All the medical teams were working around the clock to treat the infected patients and reduce the outbreak. Many medical staff have been affected by Covid-19 while treating the Covid-19 patients. Most of the recovered patients in other countries responded that their government is handling the crisis well.

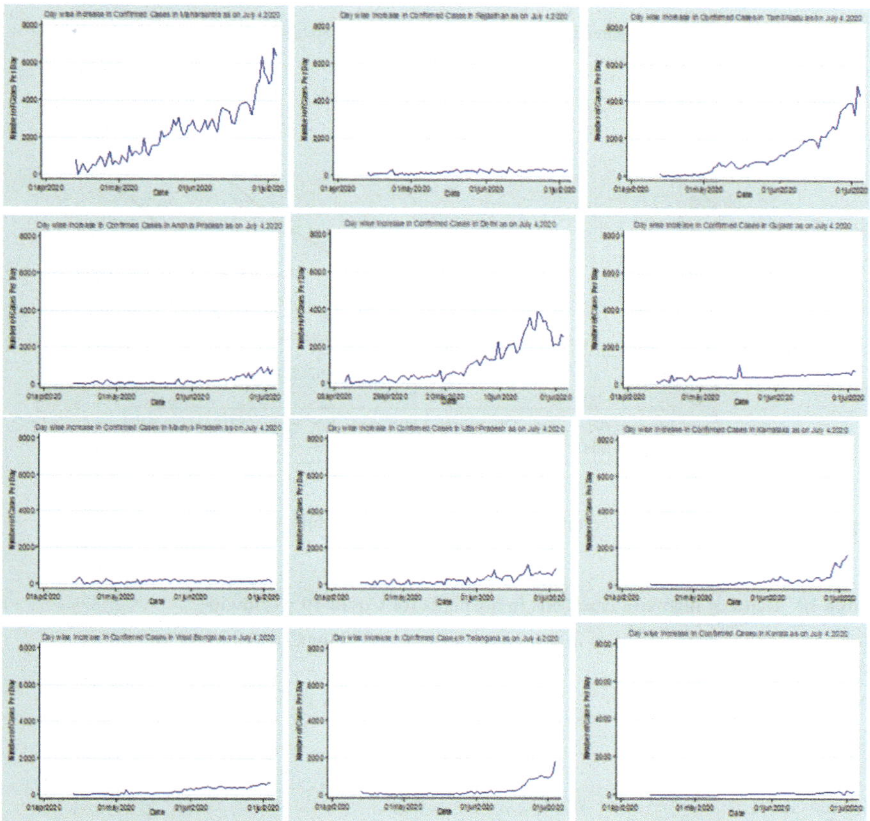

Fig. 8 Day-wise increase in confirmed cases in 12 different states in India as of July 4, 2020

Few precautionary steps to handle Covid-19 cases such as avoid crowded public places, travelling, maintain personal hygiene (hand sanitiser) [16], abstain from touching objects in public, avoid going to work, maintain social distance, avoid consumption of raw or uncooked meat, and face mask is a must when one goes out in public places were followed (https://yougov.co.uk/, p. 2).

Figure 12 shows the graph of few major Covid-19 affected states of tested samples and confirmed cases. In another report released by ICMR (June 2020) on the testing labs stated that the total number of government laboratories (642) and private (243) counted to total laboratories (885). The hospital bed per thousand in India is given as 0.53 (https://ourworldindata.org/coronavirus-testing). Figure 13 depicts the day-wise recovered cases of Covid-19 in India.

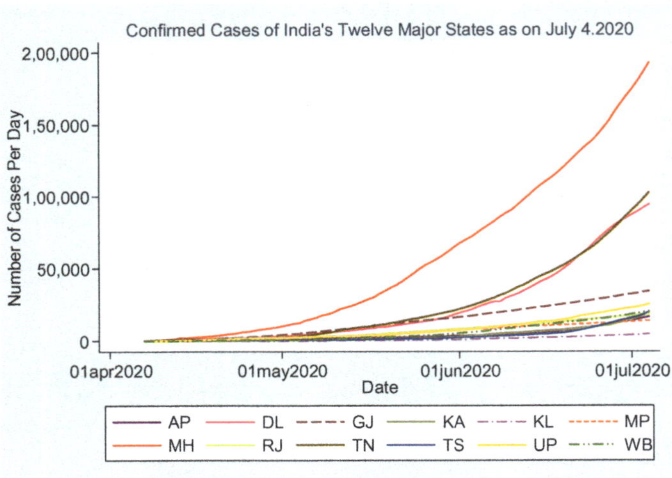

Fig. 9 Twelve different states confirmed cases in India as of July 4, 2020

Table 10 Rates of intensive care beds in hospitals for Covid-19 worldwide

Intensive care beds in hospitals in few countries as of 2020 for Covid-19 (per 100,000 population)	
Name of the country	Availability of beds
Germany	33.9
Austria	28.9
United States	25.8
France	16.3
Canada	12.9
England	10.5
Spain	9.7
Italy	8.6
Denmark	7.8
Ireland	5

Source (https://www.statista.com/statistics/1115999/intensive-care-bed-rates-hospitals-select-countries-worldwide/)

3.5 Results of GWR on Testing Samples Using STATA

3.5.1 Case-I

GWR was applied on testing samples (response) and other variables as explanatory using STATA gave the results as in Table 11 with R^2 (0.4928) and bandwidth (5.2132)

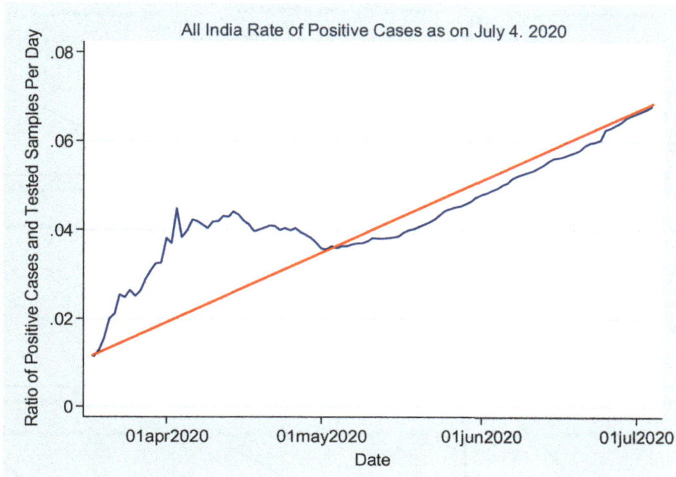

Fig. 10 The graph of all India ratio of tested samples and confirmed cases as of July 4, 2020

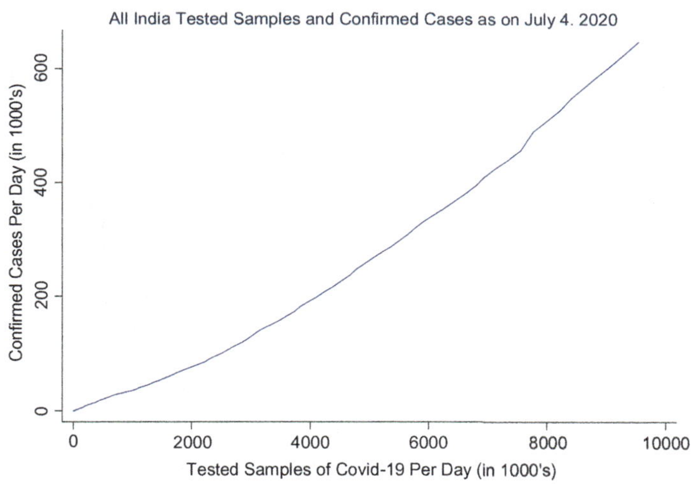

Fig. 11 All India tested samples and confirmed cases as of July 4, 2020

and Monte-Carlo simulation results were mapped through ARCGIS to get the results in Fig. 14. The significance tests (si) for CCJuly4 is 17.3650 (Table 12) and standard error (SE) (−1.03). Similarly, the variable, NumCovidtestinglabs (si) is 6.7e + 03, whereas SE is found to be 1901.701. The significance test for bandwidth showed observed (5.2132) and p-value (0.030). GWR is a statistically significantly better model with $p = 0.030$.

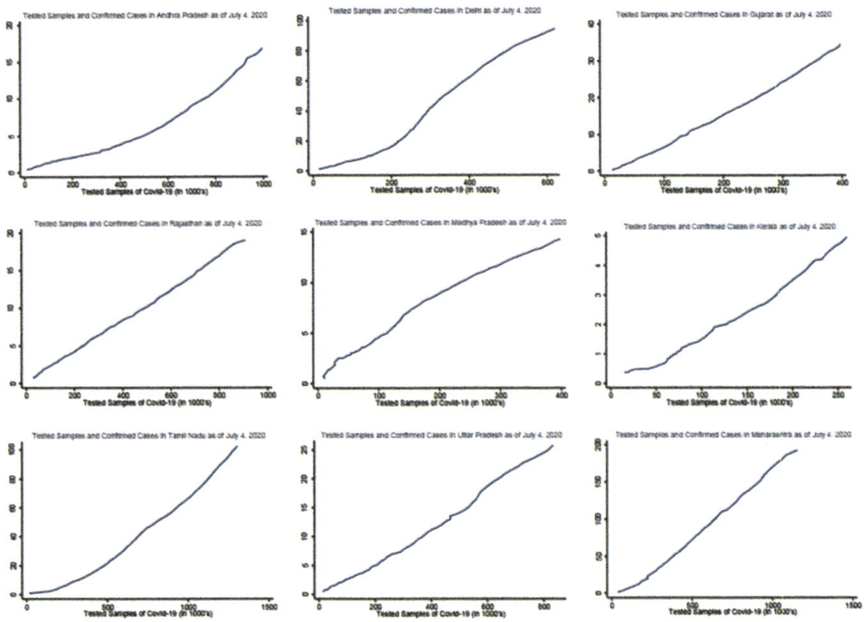

Fig. 12 Different states of India tested samples and confirmed cases as of July 4, 2020

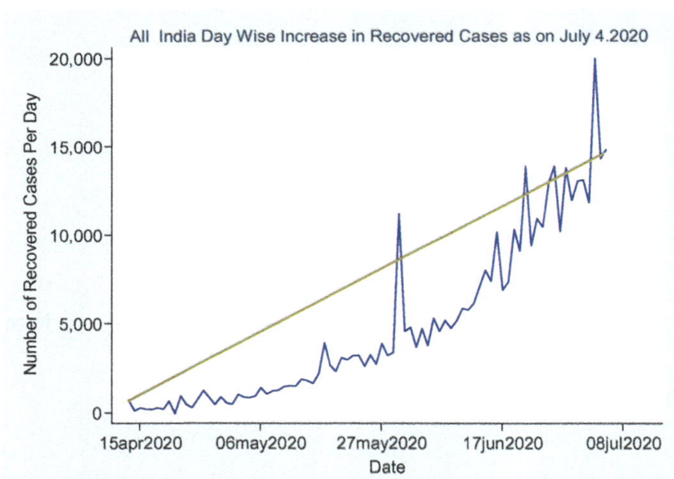

Fig. 13 All India day-wise recovered cases as of July 4, 2020

Table 11 GWR regression results of testing samples—case-I

Parameters	Coeff.	Std. error	t-value	p-value
NumCovidtestinglabs	9135.58	1901.701	4.80	0.000
CCJuly4	−1.517126	1.469029	−1.03	0.309
_cons	74,851.32	53,333.39	1.40	0.170
$R^2 = 0.4928$	Convergence: Bandwidth = 5.2132			

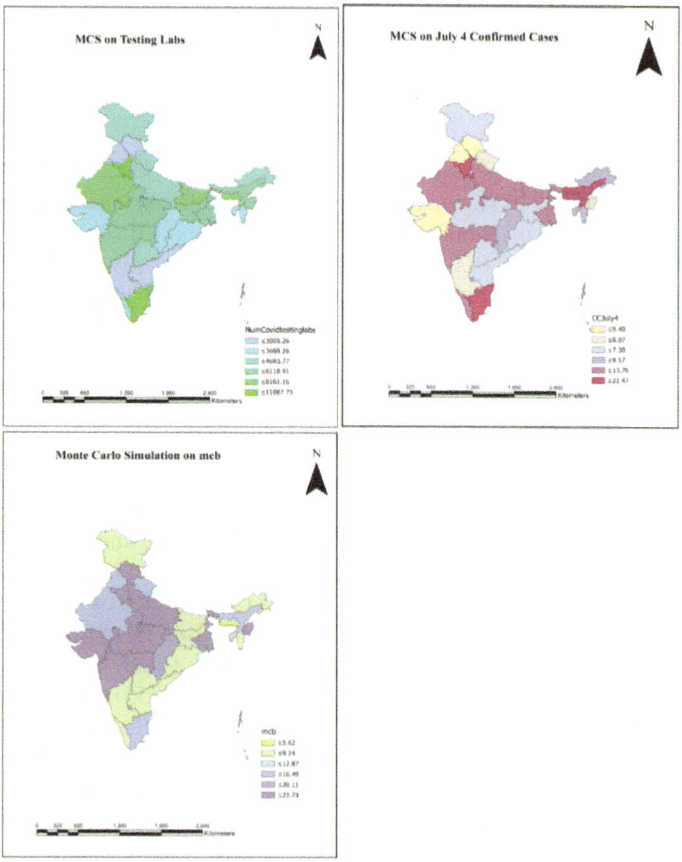

Fig. 14 The choropleth map for each explanatory variable in Testing _Sample−Case-I

Table 12 Significance tests for non-stationarity—case-I

Var	si	p-value
Constant	1.1e + 05	0.000
NumCovidtestinglabs	6.7e + 03	0.182
CCJuly4	17.3650	0.091

Table 13 GWR regression results of testing samples—case-II

Parameters	Coeff.	Std. error	t-value	p-value
TotalBeds	0.3786624	2.077251	0.18	0.857
CCJuly4	−2.08735	2.100136	−0.99	0.328
NumofHospitals	−12.76002	34.24787	−0.37	0.712
NumCovidtestinglabs	9715.962	3897.636	2.49	0.018
_cons	74,202.38	55,422.48	1.34	0.190
$R^2 = 0.4956$	Convergence: bandwidth = 8.4204			

3.5.2 Case-II

In the case-II of GWR on testing samples (response), it produced R^2 (0.4956) and bandwidth (8.4203982), as shown in Table 13. The simulation was mapped to obtain the maps in Fig. 15, si of NumofHospitals (57.9820) and SE (34.24787), whereas si of NumCovidtestinglabs (6.6e + 03) from Table 14 is nearly 3000 times greater than SE (3897.636). The significance test for bandwidth gave the observed (8.4204) and p-value (0.030). Thus, the model is statistically significant with $p = 0.030$.

3.6 *GWR and OLS Results Using ARCGIS*

The models are generated on the datasets of testing labs (response) (Appendix-I Table 18) and explanatory variables are shown in Table 15. GWR and OLS results were obtained using ArcGIS Pro. The summary of the GWR model (Fig. 16) is shown in Table 15; the model is described based on various factors, such as sigma square and sigma square MLE and DF. Finally, sigma squared is used for AICc computations. From Tables 15 and 16, R^2 (0.9207) is much closer to 1 and AICc (266.9961) in GWR is better over OLS, i.e., R^2 (0.8797) and AICc (268.4955), hence GWR model is a better model.

The results of the OLS model (Fig. 16) summarizes the coefficient, T-statistic and p-value along with VIFs on explanatory variables assumed in Table 16; the selected variables have relatively low multi-collinearity since the variance inflation factor (VIFs) for all of the explanatory variables were positively associated with confirmed cases ($p < 0.01$). The p-value for Total_Beds (0.011536) is better with VIF (12.1669) over the other explanatory, and the p-value of Number_Testing__Samples (0.033223) and VIF (3.4350).

However, in a discussion, the World Trade Organization reported that due to Covid-19 the world trade is expected to fall by between 13% and 32% in 2020, which clearly indicates the world economy is expected to face recession. The overall study concludes that due to unlock 1.0 from June 1 to June 30, 2020, the positive cases have been increased in overall India, except Lakshadweep where no Covid-19 cases were reported as of July 4, 2020. It was observed that north, south, east and west

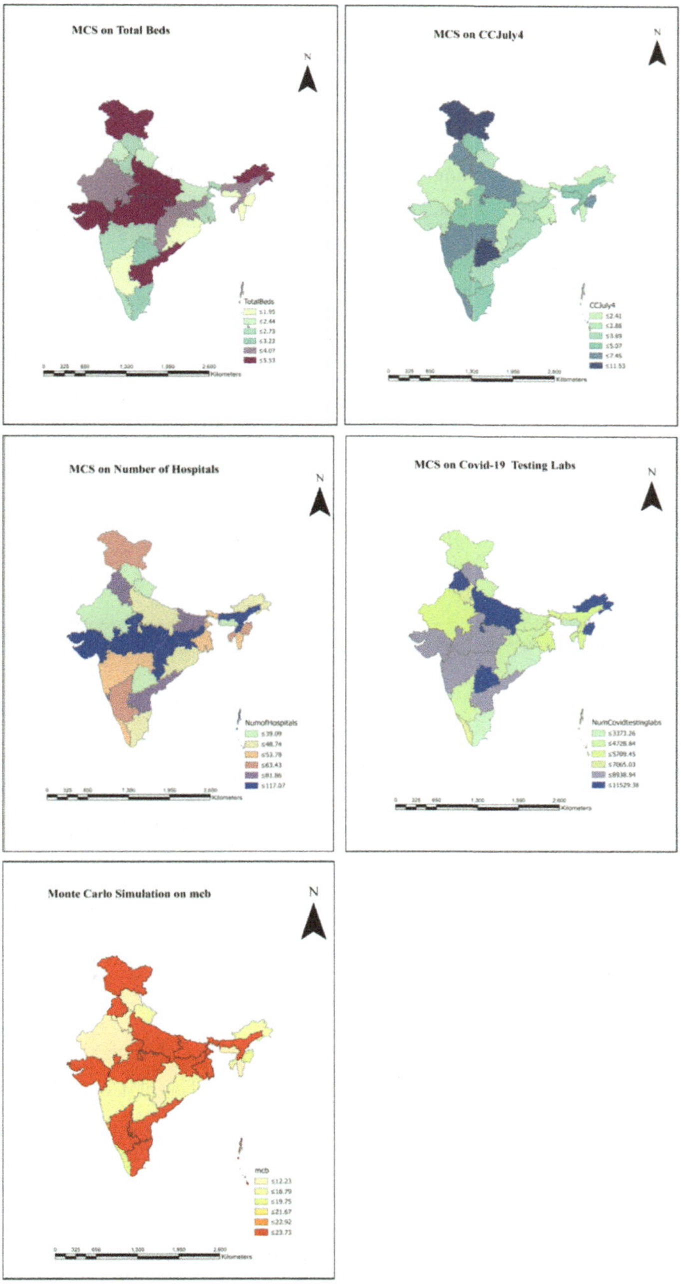

Fig. 15 The choropleth map for each explanatory variable in Testing _Sample-Case-II

Table 14 Significance tests for non-stationarity—case-II

Variable	si	p-value
Constant	6.8e + 04	0.000
TotalBeds	2.2273	0.697
CCJuly4	7.6725	0.061
NumofHospitals	57.9820	0.424
NumCovidtestinglabs	6.6e + 03	0.394

Table 15 Summary of GWR for testing samples

Dependent	Confirm
Explanatory	Number_Of_Hospitals Total_Beds Number_Testing__Samples Confirmed__July4
R^2	0.9207
AICc	266.9961
Sigma-Sq	122.2291
Sigma-Sq. MLE	68.2899
DF	17.8785

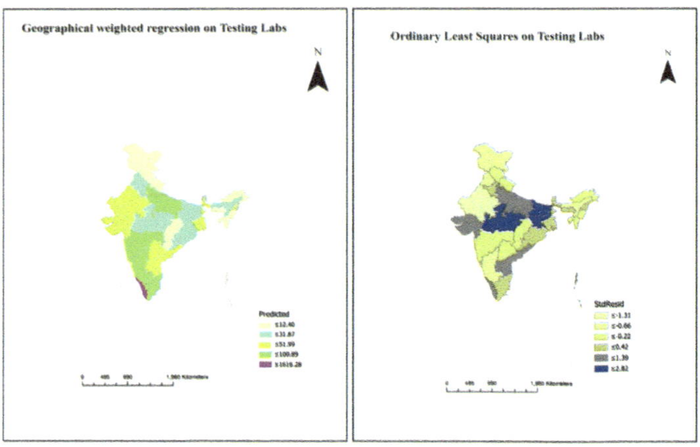

Fig. 16 GWR (left) and OLS (right) model for the variable Number_of_Covid_Testing_Labs

Table 16 Summary of OLS for testing samples

Var	Coeff.	T-statistic	p-value	VIF
Intercept	3.495449	1.352088	0.186798	–
Number_of_Hospitals	0.000266	0.176996	0.860743	7.3372
Total_Beds	0.000221	2.696957	0.011536	12.1669
Number_Testing_Samples	0.021820	2.235866	0.033223	3.4350
Confirm-July4	0.096213	1.037132	0.308247	3.5824

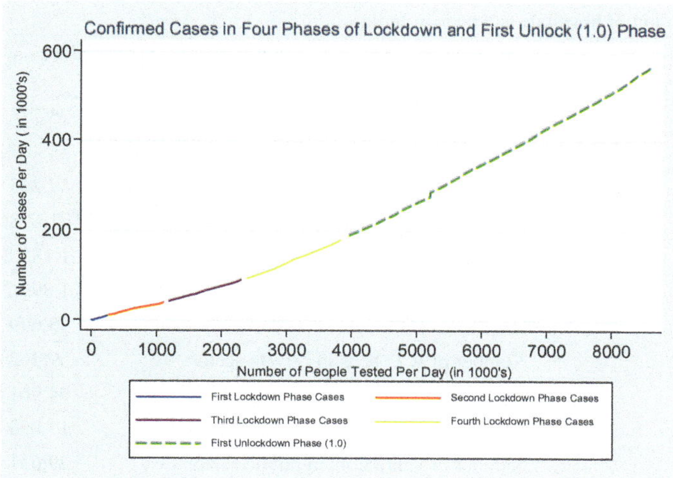

Fig. 17 The graph of all India lockdown cases

had major positive cases. Figure 17 shows the graph for all four lockdown phases and the first unlocked (1.0) phase. GWR proved that the probability distribution using Monte-Carlo simulation can be performed risk analysis for any variable that has in-built uncertainty that is useful in Covid-19 analysis.

4 Conclusions

GWR provides a significant approach in spatial analysis using a probability distribution. The Monte-Carlo helps in performing risks analysis for the number of uncertainties of Covid-19. The need for testing labs and hospitals is a must to face the situation in such a need of the hour with a sudden outbreak of Covid-19. Based on the results obtained, GWR on testing samples with R^2 (0.4928) and bandwidth (5.2132) proves that there are requirements for more testing labs compared to other higher-income countries. The testing samples are more in number with the population. GWR on confirmed cases shows R^2 (0.9998) and bandwidth (8.4203) such that the increase is due to unlockdown phase. Moreover, many teenagers should be encouraged to take up courses in the medical field to handle such a crisis in future.

Appendix-I

See (Tables 17 and 18 and Fig. 18).

Table 17 Count of hospitals and beds in India

S. no.	Total number of hospitals and beds available in India as of October 2019	
1	Number of hospitals in public sector	25,778
2	Number of hospitals in private sector	43,487
3	Total number of hospitals (1 & 2)	69,265
4	Number of hospital beds in public sector	713,986
5	Number of hospital beds in private sector	1,185,242
6	Total number of hospitals beds (4 & 5)	1,899,228
7	Number of ICU beds in public sector	35,699
8	Number of ICU beds in private sector	59,262
9	Total number of ICU beds (7 & 8)	94,961
10	Number of ventilators in public sector	17,850
11	Number of ventilators in public sector	29,631
12	Total number of ventilators (9 & 10)	47,481

Source National Health Profile of India 2019 released in October 2019

Table 18 State-wise total Covid-19 testing labs in India

State	Number of Covid-19 testing labs (includes government and private)
Maharashtra	98
Tamil Nadu	79
Delhi	41
Gujarat	42
Uttar Pradesh	108
Rajasthan	24
West Bengal	47
Madhya Pradesh	60
Haryana	21
Karnataka	72
Andhra Pradesh	58
Bihar	38
Telangana	26
Jammu and Kashmir	7
Assam	13

(continued)

Table 18 (continued)

State	Number of Covid-19 testing labs (includes government and private)
Odisha	20
Punjab	13
Kerala	34
Uttarakhand	7
Chhattisgarh	5
Jharkhand	30
Tripura	1
Ladakh	1
Goa	5
Himachal Pradesh	6
Manipur	2
Chandigarh	3
Puducherry	4
Nagaland	7
Mizoram	2
Arunachal Pradesh	2
Sikkim	2
Dadra and Nagar Haveli	1
Andaman and Nicobar Islands	3
Meghalaya	3
Lakshadweep	NA

Source ICMR, Department of Health Research released on June 2020

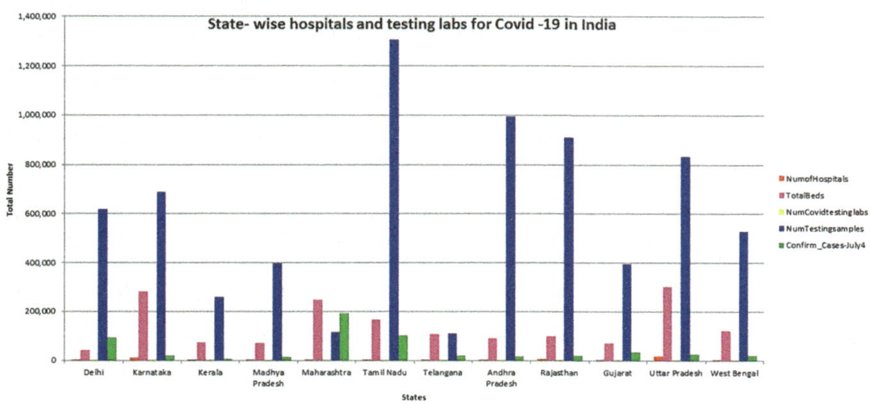

Fig. 18 Twelve states, total Covid-19 testing labs, hospitals, total beds and total tested samples in India

References

1. Brunsdon C, Fotheringham AS, Charlton ME (1996) Geographically weighted regression: a method for exploring spatial nonstationarity. Geogr Anal 28(4):281–298
2. Brunton LA, Alexander N, Wint W, Ashton A, Broughan JM (2017) Using geographically weighted regression to explore the spatially heterogeneous spread of bovine tuberculosis in England and Wales. Stoch Environ Res Risk Assess 31:339–352. https://doi.org/10.1007/s00477-016-1320-9
3. Calvo E, Escolar M (2003) The local voter: a geographically weighted approach to ecological inference. Am J Polit Sci 47(1):189–204
4. Chen Y, Jiao J, Bai S, Lindquist J (2020) Modeling the spatial factors of COVID-19 in New York City. SSRN: https://doi.org/10.2139/ssrn.3606719
5. Cui J, Li F, Shi Z (2019) Origin and evolution of pathogenic Coronaviruses. Nat Rev Microbiol 17:181–192
6. Fotheringham AS, Charlton M, Brunsdon C (1996) The geography of parameter space: an investigation of spatial non-stationarity. Int J Geogr Inf Syst 10(5):605–627
7. Fotheringham AS, Charlton ME, Brunsdon C (1998) Geographically weighted regression: a natural evolution of the expansion method for spatial data analysis. Environ Plan A 30(11):1905–1927
8. Lu B, Charlton M, Harris P, Fotheringham AS (2014) Geographically weighted regression with a non-Euclidean distance metric: A case study using hedonic house price data. Int J Geogr Inf Sci. https://doi.org/10.1080/13658816.2013.865739
9. McCarthy N (2020) The countries with the highest density of nurses. Statista. https://www.statista.com/chart/21163/nurses-per-1000-inhabitants/
10. Mollalo A, Vahedi B, Rivera KM (2020) GIS-based spatial modeling of COVID-19 incidence rate in the continental United States. Sci Total Environ 728:138884
11. Ndiath MM, Cisse B, Ndiaye JL, Gomis JF, Bathiery O, Dia AT, Gaye O, Faye B (2015) Application of geographically-weighted regression analysis to assess risk factors for malaria hotspots in Keur Soce health and demographic surveillance site. Malar J 14:463. https://doi.org/10.1186/s12936-015-0976-9
12. Oshan TM, Li Z, Kang W, Wolf LJ, Fotheringham S (2019) MGWR: a python implementation of multiscale geographically weighted regression for investigating process spatial heterogeneity and scale. ISPRS Int J Geo-Inf 8(269):31. https://doi.org/10.3390/ijgi8060269
13. Phua J, Faruq MO, Kulkarni AP, Redjeki IS, Detleuxay K, Mendsaikhan N, Sann KK, Shrestha BR, Palo JEM, Haniffa R, Wang C, Hashemian SMR, Konkayev A, Nor MBM, Patjanasoontorn B, Nafees KMK, Ling L, Nishimura M, Bahrani MJA, Arabi YM, Lim C, Fang W (2020) Critical care bed capacity in asian countries and regions. Critical Care Med 48(5):654–662. https://doi.org/10.1097/CCM.0000000000004222
14. Shereen MA, Khan S, Kazmi A, Bashir N, Siddique R (2020) COVID-19 infection: origin, transmission, and characteristics of human coronaviruses. J Adv Res 24:91–98. https://doi.org/10.1016/j.jare.2020.03.005
15. Wang C, Zhang J, Yan X (2012) The use of geographically weighted regression for the relationship among extreme climate indices in China. Hindawi Publishing Corporation Mathematical Problems in Engineering 2012(369539):15. https://doi.org/10.1155/2012/369539
16. WHO (2020) Water, sanitation, hygiene, and waste management for the COVID-19 virus: interim guidance

Chapter 7
Integration of IoT and Fog Computing for the Development of COVID-19 Cluster Tracking System in Urban Cities

H. M. K. K. M. B. Herath, G. M. K. B. Karunasena, H. M. W. T. Herath, H. D. N. S. Priyankara, B. G. D. A. Madushanka, and W. R. De Mel

Abstract In recent years, the Internet of Things (IoT) has gained compelling support for studies as a new research topic in a wide range of academic and industrial disciplines, especially in health care. The IoT revolution is reshaping contemporary healthcare systems by integrating technical, economic, and social perspectives. The emerging global pandemic challenges caused by the novel coronavirus (SARS-CoV-2) have been the biggest global public health issue since 2019. At the time this chapter was written, more than 61.8 million cases worldwide had already been diagnosed with COVID-19. In the sense of COVID-19, IoT-enabled devices or applications have the potential to recognize and monitor COVID-19 patients. Since there is no new development introduced to recognize COVID-19 clusters using the IoT and Fog computing up to date, this chapter provides a potential solution for identifying and monitoring clusters in the 1st, 2nd, and 3rd phases of the pandemic. Technologies regarding Fog computing, smart sensors, and IoT-based techniques have been used to build the framework. We have observed that the findings are very promising with high-performance accuracy.

Keywords COVID-19 · Cluster tracking · Fog computing · Internet of Things (IoT) · Network protocol · Smart health care · Urban IoT

H. M. K. K. M. B. Herath (✉) · G. M. K. B. Karunasena · H. D. N. S. Priyankara · B. G. D. A. Madushanka
Faculty of Engineering Technology, The Open University of Sri Lanka, Colombo, Sri Lanka
e-mail: kkmbh@ou.ac.lk

H. M. W. T. Herath
Faculty of Engineering, University of Moratuwa, Moratuwa, Sri Lanka

W. R. De Mel
Faculty of Technology, University of Sri Jayewardenepura, Nugegoda, Sri Lanka
e-mail: wrdemel@sjp.ac.lk

U. Kose et al. (eds.), *Computational Intelligence for COVID-19 and Future Pandemics*, Disruptive Technologies and Digital Transformations for Society 5.0,
https://doi.org/10.1007/978-981-16-3783-4_7

1 Introduction

Coronaviruses (CoV) are known as a large family of positive-stranded RNA respiratory viruses. It is named as such because of the crown-shaped tips on the surface of the virus. Coronavirus is recognized as the major pathogens that mainly affect the human respiratory system. In December 2019, a new coronavirus disease (SARS-CoV-2, which causes the COVID-19 disease) was first reported in Wuhan, China, and has spread rapidly worldwide [1, 2]. The World Health Organization (WHO) has formally announced the discovery of a new virus named 2019-nCoV in January 2020 [3]. Previous coronavirus outbreaks include the Severe Acute Respiratory Syndrome, also called SARS-CoV, and the Middle East Respiratory Syndrome Corona Virus known as MERS-CoV, which have been identified as major public health threats. Table 1 depicts the comparison between the different viruses and their characteristics.

More than 58.6 million positive COVID-19 cases were reported by the end of November 2020 at the time of this writing, resulting in more than 1.3 million deaths globally. Figure 1 illustrates the global COVID-19 spread as of November 2020.

A study carried out by Jiang et al. [4] has shown that the death rate of COVID-19 is 4.5% globally. The mortality rate for patients is 8.0% in the age range of 70–79 years, while 14.8% for patients above 80 years. Patients above the age of 50 with chronic illnesses are at the greatest risk and finding a way to diagnose illness before heading into severe conditions is increasingly necessary. As a result of the COVID-19 pandemic, countries have faced several financial and social challenges in terms of health care. Due to the extremely rapid growth of new COVID-19 patients, overwhelmed healthcare facilities are experiencing interruptions in the provision of regular health services. Besides, healthcare staff is also becoming vulnerable to COVID-19 and this is further involved in taxing healthcare resources. Ongoing health and economic crisis have been triggered by the COVID-19 pandemic. The severity of

Table 1 Comparison between coronavirus disease 2019 (COVID-19), Severe Acute Respiratory Syndrome (SARS), Middle East Respiratory Syndrome (MERS), and Ebola Virus Disease (EVD)

Description	COVID-19	SARS	MERS	EVD
Incubation period (days)	2–14	2–10	2–14	2–21
Contagious during incubation	Yes	Yes	Yes	No
Vaccine as of 2020	No	No	No	Yes
Infected body part	Blood vessels	Respiratory system		All muscles
Main transmission	Respiratory droplet secretions			Body fluids
Shape and size in nm	Spherical 80–120	Spherical 80–90	Spherical 90–125	Filament 14,000 × 80 nm
Mortality rate %	2.65	14–15	34	Up to 90
Pronounced R-naught	2–2.5	3.1–4.2	< 1	1.5–1.9

Fig. 1 COVID-19 confirmed cases as of November 2020, *Source* Johns Hopkins University Center for Systems Science and Engineering

the situation has called for an urgent response from leaders of the government, scientists, and medical professionals. The epidemic has demanded creative solutions with limited resources and an urgent need for medical equipment, healthcare assistance, and treatment [5].

The primary problem for COVID-19 is its tendency to be fast-spreading. The spreading of the disease has been classified into four stages in every country [6–9]. The first stage is called an imported stage in which all positive cases of COVID-19 have a history of travel to the COVID-19 affected country. The second stage is known as the stage of local transmission, in which a person with a history of travel comes into contact with other people and transfers the infection to them. The third stage is known as the transmission phase of the community. Here at this stage, it is not possible to detect the transmission source at all. In this case, the history of travel and the number of contaminants with the virus of COVID-19 patients will not be conceivable [10]. The last stage is the epidemic stage. At this stage, the disease has no definitive end, and there is no choice but to wait for it to be over. The proposed solution will operate effectively if the suggested approach will be used successfully during phases 1, 2, and 3 so that it will be manageable if proper measures and action are taken before getting critical.

Innovative ICT technologies such as the Internet of Things (IoT) [11], cloud computing, big data, 5G communications [12], artificial intelligence [13], and blockchain [14] have been playing a key role in promoting the security and improvement of the people and economies of the world. The capabilities of ICT for omnipresent and accessible health care are necessary to alleviate the issues related to the pandemic. Over the last decade, there has been a shift toward transferring storage, managing, and maintaining data to the cloud. In specific, the roles of computing,

Table 2 Various IoT protocols and their uses in the smart city applications

IoT Protocol	Frequency	Data rate	Transmission range	Power usage
2G/3G	Cellular bands	10 Mbps	Several miles	High
Bluetooth/BLE	2.4 GHz	1, 2, 3 Mbps	~300 feet	Low
IEEE 802.15.4	Sub, 2.4 GHz	40, 250 kbps	>100 square miles	Low
LoRa	Sub GHz	<50 kbps	1–3 miles	Low
LTE Cat 0/1	Cellular bands	1–10 Mbps	Several miles	Medium
NB-IoT	Cellular bands	0.1–1 Mbps	Several miles	Medium
SigFox	Sub GHz	<1 kbps	Several miles	Low
Weightless	Sub GHz	0.1–24 Mbps	Several miles	Low
Wi-Fi	Sub, 2.4, 5 GHz	0.1–54 Mbps	<300 feet	Medium
WirelessHart	2.4 GHz	250 kbps	~300 feet	Medium
ZigBee	2.4 GHz	250 kbps	~300 feet	Low
Z-Wave	Sub GHz	40 kbps	~100 feet	Low

storage, and network maintenance are transferred to consolidated data centers, backbone IP networks, and wireless core networks [15]. Nowadays, significant improvements in health care are expected to be made through the Internet of Things (IoT). Various portable IoT devices are being introduced to monitor the health condition of the patient. These devices can connect to the Internet directly or through a local smartphone or a gateway. The IoT architecture consists of a multi-component structure, such as cloud servers, protocols, sensors, actuators, and layers. The IoT layer consists primarily of physical, network, and application [16]. Alternative protocols for communication between IoT devices using IP (Internet Protocol) have been developed by the Internet Engineering Task Force (IETF) due to the stable and reliable IP standard [17, 18]. Table 2 shows the different IoT networks used for healthcare applications.

Fog computing is a unique benefit for healthcare applications due to its crucial characteristics such as real-time, low latency, and fast response time. The Fog computing layer provides computing, networking, storage, and other domain-specific resources for IoT systems. The health domain has a variety of requirements that specifically distinguish it from other IoT applications such as remote monitoring, requiring a high degree of reliability. Safety and privacy aspects of health care are often of vital importance compared to other domains. According to the Fog system architecture, the Fog layer continuously collects a large volume of sensory data from the sensor network over a limited period of time, so that it controls the incoming data to provide a rapid response to the different user and system conditions.

There are many investigations in the area of monitoring, contact tracking, predicting, diagnosing COVID-19 disease, and many virologists are actively working to produce a vaccine as quickly as practicable. Since there is no specific cure for the pandemic disease, the world is now trying to manage spread by en-forcing a worldwide lockdown and making people aware of wearing masks and using sanitizers [19].

Innovations, including the Internet of Things (IoT) and Fog computing, are attracting global exposure to growing technological assistance in healthcare systems, in particular for predicting, identifying, avoiding, and tracking most infectious diseases. It also helps to combat COVID-19 by monitoring, tracing contacts, and detecting COVID-19 patients by linking it to IoT-based smart solutions. This chapter aims to develop an IoT-based COVID-19 cluster tracking system with the aid of Fog computing technologies.

The second section of this chapter explains the theoretical background of the proposed model, and section three sets out the methodology for the proposed design. The result section (section four) consists of the finding of the experiment and the analysis of the results. Section five discusses the advantages and disadvantages of the proposed model and finally, Section six concludes the proposed system.

2 Background

The novel coronavirus is a new strain not previously identified in the human body. Many nations have now left the clusters undetected and this causes the immense problem of spreading the virus. This study aims to develop a COVID-19 cluster tracking system using IoT and Fog computing technologies. To achieve the research objectives, previous studies of GPS acquisition for healthcare applications, IoT frameworks for cloud-based applications, reliable network protocols used in smart cities, and Fog computing technologies have been focused on primarily. This section introduces the theoretical background and previous studies led by researchers in the field of smart health care and IoT.

IoT technology has gained tremendous attention in the field of health care where it plays an important role in the different processes of numerous infectious diseases [20]. In the current pandemic, as the contingency of COVID-19 is strong, there is a profound need to classify and monitor patients in various phases of COVID-19. The Fog computing model [21] offers a great platform for the implementation of sensing applications because it enables the unparalleled capabilities of dedicated sensor networks, smart devices, and IoT devices to monitor the physical environment to converge smoothly [22]. Different IoT technologies are currently involved in preventing the COVID-19 pandemic, particularly in smart cities. Table 3 depicts the specifications of selected IoT technologies regarding the COVID-19 pandemic.

Smartphone apps are application software designed to execute restricted functions on a handheld computer, such as a smartphone. Provided that there are 3.5 billion operating smartphones in 2020, these IoT-based smartphone applications may be very effective in a range of sectors such as health, retail, and agriculture [23, 24]. Specifically, a smartphone is a very convenient and user-friendly device with a range of built-in connectivity and data processing features. With the increasing popularity of smartphones among people, researchers are interested in developing smart IoT solutions using smartphones because of built-in sensors [25, 26]. Moreover, smart devices have built-in sensors that can be used to monitor various parameters, such as

Table 3 IoT-enabled technologies for the prevent COVID-19 pandemic

Technology	Description	Advantages	Disadvantages
Smart device applications	Application software for the acquisition and access of smart sensors on mobile devices	Monitoring and tracking, Cost-effective	Security and privacy of data, Short battery lifetime
Wearable	An app-enabled technology for receiving, processing data by wearing to the body of the consumer	Continuous monitoring of the patient, Improving the standard of the consumer's medicare, Safer and more effective use for hospitals, Lowering hospital visits and the ability to remote access	Security and privacy of data, Short battery lifetime, Wearing may be uncomfortable for the elderly

motion, location, and environmental factors (humidity, light, etc.). Sensing and data processing features of smart devices have improved over the last decade. Depending on the specifications, additional sensors may also be used. Applications may be designed on smartphones that use sensor data to generate useful results. Some of the sensors in a modern smartphone are listed in Table 4.

Usak et al. [28] have published a comprehensive literature analysis of the application of IoT in healthcare systems. This work has included a discussion of the

Table 4 Smartphone sensors used for the development of IoT-based system

Sensors in smartphones	Function of the sensor
Accelerometer [27]	It detects the motion and acceleration of the smartphone. Typically, it measures changes in the velocity of the smartphone in three dimensions
Gyroscope [27]	Detects the orientation of the smartphone very precisely. Orientation is measured using capacitive shifts as the seismic mass travels in a certain direction
Magnetometer [27]	Used to detect the magnetic field. This can be used as an electronic compass and in applications for metal presence detection
GPS [27]	The Global Positioning System (GPS) detects the geological positions of smartphones, and is one of the most important data sensing sources for the smart applications
Light [27]	Detects the intensity of ambient light. It can be used to set the brightness of the screen and other applications where intervention is required based on the intensity of the ambient light
Proximity [27]	The proximity sensors use infrared LEDs that emit IR rays. The rays are bounce back when they hit the object. It will calculate the distance, depending on the time differential. In this way, the distance between the smartphone and the different objects can be measured
Thermometer [27]	Used to measure the temperature

key problems of using IoT to provide health care and the classification of the work discussed in the literature.

Perera et al. [29] have proposed and demonstrated the DAM4GSN architecture for capturing sensor data using sensors built into mobile phones. In addition, they have developed a prototype framework that can be enabled on Android devices as well as an AndroidWrapper as a GSN middleware component. Bayat et al. [30] have suggested a recognition method in which a modern digital low-pass distinguishes the aspect of gravitational acceleration from that of body acceleration in raw data. Their method has been trained and validated in an experiment with several human subjects under real-world conditions. Several classifiers have been evaluated using different statistical features. High-frequency and low-frequency data elements have been taken into consideration. Wang and Kim have suggested a method to perform movement detection using smartphone accelerometers, a role that requires recognizing the physical activities performed by a person. To classify the data, they have used the k-Nearest-Neighbor (KNN) classifiers algorithm [31].

Ashraf et al. [32] have proposed a smart edge monitoring system that is effective in remote control, early warning, and identification of a person's fever, heart rate, body temperature, and some radiological features to identify an infected or suspected person using wearable smart gadgets. Their proposed model helped to identify and monitor the infectious individual. In addition, the patient's data record for analysis and decision-making using modern computing technology has been preserved. Negash et al. [33] have focused on implementing a smart e-health gateway for use in the Fog computing layer, connecting a network of such gateways both at home and in the hospital environment. Moreover, Negash has provided simple health scenarios to demonstrate the application of the services. In addition, the functionality of the gateway in the implementation of Fog has been discussed and evaluated in the study.

A new approach to improve the accuracy of indoor navigation using the learning to perdition model has been proposed by Jamil et al. [34]. Their proposed design has monitored the location of objects in an indoor setting where the global positioning system does not operate properly. In addition, learning to predict model-based artificial neural networks has been used to enhance prediction precision.

With a careful review, it has been discovered that not many potential systems have been developed for the COVID-19 patient tracking and in particular, IoT solutions for cluster identification. The next section of this chapter describes the design and implementation of the proposed COVID-19 cluster tracking system.

3 Proposed Design

The proposed design of this chapter consists of two sections. Section one consists of the development of smartphone applications to detect activities of the user. Since our model collects the location of the user, a significant volume of data was obtained. The activity detection system has classified the activities conducted by the user using the inbuilt accelerometer sensor data. According to the sensor data, the GPS acquisition

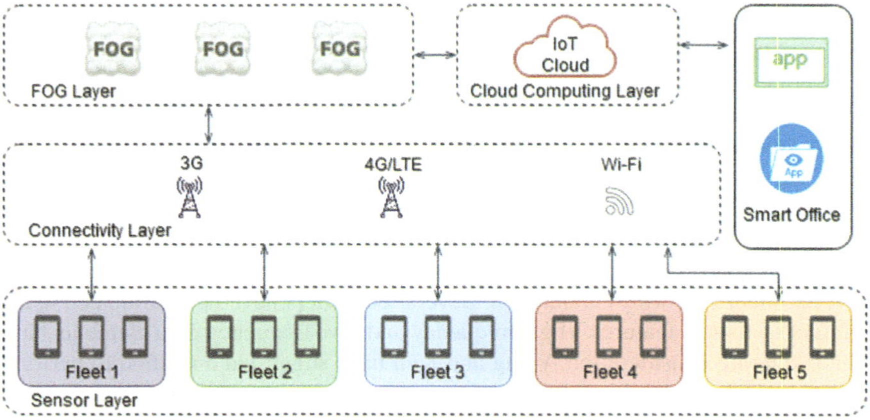

Fig. 2 Proposed IoT architecture of the COVID-19 cluster tracking system

process was distinct from a variety of tasks, such as sitting, walking, resting, stairs up, and stairs down.

The second section describes the development of a COVID-19 cluster tracking system based on the GPS history of the user. The cluster tracking system consists of the client-side (Smartphone Application) and the admin-side (Web-based Application) process to detect possible clusters. A smartphone application was developed to store GPS locations according to the user's activities. It also provides for alerting users whether they identified as a COVID-19 positive, negative, or suspicious. Once the smart office has marked the user as a COVID-19 patient, the GPS history of the past 15 days will be uploaded to the cloud database. The smartphone application frequently searches the cloud database with the local database to see how the user interacted with the prior COVID-19 patients (based on Date/Time and GPS locations). Proposed design is composed with Fog layer, sensor layer, connectivity layer, cloud commuting layer, and administrative actions. Figure 2 illustrates the proposed IoT architecture of the COVID-19 cluster tracking system with Fog layer.

3.1 Development of the Activity Detection System

Human activity recognition is an important and challenging area of research with many uses in health care, smart environments, and security [35]. When the GPS acquisition was carried out at 1 data per second frequency, 1,296,000 records were recorded in the database for 15 days. This is induced by the issue of data handling and location filtering process. Then frequency was set for 1 data per hour, and we found that there have been 21,600 records for the past 15 days. This action also created issues since the different geo locations were not recorded at the same time span. We, therefore, suggested a method for storing data on a local database based

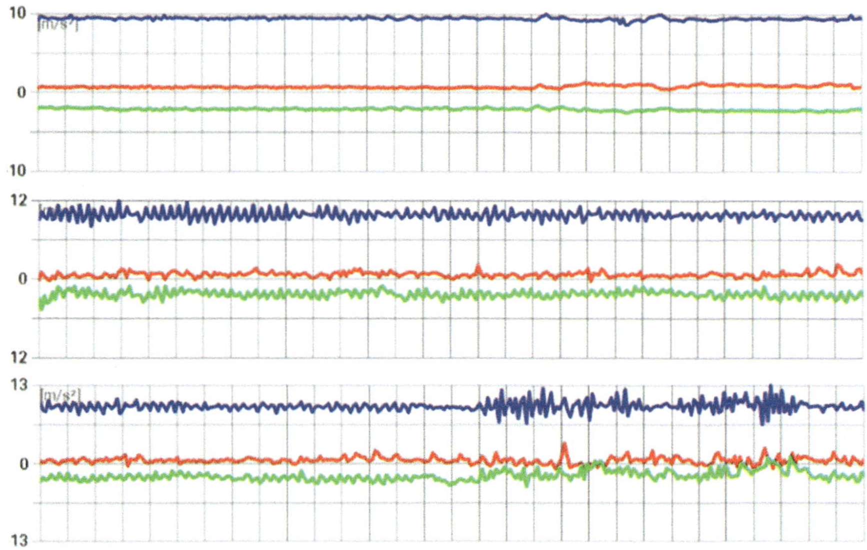

Fig. 3 Accelerometer sensor reading of the smartphone while performing activities, (From Top to Bottom: Resting, Walking, and Running)

on an activity-based approach. The data storing frequency was defined based on the activities performed by the user. After implementing this approach, we have observed that almost 7,000 records have been found in the local database for the past 15 days with nearly 2.0 MB of data. This intervention kept the program running effectively and easy to manipulate data for the database manager.

This section introduces the development of the smartphone application for accelerometer measurements and predicts user activity using an Artificial Neural Network (ANN) technique. To collect data, we used a triaxial accelerometer on the Android smart device to measure acceleration. Data from the accelerometer contains acceleration along the *X*-axis, *Y*-axis, and *Z*-axis. These axes captured the horizontal/sideway orientation (*X*-axis), up/down movement (*Y*-axis), and forward/backward movement (*Z*-axis) of the user. Figure 3 shows the variation in acceleration with the different activities of the user.

Since the proposed design uses built-in mobile accelerometer sensors for reading acceleration (ACC_X, ACC_Y, ACC_Z) and classifying them according to the activities conducted by users is highly required. The MotionSense Dataset that is publicly available in the Queen Mary University of London repository [36] was used to train and test the Artificial Neural Network (ANN) model. The dataset includes time-series data generated by the smartphone's accelerometer and gyroscope sensors. From the original dataset, we extracted accelerometer sensor data for five different activities (walking, downstairs, upstairs, sitting, and resting) for this research.

Artificial Neural Networks (ANNs) are biologically inspired computational networks. Among the different types of ANNs, this chapter focuses on Multilayer

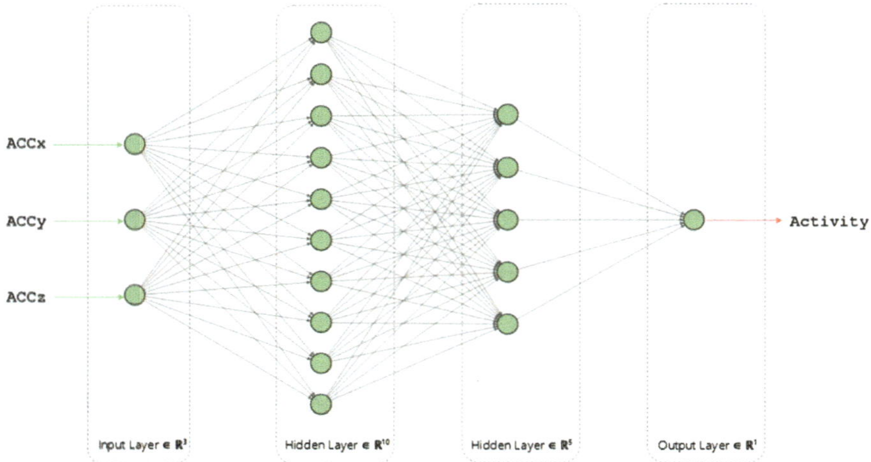

Fig. 4 Architecture of the proposed ANN model of activity prediction process

Perceptron (MLP) with backpropagation learning algorithms. The MLPs are most intensively used for a wide range of problems, depending on the supervised method, and consist of three layers known as input, hidden, and output. The proposed ANN architecture was developed with 3 nodes (ACC_X, ACC_Y, and ACC_Z) for input layer, single node for output layer (class ID), 10 nodes for first hidden layer, and 5 nodes for second hidden layer. Figure 4 illustrates the architecture of the proposed ANN model for predicting the user activities based on the accelerometer reading.

The hyperbolic tangent sigmoid transfer function (Tansig function) is widely used as the activation function for the hidden layer of a neural network. The activation function is a mathematical representation that helps to determine the output of a neural network. The activation function is associated with each neuron and determines the state of activation, whether the neuron should fire or not. This activation status is based on the neuron's input state relevant to the prediction of the model. The S-shaped anti-symmetric function was used for input to output transformation [37]. This transformation allows the output of each neuron to stay between -1 and $+1$ in the interval. The hyperbolic tangent sigmoid function is defined by Eq. (1). We used the Tansig activation function for activating the neurons.

$$a = Tansig(n) = \frac{2}{1 + e^{-2n}} - 1 \tag{1}$$

Figure 5 displays the program flow chart of the user activity detection system and the GPS acquisition system. When the program begins, the GPS and accelerometer sensors have been initialized to read the data from the sensors. Initially, the GPS acquisition sub-system checked the availability of the 3G/4G access. Then, accelerometer data (ACC_X, ACC_Y, *and* ACC_Z) were acquired to predict user

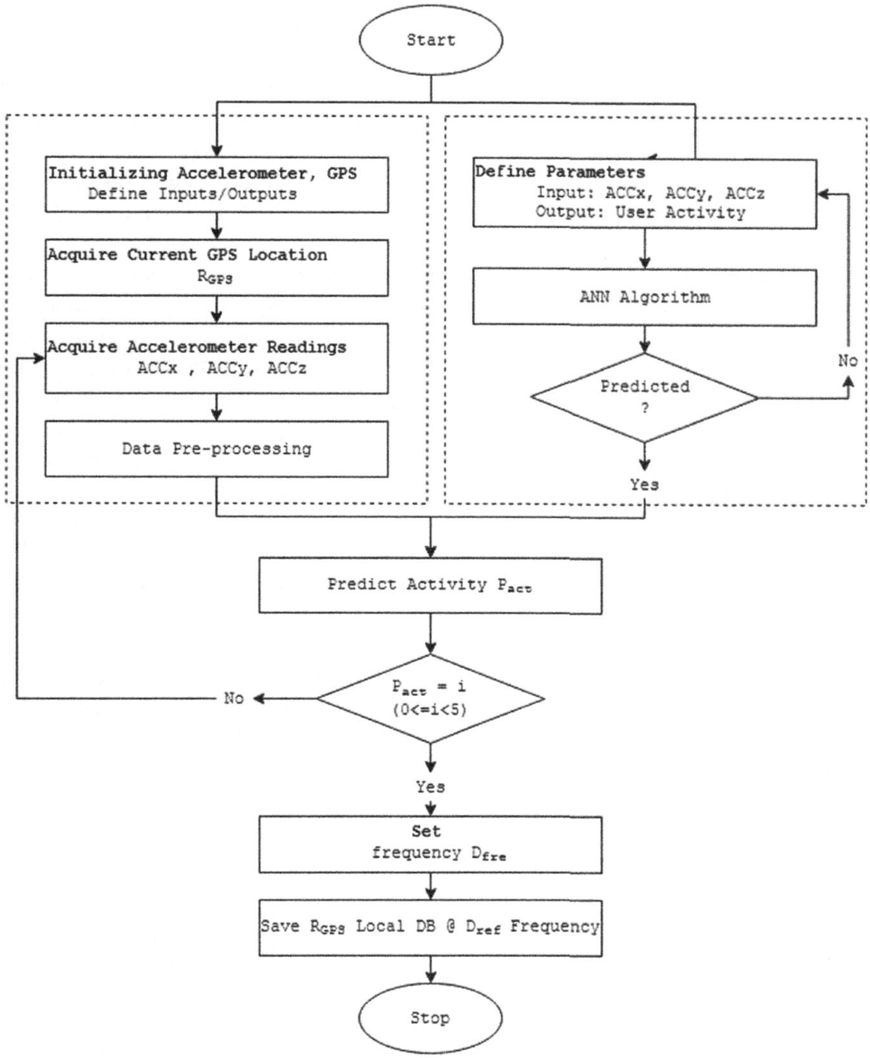

Fig. 5 Program flow chart of the user activity recognition and GPS acquisition system

behaviors. Preprocessing data was carried out to eliminate noise from the data collected.

The Activity Prediction (P_{act}) process was used to forecast activities using the ANN algorithm. The system then checked the P_{act} ($0 \leq P_{act} < 5$) for each activity in 0 to 4 indices (0-Sitting, 1-Walking, 2-Resting, 3-Stairs up, 4-Stairs down). The frequency D_{fre} was generated in accordance with the variation of P_{act}. Finally system was stored the GPS Locations (R_{GPS}) on the local database at the D_{fre} frequency.

3.2 Development of the Cluster Tracking System

This section introduces the development of the COVID-19 patient monitoring and cluster tracking system. This research used 31 test subjects throughout 12 different locations (Table 5) in the Western Province of Sri Lanka. Figure 6 shows the experimental scenario of a potential COVID-19 cluster. The cluster scenario is based on past events involving COVID-19 in the Western Province of Sri Lanka. As seen in the diagram, *P0* was the patient zero of the cluster. The sequence of the cluster was based on the time they interacted with each other and the data was sorted by time.

After *P0* was labeled as a COVID-19 patient, the test subjects (App users) at Location ID 1 (*6°58′08.6"N 79°57′15.3"E*) were the initial suspects of COVID-19, therefore it is important to conduct a PCR test to establish the critical pathways of the cluster. In this case, we assumed that users *P0*, *P1*, *P2*, and *P3* were tested positive for COVID-19, and *P4* tested negative for COVID-19. Based on this information, the system determines the critical paths of the cluster and continues to draw a critical path based on historical GPS data.

The processing of GPS data from the GPS module on a smartphone offers the ability to analyze the travel activities of smartphone owners. GPS data typically contains basic information, such as longitude, latitude, and time stamp, while additional information, including speed, acceleration, and signal quality, is often available [38]. Figure 7 illustrates the distance calculation methodology of two different GPS locations on earth. The Haversine formula [39] was used to determine the great-circle distance between two points of geographical locations using their longitudes and latitudes. For the development of the distance calculation algorithm (Algorithm 1), Eqs. (1–5) was used.

$$\Delta\varphi = (\varphi_i - \varphi_0) \times \frac{\pi}{180} \tag{2}$$

Table 5 GPS coordinates of the locations used for the experiment

Location ID	GPS coordinate
1	6°58′08.6"N 79°57′15.3"E
2	6°58′08.0"N 79°57′17.8"E
3	6°58′08.1"N 79°57′23.2"E
4	6°58′07.2"N 79°57′31.0"E
5	6°58′02.5"N 79°57′40.3"E
6	6°58′04.0"N 79°58′03.1"E
7	6°58′14.5"N 79°58′19.3"E
8	7°00′25.7"N 79°57′36.5"E
9	7°00′07.4"N 79°57′03.5"E
10	6°58′39.8"N 79°55′35.9"E
11	6°52′57.6"N 79°53′13.3"E
12	6°52′51.7"N 79°53′24.5"E

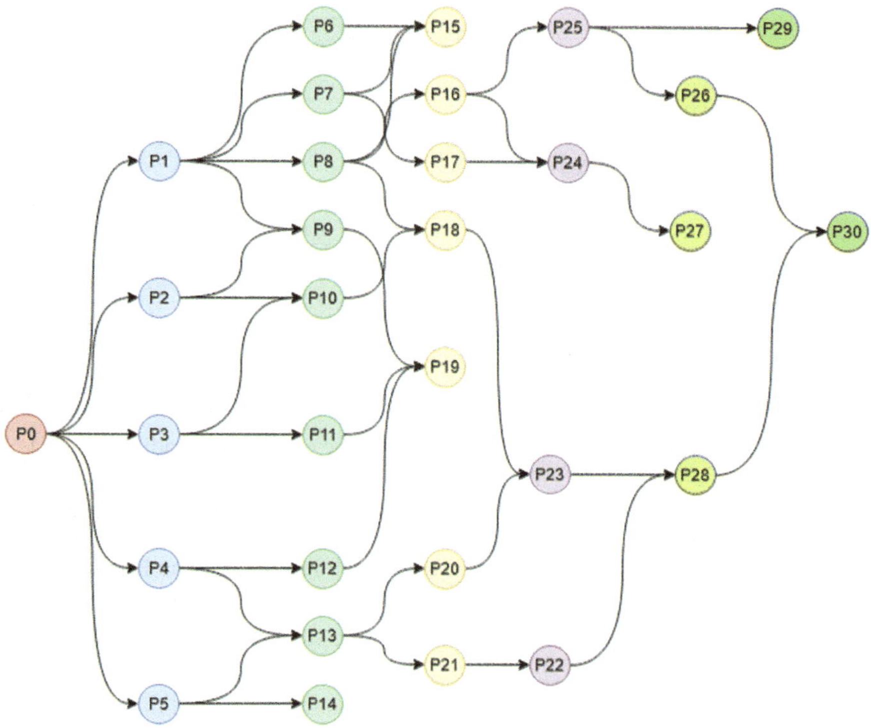

Fig. 6 An experimental scenario of a possible COVID-19 cluster

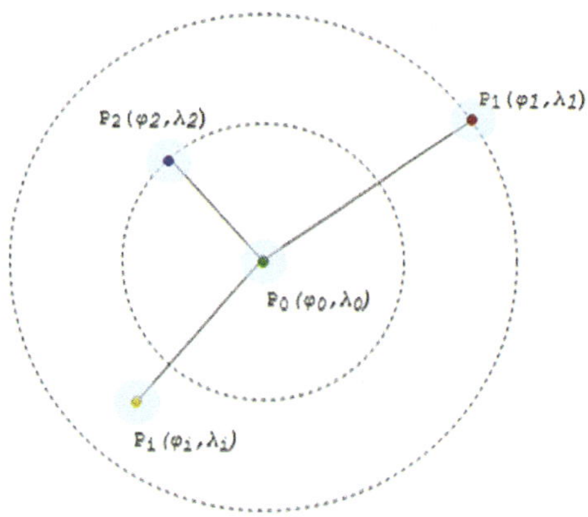

Fig. 7 Distance calculations between two different locations from the GPS coordination

$$\Delta\lambda = (\lambda_i - \lambda_0) \times \frac{\pi}{180} \quad (3)$$

Using the Haversine formula,

$$a_i = \sin^2\left(\frac{\Delta\varphi}{2}\right) + \left(\cos\varphi_0 \times \cos\varphi_i \times \sin^2\left(\frac{\Delta\lambda}{2}\right)\right) \quad (4)$$

$$C_i = 2 \times \operatorname{atan2}\left(\sqrt{a_i}, \sqrt{1 - a_i}\right) \quad (5)$$

$$d_i = R \times C_i \quad (6)$$

where

φ is latitude, λ is longitude, and R is radius of the earth (mean radius $= 6{,}371 \times 10^3$ m).

Table 5 depicts the GPS data of the twelve different locations in the Western Province of Sri Lanka. The GPS coordinates of the experiment locations were extracted using the Google Map Android app which installed on the Huawei MediaPad T1 7.0. Once the GPS coordinates were extracted, they were indexed according to the time we interacted with the location.

Algorithm 1 GPS Distance Calculation

procedure DISTANCE-CALC
    $\varphi_0 \leftarrow$ initial point latitude
    $\lambda_0 \leftarrow$ initial point longitude
    $R \leftarrow 6.371$
top:
    if φ_i, λ_i = *exist*, **then return** true
    $\Delta\varphi_i \leftarrow \varphi_i - \varphi_0$
    $\Delta\lambda_i \leftarrow \lambda_i - \lambda_0$
    $a_i \leftarrow f(\Delta\varphi_i, \Delta\lambda_i, \varphi_0, \lambda_0)$
    $C_i \leftarrow f(a_i)$
    $d_i \leftarrow f(C_i, R)$
loop:
    $\sum_{i=1}^{N} d_i$
    if $d_i <= 15$ **then**
        location $\leftarrow (\varphi_i, \lambda_i)$
        goto *loop*.
        close;
    $i = i+1$
    goto *top*.

The program flowchart of the cluster tracking system is shown in Fig. 8. When the program begins, the Android application starts reading both local and cloud databases. As dynamic variables, the time of the user carrying out an activity on a location (U_{time}), the distance between a contact individual ($U_{distance}$), and the GPS

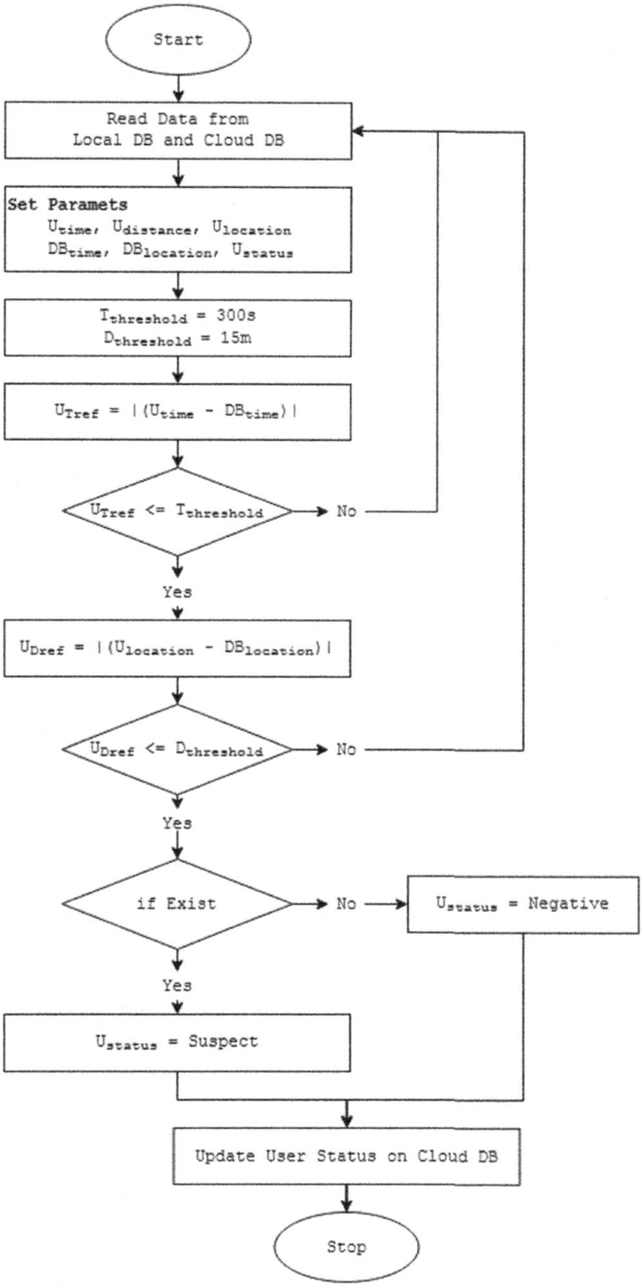

Fig. 8 Program flow chart of the cluster tracking system

coordinate of the user location ($U_{location}$) were initialized at the beginning. The threshold value for time ($T_{threshold} = 300$ s) and distance ($D_{threshold} = 15$ m) was initialized as the static variable of the program. Then several mathematical steps were taken to find possible suspects based on the date/time and distance. After data acquired by the system, reference values for time (U_{Tref}) and distance (U_{Dref}) were frequently matched with the thresholds ($T_{threshold}$, $D_{threshold}$). Once the algorithm verified all the criteria, the user was identified as a COVID-19 suspect. In addition, the system uploaded user datasets (user information, location, COVID-19 contacted individuals) to the cloud database.

4 Results and Analysis

This section describes the results and analysis of the activity detection system and the COVID-19 cluster tracking system. The activity detection system was conducted with 10 test subjects and the COVID-19 cluster tracking system was validated with 31 test subjects.

4.1 Results of the Activity Detection System

We enlisted the assistance of 10 volunteer subjects to hold a smartphone when carrying out five sets of activities. No noise cancellation was carried out with respect to the results. To prevent mislabeling, subjects were asked to stop and wait a few seconds before starting the next activity. It was essential to have subjects carry an Android-based smartphone app when conducting designated activities in order to test the model.

For better prediction, the ANN model needed a lower mean square error. The best training performance 0.20955 was observed at epoch 100, as seen in Fig. 9.

Figures 10 and 11 show the developed Android program for the user activity detection system. Once the system is started, this program must run as a background operation and serve as a prerequisite for the GPS acquisition system. As shown in the figure, different acceleration on *X*-, *Y*-, and *Z*-direction was achieved to predict user activities using the ANN activity prediction algorithm.

A confusion matrix, also known as an error matrix, is widely used in the area of machine learning and especially the issue of statistical classification [40]. A confusion matrix is a table often used to describe the performance of a classification model or classifier on a set of test data for which the true values are known. Confusion matrix was used to validate the performance of our algorithm.

The entries in the confusion metrics were calculated from the coincidence matrix by using the following hypothesis,

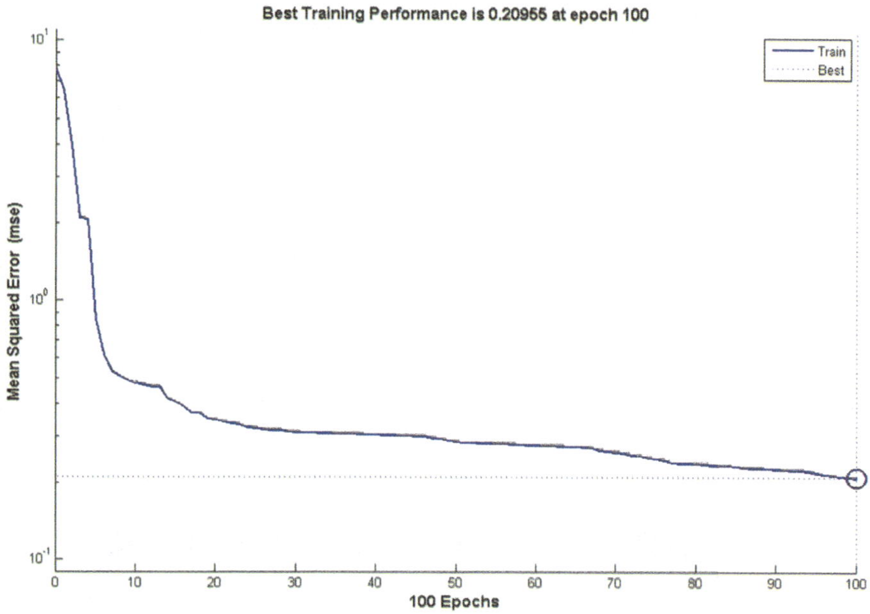

Fig. 9 Training performance of the proposed ANN model

Fig. 10 User activity recognition Android application for walking, stairs down, and stairs up activities

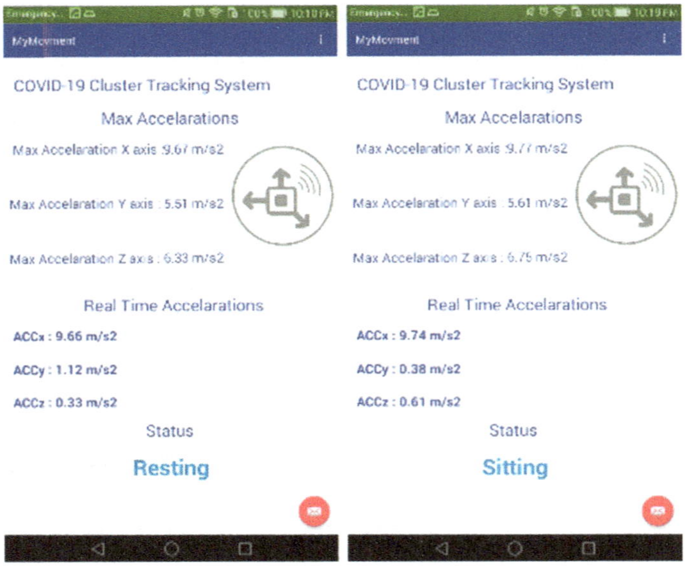

Fig. 11 User activity recognition Android application for resting, and sitting activities

- True Negative (TN) is the number of correct predictions that an instance is negative.
- True Positive (TP) is the number of correct predictions that an instance is positive.
- False Positive (FP) is the number of incorrect predictions that an instance is positive.
- False Negative (FN) is the number of incorrect predictions that an instance is negative.

Equations 7 and 8 were used to calculate the accuracy and precision of the test results based on the data extracted from the confusion matrix.

$$Accuracy = \frac{\sum(TP + TN)}{\sum(TP + FP + FN + TN)} \tag{7}$$

$$Precision = \frac{\sum(TP)}{\sum(TP + FP)} \tag{8}$$

Figure 12 illustrates the confusion matrix of the proposed activity detection system. The experiment was conducted for five activity classes with 500 test trials for ten test subjects. All the subjects were performed 10 test trials from each activity (Sitting, Walking, Resting, Stairs up, and Stairs down).

Overall accuracy of the activity recognition system was 84.8% observed at 0.83, 0.76, 0.88, 0.89, and 0.91 precision for each activity class. Table 6 depicts the accuracy

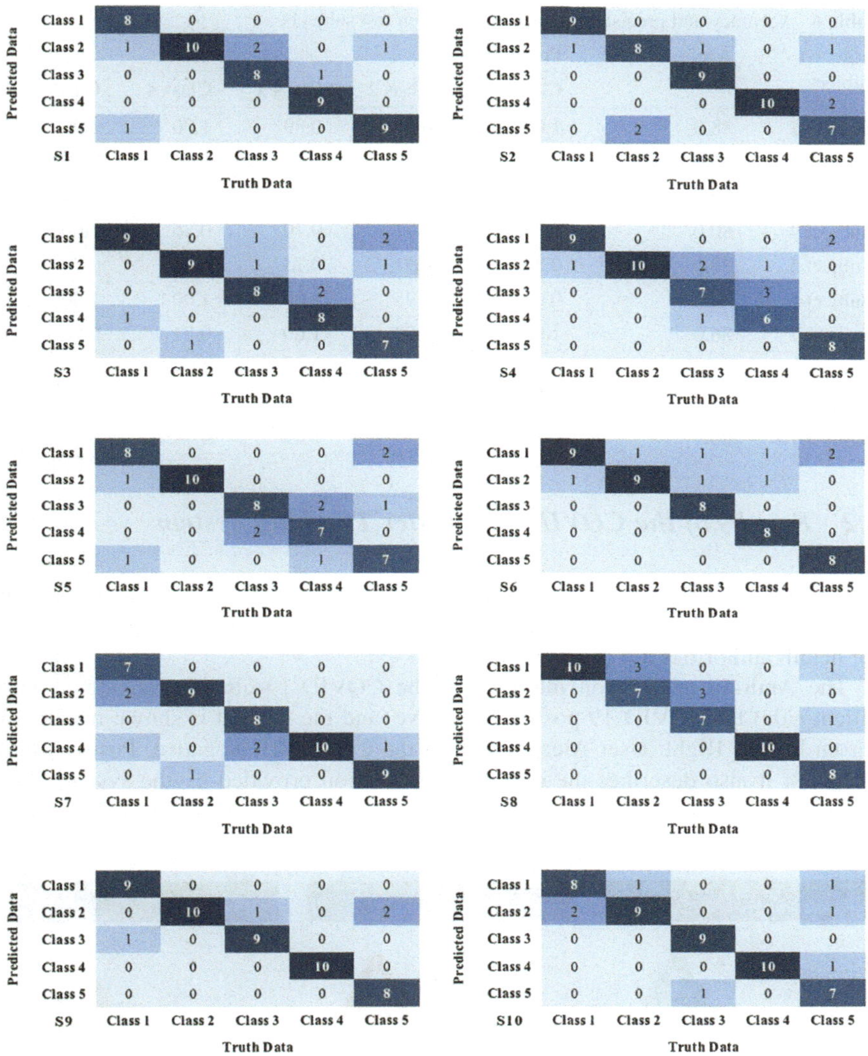

Fig. 12 Confusion matrix of the experimental results on 10 test subjects

and precision of each activity of 10 test subjects. We have observed that 85.4% overall precision was achieved by the proposed activity detection system.

Table 6 Accuracy and precision of each class for ten test subjects

Test subject	Accuracy (%)	Precision				
		Class 1	Class 2	Class 3	Class 4	Class 5
Subject 1	88.0	1.00	0.71	0.89	1.00	0.90
Subject 2	86.0	1.00	0.73	1.00	0.83	0.78
Subject 3	82.0	0.75	0.82	0.80	0.89	0.88
Subject 4	80.0	0.82	0.71	0.70	0.86	1.00
Subject 5	80.0	0.80	0.91	0.73	0.78	0.78
Subject 6	84.0	0.64	0.75	1.00	1.00	1.00
Subject 7	86.0	1.00	0.82	1.00	0.71	0.90
Subject 8	84.0	0.71	0.70	0.88	1.00	1.00
Subject 9	92.0	1.00	0.77	0.90	1.00	1.00
Subject 10	86.0	0.80	0.75	1.00	0.91	0.88

4.2 Results of the COVID-19 Cluster Tracking System

This section describes the results of the COVID-19 cluster tracking system for both the client-side and admin side. Health alerts for mobile users and patient's information for health authorities are described here.

The Android application interface of the COVID-19 cluster tracking system (client-side) for COVID-19 positive, negative, and the suspect is shown in Fig. 13 (From Left to Right: User Alert Screen of the COVID-19 Negative, Positive, and Suspect). It also describes the different information provided by the system after the user's status has been detected. Figure 14 displays the GPS location history of

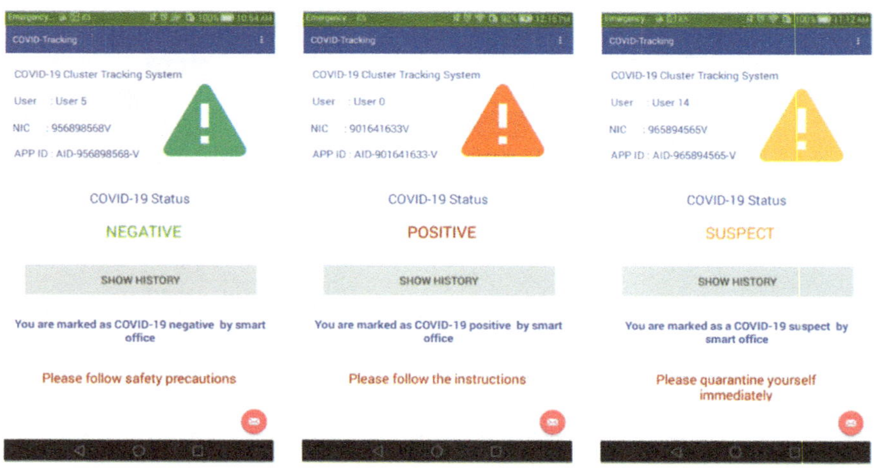

Fig. 13 Android smartphone cluster tracking system application (Client-side)

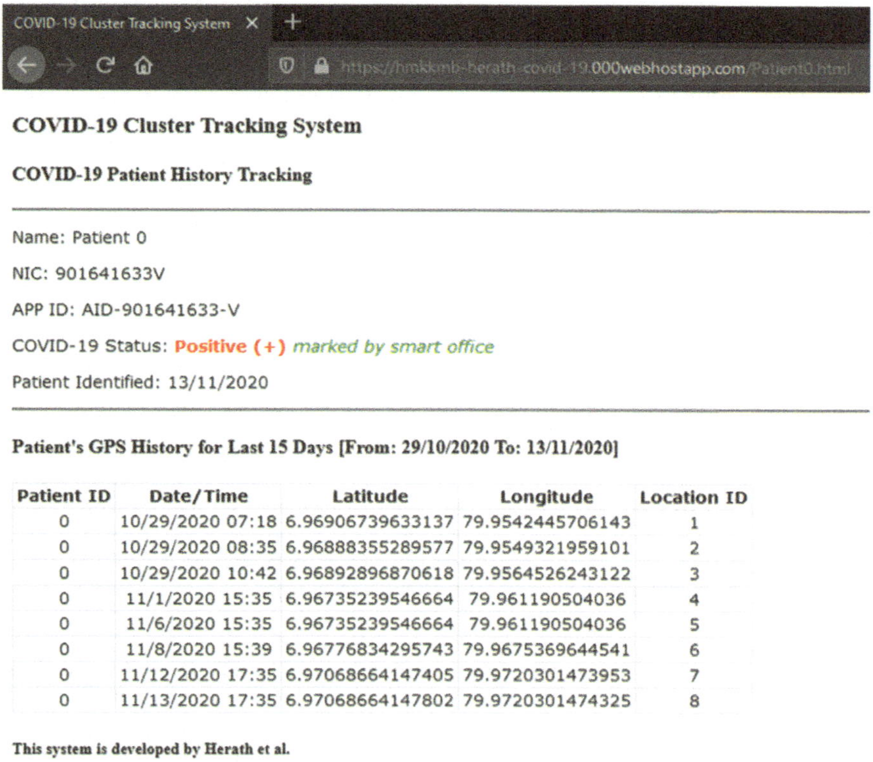

Patient ID	Date/Time	Latitude	Longitude	Location ID
0	10/29/2020 07:18	6.96906739633137	79.9542445706143	1
0	10/29/2020 08:35	6.96888355289577	79.9549321959101	2
0	10/29/2020 10:42	6.96892896870618	79.9564526243122	3
0	11/1/2020 15:35	6.96735239546664	79.961190504036	4
0	11/6/2020 15:35	6.96735239546664	79.961190504036	5
0	11/8/2020 15:39	6.96776834295743	79.9675369644541	6
0	11/12/2020 17:35	6.97068664147405	79.9720301473953	7
0	11/13/2020 17:35	6.97068664147802	79.9720301474325	8

Fig. 14 GPS history of the COVID-19 patient zero *(P0)* of the cluster from admin side of the system

patient zero (*P0*) over the last 15 days. All the unique locations that the user has interacted with were tagged by the location ID. In this case, Patient zero *(P0)* was classified as a positive COVID-19 patient in clinical trials. The system has processed all the unique locations in which the patient has dealt for the past 15 days. As shown in the image, the program extracted eight separate coordinates from the local database. In addition, the data in the data table was sorted by Date and Time.

The latitude and longitude of the locations were stored for the measurement of distances between individuals. All users who interacted with or suspected as COVID-19 patients have a background profile as shown in the figure.

Figure 15 shows the COVID-19 cluster tracking system from the admin side. This web application was developed to identify patients with COVID-19 and to illustrate the cluster using JavaScript (JS) language. As shown in the figure, the cluster was plotted based on the location ID and patient ID attribute of the DataTable. After generating the cluster, health authorities have ability to track them according to the user details such as name and NIC (ID number).

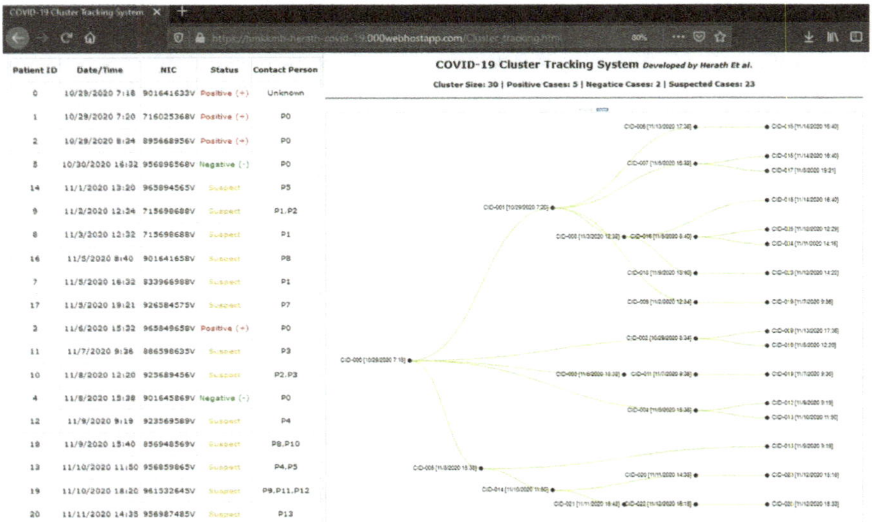

Fig. 15 COVID-19 cluster tracking system from admin-side (Left: Positive, Negative, and Suspects of COVID-19, Right: Simulation of the possible COVID-19 cluster)

The cluster nodes were indexed by patient ID and Date/Time of the user who initially interacted with a patient or a suspect. Since *P0*, *P1*, *P2*, and *P3* users are deemed to be COVID-19 positive instances; the admin would indicate their profiles as COVID-19 positive users. Once they became negative, the admin was updated their status as negative. The "Contact Person" attribute on the DataTable indicates the COVID-19 person or suspect who interacted with the smartphone App (Fig. 15) user. All data on the DataTable was sorted according to a Date/Time-based approach.

5 Discussions

COVID-19 is a highly contagious disease and thus immediate prevention action must be taken. It is important to the whole world to have a new innovative approach to coping with the COVID situation when heading into a period of technology and commerce. The technology liaison with Fog computing would help in monitoring the outbreak of the virus as well as in identifying clusters at a limited cost and resources. It has evolved as a technology which is used worldwide, in-tended to enable the implementation of the Internet of things. The proposed integrated design of IoT and Fog computing based COVID-19 cluster tracking system development could be applied successfully to identify and monitor the outbreak of COVID-19 virus cluster basis in different regions, different urban cities all over the world.

The use of conventional methods of emergency response in pandemics of rapid outbreaks, such as COVID-19, will not be appropriate for several potential reasons. Such management procedures can entail the use of tremendous valuable human resources in the health sector and also the security sector. During Pandemics as such, it is likely that it will lead to the risk of many lives. The use of the proposed system is very supportive to eliminate such contamination and also to minimize human interactions. There is no argument that conventional emergency aid systems need a high degree of financial assistance. Programs involving greater human capital entail extra costs and investments such as travel and sanitation. It is evident that such costs can be reduced to a larger extent by using a system, which is an integration of IoT and Fog computing not even to identify but also to map and control clusters of COVID-19 distribution.

One downside of the project can be mentioned as a concern of data vulnerability and misuse. It is necessary to use a lot of user's sensitive data such as personal information and location history in order to succeed the program, although it has a social and ethical impact on data privacy. By understanding the emergency condition of the pandemic, users can always contribute to the system by providing proper information. When exchanging their true information with the authorities, it is possible to provide the individual with a confidentiality arrangement directly with the government. The security dimensions should be solidified when looking at internal as well as external risks and it is essential to follow prevention measures. The implementation of encrypted access can be suggested which must be pre-measured against a range of possible security threats.

The combination of AI and Fog computing has inspired to propose a contaminated identification and cluster tracking system that can be used to handle large data and smartly predict a virus transmission and identify most possible clusters without delay. It is suitable for applications where real-time, fast response times, and low latency are required as in this COVID-19 Pandemic situation.

6 Conclusion

Since the end of 2019, the world has been faced with the threat of the COVID-19 virus emerging in Wuhan, China. Precautionary steps are likely to be required in the world to tackle the COVID-19 pandemic before an appropriate vaccine is developed. Many nations have now left the clusters undetected, and this is causing a major problem with the spreading of the virus. This chapter proposed an IoT and Fog computing-based system to mitigate the impact of COVID-19 diseases by recognition of the COVID-19 clusters. The proposed system is composed of activity detection and COVID-19 cluster recognition system. The overall accuracy of the proposed activity detection system was 85.4% observed. The cluster recognition system was tested and validated the results for 31 test subjects. The proposed system showed promising results in mitigate the impact of COVID-19 in the urban cities.

References

1. Herath HMKKMB (2021) Internet of Things (IoT) enable designs for identify and control the COVID-19 pandemic. In: Artificial Intelligence for COVID-19. Springer, Cham, pp 423–436
2. Herath HMKKMB, Karunasena GMKB, Herath HMWT (2021) Development of an IoT based systems to mitigate the impact of COVID-19 pandemic in smart cities. In: Machine intelligence and data analytics for sustainable future smart cities. Springer, Cham, pp 287–309
3. Mohammed MN, Syamsudin H, Halim S, Ramli R, Yusuf E (2020) Novel Covid-19 detection and diagnosis system using IoT based smart helmet. Am J Polit Sci 24(7)
4. Jiang F, Deng L, Zhang L, Cai Y, Cheung CW, Xia Z (2020) Review of the clinical characteristics of coronavirus disease 2019 (COVID-19). J Gen Intern Med 35(5):1545–1549
5. Vafea MT, Atalla E, Georgakas J, Shehadeh F, Mylona EK, Kalligeros M, Mylonakis E (2020) Emerging technologies for use in the study, diagnosis, and treatment of patients with COVID-19. Cell Mol Bioeng 13(4):249–257
6. Wong SH, Lui RN, Sung JJ (2020) Covid-19 and the digestive system. J Gastroenterol Hepatol 35(5):744–748
7. Chen H, Guo J, Wang C, Luo F, Yu X, Zhang W, Li J, Zhao D, Xu D, Gong Q, Liao J, Yang H, Hou W, Zhang Y (2020) Clinical characteristics and intrauterine vertical transmission potential of COVID-19 infection in nine pregnant women: a retrospective review of medical records. Lancet 395(10226):809–815
8. Chen N, Zhou M, Dong X, Qu J, Gong F, Han Y, Qiu Y, Wang J, Liu Y, Wei Y, Xia J, Yu T, Zhang X, Zhang L (2020) Epidemiological and clinical characteristics of 99 cases of 2019 novel coronavirus pneumonia in Wuhan, China: a descriptive study. The Lancet 395(10223):507–513
9. Salehi S, Abedi A, Balakrishnan S, Gholamrezanezhad A (2020) Coronavirus disease 2019 (COVID-19): a systematic review of imaging findings in 919 patients. Am J Roentgenol 215(1):87–93
10. Angurala M, Bala M, Bamber SS, Kaur R, Singh P (2020) An internet of things assisted drone based approach to reduce rapid spread of COVID-19. J Saf Sci Resil 1(1):31–35
11. Talal M, Zaidan AA, Zaidan BB, Albahri AS, Alamoodi AH, Albahri OS, Alsalem MA, Lim CK, Tan KL, Shir WL, Mohammed KI (2019) Smart home-based IoT for real-time and secure remote health monitoring of triage and priority system using body sensors: multi-driven systematic review. J Med Syst 43(3)
12. Siriwardhana Y, Gür G, Ylianttila M, Liyanage M (2020) The role of 5G for digital healthcare against COVID-19 pandemic: opportunities and challenges. ICT Express
13. Albahri AS, Hamid RA, Alwan JK, Al-Qays ZT, Zaidan AA, Zaidan BB, Albahri AO, AlAmoodi AH, Khlaf JM, Almahdi EM, Thabet E (2020) Role of biological data mining and machine learning techniques in detecting and diagnosing the novel coronavirus (COVID-19): a systematic review. J Med Syst 44(7)
14. Kaur H, Alam MA, Jameel R, Mourya AK, Chang V (2018) A proposed solution and future direction for blockchain-based heterogeneous medicare data in cloud environment. J Med Syst 42(8)
15. Chiang M, Zhang T (2016) Fog and IoT: an overview of research opportunities. IEEE Internet Things J 3(6):854–864
16. Zhao K, Ge L (2013) A survey on the internet of things security. In: 2013 ninth international conference on computational intelligence and security
17. Sheng Z, Yang S, Yu Y, Vasilakos A, Mccann J, Leung K (2013) A survey on the IETF protocol suite for the internet of things: standards, challenges, and opportunities. IEEE Wirel Commun 20(6):91–98
18. Vasseur JP, Dunkels A (2008) IP for smart objects. White Paper, pp 1–6
19. Rajeesh NVK, Arun M, Baraneetharan E, Stanly J, Kanchana A, Prabu S (2020) Detection and monitoring of the asymptotic COVID-19 patients using IoT devices and sensors. Int J Pervasive Comput Commun
20. Christaki E (2015) New technologies in predicting, preventing and controlling emerging infectious diseases. Virulence 6(6):558–565

21. Menchaca-Mendez R, Luna-Nuñez B, Menchaca-Mendez R, Yee-Rendon A, Quintero R, Favela J (2018) Opportunistic mobile sensing in the fog. Wirel Commun Mob Comput 2018:1–18
22. Lee EA, Rabaey J, Hartmann B, Kubiatowicz J, Pister K, Sangiovanni-Vincentelli A, Seshia SA, Wawrzynek J, Wessel D, Rosing TS, Blaauw D, Dutta P, Fu K, Guestrin C, Taskar B, Jafari R, Jones D, Kumar V, Mangharam R, Pappas GJ, Murray RM, Rowe A (2014) The swarm at the edge of the cloud. IEEE Design Test 31(3):8–20
23. Kamaludeen NBA, Lee SP, Parizi RM (2019) Guideline-based approach for iot home application development. In: 2019 international conference on internet of things (iThings) and IEEE green computing and communications (GreenCom) and IEEE cyber, physical and social computing (CPSCom) and IEEE smart data (SmartData)
24. Parizi RM, Guo L, Bian Y, Dehghantanha A, Azmoodeh A, Choo KKR (2018) CyberPDF. In: Proceedings of the 2018 IEEE/ACM international conference on connected health: applications, systems and engineering technologies
25. Lane N, Miluzzo E, Lu H, Peebles D, Choudhury T, Campbell A (2010) A survey of mobile phone sensing. IEEE Commun Mag 48(9):140–150
26. Khan WZ, Xiang Y, Aalsalem MY, Arshad Q (2013) Mobile phone sensing systems: a survey. IEEE Commun Surv Tutor 15(1):402–427
27. Sethi P, Sarangi SR (2017) Internet of things: architectures, protocols, and applications. J Electr Comput Eng 2017:1–25
28. Usak M, Kubiatko M, Shabbir MS, Dudnik OV, Jermsittiparsert K, Rajabion L (2019) Health care service delivery based on the internet of things: a systematic and comprehensive study. Int J Commun Syst 33(2)
29. Perera C, Zaslavsky A, Christen P, Salehi A, Georgakopoulos D (2012) Capturing sensor data from mobile phones using global sensor network middleware. In: 2012 IEEE 23rd international symposium on personal, indoor and mobile radio communications—(PIMRC)
30. Bayat A, Pomplun M, Tran DA (2014) A study on human activity recognition using accelerometer data from smartphones. Procedia Comput Sci 34:450–457
31. Wang X, Kim H (2015) Detecting user activities with the accelerometer on android smartphones 2(2):234–239
32. Ashraf MU, Hannan A, Cheema SM, Ali Z, Jambi KM, Alofi A (2020) Detection and tracking contagion using iot-edge technologies: confronting COVID-19 pandemic. In: 2020 international conference on electrical, communication, and computer engineering (ICECCE)
33. Negash B, Gia TN, Anzanpour A, Azimi I, Jiang M, Westerlund T, Rahmani AM, Liljeberg P, Tenhunen H (2017) Leveraging fog computing for healthcare IoT. In: Fog Computing in the Internet of Things, pp 145–169
34. Jamil F, Kim DH (2019) Improving accuracy of the alpha-beta filter algorithm using an ANN-based learning mechanism in indoor navigation system. Sensors 19(18):3946
35. Khan AM, Lee YK, Lee SY, Kim TS (2010) A Triaxial accelerometer-based physical-activity recognition via augmented-signal features and a hierarchical recognizer. IEEE Trans Inf Technol Biomed 14(5):1166–1172
36. Malekzadeh M, Clegg RG, Cavallaro A, Haddadi H (2019) Mobile sensor data anonymization. In: Proceedings of the international conference on internet of things design and implementation
37. Marquardt DW (1963) An algorithm for least-squares estimation of non-linear parameters. J Soc Ind Appl Math 11:431–441
38. Gong L, Yamamoto T, Morikawa T (2018) Identification of activity stop locations in GPS trajectories by DBSCAN-TE method combined with support vector machines. Transp Res Procedia 32:146–154
39. Anisya A, Swara GY (2017) Implementation of haversine formula and best first search method in searching of tsunami evacuation route. In: IOP conference series: earth and environmental science, vol 97, p 012004
40. Susmaga R (2004) Confusion matrix visualization. In: Intelligent information processing and web mining, pp 107–116

Chapter 8
Reinforcement Learning Model for Pandemic Precautions in Healthcare Environment

Ramesh Sekaran, Arun Prasath Raveendran, Ashok kumar Munnangi, and Manikandan Ramachandran

Abstract Coronavirus (COVID-19) widespread has produced a potential demand for the need for healthcare equipment and medicines as well as applications with advanced information technologies. The traditional ways of healthcare management have collapsed under these exigent circumstances. Without a question, health care offers an upper hand on conventional analytics and clinical decision-making strategies. Many scholars around the world have dedicated intense efforts to the advancement of computational intelligence methods due to rapid and widespread COVID-19. From early alerts, prompt emergency responses, and crucial decision-making, for monitoring drones, the learning model will help to fight against this deadly virus. To decide against COVID-19 based on computational intelligence, this study investigates how to fight COVID-19 and protect the public in a pandemic crisis. With proper surveillance systems, the day-wise update of the infected patients, age-wise, state-wise, and area-wise are helpful. This work discusses data that can be analyzed based on the Hybrid Ensemble-based Reinforcement Learning (HERL) algorithm which provides better prediction and detection of COVID-19.

Keywords Reinforcement learning · Computational intelligence · COVID-19 · Pandemics · Health care

1 Introduction

In the history of the world, there exist numerous outbreaks due to diseases that attacked humanity. World Health Organization (WHO), clinicians, and several national specialists worldwide fight against these pandemics till today. Initially, the disease caused by Coronavirus [1] was observed in December 2019, at Wuhan

R. Sekaran · A. P. Raveendran · A. Munnangi
Department of Information Technology, Velagapudi Ramakrishna Siddhartha Engineering College (Autonomous), Vijayawada, Andhra Pradesh, India

M. Ramachandran (✉)
School of Computing, SASTRA Deemed University, Thanjavur, India

U. Kose et al. (eds.), *Computational Intelligence for COVID-19 and Future Pandemics*, Disruptive Technologies and Digital Transformations for Society 5.0,
https://doi.org/10.1007/978-981-16-3783-4_8

District of China, which continuously and rapidly spread across several nations. WHO declared the pandemic internationally considering public health on January 30, 2020 [2]. Disease due to Coronavirus (SARS-CoV-2) [3] spread over 185 nations infecting about 7,145,800 people from which 407,067 deaths were incurred as of June 9, 2020 [4, 5]. To deal with this worldwide pandemic, WHO, medicinal experts, and scientists are trying to develop a novel approach to identify the infected individuals in different stages and minimize the spread of coronavirus, and to find out a vaccine to provide treatment for patients. Moreover, the persons contacted by these infected patients have to be traced out.

Pandemic due to COVID-19 has caused enormous difficulties in the life of humans. Till now, these exist no immunization to secure humans from this dreadful destructing disease. In this current situation, increase and rapid testing of individuals and recognizing the infected individuals can control the spread of this threatening virus. But, testing each and every person is expensive which has a direct effect on resources but not on the economy. Other than testing, another method to control the spread of this infection is to sanitize which kills this virus [6]. But, this method uses more resources and is thus expensive and it is impossible to offer sanitization for every individual. Along with the two methods discussed above, social distancing has to be maintained by the public. Further, lockdown can be followed, and quarantine centers can be established which are expensive too.

The government helps people and cares for them by providing treatment for people infected mildly and ventilators for those in critical situations. As there is a rapid spread, free communication is not permitted. Apart from the three methods discussed above, human civilization has to be protected and routine life has to be continued with no or little effect based on the life, resources, and economic conditions.

Lockdown has been imposed by Government so that the spread of this dreadful virus can be prevented. This method includes a few controlling mechanisms like social distancing as well as self-isolation. These aspects restrict the contact of individuals but affect their day-to-day lives [7, 8]. Moreover, the transmission of this disease among individuals is arrested when lockdown is strictly followed [9]. Lockdown restricts people to move unnecessarily and is permitted only at the time of emergency. Methods used to prevent the transmission of COVID-19 have a significant cause worldwide in the economic, geopolitical, and health consequences [10]. This lockdown has helped the academicians and researchers to concentrate on their higher studies. They have published numerous articles while COVID-19 was at its peak [11]. The rest of the chapters of this paper are organized into related works in Sect. 2, system model in Sect. 3, results and discussion in Sect. 4, and at last concludes in Sect. 5.

2 Related Works

Shubham Debnath et al. presented the use of evidence-based prediction tools with various clinical settings and the way to deploy similar models in this COVID-19 pandemic to assist hospitals and healthcare managers to decide on the health of patients and manage the crowd at hospitals [12].

Gloria Hyunjung et al. estimated the timing and intensity of public health policies worldwide in the COVID-19 pandemic and moreover evaluated if machine learning can assist in providing better global health schemes. This strategy generally suggested too early and high power of lockdown and limitation in traveling. Machine learning is employed as a tool for decision support implementation of public health care at the time of COVID-19 and pandemics in the future [13].

William R. Zamea et al. listed out a few challenges and recommended several ways where machine learning (ML) can assist to overcome these challenges. The symptoms are observed from the 2nd day till the 14th day, which is the development duration of Middle East Respiratory Syndrome (MERS) and SARS. The specialist faces fundamental difficulties while sharing the information during disorder recognizing and giving treatment [14]. Three challenges were identified which are ongoing clinical efforts for non-COVID-19 drugs, for re-purposing drugs, and for new drugs to provide treatment for COVID-19. Machine language can help in almost all these areas by providing invaluable support [15].

Shreyas Setlur Arun et al. analyzed the transmission of COVID-19 and predicted the measure of pandemic along with the rate of recovery rate as well as death. The dataset available at Johns Hopkins Corona Virus Resource Center was used for analysis. A few standard machine learning approaches were employed to achieve remarkable results [16].

Amrita Mathur et al. suggested the ways adopted by the authorities to analyze how people are affected mentally due to COVID-19 and suggested a few important measures to be taken to fight against this virus which affects the social welfare and economic status of the entire world [17].

Yujin Oh et al. [18] introduced a patch-based convolutional neural network method which used only a few trainable parameters for diagnosing COVID-19. This approach had motivated statistical analysis of the imaging biomarkers of the CXR radiographs. From the experiment, it was observed that the performance of this method was supported clinically which helped in detecting COVID-19 and patient triage.

Aayush Jaiswal et al. [19] employed pre-trained deep learning structures as an automatic tool for detecting and diagnosing COVID-19 in chest CT. COVID patients were classified using DenseNet201-based deep transfer learning (DTL). With this approach, features were extracted using a convolutional neural architecture where the ImageNet dataset was employed. From the several experiments carried out, the performance of this DTL model was evaluated on COVID-19 chest CT scan images. From the comparative analysis, DTL-based COVID-19 classification approach provided remarkable results than the other competitive methods.

Kumar et al. [20] predicted the COVID-19 spread in the 15 most-infected countries in the world using the autoregressive integrated moving average (ARIMA) strategy. The outcome of their prediction indicates that circumstances would worsen in Iran and Europe, especially in Italy, Spain, and France. Moreover, their prediction indicated that the number of cases in South Korea and Mainland China would become more stable. The study also indicated that COVID-19 would spread exponentially in the USA and that strict official measures would be urgently required to stop the transmission of the disease. Although the prediction of COVID-19 cases for the USA was 1 million between April 8 and 30, 2020, it reached 677,570 on April 17, 2020. Furthermore, Italy had 168,941 cases, although it was predicted to have 300,000 cases.

Pandey et al. [21] utilized two statistical algorithms—the susceptible-exposed-infectious-recovered (SEIR) and regression models—to estimate and predict the distribution of COVID-19 in India. They used a dataset retrieved from the John Hopkins University repository. The prediction results from the SEIR and regression models showed that the number of COVID-19 cases confirmed would reach 5300 and 6153 cases, respectively, by April 13, 2020. However, the confirmed cases in India on that date were 10,453 and 6153 for the SEIR and regression models, respectively.

3 System Model

The system model discusses data that can be analyzed based on the Hybrid Ensemble-based Reinforcement Learning (HERL) algorithm which provides better prediction and detection of COVID-19. Figure 1 shows the system model. The patient data is collected and the raw data is preprocessed by using the filtering method, and the preprocessed data is given to the feature selection stage that can process the data based on selecting the correlation features and the trained data can be stored as features. The classification using HERL semantically generates deep instances behind the promising infected regions. An attention-based pooling strategy is applied

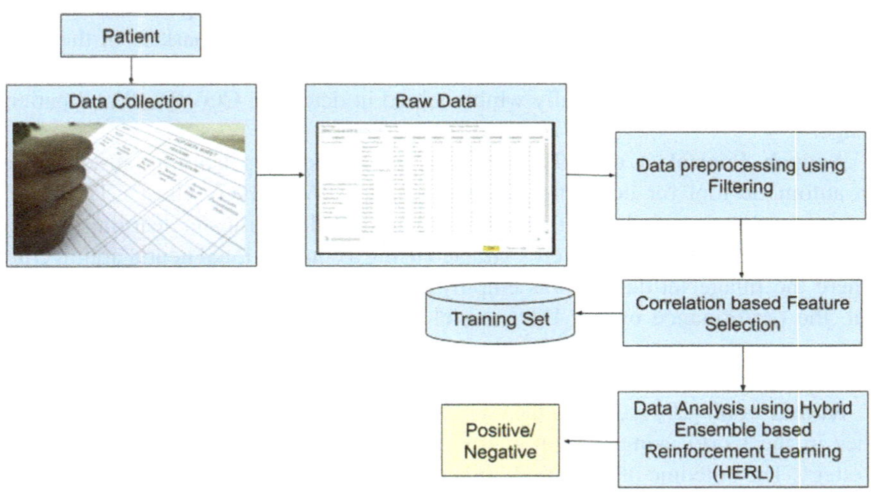

Fig. 1 System model

to the instances to provide insight into the contribution of every instance to the label. Finally, Bernoulli distributions of labels are learned for easier learning. Finally, the result is classified as COVID positive and COVID negative.

Data Collection: Data collection is the process to obtain healthcare data from a particular patient regularly for some certain duration. When this data exceeds the normal values, an alert message is sent to the doctors along with the data collected for further processing (Fig. 1).

Data Preprocessing: The data preprocessing is the initial step which utilizes the filtering method. The text is non-structured and thus requires refinement which is achieved by machine learning. This phase performs several steps; unnecessary tests are removed. Punctuation and lemmatization are carried out to refine data in a good manner. Symbols, stopwords, URLs, links are detached in order to classify data thereby achieving satisfactory accuracy.

Feature Selection: The methods employed in feature selection generate feature subspaces. The correlation-based feature selection approach estimates the effectiveness of the feature by determining Pearson's correlation (PC) between feature (f) and target class. PC is computed by using Eq. (1), where $E[*]$ represents the expected value, and μ and σ represents the mean and standard deviation.

$$PC(f, class) = \frac{E[(f - \mu_f)(class - \mu_{class})]}{\sigma_f \sigma_{class}} \tag{1}$$

This approach is a filter-based approach for feature selection. The features are selected in relation to the threshold and are ranked with the descending scores of the obtained PC values. To estimate the effectiveness of the feature, this approach considers the redundancy between features and the predictive capability of features. Feature subset S having low inter-correlation and target class with high correlation are selected. For subset S with k features, worth is computed as per Eq. (2), where C gives the relation between the correlation of two variables and not Pearson's correlation or Spearman's ρ indeed.

$$Worth(S_k) = \frac{\sum_{f_i \in s_k} \mathtt{C}(f_i, class)}{\sqrt{\sum_{f_i \in s_k} \sum_{f_i \in s_k - \{f_i\}} \mathtt{C}(f_i, f_j)}} \tag{2}$$

Here, symmetric uncertainty is considered as a measure which is an entropy-based relative measure. For two variables X_i, X_j, symmetric uncertainty is given by Eq. (3), where MI(X_i, X_j) measures the mutual information between X_i, X_j, and $H(\cdot)$ denotes marginal entropy

.

$$Uncertainity(X_i, X_j) = 2.\frac{\mathrm{MI}(\mathrm{X_i}, \mathrm{X_j})}{H(X_i) + H(X_j)} \tag{3}$$

Classification using Hybrid Ensemble-based Reinforcement Learning (HERL) Algorithm: The classification method using Hybrid Ensemble-based Reinforcement Learning (HERL) algorithm is categorized into four steps. The steps to be carried out are Deep Instance Generator (DIG), Attention-based Pooling (AP), transformation function, and Bernoulli distribution which is shown in Fig. 2.

(i) ***Deep Instance Generator (DIG):*** The proposed DIG considers entire data and automatically generates deep instances. In general, DIG is considered to be a fully 3D CNN. Practically, with the given data Xi whose shape is H x W x S, the last layer of fully 3D CNN provides a sequence of 3D feature maps whose shape is H* x W* x S* x FD, where H*, W*, S*, and FD denotes the height, width, spatial, and feature dimension of 3D feature maps, respectively. The dimension of every point of $H^* \times W^* \times S^*$ was D^*1 which means that inside the last layer of DIG, there exists a total $N = H^* \times W^* \times S^*$ instances which were produced with deep representation. A deep instance $H = \{h1, h2,.., hn\}$ was generated where n represents the number of deep instances $H \in \mathbb{R}^{n \times D}$. On the cube, with the location of deep instances, the actual location of the respective instances on data was derived easily. The mathematical formulation of this step is provided in Eq. (4)

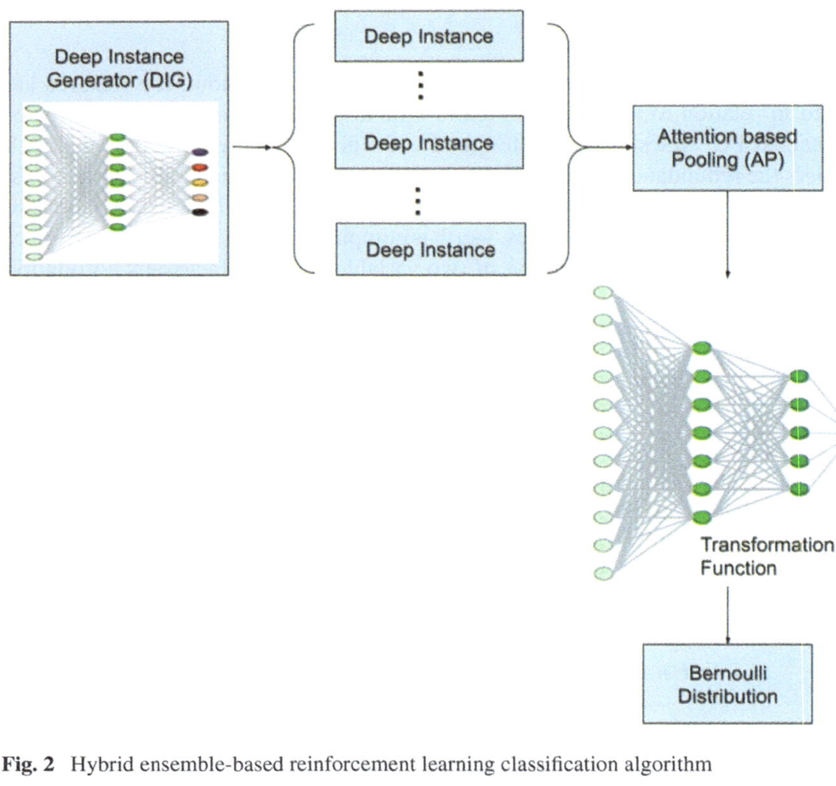

Fig. 2 Hybrid ensemble-based reinforcement learning classification algorithm

$$H = \psi X_i \tag{4}$$

where X_i is the raw input and $H \in \mathbb{R}^{n \times D}$.

(ii) ***Attention-based Pooling (AP):*** As the maximum and mean pooling operators have limitations, flexible and adaptive pooling has to be developed to achieve remarkable performance. Once a bag of deep 3D instances H is obtained, an interpretable symmetric function is applied which is termed as attention-based pooling. It is denoted as H = {h1,h2,..,hn} bag of N deep instances, and attention-based pooling is given by

$$\mathrm{Z} = \sum_{\mathrm{a}=1}^{\mathrm{n}} \mathrm{s_n h_n} \tag{5}$$

where

$$\mathrm{s_n} = \frac{\exp\{\mathrm{w^T} \tanh(\mathrm{vh_n^T})\}}{\sum_{\mathrm{j}=1}^{\mathrm{n}} \exp\{\mathrm{w^T} \tanh(\mathrm{vh_j^T})\}} \tag{6}$$

where $w \in \mathbb{R}^{n \times 1}$ *and* $v \in \mathbb{R}^{n \times D}$ are parameters that are trainable. The hyperbolic tangent tanh(.) element-wise non-linearity is applied for appropriate gradient flow.

Instead of applying the existing attention mechanism, an approach based on deep 3D instances was applied in this study. Essentially, when a deep 3D instance is given to maximum attention weight, then it is said as a key instance. This means that attention weights provide deep insight to contribute toward every instance to the label. Hence, this type of pooling provides robust interpretability for prediction. Moreover, generated instance representation z is more semantic than conventional pooling operators which use machine learning approaches. To summarize, assume σ_a with parameters θ_{σ_a} denote attention-based pooling; the above process is formulated as in Eq. (7):

$$z_i = \sigma_a(H_i) \tag{7}$$

(iii) ***Transformation Function and Bernoulli Distribution:*** Given a representation z_i of an instance *Xi*, two FC layers are used in which transformation function g transforms bag representation z_i into bag label Y_i which is distributed based on the Bernoulli distribution with the parameters θ_g, i.e., g(z_i) 2 [0; 1]. The probability $_{\mathrm{pi}}$ of $Y_i = 1$ is for the given bag representation z. The distribution used is discrete which has two possible outputs labeled as $Y = 1$ and $Y = 0$. With the probability p, Y represents 1 indicating the presence of COVID-19 with probability 1-p representing Y as 0 indicating the absence of COVID-19, where p lies between 0 and 1. The last layer provides a scalar value representing the probability of COVID-19. This step can be formulated into $pi = g(z_i)$. With no loss in generality, the final bag label $\widehat{Y}_i$ is estimated by threshold τ of 0.5.

4 Experimental Results

The existing methods used for comparison are Convolutional Neural Network (CNN) and Boosted Neural Network Ensemble Classification (BNNEC). The performance analysis is carried out using various parameters as follows.

1. **Accuracy**—Accuracy (Acc) is the ratio of the truly recognized cases to the total items and is computed as in Eq. (8). It is the measure when both COVID-19 (+) and COVID-19 (–) are equally important. It can be computed as

$$Accuracy(Acc) = \frac{TP + TN}{TP + TN + FP + FN} \tag{8}$$

2. **Prediction rate**—Prediction Rate (PR) is the ratio of the number of relevantly predicted cases that are correctly selected to the total number of predictions. It is measured in terms of percentage (%).

$$Predictionrate = \frac{\text{numberofcorrectlyselectedfeatures}}{\text{totalnumberofprediction}} * 100 \tag{9}$$

3. **Precision**—Precision (P) evaluates the exactness of the classifiers. When the P value obtained for the classifier is low, then it suffers from more FP: Mathematical P is computed by

$$Precision(P) = \frac{TP}{TP + FP} \tag{10}$$

4. **Recall**—Recall (R) also termed as sensitivity is the measure which provides the completeness of the classifier. When R has a low value, the classifier suffers from large FP and is expressed by

$$Recall(R) = \frac{TP}{TP + FN} \tag{11}$$

5. **Area Under Roc Curve (AUC)**—Area under ROC curve (AUC) is one among the outstanding measures to compare the classifiers. When the ROC curve quickly rises upwards toward the left corner of the graph, better performance is achieved. When AUC is closer to 1.0, the diagnostic test is said to be reliable, or else when closer to 0.5, the test is unreliable.

Table 1 shows the analysis of existing and proposed methods based on the number cases, i.e., patients

Table 1 Analysis of accuracy

Number of cases	PCNN	DTL	HERL
20,000	45	56	65
40,000	56	67	72
60,000	63	72	84
80,000	71	78	89
1,00,000	79	86	93

Figure 3 illustrates the comparative analysis of accuracy for existing and proposed approaches. *X*-axis and *Y*-axis provide the number of cases and values obtained in percentage, respectively. The green, orange, and pink colors indicate HERL, DTL, and PCNN classifiers, respectively. When compared to the existing methods, the proposed method achieves better results. Table 2 shows the analysis of the Prediction rate with existing and proposed methods.

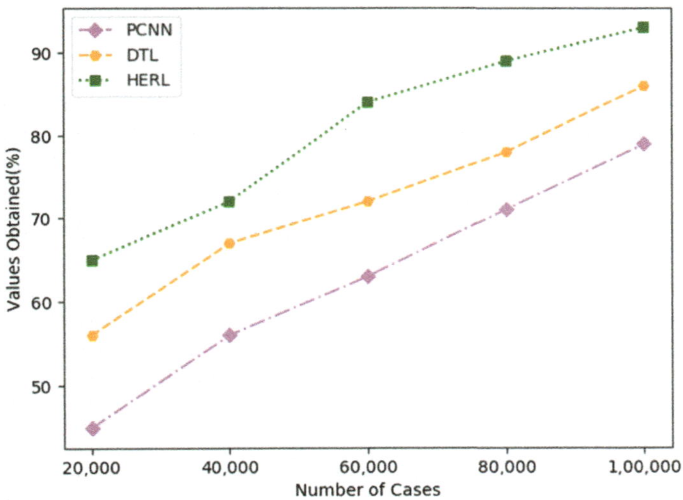

Fig. 3 Comparison of the existing and proposed methods for Accuracy

Table 2 Analysis of prediction rate

Number of cases	PCNN	DTL	HERL
20,000	37	49	56
40,000	42	51	63
60,000	53	58	74
80,000	61	69	83
1,00,000	73	79	89

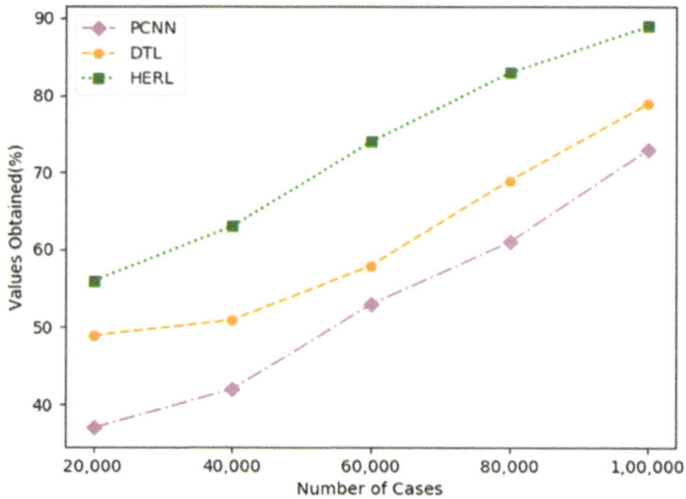

Fig. 4 Comparison of the proposed and existing methods for Prediction rate

Figure 4 illustrates the comparative analysis of the prediction rate for existing and proposed approaches. *X*-axis and *Y*-axis provide the number of cases and days, respectively. The green, orange, and pink colors indicate HERL, DTL, and PCNN classifiers, respectively. When compared to the existing methods, the proposed method achieves better results. Table 3 shows the analysis of precision with existing and proposed approaches.

Figure 5 illustrates the comparative analysis of Precision for existing and proposed approaches. *X*-axis and *Y*-axis provide the number of cases and precision in percentage, respectively. The green, orange, and pink colors indicate HERL, DTL, and PCNN classifiers, respectively. When compared to the existing methods, the proposed method achieves better results. Table 4 shows the analysis of Precision with existing and proposed approaches.

Figure 6 illustrates the comparative analysis of Recall for existing and proposed approaches. *X*-axis and *Y*-axis provide the number of cases and recall in percentage, respectively. The green, orange, and pink color indicates HERL, DTL, and PCNN classifiers, respectively. When compared to the existing methods, the proposed

Table 3 Analysis of precision

Number of cases	PCNN	DTL	HERL
20,000	46	52	61
40,000	53	64	70
60,000	58	69	73
80,000	64	71	81
1,00,000	72	76	89

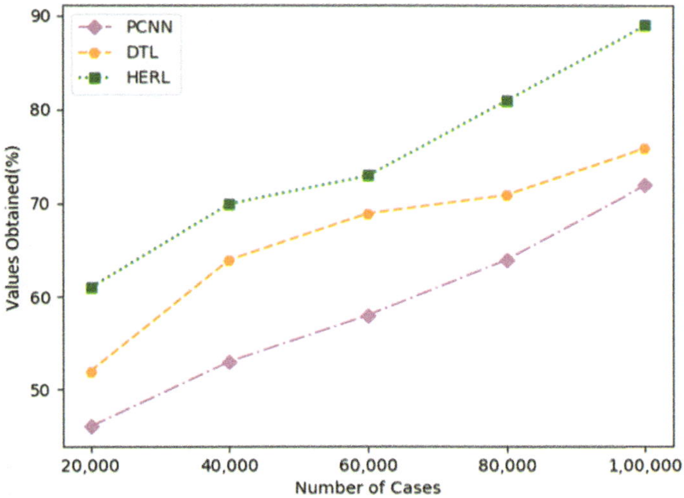

Fig. 5 Comparative analysis of Precision for existing and proposed methods

Table 4 Analysis of recall

Number of cases	PCNN	DTL	HERL
20,000	39	41	57
40,000	43	48	66
60,000	50	54	71
80,000	54	61	75
1,00,000	61	69	79

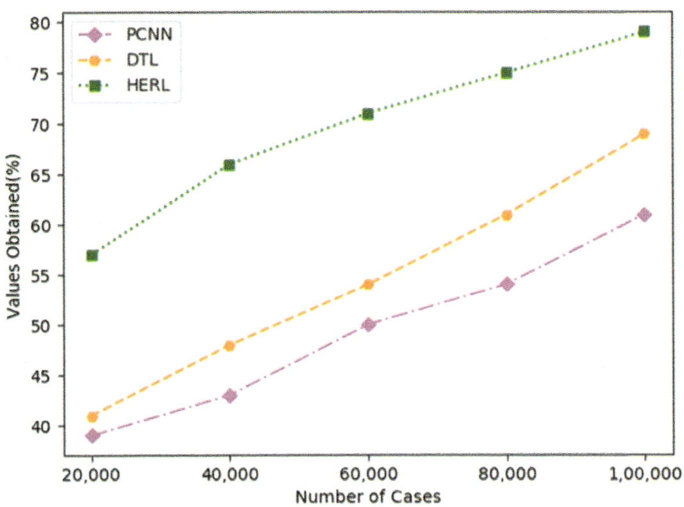

Fig. 6 Comparative analysis of Recall for existing and proposed methods

Table 5 AUC analysis

False positive rate	PCNN	DTL	HERL
0.2	0.75	0.78	0.92
0.4	0.85	0.88	0.95
0.6	0.88	0.90	0.97
0.8	0.90	0.95	0.98
1	1	1	1

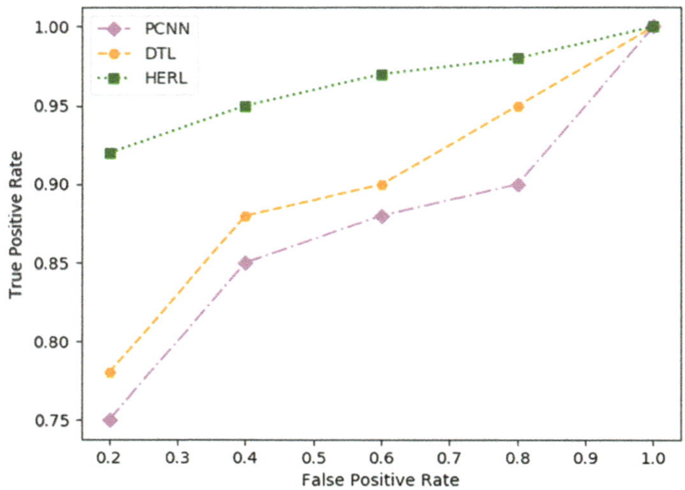

Fig. 7 Comparative analysis of AUC for existing and proposed methods

method achieves better results. Table 5 shows the analysis of AUC with existing and proposed methods.

Figure 7 illustrates the comparative analysis of AUC for existing and proposed approaches. *X*-axis and *Y*-axis provide true positive rate and false positive rate, respectively. The green, orange, and pink colors indicate HERL, DTL, and PCNN classifiers, respectively. When compared to the existing methods, the proposed method achieves better results.

5 Conclusion

Due to the COVID-19 pandemic, nearly five million humans are infected; moreover, the number still increases. With the recent dreadful epidemiological circumstance and huge socioeconomic effects of COVID-19, the important requirement is to create devices which can help in identifying or preventing the rapid spread of terrible diseases. This paper presented the analysis of data which is based on the

HERL classification technique to achieve remarkable results. This HERL model was prominent among the Reinforcement learning approaches employed in the experiment. The proposed method achieves an accuracy of 80.6%, Prediction rate is 73%, Precision is 74.8%, Recall is 69.6%, and AUC is 98%. In future, this method is enhanced by automatic labeling to avoid time consumption and apply security by using encryption methods.

References

1. Li Q, Guan X, Wu P, Wang X, ZhouL TY, Ren R, Leung KS, Lau EH, Wong JY, Xing X (2020) Early transmission dynamics in Wuhan, China, of novel coronavirus-infected pneumonia. N Engl J Med 382:1199–1207
2. Sohrabi C, Alsafi Z, O'Neill N, Khan M, Kerwan A, Al-Jabir A, Losifidis C, Agha R (2020) World health organization declares global emergency: a review of the 2019 novel coronavirus (COVID-19). Int J Surg 76:71–76. https://doi.org/10.1016/j.ijsu.2020.02.034
3. Niehus R, Salazar PMD, Taylor A, Lipsitch M (2020) Quantifyingbias of COVID-19 prevalence and severity estimates in Wuhan, China that depend on reported cases in international travelers. medRxiv, p.2020.02.13.20022707
4. WHO (World Health Organization) (2020) Coronavirus disease (COVID2019) situation reports. https://www.who.int/emergencies/diseases/novelcoronavirus-2019/situation-reports/. Accessed 9 Jun 2020
5. JHU (John Hopkins University) (2020) COVID-19 dashboard by the Center for Systems Science and Engineering (CSSE) at Johns Hopkins University (JHU). https://www.coronavirus.jhu.edu/map.html. Accessed 9 Jun 2020
6. Huang C, Wang Y, Li X (2020) Clinical features of patients infected with 2019 novel coronavirus in Wuhan, China. Lancet 395(10223):497–506
7. World Health Organization (2020) Coronavirus disease (COVID-19) advice for the public. https://www.who.int/emergencies/diseases/novel-coronavirus-2019/advice-for-public. Accessed 25 Oct 2020
8. Haleem A, Javaid M, Vaishya R (2020) Effects of COVID 19 pandemic in daily life. Curr Med Res Pract 10(2):78–79. https://doi.org/10.1016/j.cmrp.2020.03
9. Iyengar K, Upadhyaya GK, Vaishya R, Jain V (2020) COVID-19 and applications of smartphone technology in the current pandemic. Diabetes Metab Syndr 14:733–737
10. World Economic Forum (2020) The economic, geopolitical and health impacts of COVID-19. https://www.weforum.org/agenda/2020/03/the-economic-geopolitical-and-health-consequences-of-covid-19/. Accessed 25 Oct 2020
11. Srinivas SBK, Vaishya R, Vaish A (2020) Unprecedented surge in publications related to COVID-19 in the first three months of Pandemic. J Clin Orthop Trauma 11(Suppl 3):S304–S306. https://doi.org/10.1016/j.jcot.2020.04.030
12. Debnath S, Barnaby DP, Coppa K, Makhnevich A, Kim EJ, Chatterjee S, Tóth V, Levy TJ, de Paradis M, Cohen SL, Hirsch JS (2020) Machine learning to assist clinical decision-making during the COVID-19 pandemic. Bioelectron Med 6(1):1–8
13. Hyunjung KG, Ling L, Hui P (2020) Modelling global public health strategies in COVID-19 pandemic using deep reinforcement learning. Res Square, 1–14. https://doi.org/10.21203/rs.3.rs-31226/v1
14. Sampathkumar A, Murugan S, Rastogi R, Mishra MK, Malathy S, Manikandan R (2020) Energy efficient ACPI and JEHDO mechanism for IoT device energy management in healthcare. Internet of things in smart technologies for sustainable urban development, Cham, pp 131–140

15. Zame WR, Bica I, Shen C, Curth A, Lee HS, Bailey S, Weatherall J, Wright D, Bretz F, van der Schaar M (2020) Machine learning for clinical trials in the Era of COVID-19. Stat Biopharm Res 12(4):506–517
16. Arun SS, Iyer GN (2020) On the analysis of COVID19—novel corona viral disease pandemic spread data using machine learning techniques. In Proceedings of the international conference on intelligent computing and control systems (ICICCS 2020). IEEE, New York, pp 1222–1227
17. Mathur A, Kubde P, Vaidya S (2020) Emotional analysis using Twitter data during pandemic situation: COVID-19. 5th international conference on communication and electronics systems (ICCES). IEEE, New York, pp 845–848
18. Oh Y, Park S, Ye JC (2020) Deep Learning COVID-19 Features on CXR Using Limited Training Data Sets. IEEE Trans Med Imaging 39(8):2688–2700
19. Jaiswal A, Gianchandani N, Singh D, Kumar V, Kaur M (2020) Classification of the COVID-19 infected patients using DenseNet201 based deep transfer learning. J Biomol Struct Dyn 2020:1–8. https://doi.org/10.1080/07391102.2020.1788642
20. Kumar P, Kalita H, Patairiya S, Sharma YD, Nanda C, Rani M,...Bhagavathula AS (2020) Forecasting the dynamics of COVID-19 Pandemic in Top 15 countries in April 2020: ARIMA model with machine learning approach. MedRxiv
21. Pandey G, Chaudhary P, Gupta R, and Pal S (2020) SEIR and regression model based COVID-19 outbreak predictions in India. Preprint at arXiv:2004.00958

Chapter 9
Transfer Learning-Based Economical and Rapid COVID-19 Detection Using X-Rays Images

Shatakshi Singh, Keshav Jangid, Chandraprakash Sharma, Nakul Joshi, Jyoti Tomar, Manish Kumar, Dinesh Goyal, Madhushi Verma, and Sunil Kumar Jangir

Abstract Recently, Coronavirus disease (COVID-19) has spread exponentially across many countries in the world. This disease, caused by Novel Coronavirus spreads so rapidly that within a few months of its spread the World Health Organization (WHO) declared it as a Public Health Emergency of International Concern. The incubation period of this disease is 2–14 days, within which an infected person may be exposed to many people, further spreading the infection at a rapid rate.

S. Singh (✉) · S. K. Jangir
Department of Computer Science & Engineering, Mody University of Science and Technology, Sikar, Rajasthan, India

K. Jangid
EDApply Pvt. Ltd, Jaipur, Rajasthan, India
e-mail: ceo@edapply.in

C. Sharma
Wisflux Private Limited, Jaipur, Rajasthan, India
e-mail: cp@wisflux.com

N. Joshi
Celebaltech Private Limited, Jaipur, India

J. Tomar
Department of Microbiology, Pacific Institute of Medical Science, Udaipur, Rajasthan, India

M. Kumar
Department of Biomedical Engineering, Mody University of Science and Technology, Sikar, Rajasthan, India

D. Goyal
Department of Computer Science Engineering, Poornima Institute of Engineering and Technology, Jaipur, Rajasthan, India
e-mail: dinesh.goyal@poornima.org

M. Verma
Department of Computer Science Engineering, Bennet University, Greater Noida, Uttar Pradesh, India
e-mail: madhushi.verma@bennett.edu.in

U. Kose et al. (eds.), *Computational Intelligence for COVID-19 and Future Pandemics*, Disruptive Technologies and Digital Transformations for Society 5.0,
https://doi.org/10.1007/978-981-16-3783-4_9

In a few months, millions of people have been infected by this virus and thousands have lost their lives. Laboratory test Reverse Transcription Polymerase Chain Reaction (RTPCR), an antigen-based test, is used for diagnosis of this disease. This test is expensive and time-consuming. Moreover, not all labs are well equipped to perform this test. Due to the high cost and scarcity of kits, only limited and necessary tests are performed which makes it difficult to diagnose more people. The use of computational intelligence to diagnose this disease could not only reduce the cost but also make it accessible to most of the diagnostic centers, and reduce the time taken to obtain the results of the test. This work implements an approach based on Transfer Learning, where pre-trained neural networks are used to transfer their weights and bias to a newly trained model for diagnosing this disease using an X-ray image of the human lungs. Pre-trained architectures InceptionResNetV2, InceptionV3, MobileNetV2, ResNet50V2, VGG16, and DenseNet201 are applied for transfer learning to create new models for Covid-19 classification. Conclusion: When they were tested on an unseen dataset, it was found that MobileNetV2 outperformed all others by achieving the highest accuracy and precision.

Keywords Classification · COVID-19 · Deep learning · Image processing · Machine learning · Transfer learning

1 Introduction

Coronavirus is a group of spherical or pleomorphic enveloped RNA viruses, having petal-like peplomers on their surface. The reason behind this name is the fringes of surface projections surrounding the virus that resemble the solar corona [1]. There are two groups of Coronaviruses, the first is acid-labile viruses (associated with common cold-like illness) and the second is acid-stable viruses (associated with human and animal gastroenteritis). There are many serotypes, which are mostly fastidious and difficult to grow in the cell culture system. Human Coronaviruses were first isolated from the cases of the common cold. They were isolated by inoculating nasopharyngeal washings on organ cultures of human embryonic trachea. The growth of the virus is indicated by the inhibition of ciliary motility. When the Coronavirus is inoculated in humans it induces common cold-like symptoms in a period of 2–7 days. The same serotype of this virus was found to cause reinfections. After the Rhinoviruses, it is the second most common cause of the common cold. The classic Coronavirus causes common cold. A dramatic addition to the genus was recognized in 2003, when the novel Coronavirus was identified as the cause of a mysterious, new respiratory illness in China in 2003 [2]. This infection was called Severe Acute Respiratory Syndrome (SARS). SARS is characterized by lower respiratory infections that require admission to an ICU, as it may be fatal. This virus was new and different from other Coronaviruses which had been classified into three types: mammalian viruses in type 1 and 2, and avian viruses in type 3 [3]. The new SARS virus became Coronavirus type 4.

Initially, a cluster of pneumonia cases was found in Wuhan, China [4, 5]. Thousands of patients were reported with common symptoms including, cough myalgia or fatigue, difficulty in breathing, cough, and invasive lesions on both lungs, headache, hemoptysis, and diarrhea [6]. A novel beta Coronavirus was identified as the causative agent [7]. Slowly, the virus spread all across the world and the number of cases were found to be exponentially rising. On January 22, 2020, 571 confirmed cases of COVID-19 were reported in 25 provinces in China and up to January 22, 2020, China National Health Commission reported 17 deaths. By January 25, 2020, 1,975 cases were reported as confirmed in China with 56 deaths. This Figure went up to 5502 by January 24, 2020. 7734 cases were confirmed in China by January 30, 2020, and 90 cases were reported from countries including Thailand, Malaysia, Taiwan, Singapore, Republic of Korea, Vietnam, Cambodia, Japan, Nepal, Sri Lanka, the United Arab Emirates, the United States, the Philippines, India, Australia, Canada, Finland, France, and Germany [8]. On January 30, 2020, WHO declared this situation as a Public Health Emergency of International Concern. Countries with vulnerable health systems were at high risk. By March 1, 2020, 79,968 cases were reported from China, and 87,137 cases were confirmed globally [9, 9]. As of May 27, 2020, 5.61 million cases were reported all over the world, with 351,000 deaths and 2.31 million recovered cases.

Currently, no specific therapy or prophylaxis has yet been identified for treating this disease effectively. Many countries have claimed that vaccine was found to be effective. The development of a vaccine is under the trial phase. Proper Isolation and Quarantine of the infected person are being used to control the situation [11]. The virus is highly mutable so vaccine prophylaxis may not be easy. Anti-viral, anti-malarial, and herbal treatments are potential treatments for this infection [12, 12]. The incubation period of COVID-19 is 2–14 days. Symptoms may show up in an infected person after 4–5 days or more after getting infected. During this duration, the infected person comes in contact with many people and infects them, as a result of which the number of patients of COVID-19 is exponentially increasing across the world. Laboratory test Reverse Transcription Polymerase Chain Reaction (RTPCR), an antigen-based test, is used for the diagnosis of this disease [14]. The rise in titer of antibodies can also be demonstrated by ELISA or indirect immunofluorescent test in paired serum samples. RTPCR is an expensive and time-consuming test. The use of intelligence symptoms can significantly reduce the expenses and duration of disease detection. A system is required that can distinguish between an X-ray image of an infected and a non-infected person.

Various CNN architectures are available that can help in identifying the infected person by scanning the X-ray image of the suspect's lungs.

The main contribution of this research article is summarized as follows:

1) Prepared dataset for this study from 2 different open-source datasets available online.
2) Implemented the state of the art using Deep Learning architectures InceptionResNetV2, InceptionV3, MobileNetV2, ResNet50v2, VGG16, and DenseNet201, on the Lung X-ray dataset.

3) Checked the adaptiveness of the models by testing their performance on an unseen dataset and provided a comparative study of results obtained by these deep learning models on both the datasets.
4) Summarized the performance of various deep learning models and compared them with remarks on their performance to show the most accurate classification results on the lung X-ray image dataset.

The layout of the chapter is intended as follows: Literature review is given in Sect. 2. Methods and materials are presented in Sect. 3. Results discussion and conclusion of this research chapter are discussed in Sects. 4 and 5, respectively.

2 Literature Review

The use of deep learning for image identification and classification has been one of the high-priority research areas for the last few years. Many researches have tried to develop systems which leverage the potential of Artificial Intelligence for medical diagnosis. Ezz El-Din Hemdan et al. [15] applied a new deep learning framework COVIDX-Net including 7 distinct architectures of deep CNN models, such as modified VGG19 on a dataset having 50 Lung X-rays with 25 X-rays of confirmed COVID-19 cases and 25 X-ray images of healthy lungs. They found that the two architectures VGG19 and DenseNet showed similar performance with the value of F1 score 0.89 and 0.91 for healthy lung's X-ray and COVID-19-affected lung X-ray, respectively.

Muhammad Ilyas et al. [16] developed a system for detecting and differentiating COVID-19 from other infections including viral pneumonia, bacterial pneumonia, and healthy cases, using pre-trained deep learning models on a dataset containing 200 CT and X-Ray images of COVID-19, 200 images of healthy lungs, 200 images of lungs affected from bacterial pneumonia, and 200 images of lungs affected from viral pneumonia, respectively, and got the accuracy of 98.75%.

Li et al. [17] developed an automated framework for the detection of COVID-19, with chest CT scan images by developing a deep learning model that extracts visual features from chest CT scans. Their research concluded that deep learning techniques can effectively detect COVID-19 using CT scans and can also differentiate it from community-acquired pneumonia and other lung diseases that cause chest congestion.

Ozturk et al. [18] proposed a model for binary classification of images in two classes COVID-19 affected and normal, and for multi-class classification in three classes including COVID-19, pneumonia, and normal. They used DarknetYolo architecture and trained their model with 125 X-ray images of confirmed COVID-19 cases and achieved an accuracy of 98.08% and 87.02% for binary-class and multi-class classifications, respectively.

Wang et al. [19] proposed a deep learning model, COVID-Net, for COVID-19 detection, and obtained a classification accuracy of 92.4% for classifying given images in three classes including healthy lungs, non-COVID pneumonia, and

COVID-19. Narin et al. [20] trained a model using ResNet50 architecture on a chest X-ray image dataset and achieved an accuracy of 98%. Ioannis et al. [21] applied transfer learning on a dataset containing 224 X-ray images of confirmed COVID-19 cases and achieved 98.75% and 93.48% accuracy for two and three classes, respectively.

3 Material and Methods

Datasets used to conduct this study are discussed in Sect. 3.1, followed by the pre-trained architectures used for applying transfer learning in Sect. 3.2. Then evaluation metrics are discussed under model performance analysis in Sect. 3.3.

3.1 Dataset

The dataset taken for this research work is obtained from 2 different open-source datasets. One is the 'COVID-19 X-ray image dataset', provided by JP Cohen on Github, who prepared it by collecting images from different open-access sources [22]. This dataset is regularly updated with images provided by different researchers. The second was the 'Chest X-ray image dataset', provided by Paul Mooney on Kaggle [23].

Two datasets were prepared for this study. Dataset I contains 130 images of infected lungs and 134 images of healthy lungs. The dataset was divided into a ratio of 80:20 for training and testing, respectively (Fig. 1, Table 1).

To test the generalizability of the models, they were tested on an unseen dataset (Dataset II) that consists of 32 images of COVID-19-affected lungs and 39 healthy lungs. X-ray images in both the dataset belong to people of different age groups 12–65 years of age, both male and female (Fig. 2).

3.2 Transfer Learning

Neural Networks are designed such that, if they are trained on a given dataset for image classification, their weights and bias can be used for training another model for classifying images in a new dataset. When such pre-trained neural networks are used for training a new model to solve some other problem, it is known as transfer learning [24]. Deep neural networks have many layers containing millions of weights. Training such a huge network requires a large dataset and a huge computation power. Transfer Learning enables us to train deep neural networks with a small dataset and less computation power (Fig. 3).

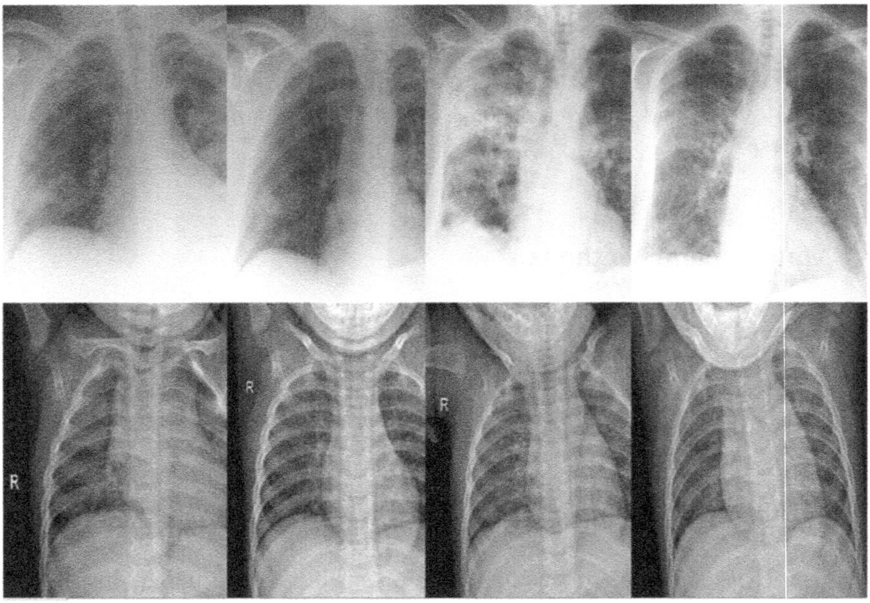

Fig. 1 Samples of images in Dataset II containing X-ray images of lungs infected by COVID-19 (First row) and images of healthy human lungs (Second row)

Table 1 Number of X-ray images for each class in both datasets

	Dataset 1	Dataset 2
COVID-19 affected X-rays	130	32
Healthy lungs	134	39

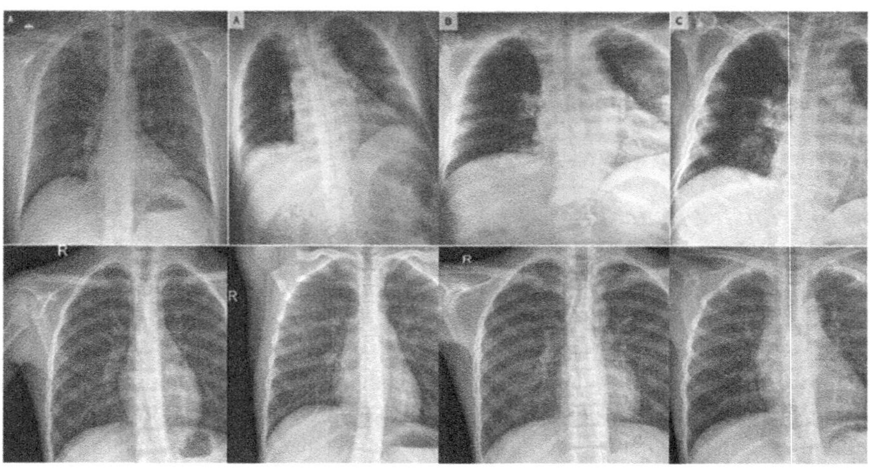

Fig. 2 Sample of images in Dataset II containing images of lungs infected by COVID-19 infection (first row) and images of healthy human lungs (second row)

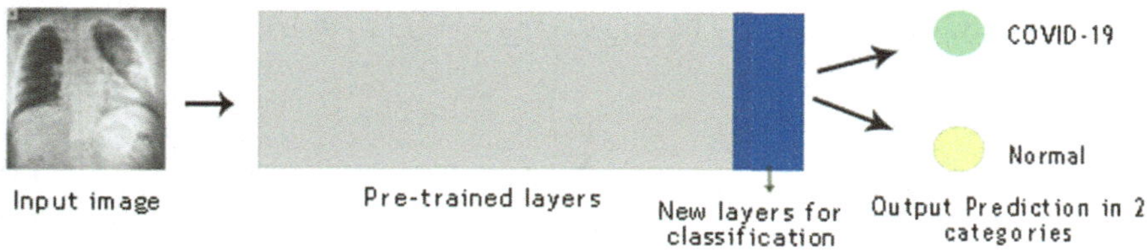

Fig. 3 Classification of given X-ray as COVID-19 affected or normal, performed using a pre-trained network with some new layers appended at the end for identifying special features

Various deep learning architectures, with pre-trained weights and bias on the ImageNet dataset, are provided by Keras. The initial layers of these pre-trained architectures are frozen while the last few layers are trained to update weights and bias to meet specific classification needs [25]. In this chapter, popular deep learning architectures—InceptionResNetV2, InceptionV3, MobileNetV2, ResNet50V2, VGG16, and DenseNet201 are used for transfer learning.

Before applying the transfer learning, image data augmentation was done using the ImageDataGenerator class, available in Keras, to expand the dataset for training. It helps to improve the generalizability of the model. We used a rotation range of 15°, which creates multiple images by transforming the original images with a random rotation from −15° to 15°.

3.3 *Deep Learning Image Classifiers*

3.3.1 ResNet50V2

ResNets consist of many stacked 'Residual Units'. Each unit performs the following computations:

$$y_l = \mathrm{h}(x_l) + f(x_l, w_l) \tag{1}$$

$$x_{l+1} = f(y_l) \tag{2}$$

where x_l and x_{l+1} are input and output of the lth unit, and f is a residual function. The function f denotes the operation after elementwise addition. The central idea of ResNets is to learn the additive residual function F with respect to $h(x_l)$; the function $h(x_l) = x_l$ is set as identity mapping [26].

If f is also an identity mapping, $x_{l+1} = y_l$, we can put Eq. (2) into Eq. (1) and obtain

$$x_{l+1} = x_l + f(x_l, W_L) \tag{3}$$

Recursively, $X_{l+2} = x_{l+1} + f(x_{l+1}, w_{l+1}) = x_l + f(x_l, w_l) + f(x_{l+1}, w_{l+1})$ and so on, so we will have

$$x_L = x_l + \sum_{i=l}^{L-1} f(x_i, w_i) \tag{4}$$

where L is the deeper unit and l is the shallower unit. Equation (4) denotes that the features of any deep unit L are the summation of x_0 and all preceding residual networks.

ResNet stacks up layers that initially do not have any significant role, and skips over them. It reuses the activations from previous layers. Skipping layers helps to increase the rate of learning as it compresses the network, and only a few layers remain. The architecture of ResNet50 is quite similar to VGG16, except that it has an additional ability of identity mapping. ResNetV2 uses batch normalization before each layer. This differentiates it from the older version ResNetV1 [27].

Initial layers of the ResNetV2 model were used as such while the last layer was removed. Three new layers were added with a dropout value of 0.5 for pooling and flattening. In the last layer, softmax activation function was used to classify the given image in one of the two classes.

3.3.2 VGG16

VGG16 is a state-of-the-art CNN model proposed by Simonyan and Zisserman [28]. It was trained for weeks using NVIDIA Titan Black GPUs. It achieved an accuracy of 92.7% on the ImageNet dataset that contains 14 million images belonging to 1000 different classes. It improved the existing AlexNet framework by replacing large kernel-sized filters with many 3 × 3 kernel-sized filters one after another.

The size of the input image is 224 × 224 RGB image. It has convolution layers with filters of 3 × 3 and a convolution stride of 1 × 1, which is the minimum required size for capturing the notion of right, left, up, and down. The spatial resolution does not change after convolution because the spatial padding of convolutional layer input is 1 pixel for 3 × 3 convolution layers. For spatial pooling, 5 max-pooling layers over a 2 × 2 pixel window with a stride of 2 are required.

This arrangement of max-pool layers and convolutional layers remains the same throughout the architecture. The stack of Convolutional layers is followed by 2 fully connected layers having 4096 channels each and 1 fully connected Softmax layer. VGG16 is named so due to the 16 layers that have weights in this network. It is a large network having approximately 138 million parameters.

All layers of this network were frozen during training, and the last layer was substituted by pooling, flattening, and classification layers with a dropout of 0.5. The learning rate was set to 0.001 so that the network converges faster. Batch size was kept at 15 and the model was trained with 16 epochs. Finally, the softmax function was used to give the final classification result.

3.3.3 DenseNet201

DenseNet201 is a 201-layer deep CNN with pre-trained weights on the ImageNet database [29]. It has 224 × 224 input image size and 20 million parameters. It supports feature reuse, as subsequent layers take as input the concatenated feature maps that were given by previous layers. This implies that feature map x_l in l^{th} layer can be projected by all the feature maps $x_0, x_1, x_2, \ldots x_{l-1}$, in the previous layers. This can be mathematically denoted as $x_l = h_l([x_0, x_1, \ldots\ldots, x_{l-1}]$, where $[x_0, x_1, \ldots, x_{l-1}]$ are, respectively, the feature maps of $0^{th} \ldots l-1^{th}$ layer and h_l is a function comprising of three operations, namely Batch Normalization, Rectified Linear unit (ReLU), and a 3 × 3 convolution. While keeping all the layers of the pre-trained network as it is, we replaced the last layer with three new layers for pooling, flattening, and classification. In the outermost layer, we applied a sigmoid activation function for the binary classification.

It provides some advantages over other architectures such as it reduces the vanishing gradient issues, encourages feature reuse, reinforces feature propagation, and reduces the number of parameters.

3.3.4 MobileNetV2

MobileNet is a pre-trained CNN architecture for image classification. In 2018, [31] proposed the MobileNetV2 model which contains 19 residual bottleneck layers, each preceded by a fully connected layer with 32 filters. Kernel size of 3 × 3 was used because, during training, it utilizes dropout and batch normalization. The depth of the network is 53 with 3.5 million parameters and the input image size for this network is 224 × 224.

In depthwise separable convolutions, a standard convolution is factorized in many depthwise convolutions. Factorization reduces the model size and the computational power to apply transfer learning. This is one of the advantages of using this architecture over others. A single filter is applied to each input channel by the depthwise convolution. In a standard CNN, filters and combines are given as input into a new set of outputs in a single step, whereas in depthwise separable networks, it splits into two separate layers, one for filtering and the other for combining [32].

$$\widehat{G}_{k,l,m} = \sum_{i,j} \widehat{K}_{i,j,m} . F_{k+i-1,l+j-1,m} \tag{5}$$

Equation (5) shows a depthwise convolution with 1 filter input channel, where $\widehat{K}$ is a depthwise convolution kernel (Fig. 4).

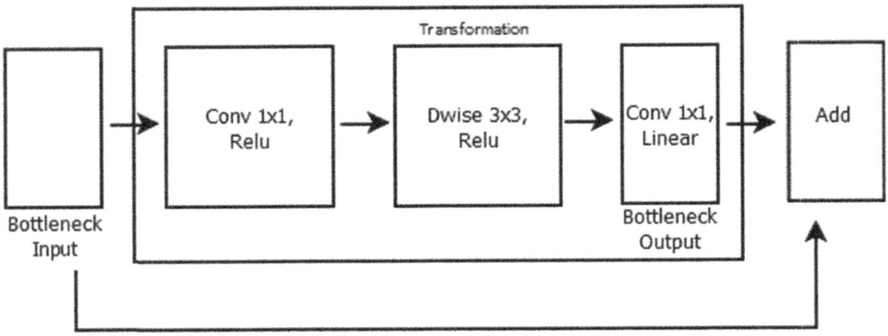

Fig. 4 MobileNetV2 building block

Table 2 Comparison of MobileNetV1 and MobileNetV2 based on MACs and Parameters

Version	Multiply accumulate operations (MACs) (millions)	Parameters (Millions)
MobileNet V1	569	4.24
MobileNet V2	300	3.47

Depthwise convolutions have a Computational cost given by $D_K.D_K.M.D_F.D_F$. Though depthwise convolutions are comparatively effective as compared to the standard convolutions, they only filter input channels and do not combine them. Therefore, another new layer is required to compute a linear combination of the output of depthwise convolution (Table 2).

All the layers of the pre-trained network were frozen during training except the last layer which was replaced with three layers, for pooling, flattening, and classification. Only the weights of the last 3 layers were updated during training, with a dropout rate of 0.5, batch size of 16. The model was trained for 14 epochs. After experimenting, we chose Adam optimizer and a rate of learning was set to 0.001 so that the network converges faster.

3.3.5 InceptionV3

Inception or GoogLeNet is a state-of-the-art architecture having a 22 layer network with pre-trained weights on the ImageNet dataset. The size of the input image for this network is 299 × 299. It is found to achieve a high accuracy for image classification [33, 34]. It was the first runner-up in the ImageNet Large Scale Visual Recognition Competition in 2015. It was initially named as InceptionV1, later its architecture was refined using batch normalization and was named as InceptionV2 [35], which was further modified by adding factorization idea in the third iteration [36]. The main aim behind factorizing convolutions was to reduce the number of connection parameters without compromising network efficiency. All the layers of the network were frozen and the last layer was replaced by pooling, flattening, and classification layers, with softmax function in the last layer for binary classification. The model was trained for 16 epochs with a batch size of 16 and a learning rate of 0.001.

3.3.6 InceptionResNetV2

InceptionResNetV2 is a 164-layer deep CNN formed by combining the Inception architecture with residual connections. It is pre-trained on millions of images from the ImageNet dataset and achieved a 3.08% top 5 error in the ImageNet challenge [27]. It has 55.9 million parameters and takes an input image of size 299 × 299 by default. It was formed after the fine-tuning of InceptionV4 and InceptionResNetV1. Unlike the residual inceptions, it uses batch normalization only on the top of the traditional layers, not on the top of summations. The last layer of the pre-trained network was replaced by three layers for pooling, flattening, and classification. The learning rate was kept to be 0.001 and a batch size of 16 was used for CPU training. Adam optimizer was used for optimization. In the last layer, the softmax activation function is used for classifying the input into two different classes.

3.4 Model Performance Analysis

To compare the performance of different methods used to classify images in the X-ray dataset, different metrics have been used. First, the confusion matrix is obtained for all the models. The outcomes obtained from a confusion matrix are True Positive (TP), i.e. the number of outcomes for which the model provided the right diagnosis, True Negative (TN), i.e. outcomes that are incorrectly classified by the model, False Positive (FP), i.e. outcomes classified as an anomaly diagnosis, and False Negative (FN), i.e. anomalies observed as an ordinary diagnosis (Table 3).

Once we have calculated the values of possible outcomes in the confusion matrix, the following performance metrics can be determined among which accuracy and precision are the most significant in the case of medical diagnosis [37].

- Accuracy: It is the sum of true positive and true negative divided by the sum of all the values in the confusion matrix.

$$Accuracy(\%) = (TP + TN) \times \frac{100}{(\mathrm{TP} + \mathrm{FP} + \mathrm{FN} + \mathrm{TN})} \tag{6}$$

Table 3 Confusion matrix

	Predictive positive	Predictive negative
Actual positive	True positive	False negative
Actual negative	False positive	True negative

- Precision: Precision gives a relation among the true positive predicted values and false positive predicted values. It is determined by dividing True Positive values by the sum of True Positive and False Positive values.

$$Presision = \frac{TP}{(TP + FP)} \tag{7}$$

- Recall: It is given by dividing the value of True Positive by the sum of True Positive and False Negative.

$$Recall = \frac{TP}{(TP + FN)} \tag{8}$$

- F1 score: It is the measure of the model's accuracy determined using precision and recall. F1 score is given as twice the ratio of the product to the summation of precision and recall metrics.

$$F1score = \frac{(2 \times Precision \times Recall)}{(Precision + Recall)} \tag{9}$$

4 Results and Discussion

All the simulation tasks are performed in Google Collaboratory using Python 3 [38]. Hardware specifications are 12.72 GB Random Access Memory (RAM) and 68 GB disk storage.

Models were first trained by applying transfer learning for all the six architectures. Figure 4 contains curves showing variation in Accuracy and Loss along with increasing epochs during the training of models. As the number of epochs increases model learns more and the training set as a result its accuracy increases and loss decreases. For all the six models, accuracy increased exponentially for the first few epochs, then it varied irregularly for the next few epochs before attaining a constant value. Similarly, loss decreased for the initial epochs at a very high rate before attaining a constant range. Accuracy and loss values cannot be further changed hence the training should be stopped after this stage (Figs. 5 and 6).

Using the TP, TN, FP, and FV values obtained from the confusion matrix, accuracy, precision, F1 score, and Recall were calculated for all the models as shown in Table 4.

The results obtained showed that all chosen pre-trained networks give an accuracy of more than 90%. To evaluate our model for its performance on unseen images, we tested it on Dataset II containing 32 images of affected lungs and 39 images of healthy lungs collected from different sources.

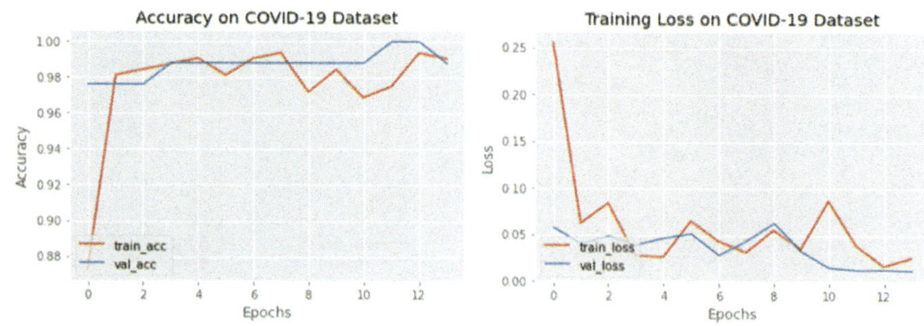

a. InceptionResNetV2

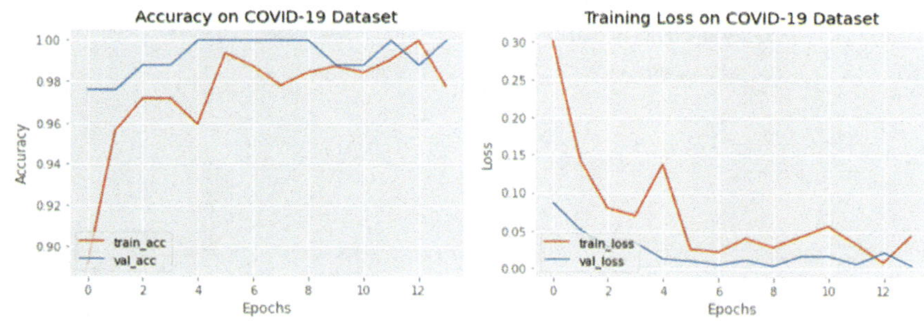

b. InceptionV3

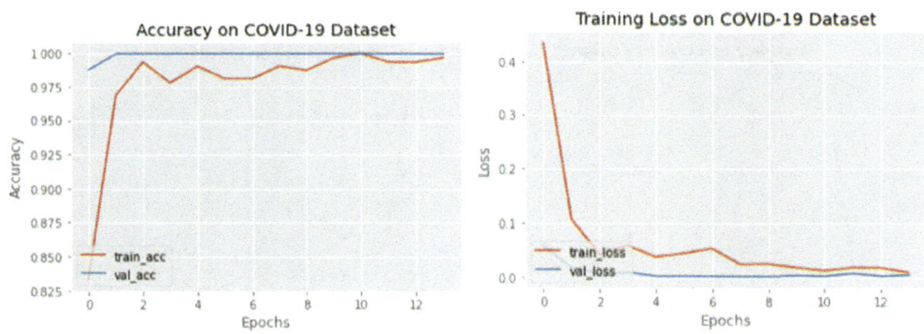

c. MobileNetV3

Fig. 5 Training loss and Accuracy Evaluation of various methods used on Dataset I

Using the confusion matrices shown in Fig. 7, the accuracy, precision, F1 score, and recall values were computed as shown in Table 5 (Fig. 8).

Precision and recall are highly significant model evaluation metrics. Precision denotes the ratio of retrieved results that were correctly classified as affected by COVID-19 and the total number of results classified as COVID-19 affected while recall gives the ratio of the results correctly classified as COVID-19 affected and the

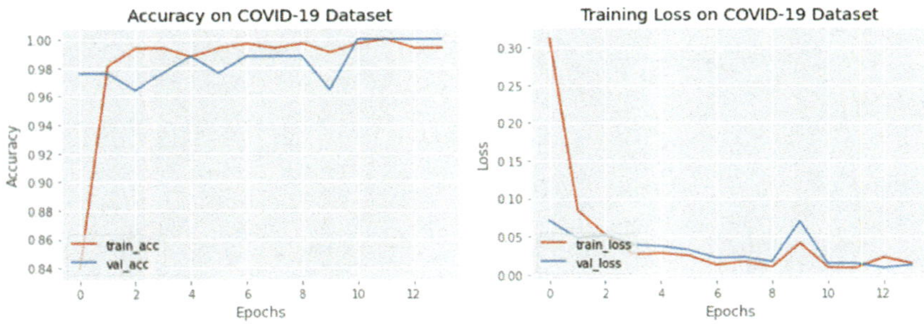

d. ResNet50V2

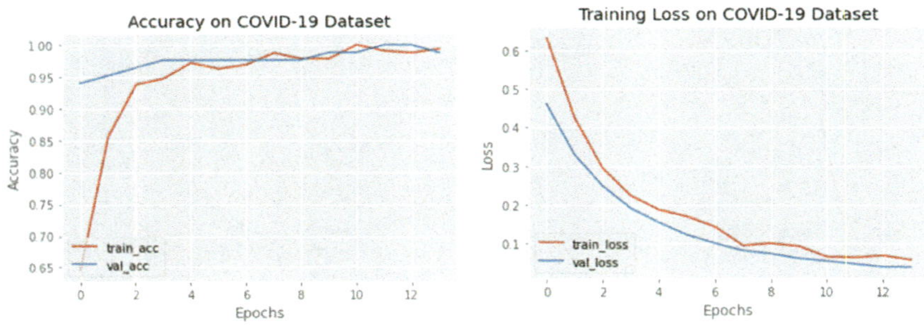

e. VGG16

f. DenseNet201

Fig. 5 (continued)

total number of X-rays in the dataset which are COVID-19 confirmed Figs. 9 and 10 show a comparison of precision values and recall obtained by different classifiers while testing the new dataset. From the graph, it is clear that MobileNetV2 obtained the highest precision rate and recall among all others.

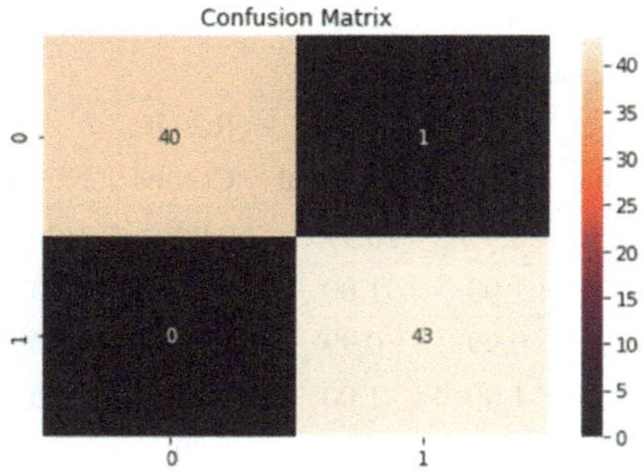

a. InceptionResNetV2

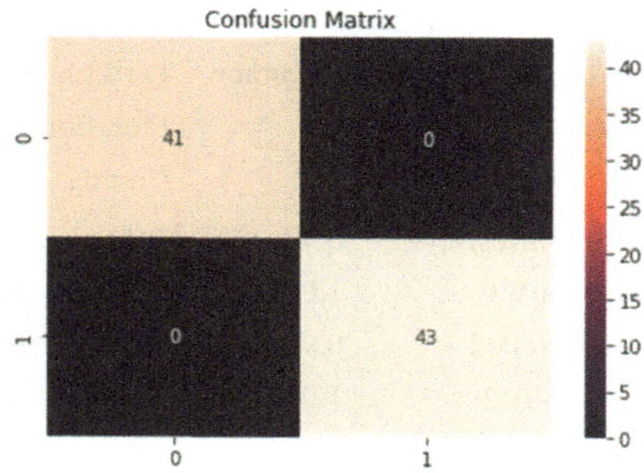

b. InceptionV3

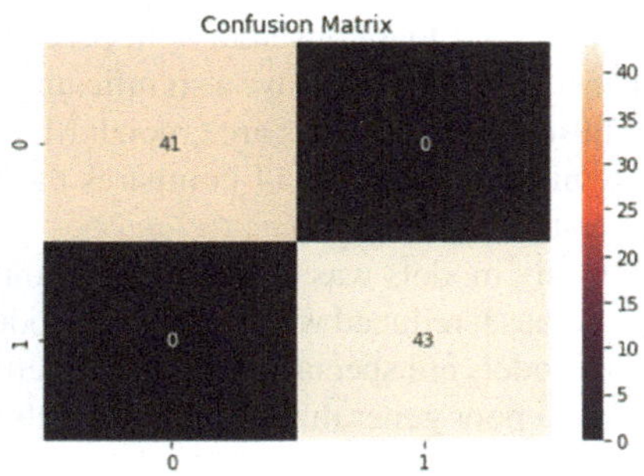

c. MobileNetV2

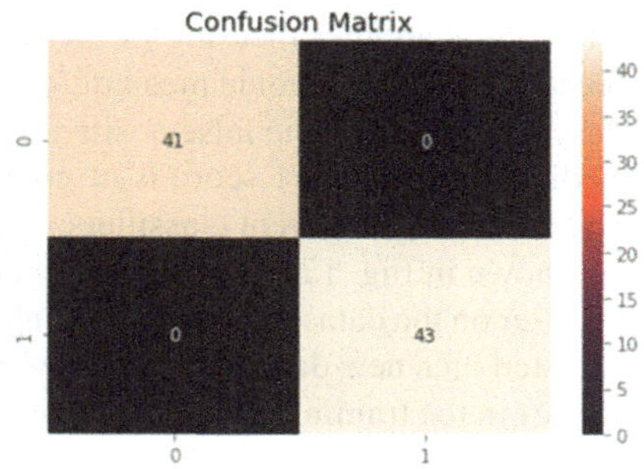

d. ResNet50V2

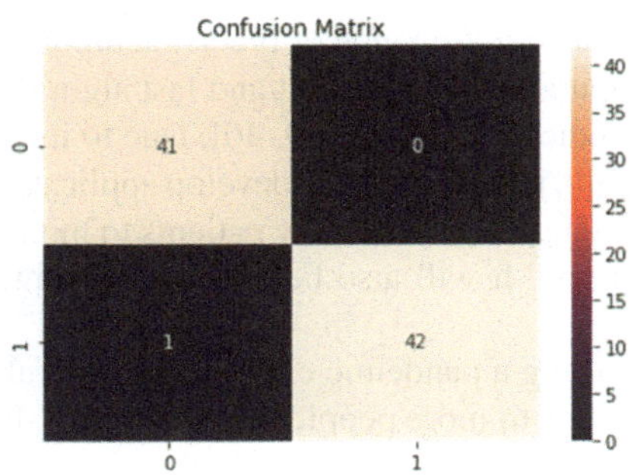

e. VGG16

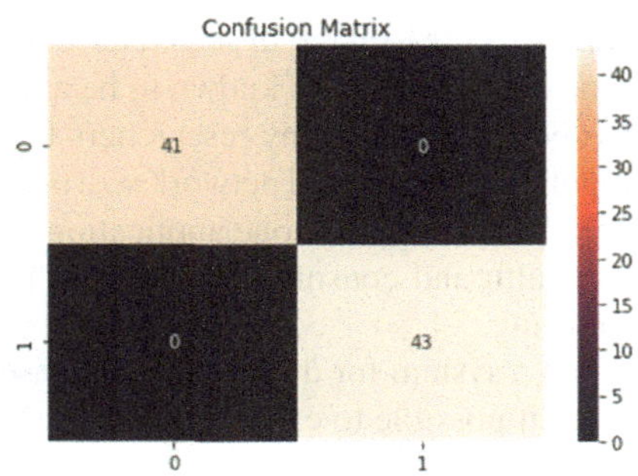

f. DenseNet201

Fig. 6 Confusion Matrix of different models for the classification of images on Dataset I

Table 4 Performance measures of various techniques on Dataset I

Classifiers	Accuracy	Performance measures					
		Precision		F1 score		Recall	
		Corona	Normal	Corona	Normal	Corona	Normal
InceptionResNetV2	0.99	1.00	0.98	0.99	0.99	0.98	1.00
InceptionV3	1.00	1.00	1.00	1.00	1.00	1.00	1.00
MobileNetV2	0.99	1.00	0.98	0.99	0.99	0.98	1.00
ResNet50V2	1.00	1.00	1.00	1.00	1.00	1.00	1.00
VGG16	1.00	1.00	1.00	1.00	1.00	1.00	1.00
DenseNet201	0.99	1.00	0.98	0.99	0.99	0.98	1.00

F1 score is another measure of the test's accuracy that considers both precision and recall. It is the harmonic mean of recall and precision and plays a significant role in classification when the false positive and false negative values are crucial. Mostly in medical diagnosis, F1 score is of great significance. Figure 11 compares the F1 score obtained by different classifiers when tested on Dataset II.

As shown in Fig. 12, the accuracy of all the six models was comparatively found to be better on the dataset used for training, whereas it reduced when the same models were tested on a new dataset. This shows that models are specialized to the structure of images in the training dataset. This results in a poor generalization power of these models.

Results obtained from the above experiments suggest that MobileNetV2 has outperformed all other pre-trained architectures for classifying COVID-19-affected lungs from the healthy lungs and shown the best generalization ability. It achieved an accuracy of 90% on the unseen dataset and the highest value of precision among all others. MobileNetV2 is known to be a computationally efficient and fast algorithm. It has been used by many researchers for medical diagnosis [39, 40]. Due to its low computation power, and network size of 17 Mb, it can be used to develop applications for cell phones. Cell-phone applications for diagnoses will allow patients to monitor their health and consult doctors when required. It will also help in post-treatment monitoring.

Such a system for diagnosis of disease during a pandemic can be very helpful in making it possible to extend diagnosis facilities to more people at a lower cost. It is more preferable for diseases which are communicable, as it reduces the probability of spreading the infection through direct contact of a patient with a doctor. This can be useful during the COVID-19 pandemic as well as in the future for most of the diseases in which diagnosis is done using lung X-ray images.

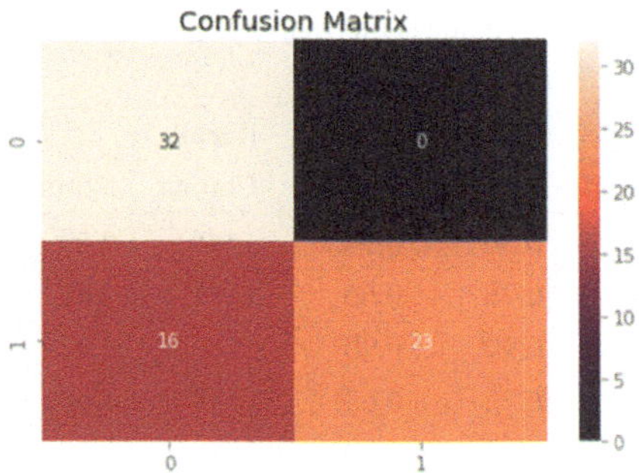

a. InceptionResNetV2

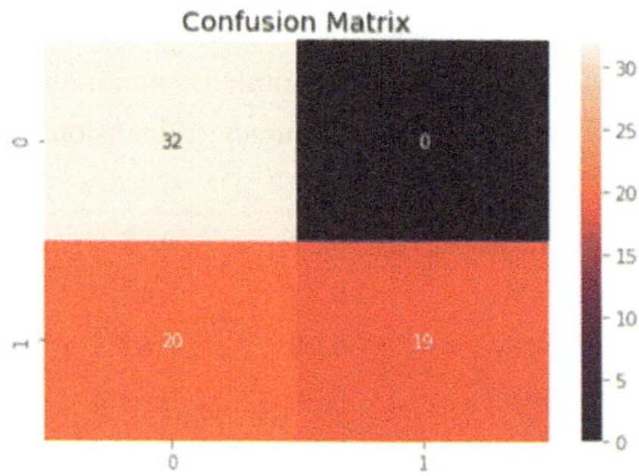

b. InceptionV3

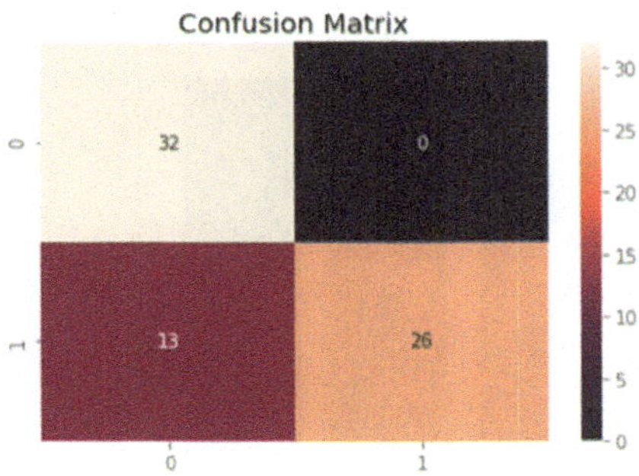

c. MobileNetV2

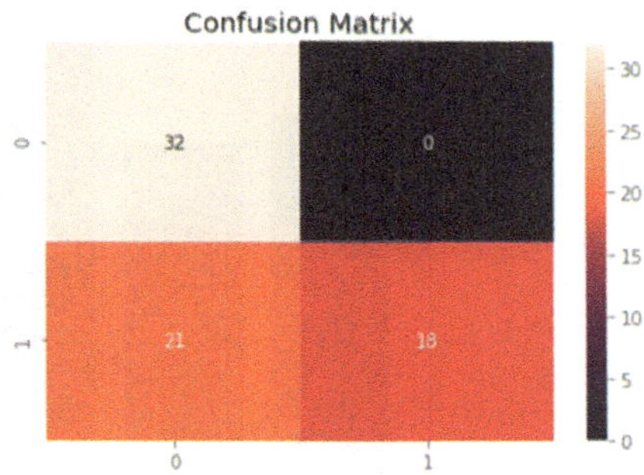

d. ResNet50V2

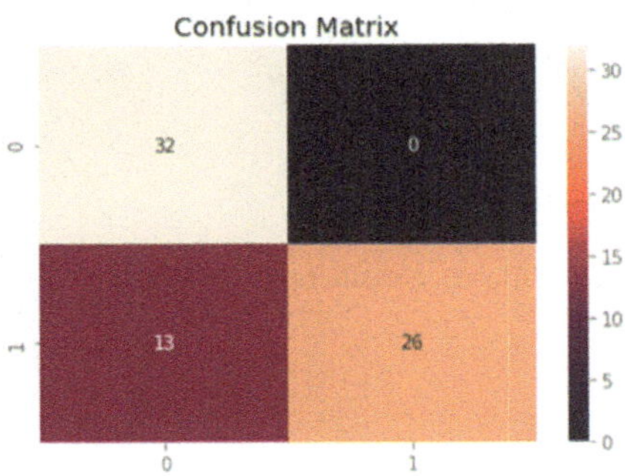

e. VGG16

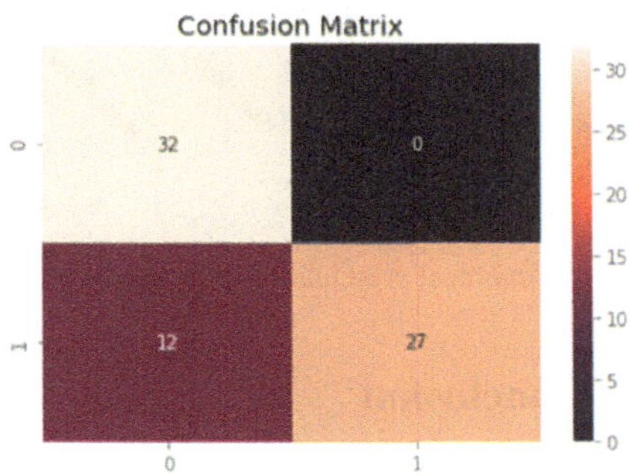

f. DenseNet201

Fig. 7 Confusion Matrix of all models when they were tested for their generalization ability on Dataset II

Table 5 Performance measures of various classification techniques on Dataset II

Classifiers	Performance measures						
	Accuracy	Precision		F1 score		Recall	
		Corona	Normal	Corona	Normal	Corona	Normal
InceptionResNetV2	0.77	0.67	1.00	0.80	0.74	1.00	0.59
InceptionV3	0.72	0.62	1.00	0.76	0.66	1.00	0.49
MobileNetV2	**0.90**	**0.82**	1.00	0.90	0.90	1.00	0.82
ResNet50V2	0.70	0.60	1.00	0.75	0.63	1.00	0.46
VGG16	0.82	0.71	1.00	0.83	0.80	1.00	0.67
DenseNet201	0.83	0.73	1.00	0.84	0.82	1.00	0.69

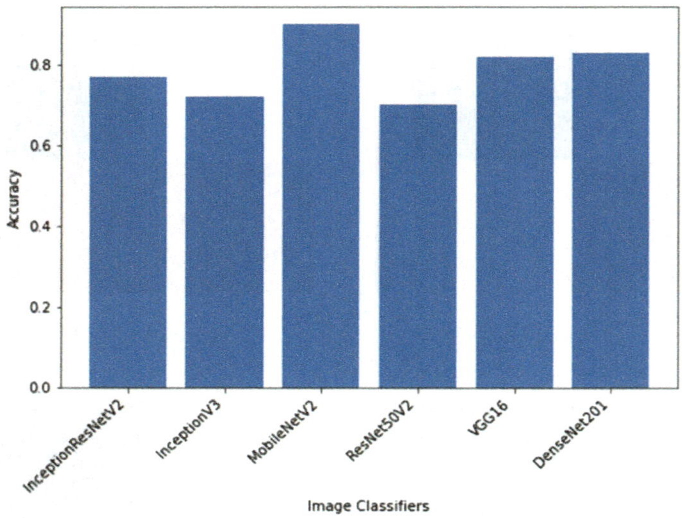

Fig. 8 Graphical presentation of accuracy obtained by different classifiers on Dataset II

5 Conclusion

COVID-19 has spread all across the world and affected millions of lives. Vaccine or any other treatment for this virus is still under research. Until we get any effective treatment for this disease, it is extremely necessary to control it from spreading further in the community. Many researchers are working on systems that can help detect COVID-19 with X-ray images of a person's Lungs. In this chapter, we trained 6 models using transfer learning on a Lung X-ray dataset with pre-processing and data augmentation techniques as discussed in Sect. 3.2. We evaluated the generalizability of these models by testing them on an unseen dataset. Experiments showed that MobileNetV2 outperformed all the classifiers in classifying COVID-19-affected

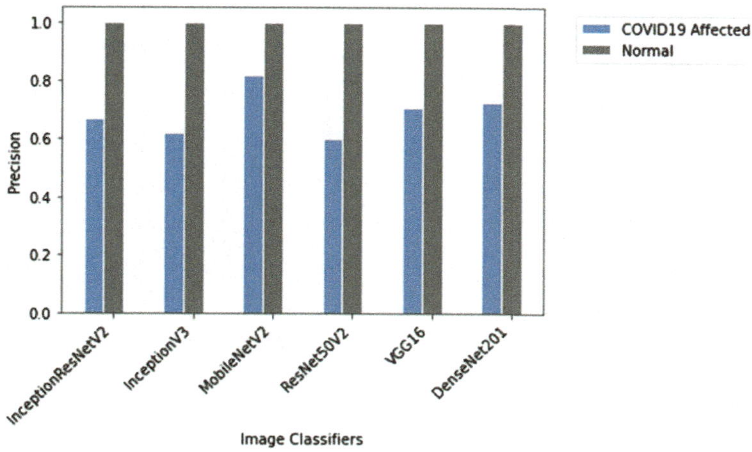

Fig. 9 Graph showing Precision values of all the classifiers for both the classes, when tested on Dataset II

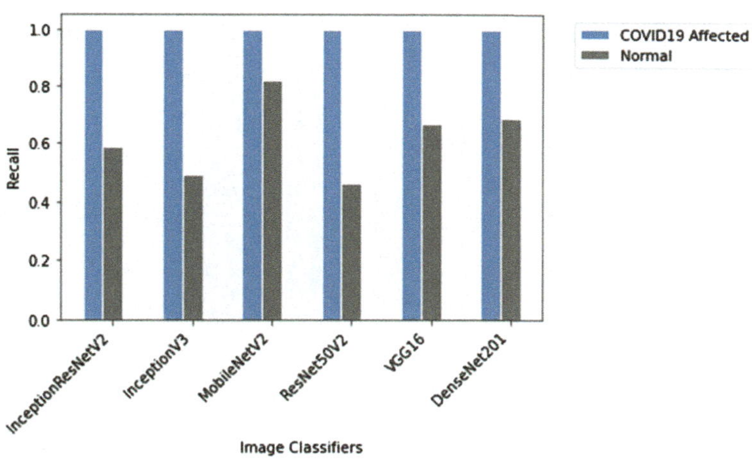

Fig. 10 Graph showing Recall values of all the classifiers for both the classes when tested on the validation dataset

lungs by obtaining an accuracy of 90% and the highest value of precision among all others. This method of screening will be very helpful as it is non-invasive and fast. Moreover, it can be deployed as an android application which will make it remotely accessible. However, using this model for real-life implementation requires more research and improved accuracy. In the future, we aim to deploy and test the model on a smartphone for its effectiveness as a diagnosis application and shall experiment with other versions of MobileNet architecture and compare their performance.

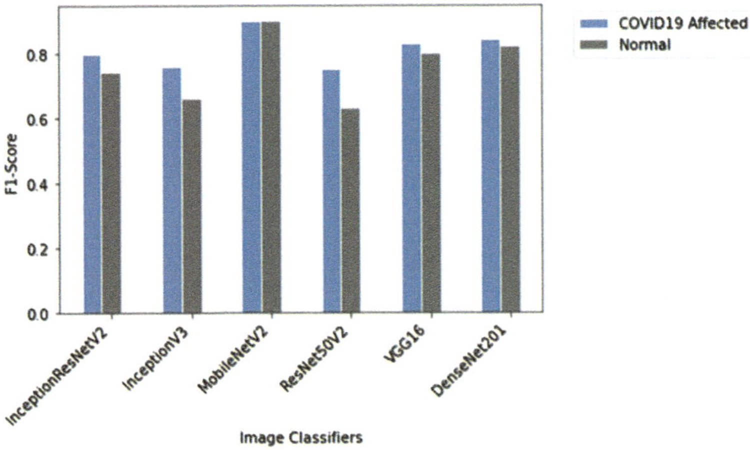

Fig. 11 Graph showing F1 score all the classifiers for both classes, when tested on Dataset II

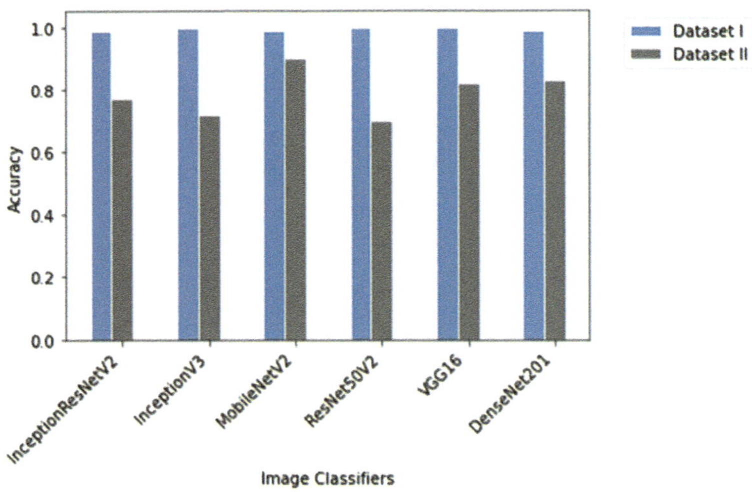

Fig. 12 Graph showing accuracy obtained by different classifiers on Dataset I and Dataset II

6 Animal/Human Involvement

There is no involvement of humans or animals in this research; data used for the simulation is available online for research purposes.

Acknowledgements We would like to express our gratitude toward JP Cohen for providing the 'COVID-19 X-ray image dataset', and Paul Mooney for the 'Chest X-ray image dataset', which were used to conduct this research.

Funding This study has been done as part of a project using open-access datasets and open-source software for simulation.

Conflict of Interest The authors declare that there is no conflict of interest.

References

1. Ananthnarayan R, Paniker KJ (2009) Ananthnarayan and Paniker's Textbook of Microbiology. Orient Blackswan, Hyderabad, India
2. Ter Meulen J, Bakker AB, Van Den Brink EN, Weverling GJ, Martina BE, Haagmans BL, Kuiken T, De Kruif J, Preiser W, Spaan W, Gelderblom HR (2004) Human monoclonal antibody as prophylaxis for SARS coronavirus infection in ferrets. Lancet 363(9427), 2139–2141. https://doi.org/10.1016/S0140-6736(04)16506-9
3. Gibbs AJ, Gibbs MJ, Armstrong JS (2004) The phylogeny of SARS coronavirus Brief Report. Arch Virol 149(3):621–624. https://doi.org/10.1007/s00705-003-0244-0
4. Nicogossian A (2012) In the news. World Med Health Policy 4(1):2020. https://doi.org/10.1515/1948-4682.1230
5. Huang WH, Teng LC, Yeh TK, Chen YJ, Lo WJ, Wu MJ, Chin CS, Tsan YT, Lin TC, Chai JW, Lin CF (2020) 2019 novel coronavirus disease (COVID-19) in Taiwan: Reports of two cases from Wuhan, China. J Microbiol Immunol Infect 53(3):481–484. https://doi.org/10.1016/j.jmii.2020.02.009
6. Huang C, Wang Y, Li X, Ren L, Zhao J, Hu Y, Zhang L, Fan G, Xu J, Gu X, Cheng Z (2020) Clinical features of patients infected with 2019 novel coronavirus in Wuhan, China. Lancet 395(10223):497–506. https://doi.org/10.1016/S0140-6736(20)30183-5
7. Shi H, Han X, Jiang N, Cao Y, Alwalid O, Gu J, Fan Y, Zheng C (2020) Radiological findings from 81 patients with COVID-19 pneumonia in Wuhan, China: a descriptive study. Lancet Infect Dis 20(4):425–434. https://doi.org/10.1016/S1473-3099(20)30086-4
8. Rothan HA, Byrareddy SN (2020) The epidemiology and pathogenesis of coronavirus disease (COVID-19) outbreak. J Autoimmun 109:102433. https://doi.org/10.1016/j.jaut.2020.102433
9. Guo YR, Cao QD, Hong ZS, Tan YY, Chen SD, Jin HJ, Tan KS, Wang DY, Yan Y (2020) (2020) "The origin, transmission and clinical therapies on coronavirus disease 2019 (COVID-19) outbreak: an update on the status." Mil Med Res 7(1):1–10. https://doi.org/10.1186/s40779-020-00240-0
10. Lai CC, Liu YH, Wang CY, Wang YH, Hsueh SC, Yen MY, Ko WC, Hsueh PR (2020) Asymptomatic carrier state, acute respiratory disease, and pneumonia due to severe acute respiratory syndrome coronavirus 2 (SARS-CoV-2): Facts and myths. J Microbiol Immunol Infect 53(3):404–412. https://doi.org/10.1016/j.jmii.2020.02.012
11. Hellewell J, Abbott S, Gimma A, Bosse NI, Jarvis CI, Russell TW, Munday JD, Kucharski AJ, Edmunds WJ, Sun F, Flasche S (2020) Feasibility of controlling COVID-19 outbreaks by isolation of cases and contacts. Lancet Global Health 8(4):e488–e496. https://doi.org/10.1016/S2214-109X(20)30074-7
12. Cunningham AC, Goh HP, Koh D (2020) Treatment of COVID-19: old tricks for new challenges. Crit Care 24(1):91. https://doi.org/10.1186/s13054-020-2818-6
13. Lai CC, Shih TP, Ko WC, Tang HJ, Hsueh PR (2020) Severe acute respiratory syndrome coronavirus 2 (SARS-CoV-2) and coronavirus disease-2019 (COVID-19): The epidemic and the challenges. Int J Antimicrob Agents 55(3):105924. https://doi.org/10.1016/j.ijantimicag.2020.105924
14. Poon LL, Chan KH, Wong OK, Yam WC, Yuen KY, Guan Y, Lo YD, Peiris JS (2003) Early diagnosis of SARS Coronavirus infection by real time RT-PCR. J Clin Virol 28(3):233–238. https://doi.org/10.1016/j.jcv.2003.08.004

15. Hemdan EE-D, Shouman MA, Karar ME (2020) COVIDX-Net: a framework of deep learning classifiers to diagnose COVID-19 in X-ray images. Preprint at arXiv:2003.11055.
16. Ilyas M, Rehman H, Nait-ali A (2020) Detection of Covid-19 from chest X-ray images using artificial intelligence: an early review. Preprint at arXiv:2004.05436.
17. Li L, Qin L, Xu Z, Yin Y, Wang X, Kong B, Bai J, Lu Y, Fang Z, Song Q, Cao K (2020) Artificial Intelligence distinguishes COVID-19 from community acquired pneumonia on chest CT. Radiology 2020:200905
18. Ozturk T, Talo M, Yildirim EA, Baloglu UB, Yildirim O, Rajendra Acharya U (2020) Automated detection of COVID-19 cases using deep neural networks with X-ray images. Comput Biol Med 121:103792. https://doi.org/10.1016/j.compbiomed.2020.103792
19. Wang L, Wong A (2020) COVID-Net: a tailored deep convolutional neural network design for detection of COVID-19 Cases from chest X-Ray images. Preprint at arXiv2003.09871
20. Narin A, Kaya C, Pamuk Z (2020) Automatic detection of coronavirus disease (COVID-19) using X-ray images and deep convolutional neural networks. Pattern Anal Appl 24:1207–1220
21. Apostolopoulos ID, Mpesiana TA (2020) Covid-19: automatic detection from x-ray images utilizing transfer learning with convolutional neural networks. Phys Eng Sci Med 43(2):635–640
22. Cohen JP, Morrison P, Dao L (2020) COVID-19 image data collection. Preprint at arXiv:2003.11597
23. chest X-ray Image Dataset. https://www.kaggle.com/paultimothymooney/chest-xray-pneumonia/chest-xray-pneumonia(Visited on - 9-9-2021)
24. Jiang Z (2019) A novel crop weed recognition method based on transfer learning from VGG16 implemented by Keras. IOP Conf. Ser. Mater. Sci. Eng. 677(3):032073. https://doi.org/10.1088/1757-899X/677/3/032073
25. Razzak I, Naz S, Rehman A, Khan A, Zaib A (2020) Improving coronavirus (COVID-19) diagnosis using deep transfer learning. medRxiv
26. He K, Zhang X, Ren S, Sun J (2016) Identity mappings in deep residual networks. In European conference on computer vision. Lecture Notes in Computer Science, vol 9908. Springer, Cham, pp 630–645
27. Szegedy C, Ioffe S, Vanhoucke V, Alemi AA (2017) Inception-v4, inception-ResNet and the impact of residual connections on learning." In Thirty-First AAAI conference on artificial intelligence (AAAI-17). AAAI, Palo Alto, pp 4278–4284
28. Simonyan K, Zisserman A (2015) Very deep convolutional networks for large-scale image recognition. In: 3rd International Conference on Learning Representations. ICLR 2015, United Kingdom, pp 1–14
29. Deng J, Dong W, Socher R, Li L-J, Li K (2009) Imagenet: a large-scale hierarchical image database. In 2009 IEEE conference on computer vision and pattern recognition. IEEE, New York
30. Yu X, Zeng N, Liu S, Zhang YD (2019) Utilization of DenseNet201 for diagnosis of breast abnormality. Mach Vis Appl 30(7–8):1135–1144. https://doi.org/10.1007/s00138-019-01042-8
31. Sandler M, Howard A, Zhu M, Zhmoginov A, Chen LC (2018) MobileNetV2: inverted residuals and linear bottlenecks. In 2018 IEEE/CVF conference on computer vision and pattern recognition. IEEE, New York, pp 4510–4520
32. Howard AG, Zhu M, Chen B, Kalenichenko D, Wang W, Weyand T, Andreetto M, Mobilenets AH (2017) MobileNets: efficient convolutional neural networks for mobile vision applications. Preprint at arXiv:1704.04861
33. .Bai J, Wang B, Chen C (2019) Inception-v3 based method of LifeCLEF 2019 bird Recognition inception-v3 Based method of LifeCLEF 2019 bird recognition
34. Xia X, Nan B (2017) Inception-v3 for flower classification. In 2017 2nd international conference on image, vision and computing (ICIVC). IEEE, New York, pp 783–787
35. Ioffe S, Szegedy C (2015) Batch normalization: accelerating deep network training by reducing internal covariate shift. In 32nd international conference on machine learning, ICML 2015, vol 1, pp 448–456.

36. Szegedy C, Vanhoucke V, Ioffe S, Shlens J, Wojna Z (2016) Rethinking the inception architecture for computer vision. In Proceedings of the IEEE conference on computer vision and pattern recognition. IEEE, New York, pp. 2818–2826
37. Sabour S, Abbasnezhad O, Mozaffarian S, Kangavari HN (2017) Accuracy and precision in medical researches; common mistakes and misinterpretations. World J Res Rev 4:58–60
38. Singh S, Gautam K, Singhal P, Jangir SK, Kumar M A survey on intelligence tools for data analytics. igi-global.com.
39. Sheikh S, Qidwai U (2020) Using MobileNetV2 to classify the severity of diabetic retinopathy. Int J Simul: Syst Sci Technol 21(2), 16.1–16.6
40. Mor N Corona Virus Disease (COVID-19) Screening with Deep Learning Using CT Images Brain-Inspired Method of Pre-processing a Deep Neural Network for Designing a Very Deep Sparse Neural Network View project Method of Pre-processing a Deep Neural Network for Addr.

Chapter 10
A Predictive Modelling of Covid-19 Reoccurrence Using Recurrent Neural Network

Olufunke Rebecca Vincent and Jesufunbi Damilola Bolariwa

Abstract This study presents an enhanced model for using a recurrent neural network to predict the possible recurrence of Covid-19. Using a mix of a recurring neural network with a Long Short Term Memory (LSTM) and its variants, such as stacked LSTM, coevolutionary LSTM, and a bi-directional LSTM, the probability of possible recurrence of covid-19 is evaluated. A type of recurrent neural network (RNN) that takes an order of information and then uses a recurring gate mechanism is long-term memory. It can be compared to feedforward and recurrent neural networks; the proposed model was implemented and validated using the data reported for the previous month and the number of cases as input to calculate the number of issues. The number of cases was grouped into two: the training set and the validation set. As anticipated, this model achieved a high result. Besides, the feasibility of the method and the predicted model structure are discussed through analysis, the outcome shows time, and the direction of backpropagation would provide an acceptable result. The purpose of this research is to predict whether or not the curve of the virus will flatten or continue to expand over time.

Keywords Long short term memory · Back propagation through time · Recurrent neural network · COVID-19

1 Introduction

The Covid-19 pandemic has begun to parodically upsurge the world's nations and expand exponentially within a period. In reported cases, the pandemic incidence began to decline in July 2020. The world hoped that the virus has dramatically decreased; in October, coronavirus cases suddenly started to rise unexpectedly, causing the World Health Organization (WHO) to declare the second stage of the pandemic surge. COVID-19 case rates are rising hourly and monthly. The different

O. R. Vincent (✉) · J. D. Bolariwa
Department of Computer Science, Federal University of Agriculture Abeokuta, Abeokuta, Nigeria

U. Kose et al. (eds.), *Computational Intelligence for COVID-19 and Future Pandemics*, Disruptive Technologies and Digital Transformations for Society 5.0,
https://doi.org/10.1007/978-981-16-3783-4_10

data can be tracked in real-time from John Hopkins University hospital Coronaviruses are encased as a positive single-stranded large virus of ribonucleic acid (RNA) affecting humans and a vast number of animals.

In 1966, Tyrell initially detected Coronaviruses. Bynoe then produced the virus as the bulb-shaped virus in patients with common colds [1] based on their birth. It has four coronavirus mini families, namely Alpha, Beta, Gamma, and Delta. Gamma and Delta viruses are common with pigs and birds, Alpha and Beta coronaviruses originate from bats. Size varies for each genome between 26 and 32 kb. Beta-coronaviruses thus cause severe illness and deaths of the seven coronavirus subtypes that can infect humans, whereas alpha-corona viruses can trigger asymptomatic or mildly symptomatic infections. Covid-19 is part of the Beta-corona virus B family and is closely associated with the Extreme Acute Respiratory Syndrome virus [2]. Nucleocapside protein (N), spike protein (S), small membrane protein (SM), and glycoprotein (M) membrane are coded by the four key structural genes with additional glycoprotein (HE) membrane occurring in HCoV-OC43 and HKU1 beta-coronaviruses [3].

A significant clinical sign that allowed case identification was pneumonia. Gastrointestinal symptoms and asymptomatic infections are often recognized in more recent studies, especially among young people [4]. To date, observations indicate a mean five-day incubation period [5] and a median three-day incubation time (range: 0–24 days) [6]. The proportion of individuals diagnosed with Covid-19 who remain asymptomatic has not yet been measured during the period of infection. Symptoms usually begin in less than a week in symptomatic cases, consisting of nausea, cough, nasal inflammation, fatigue, and other upper respiratory tract infection signs. In approximately 75% of patients, the infection can lead to severe dyspnea and significant chest symptoms leading to pneumonia, as shown by computer tomography on admission. [6] Latest studies have shown that patients aged 60 years or older are at greater risk than young individuals who may be less likely to become infected or may have milder symptoms or asymptomatic infections [7] and $\geq$34.4 of the Middle East Respiratory System (MERSCoV) [8, 9].

The African countries selected were primarily due to their environmental conditions, population, the standard of hygiene, and the poor health infrastructure in the different countries. It has shown that the virus can only reproduce itself more rapidly and twice based on the atmosphere and the environment. Thus, the countries were chosen to know when the virus would spread and whether the spread would increase due to the countries' weather conditions. The recurrent neural network was introduced because of its more significant benefit other than other neural family members. RNN has numerous models that assist in its smooth modeling. Few of the cons and pros of RNN are listed. The pros are as follows: the probability of processing input data of any length, the size of the model not increasing with the input size, historical information, and weight are exchanged over time in RNN computation. The downside of RNN is that the computation could be very slow. It is also not very achievable to have trouble accessing knowledge from a long time ago. While LSTM is taken into account, RNN might not accept any future feedback for a current state [10, 11].

This paper uses the recurrence neural network model to predict the number of patients who may get infected with the virus, using LSTM-based models for prediction, based on the disease's propagation properties at different points. On the dataset obtained from the John Hopkins Hospital and the Country Center for Disease Control, multiple LSTM models were tested and found deeper LSTM models such as stacked LSTM; convolutional LSTM bi-directional LSTM provide better accuracy than the simple LSTM model. The data curves of seven African countries (8) will reflect the proposed model: Nigeria, South Africa, Egypt, Kenya, Morocco, Madagascar, and Guinea.

Section 2 offers an overview of the meaning of long-term, short-term memory and its variants, such as stacked LSTM, convolutional LSTM, and bi-directional LSTM, as well as the recurrence neural network. This paper is classified into different sections: The context of long-term, short-term memory and its variants, such as stacked LSTM, convolutional LSTM, and bi-directional LSTM as the recurrence neural network, are explained in Sect. 2. The descriptive analysis of the data collected for each country is in Sect. 3. Section 4 covers the model architecture, the prediction of error estimates using the proposed model, and the consideration of steps to be applied in each country to curb the virus's spread. Section 5 finished with a description of the work that was done.

2 The Proposed Recurrent Neural Networks (RNN)

Recurrent Neural Networks (RNN) differ from other neural networks in the artificial world. While other networks "move" in a linear direction in the feedforward or back-propagation point, instead of a feedforward, the Recurrent Network incorporates a recurrence relationship and uses Back-Propagation over time know. The Recurrent Neural Network consists, one for each time level, of several specified activation functions. Any computer has an internal state called the machine's hidden condition. This invisible state means past data that is currently maintained at a given time level by the network. This secret state is changed any step of the way to reflect a shift in the network's awareness of the past [12, 13].

Identifying every bit of knowledge over time is one of the Recurrent Neural Network's advantages. Only because of the function to determine the previous input is it useful to predict time series. Recurrence neural network is used to carry temporal information, involving feedback links that are internal or external, while they can encode the time context. This feedback can be internal or external. The recurrent neural network can learn patterns from previous records and even generalize and predict future virus cases for unknown data. The recurrent neural network is part of the processing of sequential data by the neural network family. RNN uses information from the previous input sequence for the prediction and takes it into account; it has the drawback of not using the data later in the series [14, 15].

Several researchers, such as Jordan and Elman RNN, Long Short Term Memory (LSTM), Hopfield, Neural History Compressor, and many more, have developed various forms of RNN. In this neural recurrence network, the feedback obtained varies from network to network. Jordan RNN has a feedback connection from output to input, and Elman RNN from its hidden layer neuron has a feedback link back to its input. The neurons are referred to as "System" or "Background" neurons in the input layer that accept feedback. The background neuron's aim in the recurrence neural network is to get input from the upper layer and send output along with other program inputs to the hidden layer after processing [16].

In Covid-19 long-term forecasting, a close association exists between current and future cases. There would be a good chance of a successful prediction of the X + 1 virus if we were able to provide similar knowledge about the X-Day virus, along with other variables. Feedforward networks are faster than feedback networks because forward feed networks are guaranteed to achieve consistency by setting a solution with only one pass in feedback networks. Before the system has stabilized, the feedback network must be iterated over some cycles. The feedback network's performance is more limited. The neural recurrence network can achieve stuff that other networks cannot do. Feedback loops allow for planning and adaptability, linear networks are restricted, and the required nonlinearity is given by neural recurrence [17].

LSTM is a different type of RNN, such networks have a link between neurons and neurons and form a directed loop. An internal self-looped cell produces this type of structure that exhibits complex temporal characteristics. Figure 2 describes that RNN has chain-like repeat module structures that allow RNN to retrieve previous information that allows arbitrary sequences to be processed by RNNs. For sequence learning, RNN is effective. In this article, to predict the future number of possible cases in the time series, a long-term memory optimization algorithm will be used.

2.1 Long Short Term Memory (LSTM) and Its Variants

LSTM can understand long-term dependencies, and for a long time, their usual tendency is to recover information. To solve difficult sequence problems in machine learning and deliver a perfect result, LSTM can construct massive, recurrent networks. There are memory blocks on the LSTM network that are connected by layers rather than neurons. A block has a component that makes it smarter than a classic neuron and a chain of memory. LSTM has a modular chain, but there are more complex structures in the replication modules. One element is contained in every repeatable LSTM module. The memory block is mainly designed for storing data over a long period. Furthermore, the LSTMs have chain-like modules, but there are more complex structures for repeated modules. In addition to three different multiplicative units called 'gates,' this block includes four sections called 'constant error carousel' (CEC) [16, 18].

Consequently, information can easily flow unchanged through the cells; LSTM learns more easily from long-term dependence. The output could track the degree to which new input flows into the CEC cell, the cell processes information, and the cell output streams to the rest of the networks, respectively as described Fig. 2.

A cell that is a memory component of the LSTM and three regulators called the gates responsible for the flow of information within the LSTM network is one of the most prevalent architectures. The cell is logically responsible for keeping track of the dependencies in the input chain between the elements. The activation function of the LSTM gate is also the "sigmoid function." logistic function inside and outside the LSTM windows, links are rendered, some of which are recurrent. The weights of these relations, which must be perfected by practice, determine how the gates work.

In the LSTM neural network, the execution stage is defined as follows: the first stage determines how much of the past it can remember. Determines the role of sigmoid, looks at the previous state $(p_t - 1)$ and the new $(x - t)$ information and determines the result,

$$Z_t = \sigma(Q_f.[P_t - 1, X_t]) + O_f. \tag{1}$$

The weight matrix and the forgotten direction bias are collectively P, Z, and O, and the sigmoid function is σ. The forgotten gate's main objective is to monitor how often the previous time's (pt-1) cell state is transferred to its cell state. The gate produces a value based on (pt-1) and X_t between 0 and 1, where full reservation is usually used and denotes complete discard. The second step specifies how much to the current state this device can add. There are two elements, the sigmoid feature, and the function tanh. The sigmoid defines the values (0 or 1) to be passed through. The tan-h function weights the values passed to demonstrate their degree of importance. (−1 to 1).

$$I_t = \sigma(W_c[P_t - 1, X_t]) + O_I, \tag{2}$$

where I_t is the input gate while σ is the output gate and.

$$c_t = \tanh(W_c[P_t - 1, X_t]) + O_c, \tag{3}$$

Determines the component that makes it output from the current cell state. A sigmoid layer is one which decides what portion of the cell state makes it to the output and determines the performance [19] The state of the cell is through tanh to push the values between (−1 and 1) and multiply it by the output of the sigmoid gate

$$O_t = \sigma(W_o[O_t - 1, x_t]) + O_o \tag{4}$$

where W = weight and σ is the sigmoid function

$$P_t = O_t * \tan h(C_t). \tag{5}$$

It is shown in Fig. 1 that the LSTM neural network not only contains a recurrent external network of hidden layer elements involved in the recurrent neural network

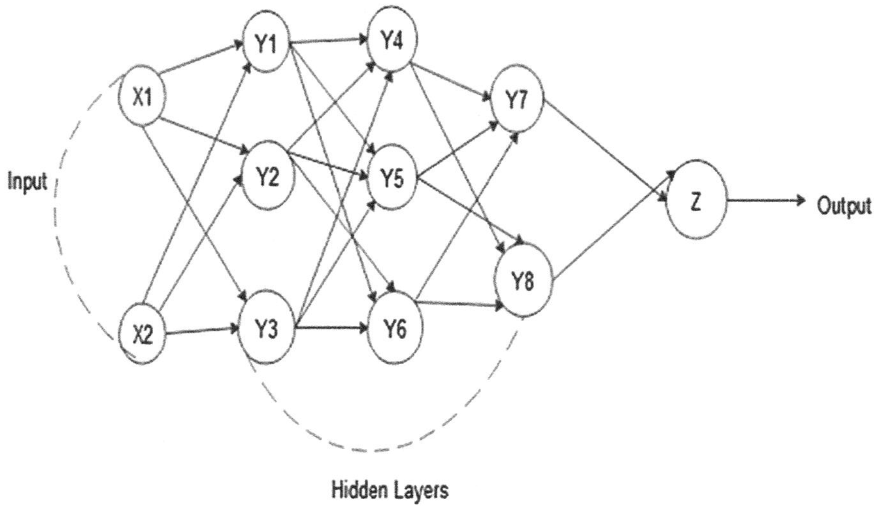

Fig. 1 A recurrent neural network model for COVID-19 analysis

but also requires neuronal self-circulation. In the threshold device unit, the input, update, and output information are managed so that the LSTM memory module can retain and analyze long-term historical details.

2.1.1 Stacked LSTM

Stacked LSTM is described as an LSTM model consisting of multiple memory cells and multiple hidden layers. A sequence output instead of a single value output is given to the LSTM by the LSTM layer. The stacked LSTM structure is shown in Fig. 2, providing an individual output for each step instead of providing a single output for all steps (Fig. 3).

2.1.2 Convolutional LSTM (Conv-LSTM)

Vector x outputs vector y in convolutional LSTM (Xianjian et al., 2015), hidden vector h and gates (I_t, F_t, O_t) are 3D tensors, and spatial dimensions are the last two dimensions. In both the transition from input to state and state to state, Convolutionary LSTM is a type of Spatio-temporal prediction RNN with a convolutionary structure. The convolutional LSTM input also specifies every cell's future state in the grid and the connecting cells' past location. In a state transformation, this can be done by using convolutionary operations rather than multiple functions.

$$I_t = \sigma(W_{xi} * X_t + W_{hi} * H_{t-1} + W_{ci} \odot C_{t-1} + B_i) \quad (6)$$

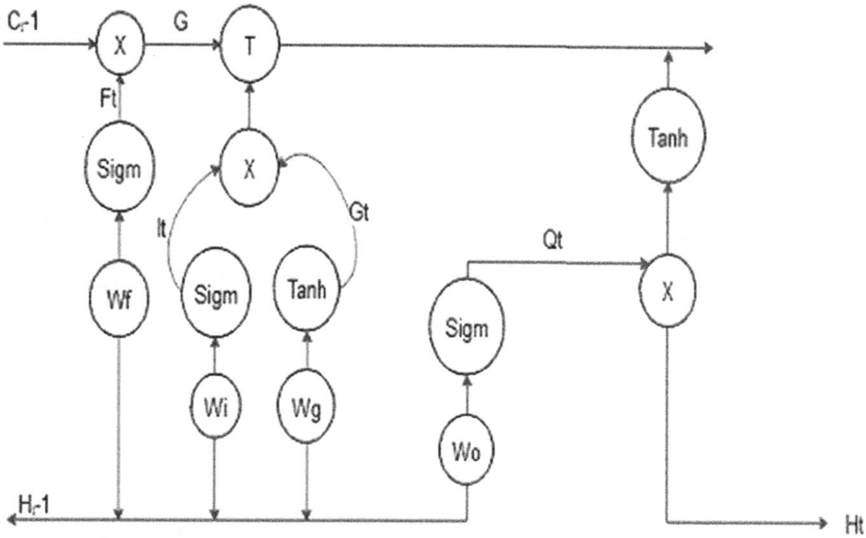

Fig. 2 A long short term memory analysis for COVID-19 patterns

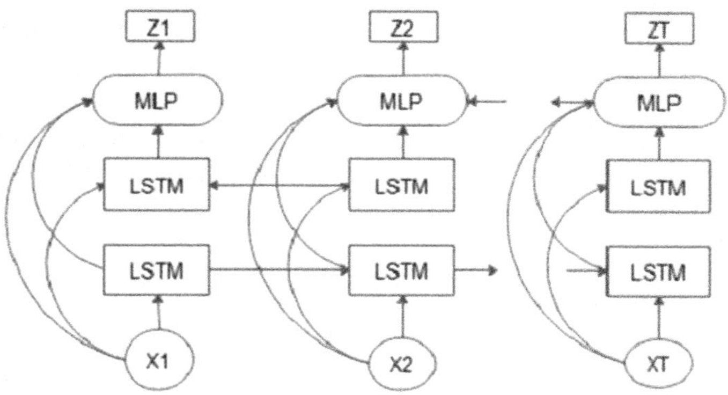

Fig. 3 Stacked LSTM architecture showing the input passage and the output passage of data

where I_t is the input gate and the projection metrics.

$$F_t = \sigma\left(W_{xf} * X_t + W_{hf} * H_{t-1} + W_{cf} \odot C_{t-1} + B_f\right), \tag{7}$$

and $(W_{xi}, W_{hi}, W_{ci}, W_{xf}, W_{hf}, W_{cf}, W_{xo}, W_{ho,} W_{co})$ are the recurrent weight matrices while

$$C_t = F_t \odot C_{t-1} + I_t \odot tan\,h(W_{xc} * X_t + W_{hc} * H_{t-1} + B_c). \tag{8}$$

In Eq. (8), $\odot$ Is the element multiplication, that is, Hadamard product

$$O_t = \sigma(W_{xo} * X_t + W_{ho} * H_{t-1} + W_{co} \odot C_{t-1} + B_o). \quad (9)$$

Besides, σ as the logistic sigmoid function of the equation

$$H_t = O_t \odot \tanh(C_t), \quad (10)$$

where $\tanh$ Is the hyperbolic tangent function, the hidden state H is being used to make predictions (Lei, L 2018).

2.1.3 Bi-directional LSTM

Bi-directional LSTM extracts temporary data by taking into account both past and future knowledge. Standard RNN hidden neurons are split into forward and backward states in the hidden neuron [20]. In which the forward-state neurons are not connected to the backward state and vice versa. The training steps of bi-LSTM over time can be summarized below: Forward pass runs input data for time $1 \leq T \leq T$ through bi-directional and evaluates the predicted output. Forward state-run from t = 1 to t = T and backward state from $t = T$ to $t = 1$. Backward pass: it evaluates the objective function derivative for time $1 \leq T \leq T$ calculated in forwarding pass. The backward pass is run for output neurons $t = T$ to $t = 1$ and backward states from $t = 1tot = T$. The bi-LSTM is useful for situations requiring context input; it has majorly been used in classification, especially text classification [6], sentimental classification [21] speech and recognition classification [22], also the bi-LSTM are used in load forecasting [23].

2.2 *Evaluation Metrics*

The neural network forecasting models were evaluated with root mean square error (RMSE), mean absolute error (MAE), and mean absolute percentage error (MAPE), variance (EV), and root mean squared log error (RMSLE).

$$RMSE = \sqrt{\frac{1}{n}\sum_{t-1}^{n}(x_t - \hat{x}_t)}, \quad (11)$$

The square root of the average square difference between prediction and actual observation is a quadratic scoring rule that calculates the error's average magnitude.

$$MAE = \sum\nolimits_{t-1}^{n} |x_t - \hat{x}_t|2, \quad (12)$$

For all instances of the tested dataset, the mean of each prediction error's absolute value is the difference between the actual values and the expected value. The measurement of two continuous variables can be referred to as MAE.

$$MAPE = \frac{100}{n}\sum_{t-1}^{n}|\frac{x_t - \hat{x}_t}{x_t}|\%, \tag{13}$$

The absolute percentage average error for each period minus the actual value is divided by the actual values.

$$EV = 1 - \frac{var(\hat{x} - x)}{var(x)}, \tag{14}$$

The variance $\hat{x} - x$ is divided by variance x, which is the real value.

$$RMSLE = \sqrt{\frac{1}{n}\sum_{t-1}^{n}(\log(x_t) - \log(\hat{x}_t))2} \tag{15}$$

where x_t is the actual value, $\hat{x}_t$ is the corresponding estimated values, and n is the measurement number; the RMSLE indicator is used because it is widely used in regression problems to assess the model quality and has a score indicator in various data science challenges [24, 25].

3 A Recurrent Neural Network Analysis of Covid-19

For Covid-19 time series forecasting, this study proposes a recurrent neural network model. To predict confirmed cases, deaths, and recovered cases, the model's testing and training are carried outputting in consideration 3 RNN model. The forecasting proposal system is illustrated in Fig. 4. The Covid-19 cases' prediction has been split into two key stages, the training and testing stages. The raw data obtained is preprocessed and standardized in the training phase and then used to create the RNN model. The RNN models' values are chosen so that the loss function is reduced during the testing process preparation, using the previously built models with the selected parameter to predict the number of cases of covid-19. Using various statistical measures such as RMSE, MAPE, MAE, and RMSLE, the model's accuracy is checked by comparing the measured data with real data. This study's sole objective is to predict days, months before another virus wave will strike again. The proposed architecture model for the study is presented in Fig. 4.

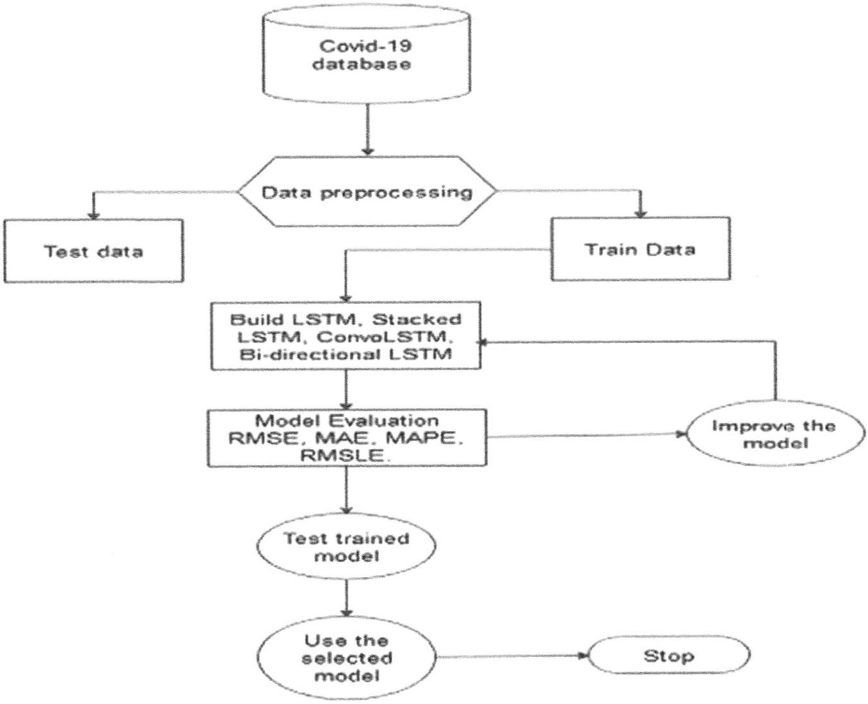

Fig. 4 The proposed model architecture

3.1 Data Analysis

In almost 218 countries and territories, the covid-19 virus has been confirmed by the WHO, and the African continent has a share of the pandemic. This study aims to forecast and predict the spread of the virus and at what specific rate. This research is focused on regular data of confirmed, death, and recovered patient cases collected from 7 Africa countries, namely Nigeria, Egypt, South Africa, Kenya, Morocco, Madagascar, and the Republic of Guinea. John Hopkins Teaching Hospital made the data readily available. This data is stochastic since the number of cases depends on other environmental and physical variables, whether it's going to increase or decrease.

3.2 Back Propagation Analysis of COVID-19

BPTT (Backpropagation through time) is conceptually more comfortable and more effective in computational time than RTR (Real-Time Recurrent Learning). BPTT starts by unfolding an RNN Feedforward neural network over time. BPTT consists of the repeated application of the chain rule. The subtlety is that the loss function of RNN depends on the activation of the hidden layer, not only by its influence on the output layer but also by its impact on the hidden layer. The square error used as the loss function is specified by.

$$et = \sum_{i=1}^{n} (X_t - Pc)^2, \tag{16}$$

where e^t represent the error, and pc represents the predicted cases per value. The back-propagation algorithm works by calculating the loss function gradient concerning each weight by the chain rule. It calculates the gradient one layer simultaneously, iterating backward from the last layer to avoid a wasteful approximation of the intermediate-term chain rule.

$$\delta^l = \nabla_{ac\varnothing\sigma^1(z^l)} \tag{17}$$

We have $\nabla_{ac\varnothing\sigma^1(z^l)}$ to be a vector whose components are the partial derivatives. $\nabla\_(ac\varnothing\sigma^(1)(z^l))$ This can be thought of as expressing the rate of change of C concerning the output activations,

$$\delta^l = \left(\left(w^{l+1}\right)Ot\delta^{l+1}\right)\sigma^1\left(z^l\right), \tag{18}$$

$\left(W^{i+1}\right)t$ The transpose of the weight matrix W^{i+1} for the $(I+1)th$ layer. Each element has an interpretation suppose we know the error (δ^l+1) at the $(I+1)th$ layer. When the transpose is being applied, the weight matrix $\left(W^{i+1}\right)$ can be thought of as moving the error back through the network, giving some measurement error at the output of the $(I+1)th$ layer. We can now take the Hadamard product $(Ot\delta^{l+1})\sigma^1(z^l)$ then move the error backward through the activation function layer giving the error δ^l in the weighted input to layer.

$$\frac{\partial c}{\partial b_j^l} = \delta_j^l, \tag{19}$$

δ^l Can denote the vector error associated with the layer, backpropagation will give a way of computing δ^l for every layer and then relating those errors to the quantities of real interest $\frac{\partial c}{\partial b_j^l}$ such that.

$$\frac{\partial c}{\partial w_J k} = ak^{\mathrm{i}-1}\delta j^z. \tag{20}$$

It is a natural expression. The first term on the right $\frac{\partial c}{\partial w_J k}$ measures how fast the cost is changing as a function of the j^k output activation. The second term on the right $ak^{i-1}\delta j^z$ measures how fast the activation function is changing. The backpropagation equation provides a means of computing the gradient of the cost function expressed in Algorithm 1.

Algorithm 1: Back Propagation Through Time
Input x: Set activation a^l for the input layer
Output: Gradient of lost function : $\frac{\partial c}{\partial b_j^l} = \delta_j^l$ *and* $\frac{\partial c}{\partial w_{jk}} = ak^{i-1}\delta j^z$

1. *Feedforward for every* $l = 1, 2, 3, \ldots\ldots\ldots z$
2. *Sum up* $z^l = w^l\, l^{z-1} + b^l$ *and* $a^1 = \sigma(z^l)$
3. *Output error*δ^z *sum up the vector* $\delta^l = \nabla_{ac}\, \emptyset\sigma^1\,(z^l)$
4. *Back propagate error for each z= z-1, z-2, z-3……z*
5. *Compute* $\delta^l = ((w^{l+1}) Ot \delta^{l+1})\sigma^1(z^l)$

   End

From the algorithm above, it is clearly stated why it is backpropagation. The error is being computed backwardly (δ^z) starting from the final layer. The backward migration is a result of the reality that the expense is a feature of the network. The study continually implements the chain law backward across the layer to get functional phrases to see how the changes with earlier weights and prejudices.

4 Analysis of Result

The Training and modeling are carried out on open sources library such as using python programming language, Numpy, Pandas, as high-level language. Jupyter notebook, MATLAB as a tool. The APIs are used to design the current model structure for different recurrent neural network variants as stacked, convo, and bi-directional LSTM. These models are used to learn the changing dependent structure in the data and map the learning sequence present to predict the number of confirmed cases present in any particular country. These models are provided with country-based historical data of the number of cases appearing daily and considering the dataset's changing structure. Hyperparameter tuning of each of these models is done rigorously.

4.1 Dataset

We used the COVID-19 dataset, which consists of date, countries, confirmed cases, recovered cases, and the number of deaths from the John Hopkins private hospital and Github repository and the center of diseases for various countries. We have put into consideration some other data which will help in predicting a good one.

4.2 Network Optimization and Training

The last 10 days' data is used for validation, and the rest is used for Training. Each data generated is trained using a maximum of 1000 iterations. 100 epoch delay on validation loss for early stopping was used. During optimization, MSE is being minimized over the validation set. Each country's data is uniquely used to generate country-specific optimized data. The dropout is chosen as zero most of the time. Relevant data from each country are educated using a case of its own. The data is validated during optimization by forecasting active cases, and the same information is used to predict death, recovery, and the current number of cases. Using a maximum of 100 epochs, the data is trained with the early stop mechanism used during optimization. It is noted that during preparation, the loss curve is an iterative reduction. Different results were obtained at various epoch stages, and Table 1 shows the values of the validity search, gradient, and Mu. We can deduct from the training that the greater the number of epochs, the lower the value of the gradient and Mu's value.

Table 1 Definition of acronyms used in Figs. 1 and 2

Acronym	Meaning	Function
X_1,X_2	Input values	It takes in data to be the process
Y	Hidden layer	Where the processing takes place
Z	Output values	It output the processed data
C_{t-1}	Cell	It is logically responsible for keeping track of the dependencies in the input chain between elements
G	Gate	They are regulators responsible for the flow of information within the LSTM network
F_t	Forget gate	It is to monitor how often the previous time's cell state is transferred to the current time's cell state
Sigm	Sigmoid	It is the activation function of the LSTM gate
W_f, W_t	Weight function and gate	It is the weight of each function in the cell f
W_i	Weight bias	It is the differences in each function's weight in the cell
W_o	Weight output	It output the weight of the processed data
Tanh	Target function	It weights the values passed to demonstrate their degree of importance
I_t	Input gate	Values are entered through this gate
Q_t	Output gate	Processed information leaves through this gate
MLP	Machine learning process	It takes data from multiple sources to process

4.3 Prediction Accuracy

The precision of the estimation of the proposed method is being debated. A suitable model is being chosen for each country, trained to predict the death rate, reported cases, and the recovered rate, whether it is going to drop or increase as the case may be The MSE, RMSE, MAPE, EV was determined on the validation data to test the methods (Table 2).

Figures 5, 6, 7, 8, 9, 10, 11 show the number of cases in each country for a specific period. Table 1 shows the values gotten from the training state in MATLAB as mentioned in (4.3), and it shows the number of the epoch, the value of the gradient, MU, and the validation check. It is seen that the values for each country differ from one another based on the data the model was trained.

Table 2 Training state table

Country	Epoch	Gradient	MU	Validation check
Nigeria	14	1,370,364	10	6
South Africa	30	652,897	1000	6
Egypt	23	7813	100	6
Kenya	15	1241	100	6
Madagascar	43	69,711	0.1	6
Morocco	16	17,412	10,000	0
Guinea Republic	22	80,715	10	6

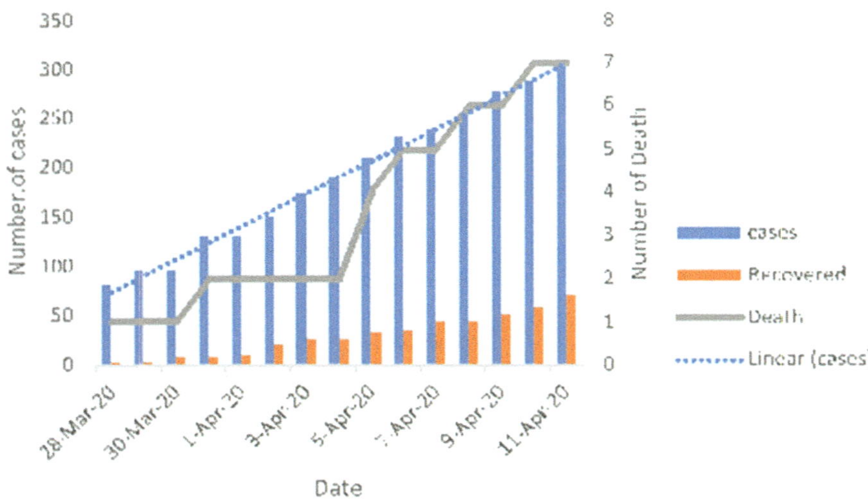

Fig. 5 Covid-19 trend in Nigeria

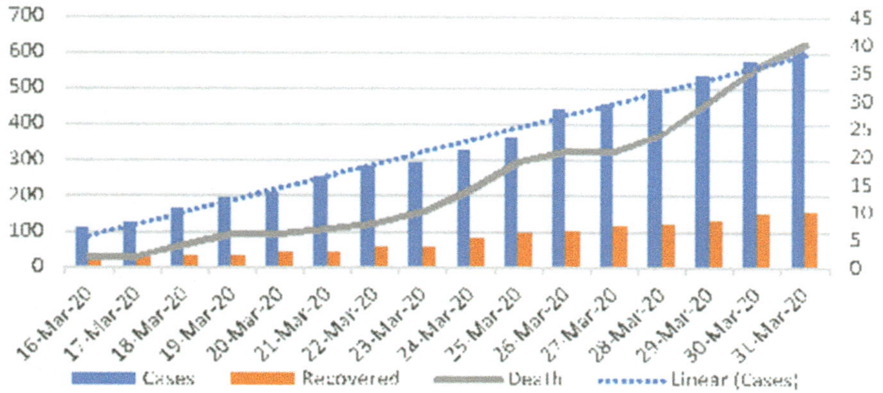

Fig. 6 The Covid-19 trend in Egypt

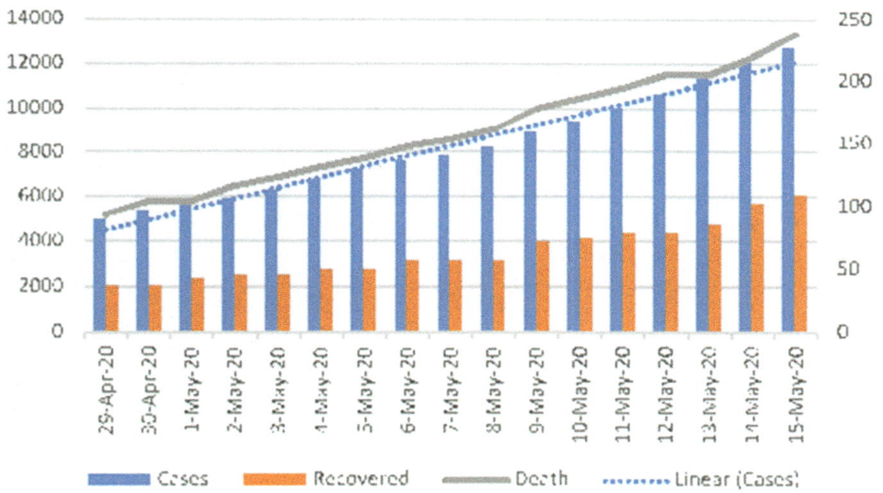

Fig. 7 Covid-19 trend in South Africa

4.4 *Model Evaluation Criteria*

In this article, the mean square error (MSE) and the coefficient of determination (R^2) are used to test the precision of the algorithm between the observed and the expected values, and R^2 tests the degree to which the effects are repeated in a process varying from (∞-1) where R2 scores close to 1 are favored for optimal model prediction Diverse types of information were measured through MSE related to the model's predictive capacities. MSE tests predictive accuracy by squaring the errors, which produces a positive value. MSE performance ranges from (0, ∞) to model prediction is optimal if MSE is 0.

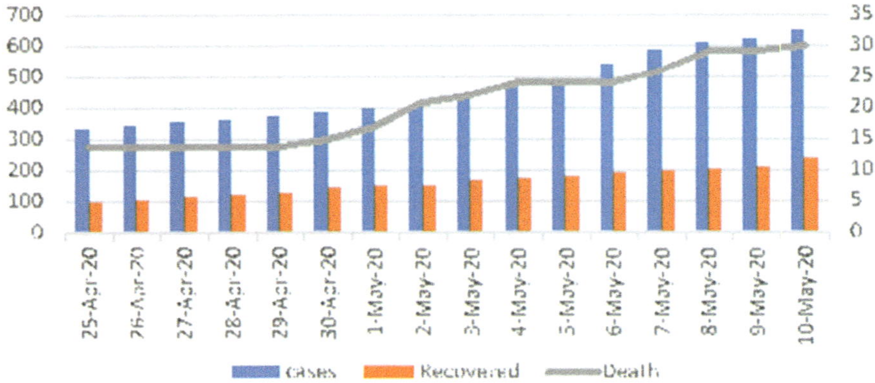

Fig. 8 Covid-19 trend in Kenya

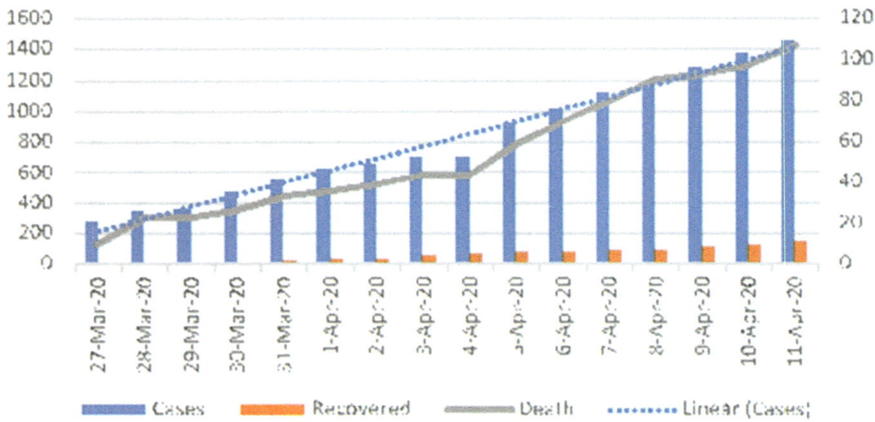

Fig. 9 Covid-19 trend in Morocco

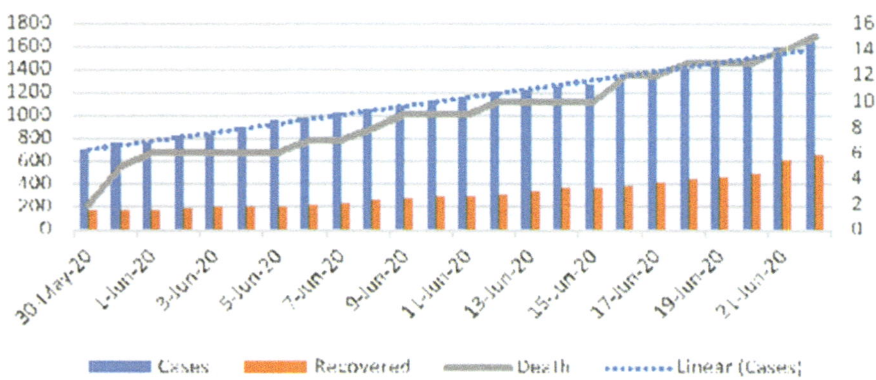

Fig. 10 Covid-19 trend in Madagascar

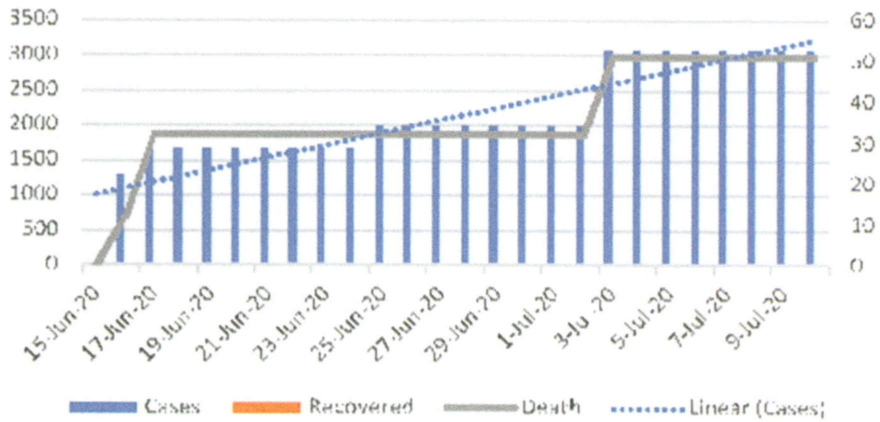

Fig. 11 Republic Covid-19 trend in Guinea

Table 3 Mean Absolute Percentage Error (MAPE) of selected countries in Africa for convolutional, stacked, and bi-directional LSTM

S/N	Country	Convo-LSTM	Stacked-LSTM	Bi-directional
1	Nigeria	12.55	11.20	8.09
2	South Africa	21.15	30.55	15.40
3	Egypt	18.00	13.05	8.95
4	Kenya	12.30	11.17	7.50
5	Morocco	7.19	7.1	0
6	Madagascar	0	0	0
7	Guinea Republic	2.77	3.5	2.15
	Average Total	10.56	10.94	6.02

Table 3 shows that there is going to be an increase in some particular countries and there is going to be a reduction in the rate of infection cause of the intending vaccine that is going to be made available which is going to reduce the spread of covid among person to person. The vaccine is not a cure but a curb to the virus. It is expected that the necessary health indicators would be in place, and also more efficiency would be shown by the health workers than when the pandemic just started (Table 4).

4.5 Discussion

In this study, the covid-19 outbreak has to be the topic of every discussion around the world. Every health organization has been finding a cure scientifically so everyone could return to their everyday life. The economic RNN model is being used to train and test to predict if there will be an increase in the number of cases. One of the

Table 4 Validation metrics for cases using RNN, LSTM, BI-LSTM, and VAE metrics

Country	Model	RMSE	MAE	MAPE	EV	RMSLE
Nigeria	RNN	1,071,465	1,060,063	4500	0211	00,030
	VAE	1,390,222	1,390,771	5911	0960	00,044
	LSTM	1,055,090	1,047,255	4455	0270	00,021
South Africa	RNN	1,672,070	167,709	7044	0260	00,050
	VAE	5,312,644	5,300,150	225	0790	00,004
	LSTM	1,260,451	1,245,810	5105	0394	00,025
Egypt	RNN	1,516,225	1,166,122	6522	1574	00,057
	VAE	1,405,217	1,359,220	5167	0710	00,007
	LSTM	1,001,200	1,410,600	5215	0399	00,029
Kenya	RNN	1,600,300	172,800	6620	0579	00,060
	VAE	1,413,222	5,400,200	5170	0721	00,017
	LSTM	1,271,550	1,244,800	5220	0410	00,031
Madagascar	RNN	32,597	1,600,100	143	0208	00,020
	VAE	18,655	1,515,000	547	0418	0
	LSTM	312,141	1,300,127	1203	0221	00,032
Morocco	RNN	50,172	1,725,000	4042	0151	0
	VAE	19,022	1,730,101	4476	0265	00,017
	LSTM	330,212	1,250,110	0128	0052	00,020
Guinea Republic	RNN	57,224	1,300,071	3112	0	00,900
	VAE	27,111	1,315,117	6214	0024	00,035
	LSTM	400,519	1,610,212	0230	0093	00,022

variants of the LSTM was able to provide a better forecast of the virus for the days. The model can help the proper authorities decide on lockdown measures and help the population comply with all security measures that the rules may provide. The forecast helps the government balance the burden of medical infrastructure and ensure that economic activities can resume, which can otherwise create many individuals' challenges. When applied to the covid-19 data from the seven African countries, the performance of the RNN, LSTM, and VAE models in terms of MAPE, MAE, EV, RMSLE, and RMSE is summarized in Table 3 that the VAE model performed well with other models that served best in their capacity. This study highlights the ability to use RNN, LSTM, and VAE techniques to predict cases of Covid-19, even when applied to small data sets of Covid-19 Training. Most importantly, the models could capture the properties of time variants and relevant past data patterns and predict the future Covid-19 trend. In the study, due to reduced hidden unit's number, VAE presents an advantage of lower dimensionality, allowing the discriminative features to be removed.

5 Conclusion

A new structure consisting of LSTM cells and well-connected cells was proposed in this study to forecast and model the rise in the number of Covid-19 cases in others to enable medical practitioners and individuals to be well informed about the virus. The method is particular to a traditional neural network and has been used mainly in fields ranging from background prediction, sequence mining, and much more. In seven African countries, the power to forecast the proposed weekly numbers of cases were assessed and discussed. The closing line is as follows: as shown, the model presented provides an effective method for providing.

an increase or decrease in the number of cases of viruses in the seven regions. Besides, in its architecture, the proposed model is logical.

The LSTM helps to retain previous details and helps to understand the time series data. The backpropagation method helps prevent the redundant measurement of the intermediate-term in the chain law. A connected layer above the LSTM helps increase the model's ability to learn and match. LSTM can learn the context needed to make a prediction, versatility in problem formation, and multiple input variables, and it can also model problems.

Thirdly, a powerful tool to estimate the number of cases involved is given by the newly developed model. This can serve as an alternative model for predicting weather-based events that can expect such a pandemic's economic effects. In the future, a larger, more comprehensive, more reliable LSTM-based model allows for more precise prediction of numbers for months and years that would be able to predict the occurrence of this kind of pandemic in the future. The given model can also be used for other time-series predictive work, such as fatality and morbidity prediction.

References

1. Joosting ACC, Head B, Bynoe ML, Tyrrell DAJ (1968) Production of common colds in human volunteers by influenza C virus. BMJ 4(5624):153
2. Zhou P, Yang XL, Wang XG, Hu B, Zhang L, Zhang W, Chen HD (2020) A pneumonia outbreak associated with a new coronavirus of probable bat origin. Nature 579(7798):270–273
3. Ubags ND, Alcazar MAA, Kallapur SG, Knapp S, Lanone S, Lloyd CM, Smits HH (2020) Early origins of lung disease: towards an interdisciplinary approach. Eur Respir Rev 29(157)
4. Chan JFW, Kok KH, Zhu Z, Chu H, To KKW, Yuan S, Yuen KY (2020) Genomic characterization of the 2019 novel human-pathogenic coronavirus isolated from a patient with atypical pneumonia after visiting Wuhan. Emerg Microbes Infect 9(1):221–236
5. Li C, Yang Y, Ren L (2020) Genetic evolution analysis of 2019 novel coronavirus and coronavirus from other species. Infect, Genet Evol 104285
6. Li Q, Guan X, Wu P, Wang X, Zhou L, Tong Y, Xing X (2020) Early transmission dynamics in Wuhan, China, of novel coronavirus–infected pneumonia. N Engl J Med
7. Holshue ML, DeBolt C, Lindquist S, Lofy KH, Wiesman J, Bruce H, Diaz G (2020) First case of 2019 novel coronavirus in the United States. N Engl J Med
8. Li P, Fu JB, Li KF, Liu JN, Wang HL, Liu LJ, Tong ZD (2020) Transmission of COVID-19 in the terminal stages of the incubation period: a familial cluster. Int J Infect Dis 96:452–453

9. Kermanshahi B (1998) Recurrent neural network for forecasting next 10 years loads of nine Japanese utilities. Neurocomputing 23(1–3):125–133
10. Arora P, Kumar H, Panigrahi BK (2020) Prediction and analysis of COVID-19 positive cases using deep learning models: A descriptive case study of India. Chaos, Solitons, Fractals 139:110017
11. Li X, Geng M, Peng Y, Meng L, Lu S (2020) Molecular immune pathogenesis and diagnosis of COVID-19. J Pharm Anal
12. Chen Y (2020) LSTM recurrent neural network prediction algorithm based on Zernike modal coefficients. Optik 203:163796
13. Cortez B, Carrera B, Kim YJ, Jung JY (2018) An architecture for emergency event prediction using LSTM recurrent neural networks. Expert Syst Appl 97:315–324
14. Guo L, Li N, Jia F, Lei Y, Lin J (2017) A recurrent neural network-based health indicator for remaining useful life prediction of bearings. Neurocomputing 240:98–109
15. Holshue ML, DeBolt C, Lindquist S, Lofy KH, Wiesman J, Bruce H, Diaz G (2020) First case of 2019 novel coronavirus in the United States. N Engl J Med
16. Montazeri M, Beigzadeh A (2016) Machine learning models in breast cancer survival prediction. Technol Health Care 24(1):31–42
17. Sigg S, Haseloff S, David K (2010) An alignment approach for context prediction tasks in ubicomp environments. IEEE Pervasive Comput 9(4):90–97
18. Lipton ZC, Kale DC, Elkan C, Wetzel R (2015) Learning to diagnose with LSTM recurrent neural networks. arXiv:1511.03677
19. Hochreiter S, Schmidhuber J (1997) Long short-term memory. Neural Comput 9(8):1735–1780
20. Schuster M, Paliwal KK (1997) Bidirectional recurrent neural networks. IEEE Trans Signal Process 45(11):2673–2681
21. Sharafudeen AM (2018). Myofascial release as a treatment choice for neuromuscular conditions: three randomized controlled trials and a systemic literature review. (Doctoral dissertation, University of Bolton in association with New York College Athens)
22. Liu W, Zhang Q, Chen J, Xiang R, Song H, Shu S, Wu P (2020) Detection of Covid-19 in children in early January 2020 in Wuhan China. N Engl J Med 382(14):1370–1371
23. Graves A, Mohamed AR, Hinton G (2013) Speech recognition with deep recurrent neural networks. In: 2013 IEEE international conference on acoustics, speech and signal processing. IEEE, pp 6645–6649
24. Huschle M, Kuhr T, Heck M, Goldenzweig P, Abdesselam A, Adachi I, Asner DM (2015) Measurement of the branching ratio of $B^{-}\rightarrow D\ (*)\ \tau-\ \nu^{-}\ \tau$ relative to $B^{-}\rightarrow D\ (*)\ \ell-\ \nu^{-}\ \ell$ decays with hadronic tagging at Belle. Phys Rev D 92(7):072014
25. Li Y, Tang WK, Chen G (2005) Hyperchaos evolved from the generalized Lorenz equation. Int J Circuit Theory Appl 33(4):235–251
26. Hammou BA, Lahcen AA, Mouline S (2020) Towards a real-time processing framework based on improved distributed recurrent neural network variants with fast text for social big data analytics. Inf Process Manag 57(1):102122
27. McInerney J, Stein S, Rogers A, Jennings NR (2013) Breaking the habit: measuring and predicting departures from routine in individual human mobility. Pervasive Mob Comput 9(6):808–822
28. McKibbin WJ, Fernando R (2020) The global macroeconomic impacts of COVID-19: seven scenarios
29. Remuzzi A, Remuzzi G (2020) COVID-19 and Italy: what next? Lancet
30. Rothan HA, Byrareddy SN (2020) The epidemiology and pathogenesis of coronavirus disease (COVID-19) outbreak. J Autoimmun 102433
31. Sajadi MM, Habibzadeh P, Vintzileos A, Shokouhi S, Miralles-Wilhelm F, Amoroso A (2020) Temperature and latitude analysis to predict potential spread and seasonality for COVID-19. SSRN 3550308
32. Wu Y, Tan H, Qin L, Ran B, Jiang Z (2018) A hybrid deep learning based traffic flow prediction method and its understanding. Transp Res Part C: Emerg Technol 90:166–180

Chapter 11
Computational Intelligence-Based Diagnosis of COVID-19

Muhammad Shoaib, Amjad Islam Aqib, Zeeshan Ahmad Bhutta, Wanxia Pu, Iqra Muzammil, and Muhammad Aamir Naseer

Abstract As the newly emerged novel coronavirus (COVID-19) spread to 210 countries worldwide, it was declared pandemic. It influences society as well as each country with a remarkable impact on the health of the people, economics, social life, and education setups at the society and country levels. Due to the increase in the cases of COVID-19 at the national and international levels, there is a need for innovative ways in initial screen out and final diagnostic evidence of COVID-19 patients in a short time. To achieve that goal, various computer-associated models belonging to the fourth industrial revolution technology have been under research and are known as computational intelligence (CI) systems. CI works as a collaborative domain where the science disciplines together with computer science work for designing systems using computer-assisted tools such as algorithms and hardware. The CI technology is an optimized tool that works on the basis of the behavior of living organisms. The

M. Shoaib
Institute of Microbiology, University of Agriculture, Faisalabad 38000, Pakistan

A. I. Aqib (✉)
Department of Medicine, Cholistan University of Veterinary and Animal Sciences, Bahawalpur 63100, Pakistan
e-mail: amjadislamaqib@cuvas.edu.pk

Z. A. Bhutta
The Royal (Dick) School of Veterinary Studies, University of Edinburgh, Easter Bush Campus, Midlothian E25 9RG, Scotland, UK

M. Shoaib · W. Pu
Key Laboratory of New Animal Drug Project, Gansu Province/Key Laboratory of Veterinary Pharmaceutical Development, Ministry of Agriculture and Rural Affairs, Lanzhou Institute of Husbandry and Pharmaceutical Sciences of Chinese Academy of Agriculture Sciences, Lanzhou 730050, PR China

I. Muzammil
Department of Veterinary Medicine, University of Veterinary and Animal Sciences, Lahore 54000, Pakistan

M. A. Naseer
Department of Clinical Medicine and Surgery, University of Agriculture, Faisalabad-38000, Pakistan

U. Kose et al. (eds.), *Computational Intelligence for COVID-19 and Future Pandemics*, Disruptive Technologies and Digital Transformations for Society 5.0,
https://doi.org/10.1007/978-981-16-3783-4_11

CI models search to get the required objective with a set of limitations to be fulfilled, and the performance index indicates an optimal solution known as the objective function. The CI saved inside the computer can exploit the required search objective using nature-inspired algorithms. Neural networks, fuzzy logic, probabilistic theorems, and artificial intelligence tools can exploit and explore the search space using the naturally supplied irregular trapping competency of animals to search out a solution for this worldwide problem. This chapter will present a review of computational models (such as neural networks, fuzzy logic, probabilistic models, evolutionary computation, computation learning theory, real-world systems and tools, big data analysis, artificial intelligence, and nature-inspired computation) for rapid diagnosis of COVID-19 cases.

Keywords COVID-19 · Diagnosis · Emerging technologies · Computational intelligence · CI Models

1 Introduction

The World Health Organization first reported in December 2019 the coronavirus disease (COVID-19), also known as a severe acute respiratory syndrome caused by SARS-CoV-2 in China. Most coronaviruses infect animals, but due to the zoonotic nature of viruses, they can also spread to humans. Severe acute respiratory syndrome (SARS-CoV-2) and Middle East Respiratory Syndrome Corona virus (MERS-CoV) in the Middle East cause severe respiratory illness and death [6]. The COVID-19 epidemic has triggered an unprecedented health and economic crisis. The disease is highly contagious and spreads from infected people to others through respiratory droplets. Mild cases of COVID-19 have symptoms such as difficulty in breathing, pyrexia, phlegm and muscle aches, but severe cases can cause pneumonia and multiple organ failure. Pandemics like COVID-19 affect the micro and macroeconomic environment as they prioritize public sector spending and reallocate it to the health sector. Due to uncertainty and declining household incomes have led to changes in consumption and production patterns. Value improves the nature of the business environment [1]. In addition, the recently identified spike in the number of COVID-19 has created an alarming condition for governments and public health professionals. These concerns are that it takes a number of infected people to participate in the health infrastructure, particularly the number of beds, ventilators, personal protective equipment (PPE) available and the COVID-19 outbreak. This is due to the possibility of an overwhelming number. In industrialized countries this has become a reality and they are known for their excellent medical infrastructure and excellent medical staff. The severity of the crisis requires an immediate and optimal response from government officials, scientists and medical professional personnel [2].

The need to control the global spread of infection will ultimately block various economies, restrict mobility, reduce social distance, use accelerated testing platforms and contact tracing. Based on the above, it is still important to identify and isolate infected cases in order to stop rapid increase or flattening the spreading curve. So, diagnosing cases in a way that matches the rate of transmission of disease between infected individuals remains important to control transmission and reduce mortality [3].

Currently, the most common method for COVID-19 diagnosis is real-time reverse-transcriptase polymerase chain reaction (RT-PCR). Chest imaging through radiological techniques, such as X-rays and computed tomography (CT), play a key role in the early detection and treatment of diseases [4]. The sensitivity of RT-PCR ranges from 60 to 70% with negative results, which can be determined by examining the X-ray of the patient [5]. Patients with active infections use saliva swabs, and patients with no previous cases use a "home" finger stick test to assess antibodies against coronavirus. However, the availability of test kits for large-scale screening remains a problem [6]. With the increase in the intensity of COVID-19 infection around the globe, life has changed significantly. Wearing gloves and masks and keeping the social distance to prevent daily basis disinfecting is a new practice around the world. Most hospitals have become COVID-19 hubs, and medical consultants are using remote health services for regular check-ups, diagnostics, prognosis and treatment. All of this is easy to do on a comfortable sofa. In addition, patients with co-morbid conditions such as heart disease, hypertension, hypothyroidism and diabetes use smart biosensors to monitor their health at home. Scientists have also presented several specified biosensors to track COVID-19, including paper-based, membrane-based and chip-based biosensors [7].

With limited resources availability and the urgent demand for medical care, support and treatment in the health sector, this crisis requires innovative solutions. New applications of newly designed technological tools have been put forward to meet the growing demand. Researchers have looked for techniques to enhance the specificity of COVID-19 case predictions, but current traditional artificial intelligence models are facing challenges and the need to consider the potential of computational models that can be useful during a crisis [6]. All existing or underdeveloped utilities are based on powerful computational intelligence-based (CI) algorithms that behave like humans but significantly reduce human intervention. As the disaster progressed, countries realized that there were still many gaps in monitoring, preventing, predicting, diagnosing and predicting large-scale diseases in this emergency. However, computational intelligence (CI) paves the way for standard health blessings that cover four major aspects in critical situations such as the current COVID-19 scenario. Therefore, scientists are promoting the idea of computational intelligence to combat the deadly disease COVID-19. Computational intelligence is a branch of computer science that combines trained algorithms developed by human intelligence to make a correct assessment of difficult problems without manual intervention. In the current COVID-19 pandemic, CI-based methods are widely used in four areas: (a) monitoring, (b) prevention, (c) diagnosis and (d) prognosis [8]. Therefore, the current chapter will highlight the possible ways that can be used for the diagnosis of COVID-19 by using different computational intelligence-based models.

2 Computational Intelligence Overview

Computational intelligence technologies are well integrated into various systems to address the potential problems of the epidemic. Computational intelligence (CI) uses computer resources to investigate complex real-world problems and gain a deeper understanding of them. It is a generic term made up of a widely used computer paradigm [7]. Bezdek formally defined computational intelligence in 1994 [9, 10]. When a system processes data at a basic level (e.g., image pixels) and includes a pattern recognition module, the system is called "smart" computing. No prior knowledge of artificial intelligence is being used. By Bezdek's definition, computer intelligence is a branch of artificial intelligence. Actually, the objectives of computational and artificial intelligence are similar. In other words, to achieve general intelligence, MARKS [11] distinguishes between artificial and computational intelligence, pointing out that the previous is made up of hard computing technology and the latter is made up of software. Judging by its existence, it can be considered that there are two types of machine intelligence: 1) artificial intelligence, developed based on the idea of hard computing tools, and 2) artificial intelligence, developed based on the idea of soft computing tools. In contrast, computing-based computer intelligence can adapt to different environments using the concept of soft computing. Hard computing technology is based on correct or incorrect values on which Boolean logic and information technology depend. The main problem with value-based logic alone is that Boolean values cannot be easily explained in natural language, but based on fuzzy logic, soft computing technology can handle dangerous situations. This logic is a unique aspect of computer intelligence. By combining data with fractional details, one can assess the behavior of the human brain [11]. As defined by the IEEE Computational Intelligence Society, an artificial neural network (ANN), fuzzy logic system (FLS) and evolutionary algorithm (EA) together form the backbone of the computing intelligent system (CIS) [7].

3 Role of Computational Intelligence Against COVID-19

The recent challenge of COVID-19 infection caused by the infamous SARS-CoV-2 virus has resulted in massive loss of life worldwide. It affected over 213 countries. The pandemic has stagnated the world and draws attention to the shortcomings of health systems and epidemics. New diseases caused by the RNA virus continue to appear and afflict us. There have been two SARS and MERS pandemics in the past, but the world is not yet ready for COVID-19. These pandemics and the devastation they cause require the use of computer intelligence in healthcare and epidemic management. Traditional medical methods require human resources exposed to patients, drugs and viruses, and it takes an undefined time limit to point out a conclusive end statement. In these emergencies, traditional treatments have collapsed. In these difficult times, it is imperative to make better use of available resources and technologies to facilitate

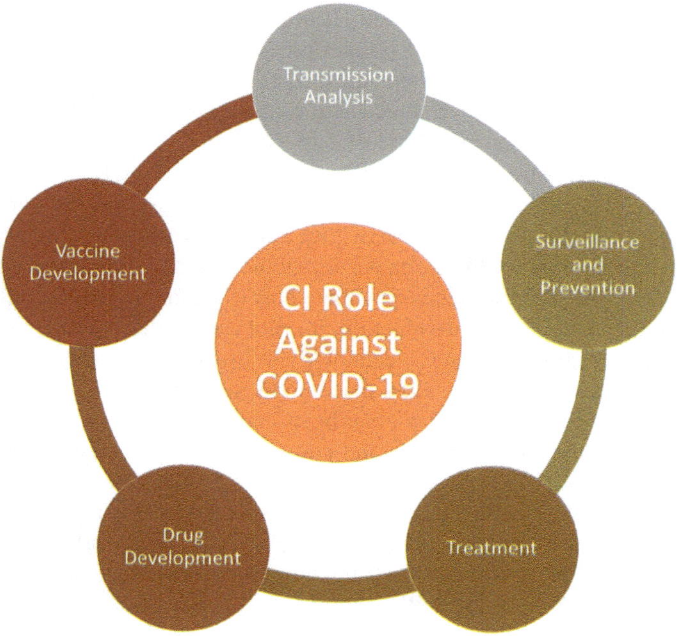

Fig. 1 Schematic diagram of role of computational intelligence (CI) against COVID-19

the fight against pandemics. Consequently, a systematic and comprehensive estimation of existing technologies, resources, tools and technologies aims to control the severity of the disease and eventually find possible solutions for its treatment (Fig. 1). Applying cutting-edge areas of computational artificial intelligence (including artificial intelligence, big data analysis and machine learning) to clinical/health data can provide an important solution in the current situation.

3.1 Role of Computational Intelligence in Transmission of COVID-19

Analysis of data from different countries has shown that direct contact is a condition favoring the spread of the COVID-19. It is known that the bat is the natural host of the coronavirus [12], and the reservoir of the bat is also the source of the human coronavirus [13]. The spread of this pathogen from the bats to animal / human populations is currently unknown. Of the 41 early cases, 14 did not come into contact with the fish market and were believed to have been infected with the virus [14]. According to Liu et al. [15] COVID-19 spread through contact with respiratory droplets. It spreads through close contact with patients or indirect contact with contaminated surfaces, according to the 29 March 2020 World Health Organization report. In the case of aerosols, we may have the following possibilities for airborne spread: bronchoscopy, tracheal incision, etc. Therefore, steps must be taken to prevent airborne spread. In addition to direct viral transfer, the feco-oral route of spread has also been in focus. Zheng et al. [16] studied that SARS-CoV-2 species were also isolated from infected stool samples. Tissues from 205 patients were tested by RT-PCR and were positive in 32% of throat swabs and 29% of stool samples [10]. Even if a nasopharyngeal swab is negative, a rectal swab can easily give a result. All research literature shows that bats are the reservoir of the coronavirus, but the lack of communication between the natural host (the bat) and the intermediate host (the mammal) ultimately spreads from person to person leading to infection. It is not easy to discover the missing link with traditional methods. Using CI technology to investigate bit rates and calculate R0 values (reproduction number) can speed up the tracking process. We can use CI to check the sent cycle and look for other hosts and operators.

We can use mathematical models and computational intelligence techniques to draw inferences about transmission chains and transmission rates that will help plan for mitigating the impact of pandemics. Basic species (R0) using various mathematical models such as bat host-reservoir-human (BHRP), network of transmission (BHRP) model, SEIR (sensitive impact-infection recovery) model can be calculated [17]. Chen et al. [18] estimated R0 value for one tank is 2.30. However, the R0 value of person-to-person transmission increased to 3.58. The author believes that SARS-Cov-2 first appeared in bats and then transferred to human beings living in the city. The SEIR model is based on public health data. The R0 value of 6.47 indicates the seriousness of the incident [19]. Tan et al. exploring a time-dependent dynamic model (TDDM) and using contact frequency as a function of time [19] recommend using an appropriate strategy to reduce R0. A conceptual model is presented by Rin et al. Lin et al. [20] estimated the R0 value as 2.8, and the emigration and migration model used by Wu et al. [21, 22] gave an R0 value of 2.68. Another study recognized potential types that can be transmitted from reservoir hosts to human hosts using ML instruments [23].

3.2 Computational Intelligence in COVID-19 Surveillance and Prevention

As the coronavirus spreads through personal contact, security measures are needed to control the spread of COVID-19. This includes trying to break this communication chain by tracking and monitoring contacts. Within general treatment, the main measures to reduce and control the transmission of diseases include case finding, isolation, diagnostic confirmation and treatment support, and contact trace out and isolation [24]. The sole purpose of monitoring is to organize the collection, assortment, analytics and dissemination of health-related data for planning, implementation and assessment of health plans [20, 25]. Established surveillance tools and systems should be checked on a daily basis to understand clear requirements for efficiency, cost and effectiveness [21]. The calculation of the monitoring effect is based on whether it computes preventive tools, treatment assessment or boosted understanding of related health mishaps. The benefit of auditing is the ability to track security incidents. Surveillance has become an important policy adopted in all countries/regions to prevent COVID-19 spread, as it helps predict long-term demographic trends associated with virus outbreaks [20]. Fighting natural disasters like the current coronavirus requires regular checks and monitoring of many geographic regions. The best method for keeping track of things is using geospatial data such as remote sensing (RS), global positioning systems (GPS) and geographic information systems (GIS). This is very effective as it allows for timely and accurate identification of natural resources in the era of the cataclysms [26]. Computational intelligence technology takes actual data input from multiple geographics and performs appropriate analysis to predict the outcome of a disaster model. In addition, high-quality geospatial data from satellites orbiting the earth can help model and understand the dynamics of the virus. Satellite imagery can be used to monitor, predict and track local diseases even after the model has been built. (https://www.planet.com/pulse/how-ssatellite-data-can-help-withcovid-19andmore/). Successful actions can be used to track contacts and spread disease. Locks are the first tool used worldwide to prevent local spread. The national blockade of India is divided into five main stages. Unfortunately, local transmission is out of control [27]. Several countries have proposed important surveillance strategies to control and track spread [28].

3.3 Computational Intelligence in COVID-19 Treatment

Telemedication or telediagnostics combines healthcare and technology to serve people in remote and difficult-to-access areas [29]. Remote assistance offers the same benefits as traditional medicine. The only difference is the use of information and communication technologies (ICT) to bridge the space between healthcare providers and consumers. In situations such as pandemics, telemedicine can offer several benefits, including improved quality of counseling, reduced viral load, low cost, comfortable living and early detection of epidemics. This technology faces a number of challenges, including a shortage of professionally trained personnel, a lack of EPR systems and data inconsistencies. Telemedicine can be used for chronic and mild diseases [30]. Qazi et al. [29] list remote healthcare services that work smoothly and raise awareness. For example, telegenomics is a neurology and cardiology consulting company. Online platforms such as *CBInsights* provide medicinal and health-related information through one portal approach. *Kanteron Systems* is another example of an IT platform that physicians can use to edit health information and deliver the most appropriate medicines to patients. It also includes remote rehabilitation and follow-ups. Many people are in home quarantine or self-quarantine, so this also applies to COVID-19. As a result, these patients find it difficult to get regular check-ups, change medications or visit a doctor. ESTRAD (Équipe Spécialisée en

Télé RéadaptationÀ Domicile) developed by Corriveau et al. [31] is an example of a rehab platform that supports oximeter, electrocardiogram, thermometer, vital signs and voice therapy technology. Some of the barriers include physical, technical and informational barriers [29]. These can be overcome with practice and further field research.

3.4 Computational Intelligence in Drug Development and Repurposing for COVID-19

Drug development requires an understanding of the manipulation of viruses. Viral manipulation involves creating drugs that exploit viral signals or the cognitive properties of host cells. Obstruction caused by either of these two conditions can reduce the risk of infection. We can also investigate signal induction suppression. Vaccines are needed to prevent the epidemic and contain its anger in a narrow range in the future. Vaccines have an important role in breaking the chain of transmission and reducing host infections. However, developing a vaccine requires personnel, time and clinical trials that can be performed on normal people. Therefore, vaccines are not the only reliable treatment. To develop antiviral drugs and vaccines, we can test various properties of existing antiviral drugs and antibiotics, including high specificity, mechanism of action, side effects, etc. Antiviral agents against HIV, SARS, Ebola, MERS and other viral diseases should be evaluated for their ability to support the medical need of COVID patients. ML also assists in drug development. In previous outbreaks such as the Ebola virus, researchers have used the Bayesian ML model in conjunction with the detection of pseudo-assays of the virus. Ebola virus replication detection data has been validated. This clarifies the potential of the molecule as a drug [32]. Research on drug development for H7N9 based on machine learning using a random forest algorithm has produced excellent results [33]. In a rapidly spreading epidemic such as COVID-19, techniques should be used to minimize causation. The study of the properties of drugs in the traditional way is time-consuming and the results may be inaccurate. ML can reuse drugs by examining knowledge graphs of the patient and predicting the interactive phenomenon between the host body and drug molecules. Richardson et al. used ML to create similar knowledge tables to study the response of the virus to baricitinib [34]. The subsequent drugs should be described utilizing the neural network, and potent drug options can be further tested in clinical runs. Wan et al. [35] have developed a system that makes it easy to create such graphs using neural networks. By training a machine learning model, we can interpret the graphics and precisely predict drug-target interactions (DTI). CVL21 was discovered using the graph topology method and is currently a drug candidate in clinical trials [36]. To make a medicine/vaccine against SARS-CoV-2, we need to understand its nature and the purpose of the danger. Epitope-based vaccines resemble the human immune system and are safer and better at preventing disease. Machine learning tools such as latent Markov models, support vector machines (SVMs) and artificial neural networks (especially deep learning) can be used to identify epitopes and design effective vaccines. Artificial intelligence can be used to search for possible treatments and vaccines. For example, the drug baricitinib with antiviral properties and six other new drugs were discovered [34, 37].

4 Computational Intelligence Techniques/Models for Diagnosis of COVID-19

As described earlier, the concept of computational intelligence has been in function for about 30 years. During this time, new concepts were continuously added to this field to strengthen the discipline. Computational intelligence technologies today can be divided into the following categories (Fig. 2).

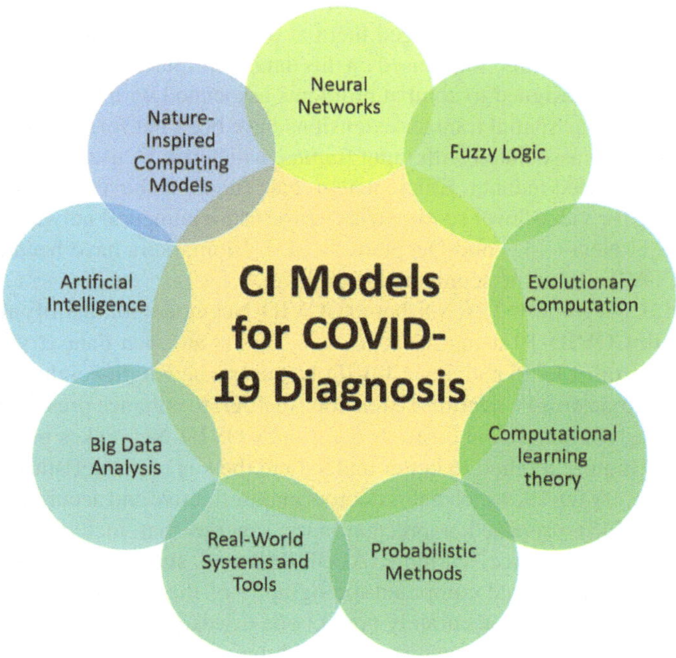

Fig. 2 Various computational intelligence (CI) models are helpful in the diagnosis of COVID-19

4.1 Neural Networks

As mentioned earlier, artificial neural networks apply deep learning principles to achieve advanced expression learning. Formula learning is a learning method that allows to automatically learn formulas from data. Learning algorithms shouldn't help people to extract features. Thus, among other things, neural networks can be easily used to extract important manifestations of a viral infection and distinctive symptoms of a viral infection. Nevertheless, if we have enough data, deep neural networks can usually pull out useful information. That is, it is difficult to efficiently handle errors and, most importantly, it is difficult to handle data that are missing values in the model. Luckily, several countries have implemented a variety of data collection methods for COVID-!9. In this way, we can build an efficient deep neural network to combat COVID-19. For example, several researchers use chest computed tomography or X-rays [38, 39] to create deep neural network models that characterize viral infections.

Roy et al. [40] suggested using an artificial neural network to analyze lung ultrasound images (LUS). They compiled a new set of fully annotated LUS images from numerous Italian hospitals and tagged them at the image, video and pixel levels to depict the status of the infection. Based on this data, multiple artificial neural network models have been designed to combat problems associated with the automatic analysis of LUS images. Spatial transfer extensions have been put forward to predict the severity of disease associated with input frames, which can be used to locate lesions in a poorly controlled manner. Efficient aggregate functions have been proposed for frame scoring for video-level scoring using three artificial neural networks: Vanille-U-Net [41], U-Net++ [42] and Deeplabv3+ [43]. Biomarkers have been created to visualize COVID-19 at the segmented pixel level.

Wang et al. [44] reported a CNN-based COVID-Net model for identifying patients infected with COVID-19 using chest X-rays. This is an open dataset with 13,800 images taken from 13,725 patients. COVID-Net uses a new lightweight residual unit (Projection-Expansion-Projection-Extension (PEPX)) to enhance presentation skills while reducing computational costs. In addition, COVID-Net makes predictions by using the GS Inquire method to better understand the key characteristics of patients infected with COVID-19, which helps doctors make effective and accurate diagnoses.

Han et al. [45] proposed a precautionary scheme that includes deep three-dimensional multi-instance learning (AD3D-MIL) to study recognition patterns based on three-dimensional computed tomography of the breast. With the AD3D-MIL warning plan, we can accurately predict one disease category such as COVID-19, frequent pneumonia or no pneumonia, and interpret the results. During training, the user does not receive a marked instance, but each package contains many instances, and each package is flagged instead of an individually marked instance. At the same time, a full 3D convolutional neural network is used to generate a functional map for each instance, and an attention-based MIL pool is used to select functional maps and combine them into a package representation. Finally, the packet representation is sent to a characteristic of a fully connected neural network to conclude the final statement.

Panwar et al. [46] established a deep learning COVID detection model. This makes it possible to identify patients with COVID-19 using X-rays within 5 s. The proposed model adds five configurable levels to the VGG-16 as the first level, the first level being the middle 2D aggregation level. Unlike maximum merge, this average merge level uses the average of all pixels with a pool size of (4, 4) to reduce the image resolution. The second layer is a flat layer that converts the 2D tensor to a vector as input to a perfectly dense tie layer (third layer), the pleasure of activating at the same time.

4.2 *Fuzzy Logic*

Fuzzy logic is one of the important tools of computer intelligence that can be used to measure and simulate complex procedures in real life. Contrary to artificial intelligence, which has a need for accurate knowledge, fuzzy logic can be used with incomplete or imprecise data applied to process models. In other words, fuzzy logic can be easily used for analysis based on limited data uncertainty. In fact, fuzzy logic can solve most of the problems in the fight against COVID-19. For example, characterizing a viral infection can confirm that each patient has different symptoms and it is difficult to diagnose COVID-19 without testing the virus.

Dhiman, Sharma [47] created a fuzzy inference system for diagnosing COVID-19 using six input elements. Utilizing the planned fuzzy reasoning system, the severity status of understudy patients can be divided into three language categories: high, medium and low. Through training and optimization, the model learned three fuzzy rules: (1) If the temperature is moderate, if the patient consumes a large amount of ethanol, or if the body temperature is low, the patient's severity may be moderate. (2) If the temperature is low, drink a small amount of ethanol. If we have a moderate body temperature and are coughing, it may be less severe. (3) If the patient consumes a small amount of ethanol, shortness of breath and insufficient pressure may increase. When these rules are applied, the proposed inference system can be used to facilitate the initial diagnosis of patients with COVID-19 infection. Plasma infusion in patients with COVID-19 was seen as a possible treatment option. It is very important to develop a donation system that can be used to determine if a donor is infected with COVID-19. However, it is not possible to test all donors on a large scale, as some infected people are asymptomatic.

Nazarov [48] used statistical data to create a fuzzy model to assess the value of the blood donation system. The system uses 12 input variables and evaluates three output variables based on three blocks of rules. Thus, it is possible to solve the issue regarding the assessment of the quality of blood supply systems in each area.

4.3 Evolutionary Computation

Evolution computing first creates several possible solutions and then refines them iteratively. The set of possible solutions at each iteration is called a collection. We can create the next generation by accidentally deleting unnecessary solutions and making small random changes to the current generation. From a biological point of view, many decisions are subject to natural or man-made selection and mutation. As a result, the health of the population gradually improved. The fitness function of the algorithm determines the learning goal. Advances in computer technology can provide highly optimized solutions to many problems. The group risk analysis has developed many options and improvements that can analyze the probable effect of COVID-19 and predict the future behavior of COVID-19.

Niazkar, Niazkar [49] predicted the occurrence of COVID-19 using the multi-gene genetic program (MGGP). Since the intensity of infection spread changes every day, predicting the occurrence of COVID-19 is difficult. Originally developed for behavioral modeling, MGGP is appropriate for modeling variable series. The proposed MGGP-based method shows very promising results. Specifically, the projected number of confirmed COVID-19 cases is similar to those seen in the seven countries studied in the study. Thus, it is assumed that the MGGP-based method is suitable for estimating the incidence of COVID-19.

Salgotra et al. [50], developers of genetic programming (GP), have proposed a predictive model for confirmed cases of illness and death in India's three most destructive states. The fitness function is designed for the root mean square error. To test the evolutionary model, use statistical parameters and metrics to assess applicability. In addition, GP-based models are supported by people with gene sizes greater than 1 using a simple linkage function. The study results found that current GP model were significantly trustable to predict number of death and confirmed cases in India.

4.4 Computational Learning Theory

There are many implementations of computational learning theory. Different principles of reasoning can be derived from different assumptions. As a result of reasoning, reasoning principles are used to develop various computational learning methods. These methods can be categorized into six types. (1) Exact guidance. (2) It is possible to properly learn machine learning with baseline mathematical analysis. (3) The theory of Vapnik–Chervonenkis, which is a statistics-based explanation. (4) Bayesian inference, which is a statistics inference based on Bayes' theorem. (5) Algorithm learning theory is a type of learning based on algorithm explanation. (6) Online machine learning technology that constantly updates data. However, advances in learning theory allow us to develop many practical learning algorithms. For example, a core idea underlies the concept of a belief network.

Duffey, Zio [51] explained a computational learning theory that uses an exponential widespread learning curve to identify the prevalence of pandemic COVID-19 infections. This method identifies infection rates to calculate erroneously predicted outcomes and time as a measure of experience or learning risk.

Barstugan et al. [52] put forward a method for early screening of COVID-19 using a machine support vector classifier. The classifier was trained from four large datasets created by capturing 16 × 16, 32 × 32, 48 × 48 and 64 × 64 color patches from 150 CT images. Perform a feature extraction process for each location to improve classification efficiency. Five algorithms for teaching theoretical calculations are used as a feature extraction method such as grayscale coexistence matrix, grayscale length matrix, local mode, discrete wavelet transform and grayscale region matrix. To avoid overfitting problems, k cross-checks are performed during training. Thanks to tenfold cross-validation with GLSZM, this classifier achieves the highest accuracy (99.68%).

Randhawa et al. [53] described the genome analysis method using computational learning theory. This method combines a decision tree with digital signal processing to classify the sequence of COVID-19 viruses and create a model that identifies the unique function of the viral genome. The suggested method can be used to analyze more than 5000 unique viral genome sequences. The dataset contains 29 sequences of the COVID-19 virus. This means that the relationship is not balanced (29: 5000).

Mei et al. [54] have proposed an integrated model for detecting COVID-19 infections. This allows COVID-19 patients to be identified as soon as possible based on clinical information from original chest computed tomography. This model combines a deep convolution network with three classifiers: multilayer perceptron, support vector machine and random forest. A deeply interconnected neural network was used to compare the symptoms of COVID-19 patients with those of the three classifiers.

4.5 Probabilistic Methods

Computer intelligence uses probabilistic methods to calculate the expected values of random variables. Probabilistic methods are widely used to analyze risk factors associated with COVID-19 and explain their importance.

Cássaro, Pires [55] believes that the number of affected people increases exponentially over time. As a result, the probabilistic model can be expressed as:

$$I(t) = I(t_0)e^{n}$$

where I (t) is the number of infections diagnosed over a certain period of time, t_0 is the onset time and r is the growth rate relative to the mean. In order to minimize the infection rate, the infection rate is determined according to the diagnosed infection schedule.

Zhang et al. [56] planned a logistic growth probability model that takes into account the law of force and the law of exponential factors to guess the number of affected patients. Contrasting exponential models, which can only handle uncontrolled prevalence, probability models of logistic growth initially approximate an exponential law but are used to constrain the model and reduce its growth rate. In accordance, the logistic growth probabilistic model is formulated as

$$I(t) = \frac{N}{1 + e^{b-c(t-t_0)}}$$

where I (t) is the number of infections diagnosed over a period of time, N is the upper limit of the prognosis, b and c are matching factors that can be derived from the under study dataset, and t_0 is the time when the first infection was detected. The forecast results depict that the logistic growth probability model can make long-term forecasts, and the forecast error is small for 3 months.

Kucharski [57] developed a dynamic random propagation model to estimate temporal variations in propagation between January 2020 and February 2020. The transmission is modeled as a geometric random run. To train the proposed dynamic random diffusion model, the model is applied to four published datasets. (1) As of January 26, 2020, the number of new cases of the disease has been confirmed daily worldwide, (2) number of confirmed COVID-19 cases in Wuhan city from December 1, 2019 to January 1, 2020, (3) number of confirmed cases in China from 29 December 2019 to 23 January 2020, and (4) daily confirmed cases on evacuation flights from January 29, 2020 to February 4, 2020.

4.6 Real-World Systems and Tools

According to the OECD (Organization for Economic Co-operation and Development) report [58], Canada has several AI-driven tools such as BlueDot [59], Epi Risk [60], CRUZR Robot [61], Canada's COVID-19 [62] and Satellite Monitor [63] that can be used to combat COVID-19. In fact, it still makes sense to showcase and promote the results of the computational intelligence community. This section focuses on these features, rather than showing how they work. BlueDot [58] is software that can be used to assess the risk of infection with more than 150 different pathogens, syndromes and toxins. In fact, COVID-19 is the most serious disease and BlueDot can determine the risk of an outbreak. The main technology behind BlueDot is a scanner that is able to scan one million media and official resources in 65 languages on daily basis. Recognizing that BlueDot can help decision-makers during the detection phase. The OECD report also gives a user-focused view and is also used to calculate individual infection potential (i.e., problems during the prevention phase). Unlike BlueDot, EpiRisk [60] is a web application that can be used to calculate potential human infections based on the airline's transport network topology. EpiRisk's data sources are so limited that some important data may be lost.

It is very easy to use, but the evaluation results are questionable. EpiRisk shows that the chances of getting infected in Taipei are greater than 45%. This is absolutely wrong. Over the past 80 days, no domestic cases have been reported in Taiwan.

When the pandemic is over, economic recovery will be a top priority for all countries. However, the question is about the timing of the implementation of economic policy. Researchers at the newly formed Chinese private bank, WeBank [64], compiled data from social networks, GPS and satellites to identify actual hotspots for steelmaking in China's factories [63]. They believe the test results may reflect China's economic recovery. Although the system is intended for internal use only, it can easily receive data from social networks, GPS and satellites, allowing computer intelligence researchers to create models based on this concept.

4.7 Big Data Analysis

The term "big data" arises from the requirement to analyze and insight large amounts of unstructured datasets, generated from various data sources every second [65]. In general, traditional analysis tools are not suitable for processing such unstructured data for understanding [66]. Big data is characterized by large amounts of data, speed and different data origins. Big data is today creating new data analysis tools coupled with machine learning models. The advantage of big data analytics lies in monitoring and predicting events in real time. "Big health data" refers to large amounts of health-related data from a variety of sources, including medical images, drug records and electronic health records. Pham, Nguyen [67] cited big data on patient care associated with COVID-19. Big data also provides information about doctors, X-rays, medical records, lists of doctors and nurses, and the location of the incident. It tracks people who are in close contact with an identified case of novel coronavirus pneumonia (COVID-19). Big data has the following characteristics:

- Volume is the amount of data to be examined. Large amounts of data range from thousands of terabytes to PB (petabytes) and EB (exabytes) [65].
- In this study, accuracy is understood as the correctness of the results of the processing system [68].
- The value is related to user interests.
- Variety refers to the structured or unstructured data types (email, call center interaction records, blog and social media posts, images, audio, video, etc. and machine data from websites, servers, networks, mobile systems applications (log files, etc.)) [69].
- Speed is the speed at which the processing system works, processes and updates data, as well as the speed at which user information is required [70].
- These features of big data apply to various problem areas, including the current coronavirus pandemic.

4.7.1 Contact Tracing with Big Data Analytics

Tracking big data contacts is a central area. We can check the occurrence using various data sources. For example, social media vehicle logs, passenger lists, subway smart cards and metadata messages for people traveling with credit cards are valuable sources of data. By defining key features in social media metadata/location tags, we can provide a tracking model to ensure that the person is in a specific location at a specific time (if not). If not, we can still track their cell phone.

Keeling et al. [71] proposed a contact tracing model for estimating transfer rates. This model identifies people who are infected before serious symptoms appear. But if we cannot do it "quickly", it will affect society. For some people, one of the most important characteristics of human social interaction is the number of secondary cases, the number of people meeting contact tracing criteria and the number and extent of contacts. The imbalance in social interactions determines the speed of transmission. The transfer rate is random and usually depends on limited and reliable data [70]. Big data can provide the availability of information needed to quickly identify high data rates. However, data architecture is critical to integrating, sharing and analyzing data in big data environments [72]. Data architecture can improve real-time insight of various public health monitoring systems. By collecting large datasets from social media, we can copy data early and spot an outbreak. In general, non-classical datasets and data flow in the early stages of development can provide information on how to design and implement a data architecture to effectively facilitate development [73].

4.7.2 Case Detection with Big Data Analytics

Qin et al. [74] use the Social Media Search Index to identify new suspects or confirmed cases. The movement of people is recognized as an important means of spreading infection. In this context, posting on social media helps to expose information about the risk of infection. However, such reports may be inaccurate and unswerving. By using keywords, the health department can create a suitable response and the required alerting mechanism. Unluckily, the COVID-19 pandemic has a wide geographic range [75]. It helps to understand the structure and handling of viruses. Therefore, the text mining algorithm is very important. However, AI interventions must be used to determine the volume and rate of data processing.

Processing streaming data requires real-time analysis. Given the geographic coverage of reported cases of COVID-19, the big data platform allows data to be collected and stored in distributed datasets [76]. Storm, Splunk and Apache are low-latency real-time streaming frameworks. Storm uses matching strategies to combine and manipulate data from diverse sources in real time. The advantage is that it provides a constant dispersed situation in which parallel processing mechanisms can be migrated to the analytic framework using "master" and "slave" nodes [77]. S4 is a common distributed computing environment that simplifies the designing of big-data-focused applications. Kafka is an open-source platform for collecting logs.

Flink is a distributed processing platform for streaming processing in which data is processed in a "circular data stream with multiple iterations". Using a series of iterations reduces the time it takes to compute the data. This architecture supports batch processing and streaming processing and integrates processing results by combining batch processing and batch processing with real-time display [78].

4.8 *Artificial Intelligence Models*

Artificial intelligence (AI) aims to create machines, called intelligent agents, that can perform actions that humans believe are rational. The term "smart agent" refers to autonomous devices that can define their environment and use the best capabilities to achieve certain goals [35]. Autonomous operations require large-scale processing of efficient heterogeneous data sources and data processing within a limited period of time [79]. Therefore, some of these agents help to create inter-layer and intra-layer functionality in the system [80]. It is a self-learning system that can cope with all forecasting tasks [79]. To learn and train, before actually using an AI system in a real environment, we need to learn how to use a dataset. This is very expensive (e.g., overfitting) as the model learns from the entire dataset and produces an inference based on the data in memory. In many areas of artificial intelligence, two important techniques are deep learning and machine learning. In general, machine learning means the capability to learn and conclude meaningful patterns from dataset, and the performance of machine learning-based algorithms and systems is highly dependent on shared functions. Machine learning architecture is simple and consists of a single layer that transforms the original input signal into the functional space of a specific problem. Unfortunately, these architectures are only effective when solving limited or simple problems. However, there are limits to how complex unstructured problems can be solved [81]. Instead, DL is a machine learning algorithm that uses a hierarchical representation structure to identify complex and unique relationships in data [82]. By scanning the representation, the machine can extract the raw data and automatically find the vector representation needed for the recognition or classification task.

Deep learning has made great strides in discovering complex multidimensional data structures that can more accurately predict the potential activity of drug molecules. This is the effect of non-coding DNA mutations on gene expression and disease [83]. Deep learning improves modern technology in many application areas such as object recognition, drug discovery and genomics [84]. This improvement can be achieved by using a backpropagation algorithm to learn from the previous layer. The internal parameters are needed to calculate a new representation for that layer. Deep learning is also used for visual tracking of medical image analysis [85] and object recognition [79, 86].

4.8.1 Case Detection with Artificial Intelligence Models

AI and ML have been used in population screening to assess the risk of infection [87]. For example, this type of AI temperature-controlled test was used in public places during the COVID-19 pandemic in China. Temperature screening helps identify symptoms and isolate suspected infections. We recently used a thermal imaging camera to take thermal images of body temperature to identify cases of COVID-19. This helps to quickly and accurately identify people with "fever", one of the main symptoms of COVID-19. Such applications predict which populations and communities are most susceptible to infection, disseminate information in real time to healthcare providers and notify people in real time of potential sources of infection to avoid such areas [87].

Other AI technologies include BlueDot (for detecting new infection risks) [46], chatbots used as virtual assistants to provide information about viruses and diagnostic robots for diagnosing diseases [88]. Predictions are done by models such as the SARIMA (seasonal auto-regressing integrated moving average) model [89] for predicting changes in influenza, and SARIMA with a vector regression model for predicting hand-foot and mouth disease (HFMD). The model is under development [90]. There are many here. However, a problem in implementing a pandemic forecasting model is the lack of historical data [85]. We can use big data and artificial intelligence models to train models and develop robust predictive models. As more and more data are published from medical websites, it is very important to include data from previously attacked countries / regions in order to accelerate the development of machine learning models [50]. By adopting an artificial intelligence model, it is possible to quickly determine the spread of the infection for the required intervention. Artificial intelligence can provide early warning and tracking, observation and prediction, diagnosis and prognosis, treatment, and social management. One way to find patients infected with COVID-19 is to take a chest X-ray. Due to a large number of patients in the hospital, it is time-consuming and difficult to see a large number of X-rays [91].

An automated framework is under development using the deep learning convolution network model (COVNet), which uses 4,356 chest computed tomography (CT) scans from 3,322 patients to detect COVID-19. The diagnostic efficiency is 0.96 [92]. The reliability of the model was tested on CT studies of community-acquired pneumonia (CAP) and non-pneumonia with a diagnostic efficiency of 0.95. Xu et al. [93] planned a deep learning early screening model to identify patients with COVID-19 with 86.7% accuracy. In a test using a single dataset with 275 CT scans, F1 was 0.85 [94]. VanBerlo, Ross [6] used a deeply entangled neural network to extract signs of COVID-19 on chest X-rays, and then went on to use the local interpretable model agnostic explanations (LIME) algorithm [95] to provide detailed information about the healthcare model. Narin et al. [96] have developed an automated convolutional neural network-based detection system with 98% accuracy as an optional rapid diagnostic option to prevent COVID-19 with a limited number of COVID-19 tests. It was launched. First aid kits were used in hospitals as more cases were reported due to COVID-19 epidemics. Numerous non-news cases have exceeded the existing capacity of the health sector. Therefore, Rahmatizadeh et al. [97] proposed an artificial intelligence system that can automatically analyze CT images to characterize COVID-19 pneumonia. The robustness of the model was tested in 1,136 case studies, of which 723 positive COVID-19 results were obtained in five hospitals with promising results. Rahmatizadeh and Valizade-Hagi proposed an AI-based control model for the wards in intensive care units [97]. Vaishya et al. [8] suggested that AI-based systems can shorten the time and process of treating patients and help control the disease. In addition, artificial intelligence-based systems can help track contacts by determining the extent of infection at a specific location [8].

4.8.2 Contact Tracing with Artificial Intelligence Models

Artificial intelligence technology is being used to determine the spread of the COVID-19 pandemic. Use smart phone technology based on artificial intelligence to track people's locations, use infrared thermal imaging cameras, use thermal detection, video surveillance and facial recognition systems to potentially be in public places. Countries like South Korea and Singapore have implemented the COVID-19 contact tracing app. It tracks COVID-19 infection with GPS phone tracking, credit card history, video surveillance and one-to-one patient surveys. At the same time, Trace Together in Singapore is a mobile app that allows you to track community-managed contacts. Every time the app discovers another phone with the app installed, the device exchanges signals via Bluetooth. For these community-driven applications, "low disk space" is a collection. However, the law requires that the person save the information on their phone so that the contact can be traced. With the rapid spread of COVID-19, contact tracing technology will help speed up the infection tracing process. In addition, an artificial intelligence model was used to track the crowd's physical guidance information. When it comes to crowd monitoring, artificial intelligence technology poses privacy concerns. This is a problem that the developers of artificial intelligence technologies must solve. Literature reviews show that the CNN model is the most common model for image recognition and classification. This may be due to its reliability (i.e., accuracy) in detection and execution. Other rarely used deep learning models are long short-term memory (LSTM) and repetitive neural network (RNN). Assembly models have been proposed that combine the predictive power of these models [98]. For example, LSTM and RNN have been used to predict life based on electronic health records [99]. In addition, LSTM and RNN models are used to predict COVID-19 epidemics [100]. AI uses for control of pandemic COVID-19 that includes early and late detection of infection, prevention, treatment, contact tracing, drug development and vaccine development, which decreases the workload of personnel and the cases of morbidity and mortality [8].

4.9 Nature-Inspired Computing Models

Nature-inspired computing (NIC) is an interdisciplinary field that joins the field of computer science with other fields of scientific knowledge to enable the development of new computer tools [101]. NIC intelligence is an optimization technology on the basis of the behavior of living organisms. The NIC model attempts to accomplish its goals within a set of limitations and uses performance indicators called goals to measure the optimal solution [101]. Natural heuristic algorithms can use the random hunting abilities that animals naturally acquire to explore the research space and use it to find the best solution on a global scale [102]. Network adapter algorithms are mainly divided into evolutionary algorithms [103] and swarm intelligence-based algorithms [104], such as ANT [105], Worm [106], particle swarm optimization (PSO) [107, 108], BAT [109], etc.

4.9.1 NIC for Contact Tracing and Case Detection

The importance of network adapters in some applications indicates the potential to solve many troubles. The NIC model has been used to diagnose and predict COVID-19 [110] and has been used to segment X-ray images to characterize COVID-19 [111]. Early detection and treatment of COVID-19 are so important that current literature reviews show that the COVID-19 pandemic contact tracing algorithm has not yet explored the computational model that was naturally inspired. Although natural computational models have been used in areas such as the automatic tracking algorithm [112], studies on the COVID-19 pandemic should still gain the attention of the research community.

5 Future Potential of Computational Intelligence and COVID-19

CI-powered technologies and methods are helping to improve the diagnosis of diseases, select candidates for vaccine and drug development, understand the evolution of the virus and conduct research to identify potential drug candidates for COVID-19. Computational intelligence plays a significant role in healthcare, and also in public health. CI-powered technologies and techniques have been established in modeling the spread of viruses and maximum usage of resources. There is a wealth of informative, high-quality data available to help you effectively explore complex real-life problems and achieve better CI governance. Researchers have used computational intelligence-based technology to generate promising preliminary results, but data exchange rules pose a major obstacle to actual implementation. CI-based techniques have helped solve important problems, but significant implications for the global healthcare industry are not yet recognized. The ability of artificial intelligence to solve problems will play an important role in making decisions about the operation of future medical systems. As far as the observations of virologists, epidemiologists or clinicians go, CI-powered tools may not be able to completely replace the human brain, but the value of CI comes from reducing the burden on each stakeholder. Decision-makers at all levels can use CI tools to suggest some preferred options that can be further explored with human intelligence before making a final decision on how to control the impact of the epidemic. In the case of epidemics, regardless of whether CI is used alone or in combination with other instruments, epidemiological characteristics, incubation time, infectivity, risk group, fertility rate, mortality, infections, pathogens, very detailed information, stability, toxicity and potential pathogenicity are keys to an effective public health feedback. It can be deduced that CI-based techniques can be used in conjunction with present routine processes that can support early diagnosis, drug development or reuse, vaccine development and early disease management. Computational intelligence interventions play an important part in understanding the global process of the COVID-19 pandemics, making decisions based on facts at all levels, from understanding pathogens to recommending effective prevention and control measures. CI-based modeling and other interventions have a significant impact on national decision-making and therefore need to be accurate. In the near future, the computational intelligence community will use limited data or combine several principles in harsh environments (visual analysis techniques, communication modeling and surveillance applications, vaccine design, drug movement, etc.). We believe that this will solve the main problems outlined earlier as well as the internet of things (IoT) for smart COVID-19 care.

6 Conclusion

COVID-19 is a global health problem that affects not only the lives of people but also the global economy. This study examined some of the important challenges in the fight against COVID-19 that could or could be solved using computational intelligence technologies. Computational intelligence is divided into different principles: neural networks, fuzzy logic, evolutionary computing, computational learning theory, probability methods, big data analysis, artificial intelligence and nature-inspired computing models. Our review showed that most of the research was based on a neural network designed to address the symptoms of a viral infection. Currently, the easiest way to collect data is with a CT scan of the breast. Thus, the current work is focused on the characterization of COVID-19 patients from breast CT images in order to create a classification model. As more and more patients recover, more treatment records are created. Such treatment records can be viewed as a time series dataset. You can then use a variety of computational intelligence techniques to analyze your treatment records. Using a model that combines natural inspiration and artificial intelligence can improve validation accuracy and reveal important details among many interactive elements in the real world. Natural computational models are used to identify relevant features, reduce remaining features and use CNN-based AI models to improve the accuracy of functional detection of diagnostic messages and responses to COVID-19. The patient's response to certain treatments improved. In addition, big data analytics tools can be used to track contacts with people infected with COVID-19, identify hotspots and alert people. We then also proposed a big data predictive contact analysis model for contact tracking. Active research is needed to overcome the limitations of CI and thus combat future trends of epidemics.

References

1. Control CfD, Prevention (2020) Centers for disease control and prevention coronavirus disease 2019 (COVID-19)
2. Gatto M, Bertuzzo E, Mari L, Miccoli S, Carraro L, Casagrandi R, Rinaldo A (2020) Spread and dynamics of the COVID-19 epidemic in Italy: effects of emergency containment measures. Proc Natl Acad Sci 117(19):10484–10491
3. Agbehadji IE, Awuzie BO, Ngowi AB, Millham RC (2020) Review of big data analytics, artificial intelligence and nature-inspired computing models towards accurate detection of COVID-19 pandemic cases and contact tracing. Int J Environ Res Public Health 17(15):5330
4. Zu ZY, Jiang MD, Xu PP, Chen W, Ni QQ, Lu GM, Zhang LJ (2020) Coronavirus disease 2019 (COVID-19): a perspective from China. Radiology 200490
5. Kanne JP, Little BP, Chung JH, Elicker BM, Ketai LH (2020) Essentials for radiologists on COVID-19: an update—radiology scientific expert panel. Radiological Society of North America
6. VanBerlo B, Ross M (2020) Investigation of explainable predictions of COVID-19 infection from chest X-rays with machine learning. Artificial Intelligence Lab
7. Raza K, Maryam, Qazi S (2021) An introduction to computational intelligence in COVID-19: surveillance, prevention, prediction, and diagnosis. In: Raza K (ed) Computational intelligence methods in COVID-19: surveillance, prevention, prediction and diagnosis. Springer Singapore, Singapore, pp 3–18. https://doi.org/10.1007/978-981-15-8534-0_1
8. Vaishya R, Javaid M, Khan IH, Haleem A (2020) Artificial Intelligence (AI) applications for COVID-19 pandemic. Diabetes Metablc Syndr: Clin Res Rev
9. Siddique N, Adeli H (2013) Computational intelligence: synergies of fuzzy logic, neural networks and evolutionary computing. Wiley
10. Bezdek JC (1994) What is computational intelligence? USDOE Pittsburgh energy technology center, PA (United States); Oregon State
11. MARKS RI (1993) Intelligence: computational versus artificial. IEEE Trans Neural Networks 4(5):737–739
12. Cui J, Li F, Shi Z-L (2019) Origin and evolution of pathogenic coronaviruses. Nat Rev Microbiol 17(3):181–192
13. Li W, Shi Z, Yu M, Ren W, Smith C, Epstein JH, Wang H, Crameri G, Hu Z, Zhang H (2005) Bats are natural reservoirs of SARS-like coronaviruses. Science 310(5748):676–679
14. Huang C, Wang Y, Li X, Ren L, Zhao J, Hu Y, Zhang L, Fan G, Xu J, Gu X (2020) Clinical features of patients infected with 2019 novel coronavirus in Wuhan China. Lancet 395(10223):497–506
15. Liu J, Liao X, Qian S, Yuan J, Wang F, Liu Y, Wang Z, Wang F-S, Liu L, Zhang Z (2020) Community transmission of severe acute respiratory syndrome coronavirus 2, Shenzhen, China
16. Zheng S, Fan J, Yu F, Feng B, Lou B, Zou Q, Xie G, Lin S, Wang R, Yang X (2020) Viral load dynamics and disease severity in patients infected with SARS-CoV-2 in Zhejiang province, China, January-March 2020: retrospective cohort study. BMJ 369

17. Qazi S, Sheikh K, Faheem M, Khan A, Raza K (2020) A coadunation of biological and mathematical perspectives on the pandemic COVID-19. Rev
18. Chen T, Rui J, Wang Q, Cui, J-A, Yin L (2020) A mathematical model for simulating the phase-based transmissibility of a novel coronavirus. Infect Dis Poverty 9 (1):24
19. Tang B, Bragazzi NL, Li Q, Tang S, Xiao Y, Wu J (2020) An updated estimation of the risk of transmission of the novel coronavirus (2019-nCov). Infect Dis Model 5:248–255
20. Lin Q, Zhao S, Gao D, Lou Y, Yang S, Musa SS, Wang MH, Cai Y, Wang W, Yang L (2020) A conceptual model for the outbreak of Coronavirus disease 2019 (COVID-19) in Wuhan, China with individual reaction and governmental action. Int J Infect Dis
21. Wu A, Peng Y, Huang B, Ding X, Wang X, Niu P, Meng J, Zhu Z, Zhang Z, Wang J (2020) Genome composition and divergence of the novel coronavirus (2019-nCoV) originating in China. Cell Host Microbe
22. Wu C, Liu Y, Yang Y, Zhang P, Zhong W, Wang Y, Wang Q, Xu Y, Li M, Li X (2020) Analysis of therapeutic targets for SARS-CoV-2 and discovery of potential drugs by computational methods. Acta Pharmaceutica Sinica B
23. Li Q, Guan X, Wu P, Wang X, Zhou L, Tong Y, Ren R, Leung KS, Lau EH, Wong JY (2020) Early transmission dynamics in Wuhan, China, of novel coronavirus–infected pneumonia. New Engl J Med
24. Tang N, Li D, Wang X, Sun Z (2020) Abnormal coagulation parameters are associated with poor prognosis in patients with novel coronavirus pneumonia. J Thromb Haemost 18(4):844–847
25. Tang B, Wang X, Li Q, Bragazzi NL, Tang S, Xiao Y, Wu J (2020) Estimation of the transmission risk of the 2019-nCoV and its implication for public health interventions. J Clin Med 9(2):462
26. Arentz M, Yim E, Klaff L, Lokhandwala S, Riedo FX, Chong M, Lee M (2020) Characteristics and outcomes of 21 critically ill patients with COVID-19 in Washington State. JAMA 323(16):1612–1614
27. Bauch CT, Lloyd-Smith JO, Coffee MP, Galvani AP (2005) Dynamically modeling SARS and other newly emerging respiratory illnesses: past, present, and future. Epidemiology 791–801
28. Pan Y, Zhang D, Yang P, Poon LL, Wang Q (2020) Viral load of SARS-CoV-2 in clinical samples. Lancet Infect Dis 20(4):411–412
29. Qazi S, Tanveer K, ElBahnasy K, Raza K (2019) From telediagnosis to teletreatment: the role of computational biology and bioinformatics in tele-based healthcare. In: Telemedicine technologies. Elsevier, pp 153–169
30. Dinesen B, Nonnecke B, Lindeman D, Toft E, Kidholm K, Jethwani K, Young HM, Spindler H, Oestergaard CU, Southard JA (2016) Personalized telehealth in the future: a global research agenda. J Med Internet Res 18(3):e53
31. Corriveau H, Tousignant M, Gosselin S, Boissy P, Azevedo L, Gelderblom G (2013) Patients satisfaction with an in-home telerehabilitation exercise program and physiotherapists' satisfaction toward technology for an acute stroke population: a pilot study. Assist Technol: Res Pract 33:753–757
32. Ekins S, Freundlich JS, Clark AM, Anantpadma M, Davey RA, Madrid P (2015) Machine learning models identify molecules active against the Ebola virus in vitro. F1000Research 4
33. Zhang L, Ai H-X, Li S-M, Qi M-Y, Zhao J, Zhao Q, Liu H-S (2017) Virtual screening approach to identifying influenza virus neuraminidase inhibitors using molecular docking combined with machine-learning-based scoring function. Oncotarget 8(47):83142
34. Richardson P, Griffin I, Tucker C, Smith D, Oechsle O, Phelan A, Stebbing J (2020) Baricitinib as potential treatment for 2019-nCoV acute respiratory disease. Lancet Publishing Group, Lancet
35. Wan F, Hong L, Xiao A, Jiang T, Zeng J (2019) NeoDTI: neural integration of neighbor information from a heterogeneous network for discovering new drug–target interactions. Bioinformatics 35(1):104–111
36. Ge Y, Tian T, Huang S, Wan F, Li J, Li S, Yang H, Hong L, Wu N, Yuan E (2020) A data-driven drug repositioning framework discovered a potential therapeutic agent targeting COVID-19. bioRxiv

37. McCall B (2020) COVID-19 and artificial intelligence: protecting health-care workers and curbing the spread. Lancet Digit Health 2(4):e166–e167
38. Huang L, Han R, Ai T, Yu P, Kang H, Tao Q, Xia L (2020) Serial quantitative chest ct assessment of covid-19: deep-learning approach. Radiol: Cardiothorac Imaging 2(2):e200075
39. Hurt B, Kligerman S, Hsiao A (2020) Deep learning localization of pneumonia: 2019 coronavirus (COVID-19) outbreak. J Thorac Imaging 35(3):W87–W89
40. Roy S, Menapace W, Oei S, Luijten B, Fini E, Saltori C, Huijben I, Chennakeshava N, Mento F, Sentelli A (2020) Deep learning for classification and localization of COVID-19 markers in point-of-care lung ultrasound. ITMI
41. Ronneberger O, Fischer P, Brox T (2015) U-net: Convolutional networks for biomedical image segmentation. International conference on medical image computing and computer-assisted intervention. Springer, pp 234–241
42. Zhou Z, Siddiquee MMR, Tajbakhsh N, Liang J (2018) Unet++: a nested u-net architecture for medical image segmentation. In: Deep learning in medical image analysis and multimodal learning for clinical decision support. Springer, pp 3–11
43. Chen L-C, Zhu Y, Papandreou G, Schroff F, Adam H (2018) Encoder-decoder with atrous separable convolution for semantic image segmentation. In: Proceedings of the European conference on computer vision (ECCV), pp 801–818
44. Wang L, Lin ZQ, Wong A (2020) Covid-net: a tailored deep convolutional neural network design for detection of covid-19 cases from chest x-ray images. Sci Rep 10(1):1–12
45. Han Z, Wei B, Hong Y, Li T, Cong J, Zhu X, Wei H, Zhang W (2020) Accurate screening of COVID-19 using attention based deep 3D multiple instance learning. ITMI
46. Panwar H, Gupta P, Siddiqui MK, Morales-Menendez R, Singh V (2020) Application of deep learning for fast detection of COVID-19 in X-Rays using nCOVnet. Chaos, Solitons, Fractals, 109944
47. Dhiman N, Sharma M Fuzzy logic inference system for identification and prevention of coronavirus (COVID-19)
48. Nazarov D (2020) Fuzzy model of digital assessment of donation systems' level in COVID-19. In: 2nd international scientific and practical conference "Modern management trends and the digital economy: from regional development to global economic growth" (MTDE 2020). Atlantis Press, pp 1201–1206
49. Niazkar M, Niazkar HR (2020) COVID-19 outbreak: application of multi-gene genetic programming to country-based prediction models. Electron J Gen Med 17(5)
50. Salgotra R, Gandomi M, Gandomi AH (2020) Time series analysis and forecast of the COVID-19 pandemic in india using genetic programming. Chaos, Solitons, Fractals 109945
51. Duffey RB, Zio E (2020) Analysing recovery from pandemics by learning theory: the case of CoVid-19. medRxiv
52. Barstugan M, Ozkaya U, Ozturk S (2020) Coronavirus (covid-19) classification using CT images by machine learning methods. arXiv:200309424
53. Randhawa GS, Soltysiak MP, El Roz H, de Souza CP, Hill KA, Kari L (2020) Machine learning using intrinsic genomic signatures for rapid classification of novel pathogens: COVID-19 case study. PLoS One 15(4):e0232391
54. Mei X, Lee H-C, Diao K-y, Huang M, Lin B, Liu C, Xie Z, Ma Y, Robson PM, Chung M (2020) Artificial intelligence–enabled rapid diagnosis of patients with COVID-19. Nat Med 1–5
55. Cássaro FA, Pires LF (2020) Can we predict the occurrence of COVID-19 cases? considerations using a simple model of growth. ScTEn 138834
56. Zhang X, Ma R, Wang L (2020) Predicting turning point, duration and attack rate of COVID-19 outbreaks in major Western countries. Chaos, Solitons, Fractals, 109829
57. Kucharski A, RT Diamond C, Liu Y, Edmunds J, Funk S, Eggo RM (2020) Centre for mathematical modelling of infectious diseases COVID-19 working group: early dynamics of transmission and control of COVID-19: a mathematical modelling study. Lancet Infect Dis 30144–30144

58. OECD Using artificial intelligence to help combat COVID-19. https://www.oecd.org/coronavirus/policy-responses/using-artificial-intelligence-to-help-combat-covid-19-ae4c5c21/. Accessed 6 Dec 2020
59. BlueDot. https://bluedot.global/. Accessed 6 Dec 2020
60. EpiRisk. https://epirisk.net/. Accessed 6 Dec 2020
61. Parrock J (2020) Coronavirus: Belgium hospital employs robot to protect against COVID-19. Euronews. https://www.euronews.com/2020/06/02/coronavirus-belgium-hospital-employs-robot-to-protect-against-covid-19. Accessed 6 Dec 2020
62. Canada's COVID-19 Chatbot. https://covidchatbot.com/. Accessed 6 Dec 2020
63. Perry TS (2020) Satellites and AI monitor Chinese economy's reaction to coronavirus. https://spectrum.ieee.org/view-from-the-valley/artificial-intelligence/machine-learning/satellites-and-ai-monitor-chinese-economys-reaction-to-coronavirus/. Accessed 6 Dec 2020
64. WeBank. https://www.webank.com/. Accessed 6 Dec 2020
65. Devakunchari R (2014) Analysis on big data over the years. Int J Sci Res Publ 4(1):1
66. Cheng S, Shi Y, Qin Q, Bai R (2013) Swarm intelligence in big data analytics. International conference on intelligent data engineering and automated learning. Springer, pp 417–426
67. Shi F, Wang J, Shi J, Wu Z, Wang Q, Tang Z, He K, Shi Y, Shen D (2020) Review of artificial intelligence techniques in imaging data acquisition, segmentation and diagnosis for covid-19. IEEE Rev Biomed Eng
68. García S, Luengo J, Herrera F (2015) Data preprocessing in data mining. Springer
69. Tsai C-W, Lai C-F, Chao H-C, Vasilakos AV (2015) Big data analytics: a survey. J Big data 2(1):1–32
70. Longbottom C, Bamforth R (2013) Optimising the data warehouse: dealing with large volumes of mixed data to give better business insights Quocirca
71. Keeling MJ, Hollingsworth TD, Read JM (2020) The efficacy of contact tracing for the containment of the 2019 novel coronavirus (COVID-19). medRxiv
72. Reyes SOL (2020) Artificial intelligence in precision health: systems in practice. In: Artificial intelligence in precision health. Elsevier, pp 499–519
73. Bragazzi NL, Dai H, Damiani G, Behzadifar M, Martini M, Wu J (2020) How big data and artificial intelligence can help better manage the COVID-19 pandemic. Int J Environ Res Public Health 17(9):3176
74. Qin L, Sun Q, Wang Y, Wu K-F, Chen M, Shia B-C, Wu S-Y (2020) Prediction of number of cases of 2019 novel coronavirus (COVID-19) using social media search index. Int J Environ Res Public Health 17(7):2365
75. Franch-Pardo I, Napoletano BM, Rosete-Verges F, Billa L (2020) Spatial analysis and GIS in the study of COVID-19. Rev ScTEn 140033
76. Mohamed A, Najafabadi MK, Wah YB, Zaman EAK, Maskat R (2020) The state of the art and taxonomy of big data analytics: view from new big data framework. Artif Intell Rev 53(2):989–1037
77. Wang J, He C, Liu Y, Tian G, Peng I, Xing J, Ruan X, Xie H, Wang FL (2017) Efficient alarm behavior analytics for telecom networks. Inf Sci 402:1–14
78. Chen CP, Zhang C-Y (2014) Data-intensive applications, challenges, techniques and technologies: a survey on big data. Inf Sci 275:314–347
79. Salah K, Rehman MHU, Nizamuddin N, Al-Fuqaha A (2019) Blockchain for AI: review and open research challenges. IEEE Access 7:10127–10149
80. Rizk Y, Awad M, Tunstel EW (2018) Decision making in multiagent systems: a survey. IEEE Trans Cogn Dev Syst 10(3):514–529
81. Elaraby NM, Elmogy M, Barakat S (2016) Deep learning: Effective tool for big data analytics. Int J Comput Sci Eng (IJCSE) 9
82. Deng L, Yu D (2014) Deep learning: methods and applications. Found Trends Signal Process 7(3–4):197–387
83. Ma J, Sheridan RP, Liaw A, Dahl GE, Svetnik V (2015) Deep neural nets as a method for quantitative structure–activity relationships. J Chem Inf Model 55(2):263–274
84. LeCun Y, Bengio Y, Hinton G (2015) Deep learning. Nature 521(7553):436–444

85. Hu D, Zhou X, Yu X, Hou Z (2015) Study on deep learning and its application in visual tracking. In: 2015 10th international conference on broadband and wireless computing, communication and applications (BWCCA). IEEE, pp 240–246
86. Xu Y, Mo T, Feng Q, Zhong P, Lai M, Eric I, Chang C (2014) Deep learning of feature representation with multiple instance learning for medical image analysis. In: 2014 IEEE international conference on acoustics, speech and signal processing (ICASSP). IEEE, pp 1626–1630
87. Petropoulos G (2020) Innovation and competition policy. www.bruegel.org/2020/03/artificial-intelligence-in-the-fight-against-covid-19/. Accessed 6 Dec 2020
88. Akhtar M, Kraemer MU, Gardner LM (2019) A dynamic neural network model for predicting risk of Zika in real time. BMC Med 17(1):171
89. Cong J, Ren M, Xie S, Wang P (2019) Predicting seasonal influenza based on SARIMA model, in mainland China from 2005 to 2018. Int J Environ Res Public Health 16(23):4760
90. Zou J-J, Jiang G-F, Xie X-X, Huang J, Yang X-B (2019) Application of a combined model with seasonal autoregressive integrated moving average and support vector regression in forecasting hand-foot-mouth disease incidence in Wuhan, China. Medicine 98(6)
91. Rahimzadeh M, Attar A (2020) A new modified deep convolutional neural network for detecting COVID-19 from X-ray images. arXiv:200408052
92. Li L, Qin L, Xu Z, Yin Y, Wang X, Kong B, Bai J, Lu Y, Fang Z, Song Q (2020) Using artificial intelligence to detect COVID-19 and community-acquired pneumonia based on pulmonary CT: evaluation of the diagnostic accuracy. Radiology 296(2)
93. Xu X, Jiang X, Ma C, Du P, Li X, Lv S, Yu L, Ni Q, Chen Y, Su J (2020) A deep learning system to screen novel coronavirus disease 2019 pneumonia. Engineering
94. Zhao J, Zhang Y, He X, Xie P (2020) COVID-CT-Dataset: a CT scan dataset about COVID-19. arXiv:200313865
95. Ribeiro MT, Singh S, Guestrin C (2016) "Why should I trust you?" explaining the predictions of any classifier. In: Proceedings of the 22nd ACM SIGKDD international conference on knowledge discovery and data mining, pp 1135–1144
96. Narin A, Kaya C, Pamuk Z (2020) Automatic detection of coronavirus disease (covid-19) using x-ray images and deep convolutional neural networks. arXiv:200310849
97. Rahmatizadeh S, Valizadeh-Haghi S, Dabbagh A (2020) The role of artificial intelligence in management of critical COVID-19 patients. J Cell Mol Anesth 5(1):16–22
98. Kaushik S, Choudhury A, Sheron PK, Dasgupta N, Natarajan S, Pickett LA, Dutt V (2020) AI in healthcare: time-series forecasting using statistical, neural, and ensemble architectures. Front Big Data 3:4
99. Beeksma M, Verberne S, Van den Bosch A, Das E, Hendrickx I, Groenewoud S (2019) Predicting life expectancy with a long short-term memory recurrent neural network using electronic medical records. BMC Med Inf Decis Making 19(1):36
100. Kolozsvari LR, Berczes T, Hajdu A, Gesztelyi R, TIba A, Varga I, Szollosi GJ, Harsanyi S, Garboczy S, Zsuga J (2020) Predicting the epidemic curve of the coronavirus (SARS-CoV-2) disease (COVID-19) using artificial intelligence. medRxiv
101. Siddique N, Adeli H (2015) Nature inspired computing: an overview and some future directions. Cognit Comput 7(6):706–714
102. Agbehadji IE, Millham RC, Fong SJ, Yang H (2018) Bioinspired computational approach to missing value estimation. Math Probl Eng
103. Said GAE-NA (2016) Nature inspired algorithms in cloud computing: a survey. Int J Intell Inf Syst 5(5):60–64
104. Tang R, Fong S, Yang X-S, Deb S (2012) Wolf search algorithm with ephemeral memory. In: Seventh international conference on digital information management (ICDIM 2012). IEEE, pp 165–172
105. Meuleau N, Dorigo M (2002) Ant colony optimization and stochastic gradient descent. Artif Life 8(2):103–121
106. Agbehadji IE, Millham R, Thakur S, Yang H, Addo H (2018) Visualization of frequently changed patterns based on the behaviour of dung beetles. International conference on soft computing in data science. Springer, pp 230–245

107. Agbehadji IE, Millham R, Fong SJ, Yang H (2018) Kestrel-based search algorithm (KSA) for parameter tuning unto long short term memory (LSTM) Network for feature selection in classification of high-dimensional bioinformatics datasets. In: 2018 federated conference on computer science and information systems (FedCSIS). IEEE, pp 15–20
108. Fong S, Wong R, Vasilakos AV (2015) Accelerated PSO swarm search feature selection for data stream mining big data. IEEE Trans Serv Comput 9(1):33–45
109. Gandomi AH, Yang X-S (2014) Chaotic bat algorithm. J Comput Sci 5(2):224–232
110. ELGhamrawy SM (2020) Diagnosis and prediction model for COVID19 patients response to treatment based on convolutional neural networks and whale optimization algorithm using CT images. medRxiv
111. Abdel-Basset M, Mohamed R, Elhoseny M, Chakrabortty RK, Ryan M (2020) A hybrid COVID-19 detection model using an improved marine predators algorithm and a ranking-based diversity reduction strategy. IEEE Access 8:79521–79540
112. Chen J, Cai H, Wang W (2018) A new metaheuristic algorithm: car tracking optimization algorithm. Soft Comput 22(12):3857–3878

Chapter 12
Short-term Forecasting of COVID-19

Hosam H. A. Mukhairez and Ahmed J. I. Alaff

Abstract Coronavirus started to spread in December 2019, especially in China, then rapidly it began to spread worldwide. COVID-19 mainly targets the elderly, the weak immune and the chronically ill people. Until now scientists could not determine the behavior of the virus to fight it and limit its reproduction and spread among humans. This steady increase in infected and dead people has exhausted the medical staff and led to the collapse of health sectors in the major countries and made them suffer from acute shortages of places and medical supplies. In order to anticipate the number of future infected and dead people, we have used in this paper short-term forecasting algorithms to predict the number of infected and deaths in the short future. The forecasting is conducted using the following forecasting algorithms: auto-regressive integrated moving average (ARIMA), hybrid ARIMA, HoltWinters and exponential smoothing with horizon value equalling 57 days. After evaluating the tested algorithms using MAPE, it is noticed that exponential smoothing was the best forecasting algorithm to be used to predict recovered cases with MAPE 2.66% and to forecast confirmed cases with MAPE equal to 1.77%, and HoltWinters is the best algorithm to be used to predict dead cases with MAPE equal to 5.33%.

Keywords Short-term forecasting · Forecasting · COVID-19 · ARIMA · Hybrid ARIMA · Exponential smoothing · HoltWinters · Mean absolute percentage error

1 Introduction

In Wuhan, China a city of 11 million people, in December 2019, the first case of COVID-19 has been detected. Thousands have been confirmed and dead in China. At that point on 30 January 2020, the World Health Organization (WHO) [1] reported that this flare-up is a worldwide pandemic. After very nearly 2 months of finding the main case in China, it sent a notice to all legislatures to take all essential measures to constrain COVID-19 from spreading [2]. Al Jazeera and news offices

H. H. A. Mukhairez (✉) · A. J. I. Alaff
Institute of Information Technology, Gazi University, Ankara, Turkey

U. Kose et al. (eds.), *Computational Intelligence for COVID-19 and Future Pandemics*, Disruptive Technologies and Digital Transformations for Society 5.0,
https://doi.org/10.1007/978-981-16-3783-4_12

have revealed step-by-step reports portraying numbers, affirmed cases and how the pandemic crossed the nation's borders. Most of the contaminated individuals were working at the city's Huanan Seafood Wholesale Market [3]. In this report, Al Jazeera mentions the most important events, statements and news issued for the first time that were documented by the leaders and officials.

To help detect coronavirus infection, thin slice chest CT images were used to help identify infected people [4]. Real-time reverse-transcriptase polymerase chain reaction (RT-PCR) tests for COVID-19 nucleic acid were performed using nasopharyngeal swabs (Novel Coronavirus PCR Fluorescence Diagnostic Kit, BioGerm Medical Biotechnology) [5]. With the passage of time, the virus has spread outside China, and it has become more cases and deaths than the country that is considered the starting point of the virus, as shown in Fig. 1.

Figure 1 shows that the USA was almost in the lead from April 2020 to December 2020. Also, Russia was almost in the second rank according to the five countries mentioned in the figure.

In conjunction with the efforts of scientists and doctors in finding a drug for the virus, there was also a role for computer scientists and artificial intelligence to develop computer models that can read images of radiation from medical centers and hospitals and try to read and understand them by taking advantage of the previously stored images. Forecasting task is understanding and recognizing the old behavior of an attribute to predict its future behavior or pattern. Forecasting is one of the most common techniques of time series data analysis. The basic idea is to use the historical timely dataset to make training to predict future data. It is used to expect and forecast future trends in water demands, retail sales, economic indicators, weather forecasting, stock markets and many other application scenarios.

Depending on the forecasting time horizon term, oil production forecasts can be classified into long term which forecasts more than 10 years, medium term which

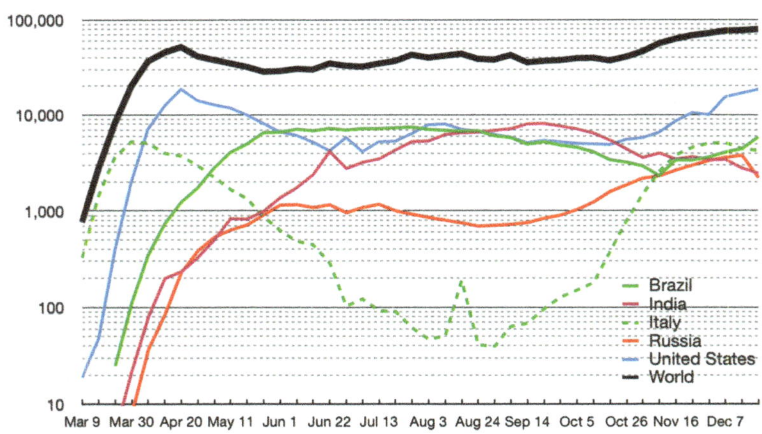

Fig. 1 Semi-log plot of daily deaths due to COVID-19 (three-day average) in the world and top five countries (mean with cases) [2] (*Source* https://en.wikipedia.org/wiki/COVID-19_pandemic)

forecasts years to decade and short term which forecasts up to a year [6]. In this paper, some forecasting algorithms were used and chosen the best depending on the mean absolute percentage error 'MAPE'. Forecasting, in general, helps specialists and decision-makers to have a better view of the future, so they can take the necessary measures according to the numbers and results.

The rest of this paper is organized as follows: Sect. 2 reviews some related works. Section 3 shows the experimental process and results. Section 4 presents the discussion and Section 5 the conclusion.

2 Related Works

Since the outbreak of COVID-19, advanced machine learning models and AI approaches have been used in the taxonomic classification of COVID-19 genomes, CRISPR-based COVID-19 detection assay, survival prediction of severe COVID-19 patients, and discovering potential drug candidates against COVID-19 [7].

Barstugan et al. [8] depending on machine learning algorithms presented an early-phase detection of coronavirus (COVID-19). They used CT images that COVID-19 shows different behaviors from other viral pneumonia. According to their claim, the model accuracy was obtained as 99.68% with tenfold cross-validation and GLSZM feature extraction method.

A mixed team of doctors and computer specialists [9] built an artificial intelligence (AI) model depending on 2003 SARS data. They observed that the epidemic of China should peak in February, showing a gradual decline by end of April. They also said that their dynamic SEIR model was effective in predicting the COVID-19 epidemic peaks and sizes. Ahmad Alimadadi et al. [7] by using artificial intelligence and machine learning built a model to force COVID-19. Their model depends on collecting data about virus information, epidemiological data, clinical data and genetic data. Then these data are managed and processed using deep learning and AI algorithms to get the final result.

AlphaFold is an AI algorithm that predicts the protein structures of COVID-19 [10]. It is made out of three layers of profound neural organizations. The principal layer is made out of a variational auto-encoder stacked with a consideration model. The subsequent layer is split into two sublayers. The first sublayer optimizes inter-residue distances using a 1D CNN on a contact map. The second sublayer optimizes a scoring network, which is how much the generated substructures look like a protein using a 3D CNN. The third layer scores the generated protein against the actual model.

A data-driven analysis [16] is done for the COVID-19 pandemic in order to give speculative forecasts of the pestilence top for Russia, Brazil, India and Bangladesh. Hospitalized-Recovered-Dead (SEIDIUQHRD) deterministic compartmental model has been proposed and aligned for deciphering the transmission elements of the novel Covid sickness (COVID-19). This study investigation likewise recommends that loosening up social movement and removing limitations excessively fast could worsen the plague episode in the previously mentioned nations.

By using a dynamic tracking model, the study [17] forecasts the spread of the COVID-19 pandemic. In this study, susceptible-infected-removed (SIR) model is being used to track the spread of COVID-19 in four countries: Italy, India, South Korea and Iran. The authors show that it is conceivable to identify the beginning stage of auxiliary spikes in diseases or the advancement of optional waves. They thought that from March to August 2020 various networks were influenced harshly and exhibit forecasts relying upon the model's boundaries identified with the spread of COVID-19 until the finish of December 2020. In contrast to the distributed information and model outcomes, they gave the reason that it could be conceivable to reflect better achievement or disappointment of the satisfactory estimates executed by governments and specialists to alleviate and control the current pandemic.

3 Methodology

Our paper methodology goes through four main steps in order to select the most suitable forecasting algorithm from the four selected algorithms over our dataset attributes, as shown in Fig. 2.

In order to achieve a high quality of forecasting results, an evaluation process must be done over timely datasets. This is done by choosing the most proper forecasting

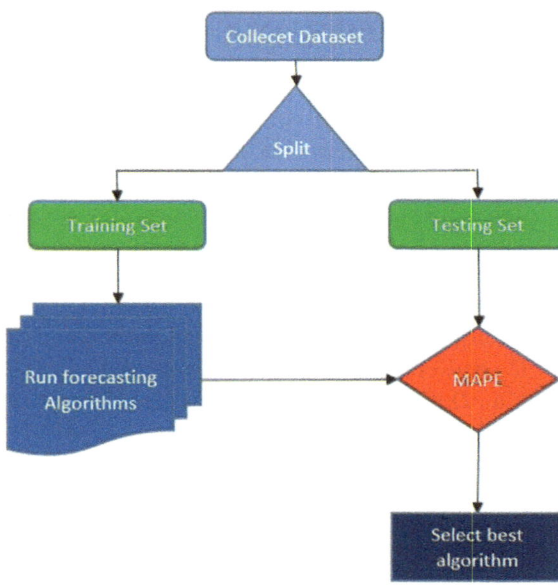

Fig. 2 Main steps of forecasting COVID-19 methodology

algorithms sucn as ARIMA [11], Hybrid ARIMA [12], HoltWinters [13], ES [14], and split the available datasets into two parts (training set and testing set).

Next, these selected algorithms run over the training sets and configure their forecast horizon to the end of the testing set. Then, comparing the results of the algorithms with the original testing set, mean percentage errors (MPE) are calculated. The algorithm with the smallest MPE is the best one to be chosen for real forecasting.

4 Experimental Results

The three successfully evaluated forecasting algorithms are considered famous algorithms used in the time series analysis field. These algorithms are auto-regressive integrated moving average (ARIMA), exponential smoothing (ES) and HoltWinters algorithm. These algorithms were run over datasets to calculate the mean absolute percentage error (MAPE) to detect which algorithm is more accurate to be used in the forecasting process depending on MAPE measure.

The dataset is a real dataset collected from Johns Hopkins University Center for Systems Science and Engineering (CSSE) [15]. The dataset includes time series data tracking the number of people affected by COVID-19 worldwide. The data were collected from 22 January 2020 to 27 December 2020, and it is constantly updated.

The original data consists of four columns (date, confirmed, recovered and death). The dataset containing 341 days was divided into a training set that contains the first 284 days of data from 22 January 2020 to 31 October 2020 (83.3% of the original data), and a testing set holding the remaining data from 1 November 2020 to 27 December 2020 (16.7% of the original data). Models were trained on the training sets before being applied to the unseen data in the test sets. Table 1 shows a sample of the training set, and Table 2 shows a sample of the testing set.

To choose the appropriate forecasting algorithms, we first evaluated and compared the results from each algorithm with the original data that we have. Then four forecasting algorithms were run on the training set with a horizon value of 57 to reach 27 December 2020 and compare the data with the actual data in the testing set.

4.1 Evaluating Algorithms Over Dataset

As a first step, the evaluation process of the selected four algorithms is done over the datasets and the mean absolute percentage error (MAPE) is computed for each algorithm.

$$\text{MAPE} = (((\text{Actual} - \text{Predicated})/\text{Actual}) * 100) \quad (1)$$

This evaluation provides insights in recommending changes to existing models to reduce deviations in future forecasts (Figs. 3 and 4).

Table 1 Sample of the training set from 22 January 2020 to 09 February 2020

Date	Confirmed	Recovered	Deaths
22/1/2020	555	28	17
23/1/2020	654	30	18
24/1/2020	941	36	26
25/1/2020	1434	39	42
26/1/2020	2118	52	56
27/1/2020	2927	61	82
28/1/2020	5578	107	131
29/1/2020	6166	126	133
30/1/2020	8234	143	171
31/1/2020	9927	222	213
1/2/2020	12,038	284	259
2/2/2020	16,787	472	362
3/2/2020	19,881	623	426
4/2/2020	23,892	852	492
5/2/2020	27,635	1124	564
6/2/2020	30,794	1487	634
7/2/2020	34,391	2011	719
8/2/2020	37,120	2616	806
9/2/2020	40,150	3244	906

Table 2 Sample of the testing set from 1 November 2020 to 10 November 2020

Date	Confirmed	Recovered	Deaths
1/11/2020	46,575,736	31,105,741	1,201,152
2/11/2020	47,136,486	31,385,632	1,207,222
3/11/2020	47,689,293	31,637,087	1,216,564
4/11/2020	48,199,622	31,916,319	1,226,752
5/11/2020	48,794,945	32,219,865	1,234,763
6/11/2020	49,438,618	32,508,227	1,244,381
7/11/2020	50,036,298	32,807,319	1,251,920
8/11/2020	50,515,513	33,060,197	1,257,688
9/11/2020	51,017,685	33,316,917	1,264,797
10/11/2020	51,575,582	33,573,921	1,274,783

Hata! Başvuru kaynağı bulunamadı., and Fig. 5 presents the values of MAPE for the four selected forecasting algorithms. It is observed that exponential smoothing (ES) was the best forecasting algorithm to be used over datasets (MAPE 1.77%) in forecasting confirmed cases. And also, exponential smoothing (ES) is the best in

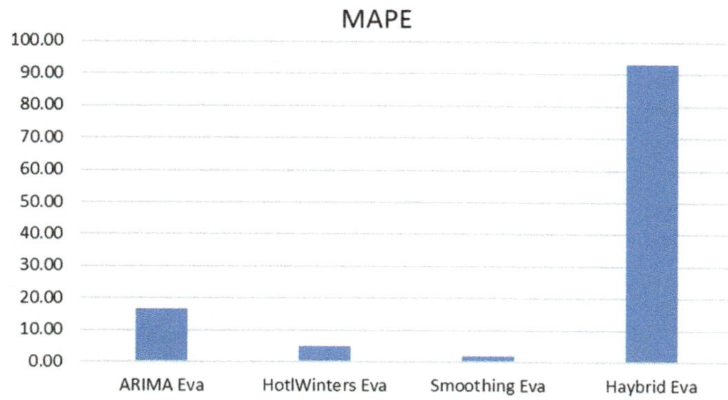

Fig. 3 Calculated MAPE for four algorithms over confirmed cases

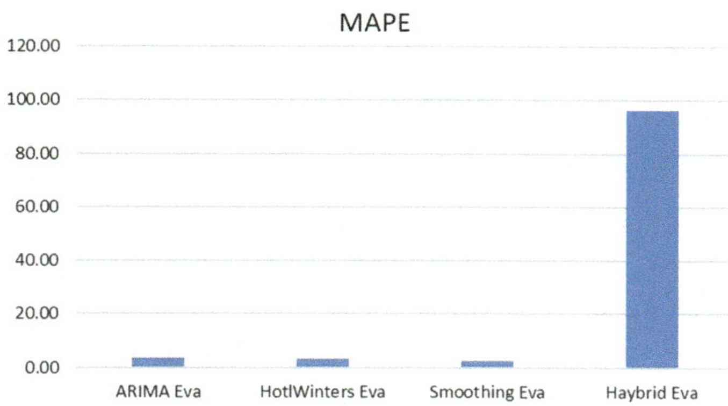

Fig. 4 Calculated MAPE for four algorithms over recovered cases

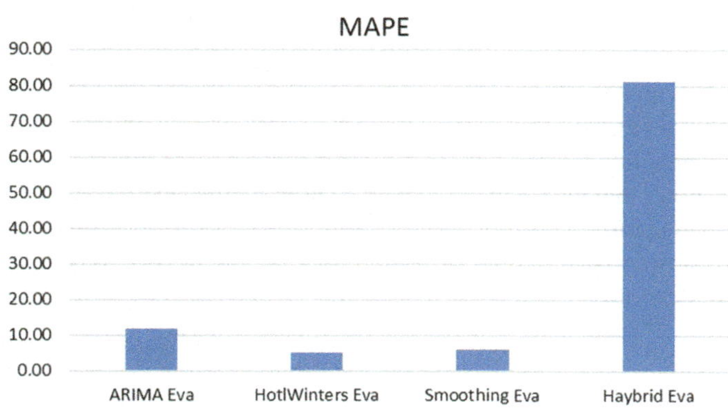

Fig. 5 Calculated MAPE for four algorithms over death cases

Fig. 6 Forecasted results for confirmed cases from 28 December 2020 to 28 February 2021

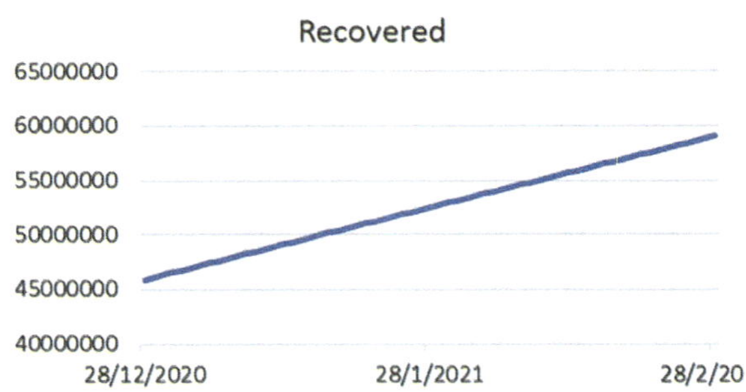

Fig. 7 Forecasted results for recovered cases from 28 December 2020 to 28 February 2021

forecasting recovered cases (MAPE 2.66%), and finally, HoltWinters is the best in forecasting death cases (MAPE 5.33%).

After detecting the most appropriate algorithm for each type of information, we are going to forecast 63 days in advance to predict number of confirmed, recovered and death cases all over the world.

Figure 6 shows that the trend of confirmed cases for the next 63 days is constantly increasing in general.

Figure 7 shows that the trend of recovered cases for the next 63 days is constantly increasing in general.

Figure 8 shows that the trend of death cases for the next 63 days is constantly increasing in general.

5 Discussion

From previous results, we noticed that the rate of all cases (confirmed, recovered and death cases) was constantly increasing. This means the pandemic is still spreading

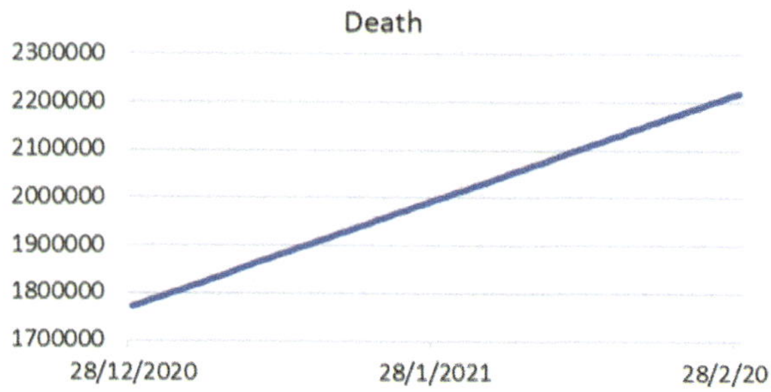

Fig. 8 Forecasted results for death cases from 28 December 2020 to 28 February 2021

and infecting strongly the human life and safety, especially after the news of the development of the coronavirus and the emergence of new genetic mutations. By continuous application of social distance and other necessary measurements by the governments, this pandemic can be restricted and the danger will be removed.

By comparing numbers with the last day of original data, we find that the percentage of recovered people according to confirmed cases is 56.5%, while at the end of the forecasting period it is about 52.5%, which gives a slight indicator that COVID-19 is going to be restricted by time.

In this paper, we used exponential smoothing to forecast both confirmed and recovered cases, because it has the best MAPE. But that doesn't mean exponential smoothing is always the best in our case or for other cases. Because the evaluation process depends on a timely dataset, this means if we got a new dataset after few months and run the model over the new dataset, it may give another algorithm with better MAPE. So, the suitable algorithms depend on the nature of the dataset values. At the same time, we cannot say that hybrid ARIMA is the worst algorithm in the forecasting field. It may give a better MAPE if it is run over another type of timely dataset.

6 Conclusion

In this paper, the number of COVID-19 victims was forecasted in the short term. Four main forecasting algorithms were used: ARIMA, hybrid ARIMA, ES and HoltWinter. We applied them to a dataset collected by the Johns Hopkins University Center for Systems Science and Engineering (CSSE).

After the experiment, we noticed that exponential smoothing is in the lead to forecast confirmed cases with the MAPE value equal to 1.77%. Also, exponential smoothing was the best forecasting algorithm to be used to predict recovered cases with MAPE of 2.66%, while HoltWinters is the best algorithm to be used to predict dead cases with MAPE equal to 5.33%.

References

1. WHO Director-General's remarks at the media briefing on 2019-nCoV on 11 February 2020 [updated 11/02/2020]. https://www.who.int/dg/speeches/detail/who-director-general-s-remarks-at-the-media-briefing-on-2019-ncov-on-11-february-2020
2. 2019–2020 coronavirus pandemic. https://en.wikipedia.org/wiki/2019%E2%80%9320_coronavirus_pandemic
3. Confirmed coronavirus cases top 3m worldwide. https://www.aljazeera.com/news/2020/04/italy-loosen-lockdown-4-coronavirus-live-updates-200426233610783.html
4. Goo N H (2020) Committee [updated 26/02/2020]. http://bgs.satcm.gov.cn/zhengcewenjian/2020-02-06/12847.html.
5. Bai Y, Yao L, Wei T, Tian F, Jin D-Y, Chen L, et al (2020) Presumed asymptomatic carrier transmission of COVID-19 323(14):1406–1407
6. Mukhairez HHA (2018) Long-term forecasting for world oil production. Int J Intell Comput Res 9
7. Alimadadi A, Aryal S, Manandhar I, Munroe PB, Joe B, Cheng X (2020) Artificial intelligence and machine learning to fight COVID-19. American Physiological Society Bethesda, MD
8. Barstugan M, Ozkaya U, Ozturk S (2020) Coronavirus (covid-19) classification using ct images by machine learning methods
9. Yang Z, Zeng Z, Wang K, Wong S-S, Liang W, Zanin M, et al (2020) Modified SEIR and AI prediction of the epidemics trend of COVID-19 in China under public health interventions 12(3):165
10. Gulamali F (2020) AlphaFold algorithm predicts COVID-19 protein structures [updated 31/3/2020]. https://www.infoq.com/news/2020/03/deepmind-covid-19/
11. Broomhead DS, King GP (1986) Extracting qualitative dynamics from experimental data 20(2–3):217–236
12. Khashei M, Bijari MJM (2011) Engineering Si: a new hybrid methodology for nonlinear time series forecasting
13. Chatfield CJJotRSSSC (1978) The Holt-winters forecasting procedure. 27(3):264–279
14. Gardner Jr ES (1985) Exponential smoothing: the state of the art 4(1):1–28
15. Novel Coronavirus 2019 time series data on cases. https://github.com/datasets/covid-19/blob/master/data/time-series-19-covid-combined.csv
16. Nabi KN (2020) Forecasting COVID-19 pandemic: a data-driven analysis. Elsevier
17. Cooper I, Mondal A, Antonopoulos CG (2020) Dynamic tracking with model-based forecasting for the spread of the COVID-19 pandemic. Elsevier

Chapter 13
Internet of Health Things (IoHT) Against COVID-19: A Review of Recent Development

Geetha Velliyangiri, Anbumani Venkatachalam, Manikandan Ramachandran, Ambeshwar Kumar, and Murugan Subramanian

Abstract Nowadays, the Internet of Things (IoT) has authentic grounds for research in a wide range of industrial and academic fields, mainly in healthcare. By employing an interconnect network for properly monitoring the individuals affected by COVID-19, IoT enabled the healthcare system. To find symptoms and give better treatment, IoT is helpful for COVID-19 infected patients and useful for a hospital management system, surgeon, physician, and patient. For enhancement of patient satisfaction and minimization of readmission rate in hospitals, this method is used. By capturing data from patients, IoT devices speed up the process of detection. This can be executed by observing the body temperatures with various devices and taking samples from wary cases and so on. To control and mitigate the virus spread, another widely used method is called tracking patients by using smartphones during quarantine time. The main concern about this work is to use IoT devices in various stages of this pandemic which is a privacy problem as patients are requested to provide their information.

Keywords Internet of Health Things (IoHT) · COVID-19 · IoT device · Smartphone · Patients information

1 Introduction

With the increase in using mobiles and smart devices, the healthcare sector has significant impacts on the world. Developing potential and powerful smart devices assists and provides benefits to observe the health of individuals, thereby providing

G. Velliyangiri · A. Venkatachalam
Department of Electronics and Communication Engineering, Kongu Engineering College, Perundurai 638 060, India

M. Ramachandran (✉) · A. Kumar
School of Computing, SASTRA Deemed University, Thanjavur, India

M. Subramanian
Department of Computer Science, Sri Aravindar Engineering College, Sedarapet, TN, India

U. Kose et al. (eds.), *Computational Intelligence for COVID-19 and Future Pandemics*, Disruptive Technologies and Digital Transformations for Society 5.0,
https://doi.org/10.1007/978-981-16-3783-4_13

improvement in healthcare sectors. IoT permits the integration of physical devices that are capable of having internet connectivity and also helps doctors by proving the actual health status of patients [1].

IoT is a distinctive strategy of interrelated computing schemes, mechanical, and digital devices that transmits data over the network with no human intervention at any level. Every device is identified with the help of its unique id. Now, IoT is defined and established in a better manner and proved that this technology acts as an intersection for numerous schemes, instantaneous analysis, machine learning, sensual products, etc. [2]. In healthcare, IoT is a challenging one due to poor internet coverage, particularly in few nations like India. Moreover, the technical ability of the hospitals, cost factors, and issues related to compromising data because of cybersecurity causes hindrances to using it to a greater extent.

IoT deployed a huge volume of data [3] and is used in various applications to achieve essential needs of alleviating impacts of COVID-19. When the data is captured more precisely, the forthcoming situation can be predicted easily. IoT services can be properly utilized by observing heart rate, monitoring blood pressure, and other possible tasks. Moreover, the health conditions of older people can be tracked easily [4]. IoT is an advanced device that assures quarantine of all infected people due to this infection. During quarantine, a good surveillance system is useful. Using an internet-based network, all high-risk patients are monitored quickly.

The critical merits of the IoT for the COVID-19 pandemic are shown in Fig. 1. This technology significantly helps in healthcare for tracking the real-time location of medical devices and equipment used for providing smooth treatment with no delay. Healthcare insurance firms can make use of this technique for detecting bogus claims, thereby transparency can be introduced in the system. Moreover, this efficiently

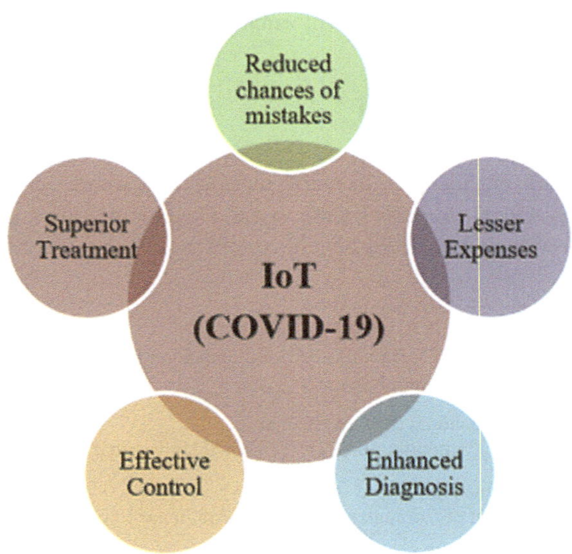

Fig. 1 IoT for fighting COVID-19 pandemic key merits

improves the way treatment is given to the patients and also helps in taking decisions while dealing with complex cases. Several recent examples of the use of technologies to monitor the spread of COVID-19 exist. Solutions to handle large numbers of people to restrict infected cases are suggested.

Coronavirus is transferred to human beings, animals as well as birds. Coronavirus HKU2 that originated from bats caused serious diarrhea syndrome in pigs in 2018. But the exposure of COVID-19 is more than ever. Nevertheless, the advantages due to care, costs, and functions balance the risks when evaluating its cause. Coronavirus (SARSCoV-2) pandemic has spread its impact worldwide whose origin was Wuhan, China, on December 31, 2019. Now it has been called coronavirus disease 2019 (COVID-19). This infectious virus has spread rapidly resulting in a global threat to human health affecting nearly 213 nations in the world. At present, the cases infected due to COVID-19 across the world are nearly 5,555,749 out of which deaths reported are 348,220 and active cases are still 2,875,734 [5]. The spreading of this coronavirus has to be controlled by the authorities who have to take suitable measures and provide awareness to the public [6]. Both nations, including India, are struggling with COVID-19 in the current pandemic situation and are also searching for a realistic and cost-effective approach to face problems that occur in various ways. At present, pandemic due to COVID-19 is considered to be a hotspot in medical research. The new technologies that are emerging can be a remedy for this crisis across the world. IoT and specifically IoHT offer a solution to the issues faced during detecting, monitoring, controlling, and contact tracing to avoid the spread of coronavirus [7]. IoHT refers to the interconnection of medical components with software products and the data related to healthcare sectors. A recent IoT application that has attracted interest is the monitoring of healthcare smartly which is best suited and applicable at this COVID-19 pandemic situation.

COVID-19 pandemic of 2019–2020 has contributed to worldwide suffering [8] and unprecedented shifts in ways in which physicians and scientists conduct their research. Immigration bans and work losses [9], psychiatric and physical illness [10, 11], and death[11] have been among the pervasive effects of this worldwide epidemic. Global stay-at-home orders have been imposed in response to the epidemic, and health visits have been supplemented by telemedicine appointments [12, 13]. In the biomedical research world, except for those studying COVID-19 and COVID-19 therapies, several buildings have been closed and research operations have ceased.

This chapter is organized into the background of the study in Sect. 1, related works discussion in Sect. 2, system model in Sect. 3, experimental results and discussion in Sect. 4, and Sect. 5 discusses the conclusion along with future work.

2 Related Works

In the past decade, digital technologies play vital roles in the major health sector, for example, the prevention of diseases. Moreover, the current health emergency sector also seeks technological support to handle COVID-2019.

Gupta et al. developed a novel architecture and defined many use-cases which helped in creating a smart city, and ITS stimulated data-driven system which enforces communal distancing at this pandemic situation and high-impact low-frequency (HILF) events [14].

Bayram M et al. focused on developing a novel critical approach to diagnosing COVID-19, especially at times with uncertain facts, high risks, and requires immediate decisions due to pandemic, and provide a potential remedy for democratizing innovation policies in this pandemic situation.

Moreover, it is proposed that not only scientific knowledge is concentrated with epistemic integrity but also its framing is generally valued, helping to minimize the gap between technological improvement and digital health investments against insufficient concerns of dimensions of developed technologies, such as AI, IoT, etc. [15].

Patil P et al. presented a recent three-dimensional printing (3DP) application that helped to develop innovative products for saving several million people across the world. Further, the benefits of 3DP technologies were highlighted. From the review, it is observed that fused filament fabrication (FFF) is one of the preferable technologies whose contribution is nearly 62% in producing protective gears through 3DP [16].

Celik HK et al. described the original design of the face shield using <10 g material per frame. Further, more useful product design was provided for AM [DfAM] with simple medical devices like face shields using advanced engineering design, simulation, and additive manufacturing (AM) applications as required to fight against coronavirus [17].

Dibyendu Roy et al. are of the view that lessons learned and knowledge acquired from these cities are highly useful and can quickly be repeated in other cities within a relatively short period, thereby offering a standard and uniform approach for controlling potential epidemics around the country [18].

Talha Burak Alakus et al. developed predictive clinical models that used deep learning approaches on laboratory data where patients may be infected with COVID-19 [19]. It is proved that this model trained with the laboratory discoveries helped in predicting COVID-19 and can assist medical experts to appropriately assign priorities for the resources.

Applications on smartphones are a kind of application software performing limited tasks in the mobile [20]. In 2020, as 3.5 billion active smartphones are available. Smartphone applications based on IoT perform tasks efficiently in different domains like agriculture, retail, and healthcare [21–23].

Several applications for smartphones have been designed and few are involved in testing COVID-19, namely nCapp [24], Detect a Chem [25], Stop Corona [26], Social Monitoring [27], Selfie app [28], Civitas [29], Stay Home Safe [30], AarogyaSetu [31], TraceTogether [32], Hamagen [33], Coalition [34], BeAware Bahrain [35], eRouska [36], and Whatsapp [37].

Eric A. Storch et al. [38] report on clinical views of effects of COVID-19 pandemic on OCD patients seeking sensitivity and reaction prevention (ERP) therapy before and after the pandemic. A study containing questions adapted by NIMH-GOCS (National Institution of Mental Health-Global Obsessive–Compulsive Scale) and Y-BOCS (Yale-Brown Obsessive–Compulsive Scale) was conducted by participating clinicians. At care beginning, just before the pandemic, and mid-pandemic (July/August 2020), clinicians rated clinical characteristics.

Addie B. Spier et al. [39] profile experiences of an early-career investigator who worked during the COVID-19 pandemic and a COVID-19 diagnosis in thrombosis science. The aims of this paper are to normalize the research community's anxiety about COVID-19, to provide a viewpoint on sustaining efficiency during stay-at-home times, and to address how the COVID-19 pandemic in the future could affect traditional research activities. While the epidemic of COVID-19 was devastating and crippling on a global basis, for many years to come, certain testing practices that were extensively utilized during the pandemic could continue to be used in scientific research.

3 System Model

A system model which is used for COVID in IoHT is shown in Fig. 2. The patient having symptoms can self-test and upload their results on the Internet of Health

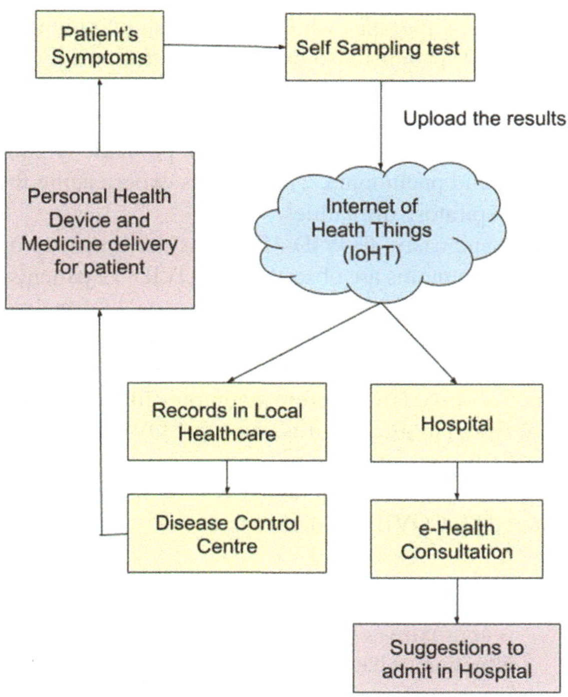

Fig. 2 System model using COVID in IoHT

Things (IoHT). The results can be categorized into two ways: one is the local healthcare which can store the records and the other one is hospitals where e-health consultation is given to the patient to know whether they are to be admitted to the hospital or not. The disease control center provides the medicine to the patient who suffers from normal fever or mild disease attack.

Patients' Symptoms

Symptoms of the disease caused by coronavirus-2019 (COVID-19) are observed from 2 to 14 days after disclosure. The duration after disclosure and before symptoms is known as the incubation period. The symptoms that are commonly observed are fever, cold, cough, and tiredness. Tastelessness and loss of smell are the early symptoms of COVID-19.

Few individuals may feel aches, nasal congestion, sore throat, running nose, or even diarrhea. These are also considered symptoms that begin gradually and are mild usually. In some cases, people get infected with COVID-19 but no symptoms are identified and even they feel well.

Nearly 80% of the people affected with COVID-19 are recovered without providing any special medical treatment. About one-sixth of the COVID-19 persons are serious and face difficulties in breathing. People who are already undergoing treatment for diabetes, blood pressure, or cardiac disorders and even older people are the most who are affected more seriously. Individuals having fever, sore throat, cough, and facing difficulties while breathing should immediately seek medical assistance. COVID-19 patients also experience neurological or gastrointestinal (GI) problems or even both.

COVID-19 symptoms vary from moderate symptoms such as fever, coughing, and shortness of breath to extreme and symptomatic symptoms such as organ failure, renal failure, and pneumonia. The patients experiencing these problems may or may not have respiratory difficulties.

In some cases, COVID-19 affects the functions of the brain also. When neurological symptoms are observed in COVID-19 patients, they exhibit loss of smell and taste, weakness in muscle, irritation or lack of feeling in hands and feet, giddiness, and even stroke. Moreover, diarrhea, vomiting, loss of appetite, abdominal pain, or uneasiness are experienced by the COVID-19 patients having GI symptoms which can be observed even before symptoms like fever, cough, and ache are experienced. Even from stools, the virus causing COVID-19 can be detected which emphasizes the significance of washing hands after every visit to the restroom and disinfect the equipment present there regularly. The following are the most common symptoms endured by COVID-19 patients.

A. Fever

   When a person is well, under normal circumstances, his/her body temperature is approximately 37 °C. However, the temperature of the body when met by an infection or bacteria is always higher and the result is fever.

   Fever can be said to be one of the main indicators of sickness or illness. Standard medical thermometers used in many countries to track the health needs of people do not seem to be an accurate tool, provided that many people need to be weighed. Thus, unexplained changes in body temperature can be observed, especially during sleep cycles. Because certain body temperature parameters need to be measured, using infrared thermopile sensors or high-precision digital temperature sensors, as found in many consumer items on the market, will provide a simplified structure.

B. Dry Cough
Cough is one of the human body's major defense systems for clearing the airway, and dry cough is associated with upper respiratory tract virus infections. [40]. One of the most frequent symptoms caused by the 2019-nCoV virus is dry cough, according to the studies on COVID-19. Infection emerges through cough of an infected person by droplets. Informing patients at the early stages of the disease by identifying irregular changes in cough is also very important.

C. Fatigue
A loss of energy can be identified as fatigue, and it can be a symptom of many diseases. There are also COVID-19 patients suffering from exhaustion. Fatigue calculation may be a complicated procedure, and there might not be a simple technique. Monitoring of physical activity during daytime can also provide an indication of the extent of fatigue instead of explicitly calculating fatigue.

D. Anorexia
Anorexia defined as a reduced appetite may be a symptom, suggesting that one may not feel hungry or lack the urge to eat. Often people don't even know that they have no appetite, so it is vital to detect this situation and alert the person as soon as possible if this occurs as a result of a disease. Anorexia cannot be accessed directly; however, it can be inferred that person displays anorexia as a symptom by following the person's normal eating patterns, anytime some drop in food consumption happens.

E. Anosmia
Lack of ability to smell one or more odors is known as anosmia. Upper respiratory tract infection due to COVID-19 results in olfactory disorders that have been documented in several European countries. It is very difficult to diagnose anosmia without taking notification from individuals.

F. Dyspnea

Dyspnea is a sensation of not breathing properly, which is often described as shortness of breath. Shortness of breath is one of the other major symptoms caused by 2019-nCoV. For example, one of their members began to exhibit shortness of breath after infection on one of the commercial fitness tracker device websites (Whoop). Through the fitness tracker device, respiratory changes are monitored in his/her, from 18.1 rpm to 15.4 rpm is changed from night to night.

Shortness of breath due to the 2019-nCoV cannot be observed in anyone, but respiratory failure is normally caused by this circumstance and the patient should be taken under medical observation if the condition persists. Depending on the case, if a person has underlying ailments, the risk of developing respiratory failure is much higher. For the measurement of shortness of breath, respiratory rate is a significant parameter. About 12 and 20 breaths per minute is the typical respiratory rate estimated for a healthy person.

Self-Sampling Test

People who think that they may be infected by COVID-19 are permitted to test their sample which is beneficial. Kids to collect samples may be distributed widely and hence numerous individuals to be tested. The advantage of using this kid is that they need not to the testing location thereby avoid the risk of disease transmission to the healthy people and to whom they are in contact.

The evidence of communicating the accuracy of the test is very complex due to several reasons. Recent research works focus on communicating the possible significances of positive and negative results of tests and the importance of providing healthcare setup, for example, treatments given for in-patients and out-patients in hospitals. Performing tests at frequencies instead of probabilities can enhance understanding. Few symptoms of flu and COVID-19 are alike and difficult to differentiate them only with the symptoms experienced. Testing is done to assist in diagnosis.

Internet of Health Things (IoHT)

To prevent people from unnecessary wandering in healthcare spaces, IoT-based heath control system help streamline services. Nowadays, robots related to medical observations are involved to detect COVID-19, for measuring the temperature, and provide medicines to isolated patients.

To encourage staff to return to their offices and to reopen public areas, buildings, stores, and all other services securely, IoT-based screening systems would also be effective in identifying potentially infectious persons.

New norms such as WFH workforce, physical distancing, and no-touch delivery are followed in most organizations in the world. Educational institutions are influenced also during these pandemic periods and are exploring the integration of online classes and distance learning. This pandemic presented many possibilities for learning from various studies and social developments spanning from home to surveillance.

Disease Control Centre

DCC suggests some policies to be followed by the employees like covering nose and mouth (wearing masks), daily monitoring of the symptoms, following the rules for screening people with and without symptoms, and even post messages to educate and provide awareness. While the rate of infection remains higher across several countries, the guidance of the CDC has highlighted that students face the risk at a higher rate as they have not attended the school/college for a longer period.

e-Health Consultation

World Health Organization has advised hospitals and healthcare clinics to promote the use of telemedicine technologies like virtual doctor visits over a video call or a phone call. e-Healthcare benefits doctors and nurses by potentially reducing the burden on already overstressed medical infrastructure. The goal is to keep the worried-ill out of medical facilities, where they might be exposed to illnesses. Face-to-face consultations potentially increase the risk for doctors and medical staff as well. Infectious diseases like seasonal flu have a high rate of transmission, and patients may infect medical practitioners too in a live clinical consultation.

4 Results and Discussion

The experimental results of the proposed system show that the Baidu search engine appropriately revealed the behavior of user's search, prevalence, and symptoms which assists in understanding COVID-19. The Baidu index was analyzed daily to confirm the number of new suspected cases. Table 1 presents the values obtained from the confirmed and non-confirmed cases. Values are obtained from the proposed model.

Figure 3 shows the Baidu index based on confirmed and suspected cases. The X-axis provides the date in 2020 and the Baidu index value in the Y-axis. Orange and brown color indicates the newly suspected cases and newly confirmed cases, respectively. The curves in the graph showing the newly confirmed and suspected cases are closely related to Baidu index.

Table 2 shows the Baidu index value for confirmed and non-confirmed cases and patient's symptoms of the proposed model. COVID-19 patient's most common symptoms are mentioned below. The patient's symptoms are cough, fever, fatigue, diarrhea, and chest pain. The confirmed and non-confirmed cases are evaluated based on the patient's symptoms. The patients experiencing these problems may or may not have respiratory difficulties.

Figure 4 shows the Baidu index based on the patient's symptoms performance. The X-axis provides the symptoms and the Y-axis shows Baidu index values. Orange and brown color indicates the newly suspected cases and newly confirmed cases, respectively. Symptoms observed not only reflect the actual needs of the people but provide significant information to the health centers to identify the characteristics of

Table 1 Baidu index for confirmed and suspected cases of the proposed model

Date in 2020	Confirmed cases	Suspected cases
30 March	38	42
28 April	42	48
31 May	46	54
23 June	48	59
27 July	53	67
29 August	56	69
30 September	66	74
18 October	68	78
20 November	73	82

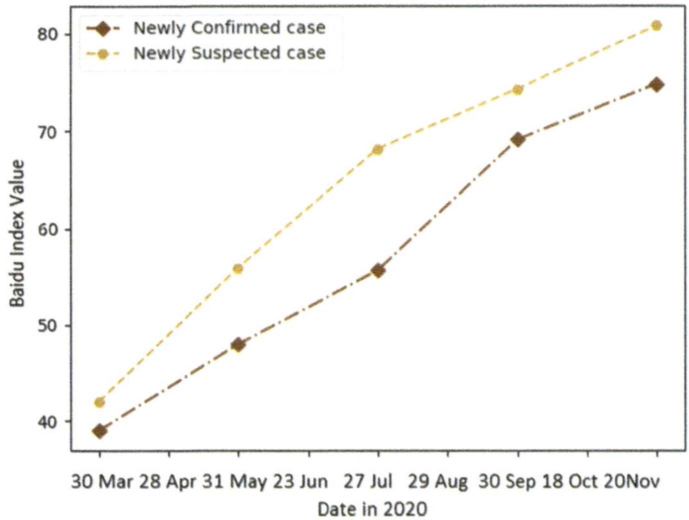

Fig. 3 Baidu index for confirmed and suspected cases

Table 2 Baidu index for patient's symptoms of the proposed model

Patient's symptoms	Confirmed cases	Non-confirmed cases
Cough	28	34
Fever	35	52
Diarrhea	43	64
Fatigue	54	78
Chest pain	57	81

COVID-19, predict the pandemic situation, and develop guidelines for diagnosing and providing treatment at the earliest.

5 Conclusion

Identifying a person infected with COVID-19 from a crowd is very tedious. To restrict the spread of this coronavirus, people have to be isolated from infected individuals. Using IoHT which involves smart sensors to estimate the temperature of individuals helps in identifying the people infected with COVID-19. Moreover, social distance has to be maintained. Doctors, as well as clinical staff, are mostly infected if safety measures are not followed and considered as significant factors to avoid infection. When not done so, infected people can spread the virus to other people who go to hospitals for other regular checkups. People affected by other medical issues

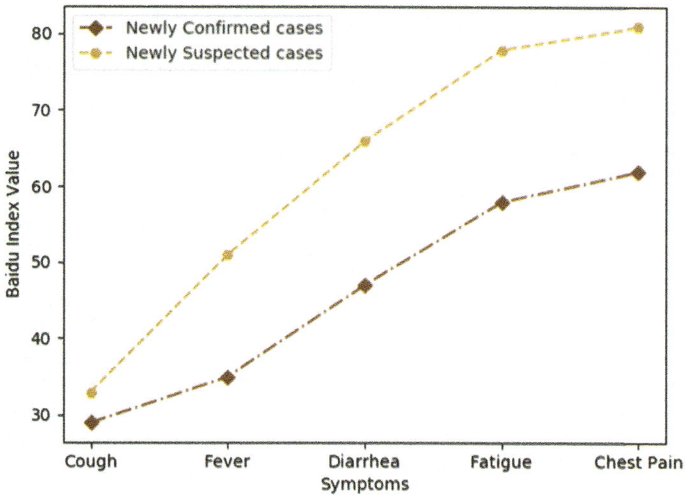

Fig. 4 Baidu index for Patient's symptoms

visiting hospitals for treatment are easily infected. This study is based on the recorded symptoms of patients infected by COVID-19 and the Baidu index calculation.

In the future, research work has to focus on improving the accuracy of searching data via the internet and explore the way of integrating these data with the actual data from medical centers. To be more precise, big data analytics can support medical development. In the present pandemic case, this study using IoT COVID-19 is about the protection and privacy of data collected, which is special and crucial from point of view of patient welfare. The second thing is about taking caution when implementing data networks between systems and protocols involved. This method is tested with people with the disease at high risk like healthcare professionals by making compact devices. Also, all data showing health status will be reached by the mobile application that will be created.

References

1. Kumar K (2020) Role of IoT to avoid spreading of COVID-19. Int J Intell Netw 1:32–35
2. Singh RP, Javaid M, Haleem A, Suman R (2020) Internet of things (IoT) applications to fight against COVID-19 pandemic. Diabetes Metab Syndr: Clin Res Rev 14:521–524
3. Sampathkumar A, Maheswar R, Harshavardhanan P, Murugan S, Jayarajan P, Sivasankaran V (2020) Majority voting based hybrid ensemble classification approach for predicting parking availability in smart city based on IoT. In: 11th international conference on computing, communication and networking technologies (ICCCNT). ,IEEE, pp 1–8
4. Li J, Zhao H, Hafid AS, Wei J, Yin H, Ren B. A bio-inspired solution to clusterbased distributed spectrum allocation in high-density cognitive internet of things. IEEE Int Things J 6(6):9294e307
5. https://www.worldometers.info/coronavirus/. Accessed 25th May 2019

6. Haleem A, Javaid M, Vaishya R (2020) Effects of COVID 19 pandemic in daily life. Curr Med Res Pract 10:78e9. https://doi.org/10.1016/j.cmrp.2020.03.011
7. Li B, Dawei Y, Xun W, Tong L (2019) Chinese experts' consensus on the internet of things-aided diagnosis and treatment of coronavirus disease 2019 (COVID-19). Clin eHealth 3(1):7e15
8. Xu B, Gutierrez B, Mekaru S, Sewalk K, Goodwin L, Loskill A, Cohn EL, Hswen Y, Hill SC, Cobo MM, Zarebski AE, Li S, Wu CH, Hulland E, Morgan JD, Wang L, O'Brien K, Scarpino SV, Brownstein JS, Pybus OG, Pigott DM, Kraemer MUG (2020) Epidemiological data from the COVID-19 outbreak, real-time case information. Sci Data 7:106
9. Iacus SM, Natale F, Santamaria C, Spyratos S, Vespe M (2020) Estimating and projecting air passenger traffic during the COVID-19 coronavirus outbreak and its socio-economic impact. Saf Sci 104791
10. Wang C, Pan R, Wan X, Tan Y, Xu L, Ho CS, Ho RC (2020) Immediate psychological responses and associated factors during the initial stage of the 2019 coronavirus disease (COVID-19) epidemic among the general population in China. Int J Environ Res Publ Health 17
11. Goyal P, Choi JJ, Pinheiro LC, Schenck EJ, Chen R, Jabri A, Satlin MJ, Campion TR Jr, Nahid M, Ringel JB, Hoffman KL, Alshak MN, Li HA, Wehmeyer GT, Rajan M, Reshetnyak E, Hupert N, Horn EM, Martinez FJ, Gulick RM, Safford MM (2020) Clinical characteristics of covid-19 in New York city. N Engl J Med 382:2372–2374
12. Randhawa AK, Fisher LH, Greninger AL, Li SS, Andriesen J, Corey L, Jerome KR (2020) Changes in SARS-CoV-2 positivity rate in outpatients in seattle and Washington state. J Am Med Assoc 323(22):2334–2336
13. Peters AL, Garg S (2020) The silver lining to COVID-19: avoiding diabetic ketoacidosis admissions with telehealth. Diabetes Technol Therapeut 22(6):449–453
14. Gupta M, Abdelsalam M, Mittal S (2020) Enabling and enforcing social distancing measures using smart city and its infrastructures: a COVID-19 use case. arXiv:2004.09246. Accessed 13 Apr 2020
15. Bayram M, Springer S, Garvey CK, Özdemir V (2020) COVID-19 digital health innovation policy: a portal to alternative futures in the making. OMICS. 2020 Aug 24(8):460-469. https://doi.org/10.1089/omi.2020.0089. Epub 8 Jun 2020. PMID: 32511054
16. Patel P, Gohil P (2020) Role of additive manufacturing in medical application COVID-19 scenario: India case study. J Manuf Syst. https://doi.org/10.1016/j.jmsy.2020.11.006. Online ahead of print
17. Celik HK, Kose O, Ulmeanu ME, Rennie AEW, Abram TN, Akinci I (2020) Design and additive manufacturing of medical face shield for healthcare workers battling coronavirus (COVID-19). Int J Bioprint 6(4):286. https://doi.org/10.18063/ijb.v6i4.286. PMID: 33088997; PMCID: PMC7557567
18. Roy D, Agarwal N, Mukherjee A, Angne J, Farhan A, Sorathiya D, Devarajan R, Nair MV (2020) Hewlett packard enterprises, utilizing smart city cyber-physical infrastructure for tracking and monitoring pandemics like Covid-19 with The ICCC As The Nerve Cetre, Digital Government: Research and Practice, vol 2, No 1, Article 8, Publication Date: Nov 2020
19. Alakus a TB, Turkoglu I (2020) Comparison of deep learning approaches to predict COVID-19 infection, Chaos, Solitons and Fractals, 140
20. Kamaludeen NBA, Lee SP, Parizi RM (2019) Guideline-based approach for IoT home application development. In 2019 international conference on internet of things (iThings) and IEEE green computing and communications (GreenCom) and IEEE cyber, physical and social computing (CPSCom) and IEEE smart data (SmartData). pp 929–936
21. El Khaddar MA, Boulmalf M (2017) Smartphone: the ultimate IoT and IoE device. Smartphones Appl Res Perspect 137
22. Sinha D (2019) IoT-based mobile applications and their impact on user experience. https://www.iotforall.com/mobile-iot/. Accessed 04 July 2020
23. Parizi RM, Guo L, Bian Y, Azmoodeh A, Dehghantanha A, Choo K-KR (2018) CyberPDF: smart and secure coordinate-based automated health PDF data batch extraction. In: 2018 IEEE/ACM international conference on connected health: applications, systems and engineering technologies (CHASE). IEEE, pp 106–111

24. Bai L, Yang D, Wang X, Tong L, Zhu X, Zhong N, Bai C, Powell CA, Chen R, Zhou J et al (2020) Chinese experts' consensus on the internet of things-aided diagnosis and treatment of coronavirus disease 2019 (COVID-19). Clin eHealth 3:7–15
25. COVID-19 smartphone testing kit (2020) https://www.detectachem.com/index.php?p=COVID19. Accessed 27 June 2020
26. United against coronavirus! Stopcorona App (2020) https://stopcorona.app/. Accessed 06 June 2020
27. Kelion L (2020) Coronavirus: Moscow rolls out patient-tracking app. https://www.bbc.com/news/technology-52121264. Accessed 19 June 2020
28. 'Selfie app' to keep track of quarantined poles (2020) https://www.france24.com/en/20200320-selfie-app-to-keep-track-of-quarantined-poles. Accessed 19 June 2020
29. Wright T (2020) Blockchain app used to track COVID-19 cases in Latin America. https://cointelegraph.com/news/blockchain-app-used-to-track-covid-19-cases-in-latin-america. Accessed 26 June 2020
30. Hui M (2020) Hong kong is using tracker wristbands to geofence people under coronavirus quarantine. https://qz.com/1822215/hong-kong-uses-tracking-wristbands-for-coronavirus-quarantine/. Accessed 19 June 2020
31. Aarogyasetu (2020) https://www.mygov.in/aarogya-setu-app/. Accessed 06 June 2020
32. TraceTogether, safer together (2020). https://www.tracetogether.gov.sg/. Accessed 06 June 2020
33. Stub ST (2020) Israeli phone apps aim to track coronavirus, guard privacy. https://www.usnews.com/news/best-countries/articles/2020-04-20/new-tech-apps-in-israel-aim-to-track-coronavirus-guard-privacy. Accessed 27 June 2020
34. Tokenpost (2020) IoT blockchain platform launches a COVID-19 contact tracing app. https://bit.ly/3eS2VGt. Accessed 27 June 2020
35. BeAware Bahrain (2020) https://bit.ly/2MOQyyt. Accessed 06 June 2020
36. I protect both you and me (2020). https://erouska.cz/en. Accessed 10 June 2020
37. How WhatsApp can help you stay connected during the coronavirus (COVID-19) pandemic (2020). https://www.whatsapp.com/coronavirus/. Accessed 06 June 2020
38. Storch EA, Sheu JC, Guzick AG, Schneider SC, Cepeda SL, Rombado BR, Gupta R, Hoch CT, Goodman WK (2020) Impact of the COVID-19 pandemic on exposure and response prevention outcomes in adults and youth with obsessive-compulsive disorder. Psych Res 113597
39. Spier AB, Evans CE (2020) Surviving and thriving in thrombosis research during a global pandemic: experiences of a vascular scientist diagnosed with COVID-19. Thrombosis Update 2:100028
40. McEwan JR, Choudry NB, Fuller RW (1990) The effect of sulindac on the abnormal coughreflex associated with dry cough. J Pharm Exp Ther 255 (1):161–164

Chapter 14
Drug Discovery with Computational Intelligence Against COVID-19

Godwin M. Ubi, Joseph K. Ebigwai, Udensi O. Ugorji, and Imaobong S. Essien

Abstract Since the advent of the current pandemic of corona virus-induced disease in December 2019 to date, the search for knowledge has specifically been focused on finding prophylactic and therapeutic solutions for COVID-19 which has murdered approximately 1,073,675 globally as of October 11, 2020 thus intensifying the global search for a cure to save humanity. The global threat posed by COVID-19 has created the urgent search for an alternative therapy and the application of a quick check on the efficacy of other available drugs on SARS-COV 2 pathogens using biotechnological tools such as molecular docking and bioinformatics on genomic data with SWISSDOCK and AUTOVINA. In the current dispensation, plant-based antimalarial compounds like quinine-, chloroquine-, hydroxychloroquine-, and artemisinin-related compounds have been widely used as antimalarial drugs for the prophylactic and therapeutic treatment of malaria and related diseases. Thus, the application of artificial intelligence through bioinformatics and molecular docking as a quick check on the efficacy of the plant-based antimalarial compounds on SARS-COV 2 pathogens becomes indispensable, with a view to discover and repurpose these drugs as potential therapy for COVID-19 and related diseases. The results of bioinformatics analysis of genomic sequence data using in silico molecular docking revealed that the antimalarial ligands, in artemisinin, chloroquine, and quinine fixed between 3272 of (3301), 2484 of (2533) and 1660 of (1709) damaged proteins of the SARS-COV 2 tumor necrosis factor receptor proteins representing about 99.12, 97.13, and 98.06% fixation, respectively. Thus, it is expected that similar results or more with the antimalarial compounds will be validated with in vitro efficacy trials on cell cultures of SARS-COV 2 in the laboratory and in vitro on experimental models like infected patients through clinical trials. Hence, there is a need for the

G. M. Ubi (✉) · U. O. Ugorji
Department of Genetics and Biotechnology, Calabar, Nigeria
e-mail: ubiology.gu@unical.edu.ng

J. K. Ebigwai
Department of Plant and Ecological Studies, Calabar, Nigeria

I. S. Essien
Department of Zoology and Environmental Biology, Faculty of Biological Sciences, University of Calabar, 1115, Calabar, Nigeria

U. Kose et al. (eds.), *Computational Intelligence for COVID-19 and Future Pandemics*, Disruptive Technologies and Digital Transformations for Society 5.0,
https://doi.org/10.1007/978-981-16-3783-4_14

constant application of artificial intelligence in drug validation and optimization. Thus, artificial intelligence using bioinformatics can therefore provide the much needed in silico validation for small molecules as potential drugs for the treatment of all health-related diseases in this twenty-first century and beyond. It is therefore recommended that computer-based knowledge and algorithms like artificial intelligence, bioinformatics, machine learning, and deep learning be applied in genomic data to identify and validate suitable candidate drugs/ligands/small molecules for specific disease protein targets for drug discovery and optimization with the sole aim of eliminating trial-and-error methods, waste of time and resources, and loss of lives.

Keywords Artificial intelligence · Molecular docking · Target ligand · Drug discovery · Drug–protein interactions

1 Introduction

1.1 Computer Intelligence in Biological Data

The study of genomics is strictly based on the availability of genomic data such as nucleotide and protein sequences usually generated from laboratory sequencing of amplified products of polymerase chain reactions after nanodrop spectrophotometer quantification of DNA or RNA isolates. The availability of genomic data which is big data has made it possible to study wide range of processes, biochemical, physiological, and molecular systems in living organisms and their interactions. These cannot be possible without the availability of computer-based algorithms or written programs such as bioinformatics that makes such analyses easier.

Bioinformatics adopts computer-based algorithms in analyzing and management of biological information coded within the human genome in form of DNA and protein sequences in which composition originates from the genetic code and central dogma template of living organisms. The computer-based algorithms enable for the determination of the characteristics, probable functions of genes, expression patterns, motifs, domain architectures, guanine cytosine contents, protein structure characteristics, active or exon sites of proteins, size of genes, evolutionary pathways and relationship, and mutations and types expressed. Bioinformatic analysis of genomic data also make possible the study of system biology, gene therapy, analysis of genes, and their derivatives as an integrated and complete system.

The applications of bioinformatics in structural- and ligand-based drug design and discovery also provide an excellent foundation for in vitro and in vivo validation of potential drugs for the much dreaded SARS-COV 2 corona viruses and other diseases.

The intervention and applications of molecular tools in the pharmaceutical industry have given rise to a new aspect of pharmacogenomics. A conjugate of pharmaceutical sciences and genetics, pharmacogenomics combines conventional sciences like biochemistry, pharmacy, genetics, the molecular structure of genes as well as the functionality and behavior of the genetic architecture.

It basically involves the evaluation and assessment of the response of a person to particular drugs in relation to their genetic constitution. This field shows great promise in contemporary medicine as it is now possible to design drugs for individual patients in relation to their genetic architecture with the aid of computer-based algorithms. This has found great applicability in the current search for prophylactics and therapeutics for the elusive COVID-19 pandemic [1].

Late in December 2019, sick people were diagnosed with viral pneumonia the cause of which was attributed to an uncultured pathogen was reported in Wuhan, China [1]. The current novel corona virus was consequently confirmed as the causative pathogenic entity, and was temporarily called the 2019 novel corona virus or COVID-19, which is the source of the pandemic currently ravaging humanity. As on August 27, 2020, total global confirmed cases had reached an alarming 24,339,123 with a global death rate of 829, 742 from 2019-nCoV (COVID-19) infections and 16,034,876 recovered cases some of which included people traveling and visiting or are visiting endemic countries and thereby increasing human-to-human transmission [2].

Since the advent of the current pandemic of corona virus-induced disease in December 2019 to date, the search for knowledge has specifically been focused on finding prophylactic and therapeutic solutions for COVID-19 which have murdered approximately 1,073,675 globally and 1,115 in Nigeria as of October 11, 2020 [3] thus intensifying the global search for a cure to save humanity. Several attempts in combating the dreaded corona viruses globally have not yielded positive results as the incidence and epidemiological statistics keep surging, shutting down lives, businesses, economies, and government. The fate of scientists globally has also been put to test and challenged with finding an urgent cure and vaccines for the pandemic. This has also not been fruitful due to several challenges such as funding, non-availability of 3-4 level bio-safety laboratories especially in developing countries, time factor in drug development, non-standardization of traditional medicines, deadly nature of the viral pathogen, and the vulnerability of all and sundry to the viral infections [4].

Recent reports [5] have shown that many drugs and vaccines have been developed using various scientific and non-scientific approaches some of which are currently undergoing clinical trials for validation before such drugs can be made publicly available [6]. These still remain elusive as it is currently seen by the helpless humanity as trial-and-error methods. The global threat posed by COVID-19 has created a heightened need for an urgent search for therapy and the application of a quick check on the efficacy of other available drugs and natural compounds of plant origin on SARS-COV 2 pathogens. For centuries and decades, quinine-, chloroquine-, hydroxychloroquine-, and artemisinin-related compounds have been widely used as an antimalarial drugs [7] for the prophylactic and therapeutic treatment of malaria caused by *Plasmodium falciparum* (a protozoan transmitted by female anopheles mosquitoes to the host through bite) [8].

In contemporary era, secondary metabolites of plant origin with previously unknown medicinal activities have been widely evaluated as major recipe for phytomedicines and chemical drugs [9]. These have consequently brought about a major renewal of interest in the utilization of medicinal plants especially in most third world and developing nations due to their safety and absence of allergy especially when compared to chemically synthesized drugs [10]. Over the years, man has used plant (whole or parts) for food, aesthetic beautification, medicine, and for the feeding of livestock. Plants have economic values and are not only used for mere satisfaction or sustenance of hunger but also for the maintenance of good health [11]. Plants have remained best sources of natural products where they form the major parts of recipe for all human endeavors of drug discovery [12]. Research institutes and companies are carrying out genuine inquiry on the phytomedicinal potentials of herbs found in the rain forest and climes for their potential medicinal properties. Contemporary allopathic method of therapeutic dispensation is said to be based on phytomedicines [13]. Plants with medicinal properties are important in both developing and developed nations of the world as recipe for drug production from herbal extracts for the various disease treatment purposes. This has continued to play a significant role in maintaining good human health from time immemorial [14].

More than 50% of all contemporary medicines are of natural product based and natural products play an important role in drug discovery, lead, and optimization [15]. In view of this approach toward drug discovery, emphasis has been laid on parts such as the leaves, bark, and root of most plants for their antimalarial, antiviral, antibacterial, antifungal, and anti-oxidant properties [16]. Klebbe [17] has posited that contemporary clinical medicines from plants play a significant role in drug-manufacturing companies. Secondary metabolites which are naturally present in plants are transformed biotechnologically and otherwise into substances that controls biochemical processes in humans [18, 19]. Some secondary metabolites or compounds produced by plants such as *Portulaca oleracea, Psidium guajava, Artemisia annua, Garcinia nitida, Azadirachta indica, Vernonia amygdalina. Gongronema latifolium, Ageratum conyzoides*, and *Emilia sonchifolia* among others have been very effective against bacteria, fungi, protozoans, nematodes, viruses, parasites, helminthes, and other pathogens causing a variety of human diseases [20].

Hence, this research is specifically designed to adopt artificial intelligence approach, an in silico molecular approach for the in silico validation of the efficacy of plant-based antimalarial compounds like artemisinin, chloroquine, and quinine on SARS-COV 2 TNF rp2 protein targets and to advocate for the repacking and repurposing of these effective plant-based antimalarial bioactive compounds as potential drug therapy for COVID-19 and related diseases.

2 Bioinformatics and in Silico Approaches

2.1 Determination of SARS-COV 2 (Pathogen) Protein Target Characteristics

Nucleotide sequences and amino acid sequences of the RNA-dependent RNA polymorphism region of SARS-COV 2 (COVID-19) spike protein were retrieved from the National Centre for Biotechnology Information (NCBI) database of the human genome. The FASTA format approach was adopted for retrieving the SARS-COV 2 gene sequences with BLASTN option. SARS-COV 2 gene sequence accession numbers from Nigeria and index country China as reference sequence were retrieved, tabulated, and used for the study. The Expasy.org. Proteomics, protparam option of the SWISS Institute for Bioinformatics (SIB), GENScan, SMART interactive, and MEGA software were used to determine the requisite characteristics of the SARS-COV 2 corona viruses required by the researchers for drug design and development. Some of the characteristics ascertained included the following:

i. Molecular weight of the corona viruses (g/mol).
ii. Size of the corona viruses (bps).
iii. Coding region of the s-gene (KDa).
iv. Beginning and end codons of the s-gene (bps).
v. Total number of amino acids.
vi. Percent aliphatic index.
vii. Percent instability index.
viii. Coefficients of extinction/attenuation of the viruses (g/l).
ix. Hydropathocity of the corona viruses.
x. G-C contents of the viruses (%).
xi. Half-life of the virus in human reticulocyte cells (hours).
xii. Secondary protein structure (%).
xiii. Tertiary protein structure.
xiv. The theoretical (pl) isoelectric point or pH.
xv. Total negatively and positively charged amino acid residues.
xvi. Mutation rate of the corona viruses (TYA).

2.2 Determination and Identification of Protein Target Sites for SARS-COV 2 Tumor Necrosis Factor Receptor Associated Proteins 2 (TNF rp2)

The SWISS target predictor online interactive program of expasy.org. Swiss target predictor was employed in the prediction of the protein target sites for the Nigerian SARS-COV 2 isolate with gene bank identification number MT159788.1 retrieved from the national center for biotechnology information (NCBI) database hosting the human genome. Structural analysis was also performed to determine the 3-D PDB structures for docking. This identifies the potential targets to binding sites on protein surfaces. Surface scanning also allows for the recognition of domains that are most likely involved in interactions with other protein or drug molecules.

2.3 Molecular Homology Modeling of Structures

The Expasy Swiss model online interactive program was used to develop the different SARS-COV 2 target models of 3-D tertiary protein structure revealing the different types, such as bad bonds; bad angles; coordinates X, Y, and Z; Ramachandran favored; Ramachandran outliers; C-Beta deviations; quality estimate; residue quality; target–template alignment; ligands; matching templates; heterotetramers; and sequence mutations.

3-D tertiary protein folding structures of partial cds RdRP region of novel corona virus sequences obtained for an SARS-COV 2 corona virus pathogen (Fig. 1).

2.4 Macromolecule (Target) Selection

Crystal protein structures of the 3-D tertiary structure of the tumor necrosis factor receptor associated protein TNF rp2 of SARS-COV 2 were obtained from the protein data bank with PDB IDs of AYI, 1CZZ, ICZY, IDOA, IDOJ, and IBCB. Also, the macromolecules and PDB IDs for the SARS-COV 2 targets with ideal sites were identified, selected using the SWISSDOCK target predictor, and selection suite. For each macromolecule, the small molecules (ligands) and water molecules were removed, while polar hydrogen and Kollman charges were added [22] (Table 1).

2.5 Ligand Selection

The chemical structures of artemisinin, quinine, and chloroquine were selected using the CHEMBl ligand databases. The derivative compounds were built using the SWISSDOCK ligand selection suite programs. Solvations, energy affinities, coordinates, bonds, atoms, fixation, binding site energies, binding modes, root mean square deviations, geometry optimization, and density functional theory calculations were performed in Swiss Dock CHARMM using the Swiss institute of bioinformatics online interactive suites (Fig. 2 and Table 2).

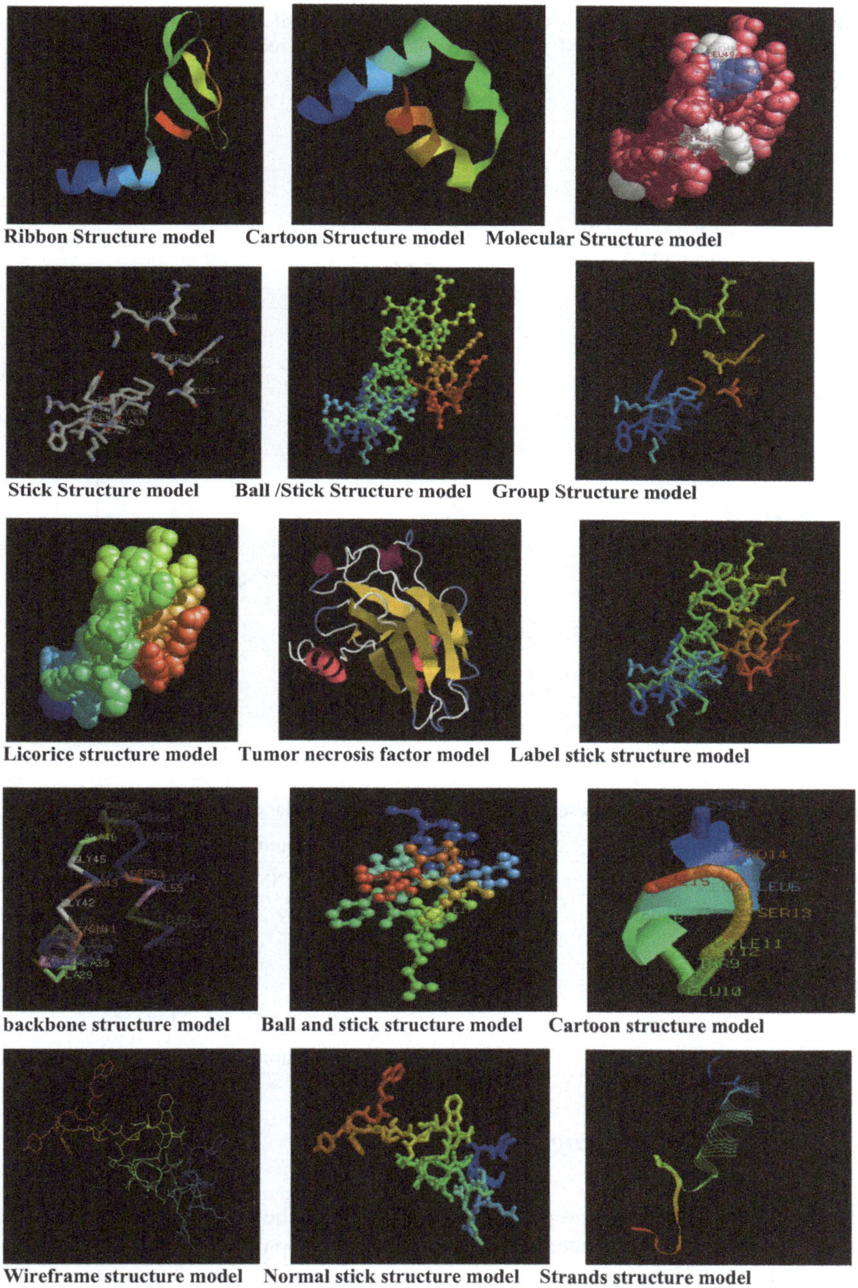

Fig. 1 3-D structures of SARS-COV 2 showing different homology modeling used for computer-aided structure-based drug design

Table 1 Target sites of the SARS-COV 2 corona virus showed 7 sites with maximum of 213 amino side chains. aa = amino acids. TNF rp 2 = tumor necrosis factor receptor associated protein 2

S/N	Target sites	Amino acid	Reaction sites
1	1 AYI	0 aa	0
2	1 BCB (DNA 5'–3')	93 aa	93
3	1 ICZY (TNF rp2)	168 aa	168
4	1 ICZZ (TNF RP2)	187 aa	187
5	IDOI (TNF rp 2)	9 aa	9
6	IDOA (TNF rp 2)	168 aa	168
7	IDOJ (TNF rp 2)	213 aa	213

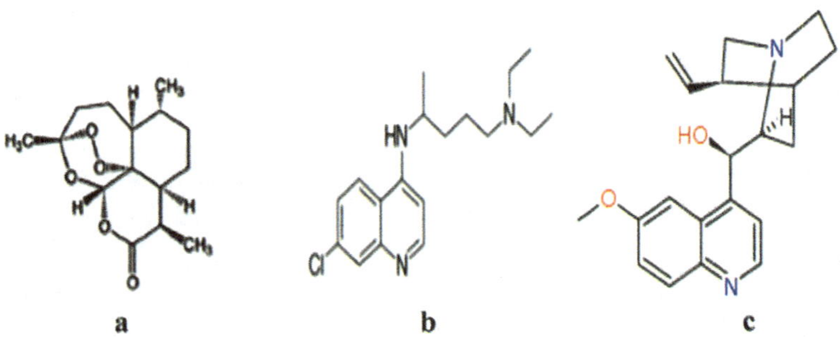

Fig. 2 **a** Artemisinin. **b** Chloroquine. **c** Quinine

Table 2 Properties of ligands (drug molecules) used for molecular docking

S/N	Properties	Artemisinin	Chloroquine	Quinine
1	Chemical formula	$C_{15}H_{22}O_5$	$C_{18}H_{25}ClN_3$	C20H24N2O2
2	Compound ID	CID68827	CID2719	CID8549
3	Molar mass g/mol	282.332	318.5	324.42
4	Chem ID	62,060	2618	84,989
5	Melting point	152–157 °C	150–155 °C	148–152 °C
6	ZINC entry	C6827	Chloroquine	Quinine

2.6 *Molecular Docking*

Each ligand molecule was prepared for docking using the AutoVina AutoDock and SWISSDOCK tools. Hydrogen atoms were added, and partial charges of each atom resulting from the root mean squared deviation calculations were incorporated. Grid maps were created by centering the grid box at the position of the ligand of each SARS-COV 2 target molecule with a spacing of 0.375 Å [23] and size covering the binding cavity of each target. Genetic algorithm and 30 docking runs were used for each docking. The resulting complexes formed revealed the site of adsorption, digestion, metabolic, and excretion pathways fixed, inhibited, or blocked by the different plant-based bioactive compounds of artemisinin, chloroquine, and quinine ligands (Fig. 3).

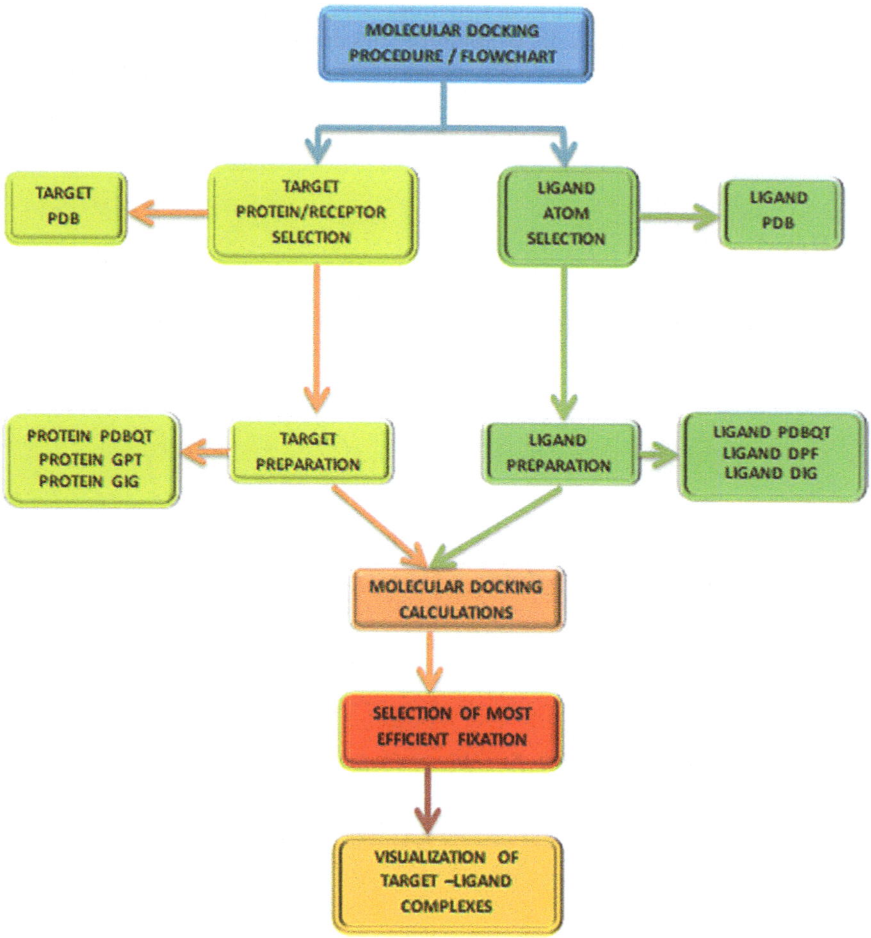

Fig. 3 Flowchart showing procedures adopted for molecular docking of antimalarial compounds with tumor necrosis factor receptor proteins of SARS-COV 2 isolate

2.7 Drug Design

The Swiss drug design online interactive program was used to access the in silico chemical protein interactions of the antimalarial drugs, artesunate, chloroquine, and quinine with the several receptor proteins of the malaria parasite *Plasmodium falciparum*. The Swiss drug design—click to drug—stitch program was used to develop the interaction network via in silico bioinformatics approach. The bigger circles with thick and bold lines indicate the major effect and direct interaction of drug with the proteins while the small circles with thin and normal lines indicate minor effect and indirect interactions between the drugs and the protein targets.

3 Results

3.1 Physiochemical Characteristics of SAR-COV 2 Isolates Protein Targets

See Table 3.

3.2 Number of Reaction Sites for Protein and Drug Targets for SARS-COV 2 Isolate

See Figs. 4, 5, 6, 7, 8, 9, 10, 11, 12, 13, 14, 15, 16, 17, 18 and Table 4.

3.3 Results of Molecular Docking Calculations and Implication on Drug Discovery

3.3.1 Cluster Groupings and Size

The results presented in Table 5 showed that the three different ligands showed variable levels of clustering in the complexes formed. Artemisinin ligands showed 36 clustering patterns with 8 sizes, chloroquine ligand showed 48 clustering patterns with 8 cluster sizes while quinine showed 31 clustering patterns with 8 cluster sizes also. The clustering patterns revealed the aggregation of different atoms and bonds relative to the formation of complexes between the target and ligand molecules. Same size reveals the highest level corresponding to the acceptable cut-off point for ligand selection.

Table 3 Results of physiochemical characteristics of two SARS-COV 2 isolates

S/N	Characteristics/properties of coronavirus isolates	Isolate 1 Accession No. (MT159778.1)	Isolate 2 Accession No. (MT042777.1)
1	Molecular weight of the SARS-COV 2 (g/mol)	16,289.82	9643.92
2	Total size of the corona viruses (bps)	29,760	29,903
3	Coding region of the s-gene	22–275	9–227
4	Start codons of the s-gene	22	9
5	End codon of s-gene	275	227
5	Total number of amino acids in the protein coat	2266	1328
6	Aliphatic index (side chain amino acids) (%)	95.45	77.36
7	Alanine (%)	6.2	8
8	Valine (%)	7.6	6.9
9	Isoleucine (%)	2.8	1.1
10	Leucine (%)	14.5	11.5
11	Instability index (%)	49.67	46.24
12	Extinction/attenuation coefficients of the viruses (m/M^3)	15,065	9190
13	Hydropathocity of the corona viruses	0.106	− 0.255
14	Half-life of cells (hours)	1.2	4.4
15	G-C contents of the viruses (%)	38.51	38.78
16	Secondary protein structure (Alpha helix)	49.66	56.32
17	Extended strand	19.31	13.76
18	Random coil	22.07	20.06
19	Beta turns	8.79	9.20
20	Nucleotide bps	483	294
21	The theoretical (pl) isoelectric point or pH	8.33	8.37
22	Total negatively charged amino acid residues	11	7
23	Total positively charged amino acid residues	14	9
24	Mutation rate of the corona viruses (%)	2/day	1/day

(continued)

Table 3 (continued)

S/N	Characteristics/properties of coronavirus isolates	Isolate 1 Accession No. (MT159778.1)	Isolate 2 Accession No. (MT042777.1)
25	Time of genetic divergence of S-gene (MYA)	0.4	0.4
26	Mutation site	1218	1098
27	Molprobity score (%)	1.48	1.21
28	Ramachandran forward (%)	91.20	90.75
29	C-beta deviations (%)	2.00	2.00
30	Bad bonds	1/1021	1/923
31	Bad angles	12/1386	10/098
32	Nucleotide diversity	0.4221	0.4042
33	Gamma distribution	15.4742	14.3812

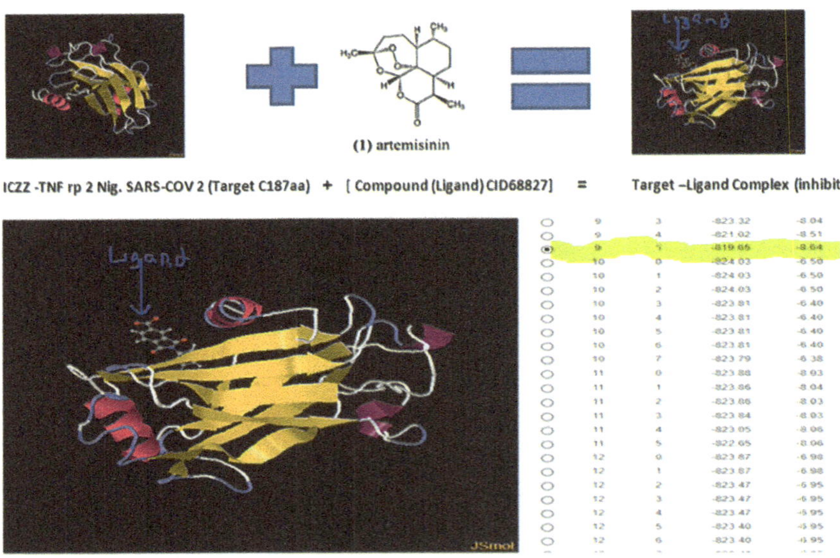

Fig. 4 Artemisinin against SARS-COV 2 TNFrp 2 target showing docking complex at cluster 9 rank 5

3.3.2 Energy Affinities and Gibb's Free Energy

Due to the formation of target ligand complexes during molecular docking, high amount of energy is usually required to form bonds in the complex and to break bonds in other to form new ones. These energies are usually generated within the system during docking. The energy requirement for each docking varies and may depend on

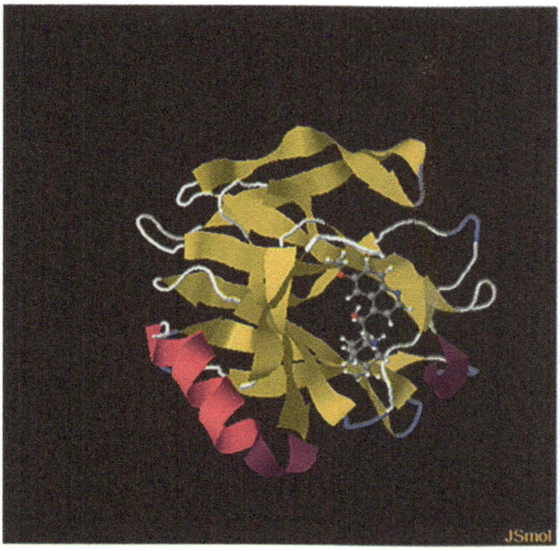

Show	Cluster	Element	FullFitness (kcal/mol)	Estimated ΔG (kcal/mol)
○	0	0	-551.47	-7.08
○	0	1	-551.47	-7.08
○	0	2	-551.47	-7.08
○	0	3	-551.47	-7.08
○	0	4	-551.11	-7.00
○	0	5	-549.19	-7.00
○	0	6	-549.19	-7.00
○	0	7	-549.19	-7.00
○	1	0	-547.99	-6.76
○	1	1	-547.99	-6.76
○	1	2	-547.98	-6.76
○	1	3	-547.98	-6.76
○	1	4	-547.98	-6.76
○	1	5	-547.00	-6.63
○	1	6	-547.00	-6.63
○	1	7	-547.00	-6.63
○	2	0	-547.35	-6.66
○	2	1	-547.35	-6.66
◉	2	2	-547.35	-6.66
○	2	3	-547.35	-6.67
○	2	4	-547.35	-6.67

Fig. 5 Quinine against SARS-COV 2 TNFrp 2 target showing docking complex at cluster 2 rank 2

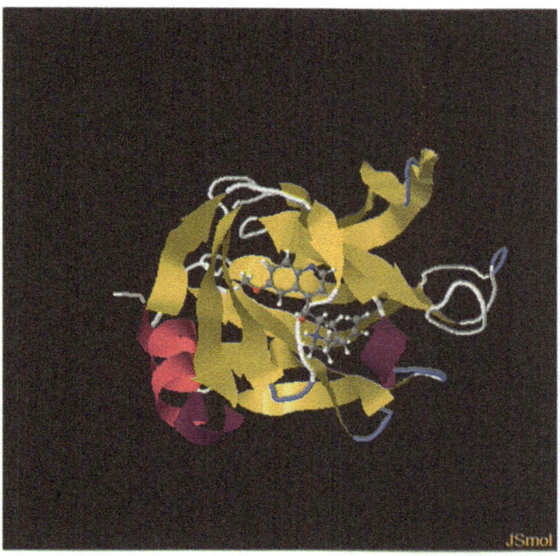

○	2	2	-547.35	-6.66
○	2	3	-547.35	-6.67
○	2	4	-547.35	-6.67
○	2	5	-547.35	-6.67
○	2	6	-546.87	-6.62
○	2	7	-546.87	-6.62
○	3	0	-547.26	-6.54
○	3	1	-547.26	-6.54
○	3	2	-547.26	-6.54
○	3	3	-547.22	-6.55
○	3	4	-547.22	-6.55
○	3	5	-547.22	-6.55
○	3	6	-547.22	-6.55
○	3	7	-546.79	-6.51
○	4	0	-546.63	-6.18
○	4	1	-546.63	-6.18
◉	4	2	-546.57	-6.17
○	4	3	-546.57	-6.17
○	4	4	-546.57	-6.17
○	4	5	-546.57	-6.17
○	4	6	-546.49	-6.16
○	4	7	-546.49	-6.16
○	5	0	-545.94	-7.46

Download CSV file

Fig. 6 Quinine against SARS-COV 2 TNFrp 2 target showing docking complex at cluster 4 rank 2

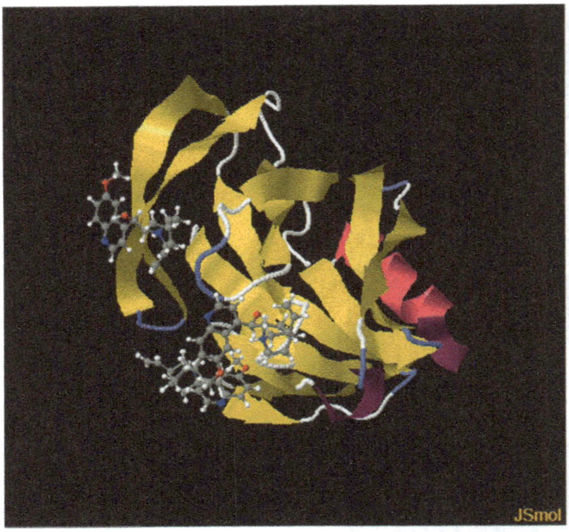

○	29	2	-539.67	-5.75
○	29	3	-539.67	-5.75
○	29	4	-539.64	-5.75
○	29	5	-539.64	-5.75
○	29	6	-539.64	-5.75
○	29	7	-539.53	-5.75
○	30	0	-539.65	-6.68
○	30	1	-539.56	-6.68
○	30	2	-539.56	-6.68
○	30	3	-539.56	-6.68
○	30	4	-539.56	-6.69
○	30	5	-539.56	-6.69
○	30	6	-539.56	-6.69
○	30	7	-539.56	-6.69
○	31	0	-539.03	-6.82
○	31	1	-539.03	-6.82
○	31	2	-538.97	-6.83
○	31	3	-538.97	-6.83
○	31	4	-538.97	-6.83
○	31	5	-538.76	-6.83
○	31	6	-538.76	-6.83
○	31	7	-538.76	-6.83
◉	31	7	-538.76	-6.83

Download CSV file

Fig. 7 Quinine against SARS-COV 2 TNFrp 2 target showing docking complex at cluster 31 rank 7

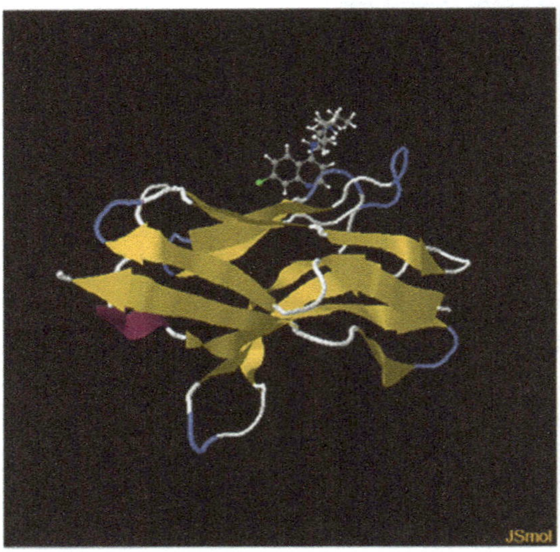

Show	Cluster	Element	FullFitness (kcal/mol)	Estimated ΔG (kcal/mol)
○	0	0	-604.03	-6.51
○	0	1	-604.03	-6.51
○	0	2	-604.03	-6.51
○	0	3	-604.03	-6.51
○	0	4	-603.25	-6.51
○	0	5	-603.25	-6.51
○	0	6	-599.52	-6.36
○	0	7	-599.52	-6.36
○	1	0	-603.37	-7.36
○	1	1	-603.37	-7.36
○	1	2	-603.35	-7.36
○	1	3	-603.35	-7.36
○	1	4	-603.35	-7.36
○	1	5	-600.11	-7.16
○	1	6	-600.11	-7.16
○	1	7	-600.11	-7.16
◉	2	0	-602.91	-6.32
○	2	1	-597.53	-6.31
○	2	2	-597.53	-6.31
○	2	3	-597.49	-6.22
○	2	4	-595.72	-6.06
○	2	5	-591.06	-6.16

Fig. 8 Chloroquine against SARS-COV 2 TNF rp 2 target showing docking complex at the second cluster rank 0

○	38	2	-586 68	-6 03
○	39	0	-591 40	-5 57
○	40	0	-590 98	-6 74
○	40	1	-589 14	-6 57
○	40	2	-582 94	-5 81
○	40	3	-581 23	-5 70
○	41	0	-590 24	-5 19
○	41	1	-590 11	-5 14
○	41	2	-589 91	-5 17
○	41	3	-585 75	-5 55
○	41	4	-584 42	-5 37
○	42	0	-589 36	-5 30
○	43	0	-588 34	-5 34
○	43	1	-587 70	-5 40
○	43	2	-587 70	-5 40
○	43	3	-587 28	-5 38
○	44	0	-588 34	-6 15
○	44	1	-577 57	-5 27
○	45	0	-586 97	-5 12
○	46	0	-585 08	-4 69
○	46	1	-584 18	-5 08
○	47	0	-583 43	-4 96
○	48	0	-583 23	-5 40
◉	48	0	-583 23	-5 40

Fig. 9 Chloroquine against SARS-COV 2 TNF rp 2 target showing docking complex at cluster 48

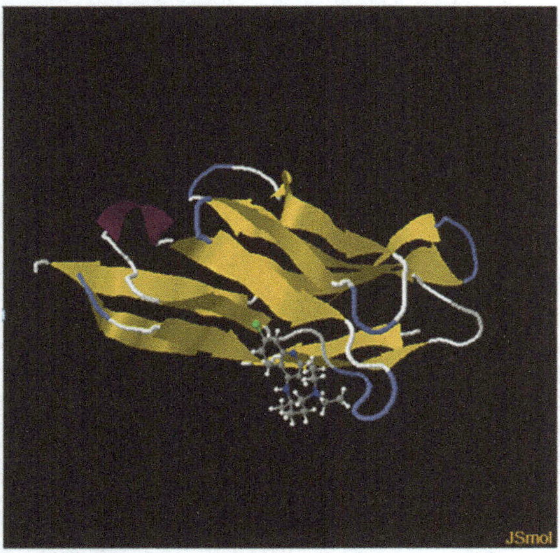

○	38	2	-586 68	-6 03
○	39	0	-591 40	-5 57
○	40	0	-590 98	-6 74
○	40	1	-589 14	-6 57
○	40	2	-582 94	-5 81
○	40	3	-581 23	-5 70
○	41	0	-590 24	-5 19
○	41	1	-590 11	-5 14
○	41	2	-589 91	-5 17
○	41	3	-585 75	-5 55
○	41	4	-584 42	-5 37
○	42	0	-589 36	-5 30
○	43	0	-588 34	-5 34
○	43	1	-587 70	-5 40
○	43	2	-587 70	-5 40
○	43	3	-587 28	-5 38
○	44	0	-588 34	-6 15
○	44	1	-577 57	-5 27
○	45	0	-586 97	-5 12
○	46	0	-585 08	-4 69
○	46	1	-584 18	-5 08
○	47	0	-583 43	-4 96
○	48	0	-583 23	-5 40
◉	48	0	-583 23	-5 40

Fig. 10 Chloroquine against SARS-COV 2TNF rp 2 target showing docking complex at cluster 48 at different coordinates

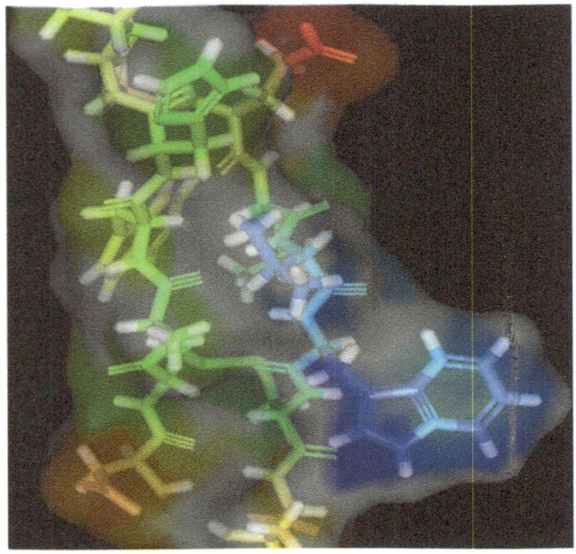

Fig. 11 Molecular simulation of ligands inside target at lower resolution

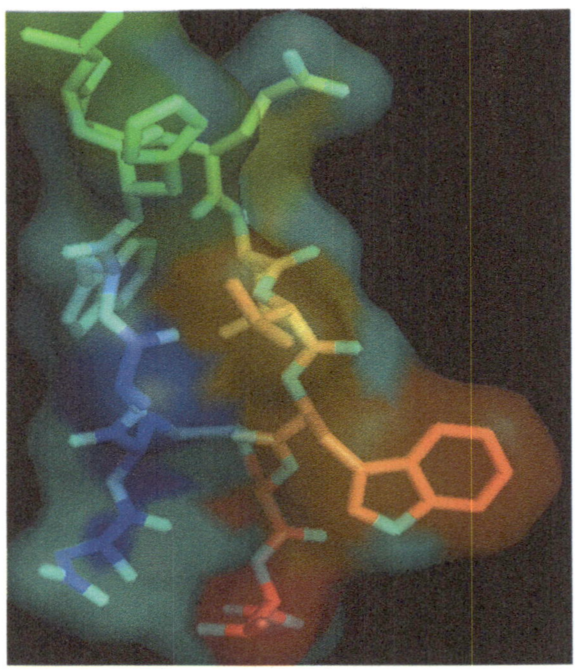

Fig. 12 Molecular simulation of ligands inside target at medium resolution

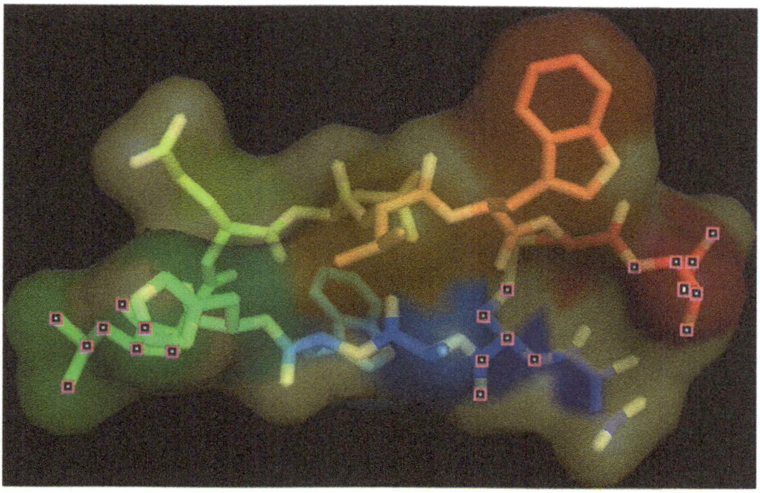

Fig. 13 Molecular simulation of ligands inside target at higher resolution

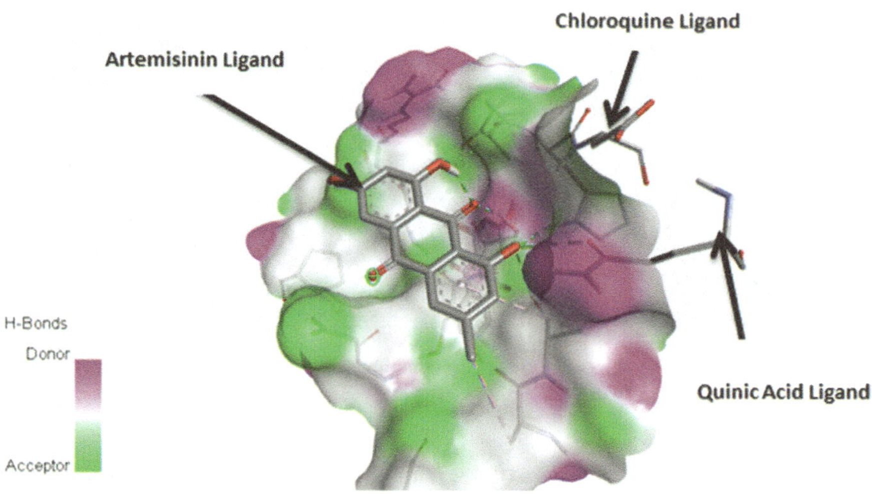

Fig. 14 AutoVina model of molecular docking of antimalarial ligands against SARS-COV 2 TNF rp 2 target

a number of factors. Also energy is also released after docking as free energy accompanying the target ligand complexes. These energies are summed up for each docking and calculated as the binding energy of the target ligand cocking. For artemisinin ligand, the energy requirement for complex formation in 36 clusters was 35.68 and free energy of −5.02 kJ/mol. Chloroquine ligand required 26.49 kJ/mol of binding energy and liberated −5.39 kJ/mol of free energy after complex formation. Quinine

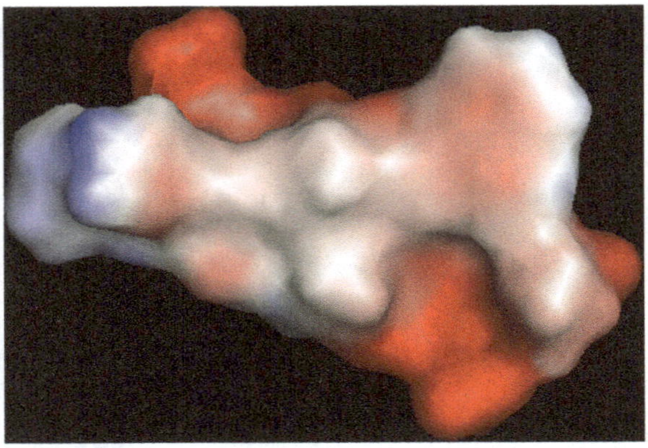

Fig. 15 AutoDock model of molecular docking of antimalarial ligands against SARS-COV 2 TNF rp 2 target

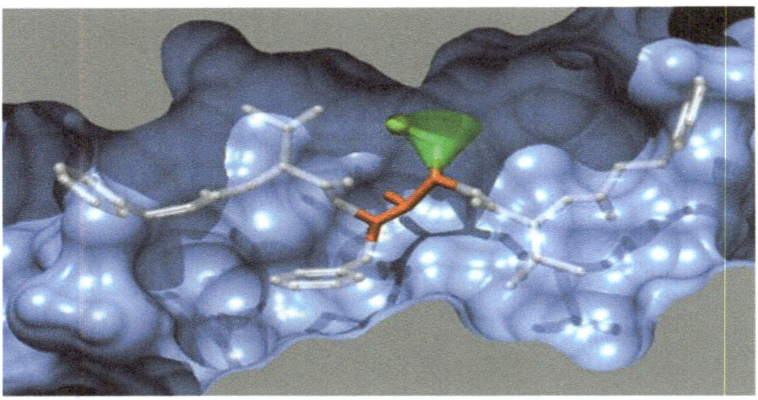

Fig. 16 SwissDock model of molecular docking of antimalarial ligands against SARS-COV 2 TNF rp 2 target

ligand showed the largest clusters of 48, with binding energy of 36.249 kJ/mol and a free energy of −6.83 kJ/mol released after complex formation (Table 5).

3.3.3 Total Systems on Target–ligand Complex Binding Modes

A total of 29 derivatives of the artemisinin ligand atoms were able to fix and form 3272 complexes with the TNF rp2 receptor proteins of the SARS-COV 2 target out of 3301 total target–ligand binding sites in the systems. This was higher and more efficient when compared to 49 derivatives of chloroquine ligand atoms which were able to fix and form 1660 complexes with the TNF receptor proteins of the SAR-COV

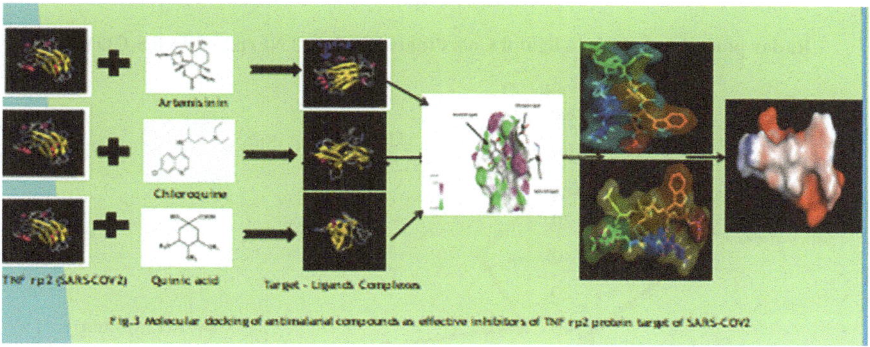

Fig. 17 Pictorial presentation of molecular docking processes of three ligands with TNF rp2 of SARS-COV 2

2 target out of 1709 total system identified and 49 derivatives of quinine ligand atoms which was able to fix and form 2484 complexes with the TNF receptor proteins of SARS-COV 2 out of 2533 total target–ligand sites identified (Table 5).

3.3.4 Percentage Fixation of TNF Receptor Proteins by the Antimalarial Ligands

The results presented in Table 5 revealed that the three antimalarial compounds of artemisinin, chloroquine, and quinine ligands were able to fix and form molecular docking complexes with the tumor necrosis factor receptor proteins of SARS-COV 2 with greater efficacy. Artemisinin fixed and formed docking complexes with 99.12% of the TNF receptor protein targets of SARS-COV2, chloroquine fixed and formed docking complexes with 97.13% of the TNF receptor protein targets of SARS-COV 2 and quinine fixed and formed docking complexes of 98. 06 % of the TNF receptor protein targets of SARS-COV 2. These results are well above the 95 percent confidence limit and probability and acceptable for further in vitro and in vivo trials with probability of obtaining same or even higher results.

3.3.5 Root Mean Square Deviations and Elapsed Time

The root mean square deviation (RSMD) presented in Table 5 from the target—ligand docking is a measure of the differences between energy values predicted by the molecular docking model and the observed values. Artemisinin, chloroquine, and quinine ligands showed RMSD of 26.37, 59.63, and 53.18, respectively, with a corresponding elapsed or completion time of 9.89 m/s, 7.58 m/s, and 8.89 m/s, respectively, with chloroquine ligand requiring lesser time to mend the broken bonds though with fewer total systems.

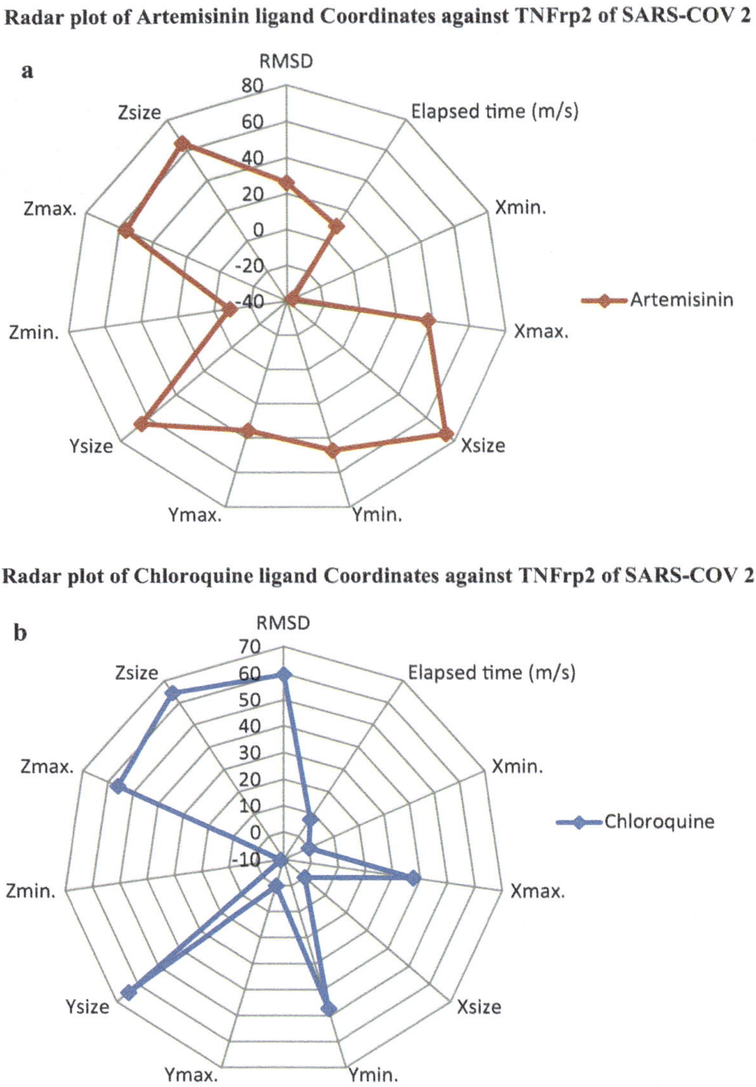

Fig. 18 a–d Radar plots showing X-, Y-, Z-coordinates of artemisinin, chloroquine, and quinine ligands with TNF rp2 targets, respectively

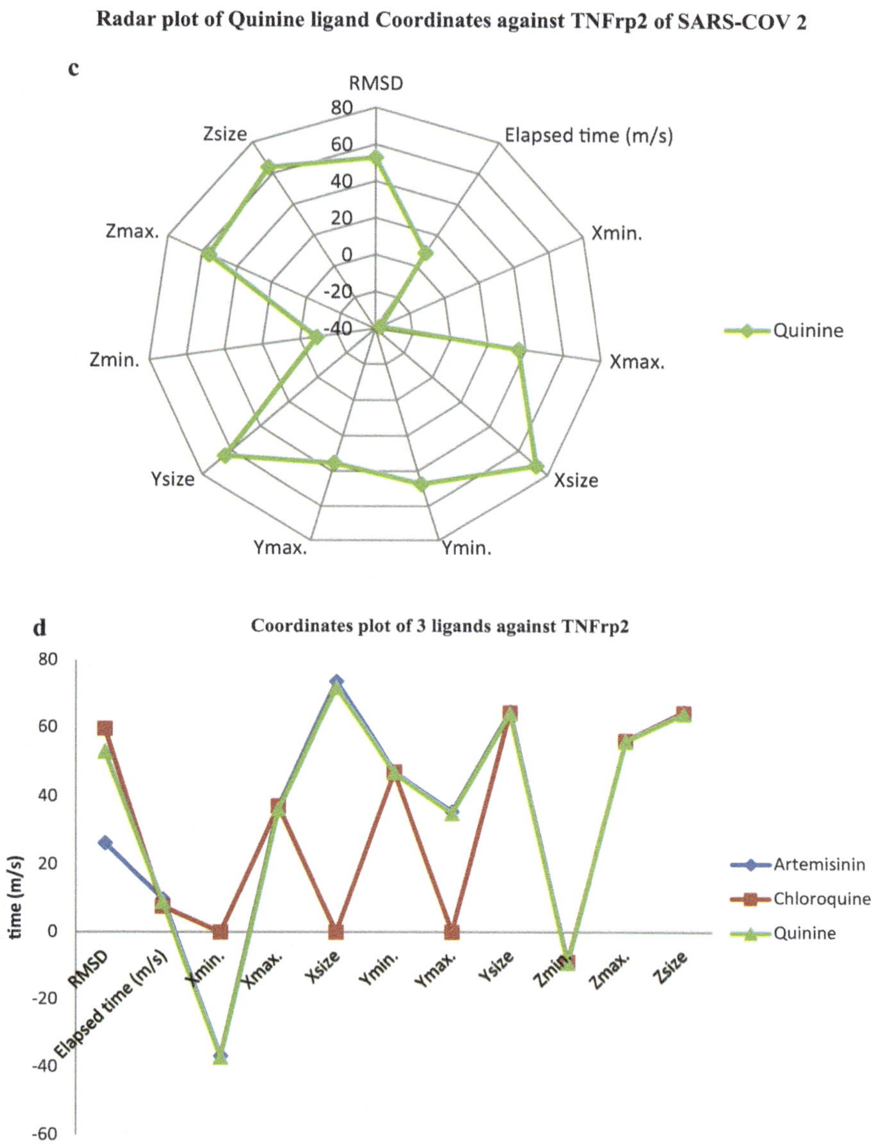

Fig. 18 (continued)

Table 4 Results of protein targets and reaction sites of the corona virus isolate (MT159778.1)

S/N	Protein type identified in the SARS-COV 2 isolate	No. of protein target sites	Drug target sites
1	3C-like proteinase (3CL-PRO)	7	7bzf
2	Non-structural proteins (1–9 proteinase)	9	9yyt
3	ORF1a polyproteins	1	1 ml
4	ORF7b polyproteins	2	2bv2
5	Papain-like proteinase (PL-PRO)	7	7bw4
6	Helicase (Hel) proteins	6	6m7l
7	Spike-glycoproteins (S-glycoproteins)	7	7c2k
8	Nucleoproteins	6	6nut
9	ORF 9b proteins	9	7m9l
10	Leader proteins	1	1bvl
11	Replicase polyproteins	6	6yyt
12	Q9BYF1–ACE2	1	1bw4
13	Envelope proteins	1	1 bzf
14	Membrane proteins	1	1c2k
15	Non-structural proteins (10–15)	7	7bvl

3.4 Drug Discovery Using Molecular Docking Calculations and Results

As stated in the previous paragraph, the three ligands used in this study are different compounds obtainable from different sources (artemisinin from *Artemisia annua* plant and chloroquine and quinine from barks/roots of the tree (*Cinchona officinalis*) as an alkaloid. The derivatives from these plants have shown different propensities and potential to fixed and repair broken bonds and atoms in TNF receptor protein targets of SARS-COV 2. Results of molecular docking of new compounds used as ligands with the TNF rp2 of SARS-COV 2 or any other target sites for SARS-COV 2 and/or other related diseases can be used to discover new ligands with higher potency and potential to fixed and repair damaged bonds and atoms in protein target sites.

Results of molecular docking can also be used as a screening tool for identification and discovery of lead drug compounds as well as their optimization. It can serve as a quick check method for screening, identifying, and discovering good potential drugs/compounds before subjecting them to in vitro and in vivo trials, thus reducing waste of time and resources on less efficacious drugs or compounds.

Table 5 Results of calculations of molecular docking of SARS-COV 2 TNF rp 2 with plant-based antimalarial compounds. TNF rp 2 = tumor necrosis factor receptor associated protein 2

S/N	Drug target sites	Bioactivity parameter estimates	Artemisinin ligand	Chloroquine ligand	Quinine ligand
1	All TNF rp 2 sites	Full fitness	−751.9844	−583.2335	−539.0270
2	All TNF rp 2 sites	Simple fitness	38.6855	26.4694	36.2266
3	All TNF rp 2 sites	ΔG (kj/mol)	−5.0528	−5.3959	−6.8323
4	All TNF rp 2 sites	Energy affinity	35.6855	26.4694	36.2490
5	All TNF rp 2 sites	Core atoms found	19	5	22
6	All TNF rp 2 sites	Bonds found	31	50	52
7	All TNF rp 2 sites	Dihedral angles	3	11	7
8	All TNF rp 2 sites	Amino acids	209	209	209
9	All TNF rp 2 sites	Cluster groups	36	48	31
10	All TNF rp 2 sites	Cluster size	8	8	8
11	All TNF rp 2 sites	Total systems	**3301**	**1709**	**2533**
12	All TNF rp 2 sites	Receptor atoms	3272	1660	2484
13	All TNF rp 2 sites	Ligand atoms	29	49	49
14	All TNF rp 2 sites	Fixed atoms	**3272**	**1660**	**2484**
15	All TNF rp 2 sites	% fixation	**99.12**	**97.13**	**98.06**
16	All TNF rp 2 sites	RMSD	26.3723	59.6396	53.1820
17	All TNF rp 2 sites	Elapsed time (m/s)	9.8941	7.5850	8.8985
18	All TNF rp 2 sites	Binding mode	250/5000	250/5000	250/5000
19	All TNF rp 2 sites	Xmin	−36.595	−36. 668	−36.990

(continued)

Table 5 (continued)

S/N	Drug target sites	Bioactivity parameter estimates	Artemisinin ligand	Chloroquine ligand	Quinine ligand
20	All TNF rp 2 sites	Xmax	37.343	37.245	36.222
21	All TNF rp 2 sites	Xsize	73.941	73. 804	72.123
22	All TNF rp 2 sites	Ymin	47.207	47.195	46.950
23	All TNF rp 2 sites	Ymax	35.667	35. 459	35.102
24	All TNF rp 2 sites	Ysize	64.855	64.569	64.205
25	All TNF rp 2 sites	Zmin	− 8.632	−8.742	−8.996
26	All TNF rp 2 sites	Zmax	56.223	56.202	56.100
27	All TNF rp 2 sites	Zsize	64.855	64.600	64.201
28	All TNF rp 2 sites	X-rays/ Resolution @ ≤ 3Á	0.375	0.375	0.375

4 Antimalarial Drug–Protein Interactions in Malaria Parasite (Plasmodium Falciparum)

4.1 Results of Chemical–protein Interactions of Antimalarial Drug, Artesunate Ligand, and Receptors Proteins of the Malaria Parasite (Plasmodium Falciparum)

The results of artesunate drug chemical and the malaria parasite receptor proteins show major and direct interactions of artesunate drug with Thymidylate synthesis CAT1 and Thymidylate synthesis AGA. The network of interactions also shows three indirect effects and minor interactions with AGAP003929_PA, AGAP0011867, and AGAP006904 as shown above. The artesunate drug takes part in carboxylic acid metabolism, drug transmembrane transporter activity, target ligand binding, and ion binding and catalytic activities. The structural models of the two major protein interactions with artesunate drug structure are shown in Fig. 19a–c. This result shows the protein target sites that interact with the artesunate drug target for complex formation, prevention, treatment, normalization, and optimization.

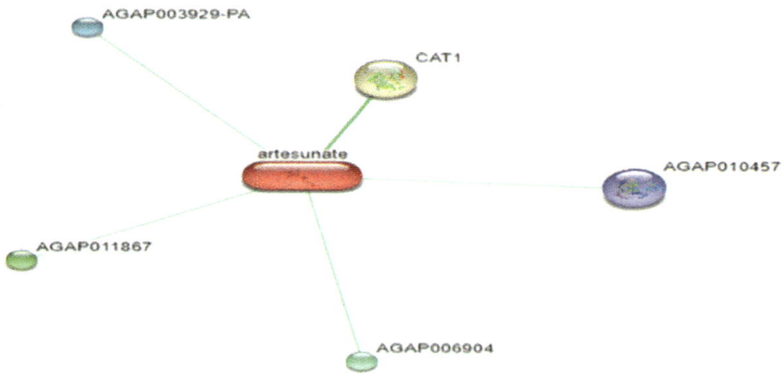

a b c

Fig. 19 **a** Artesunate drug. **b** AGA-Thmidylate synthesis. **c** Thymidylate synthesis CAT1

*4.2 Results of Chemical–Protein Interactions of Antimalarial Drug, Chloroquine Ligand, and Receptors Proteins of the Malaria Parasite (*Plasmodium Falciparum*)*

Chloroquine is a 4-aminoquinoline drug used in the treatment and prevention of malaria and related diseases. The results of chloroquine drug chemical and the malaria parasite receptor proteins show four major and direct interactions of chloroquine drug with PF14_0187, PF14_0192, PF10_0155, and PfLDH representing Glutathione S-transerase, Glutathione reductase, and L-Lactase dehydrogenase proteins, respectively. The network of interactions also shows six (6) indirect effects and minor interactions with histidine-rich proteins (HRP II), PfMDR1, PBGD, PfuroD, P1CRT, PfLDH, and the gamma-glutamylcysteine synthesis (PFE082w) as shown above. The chloroquine drug takes part in biosynthesis of secondary metabolite, glutathione metabolism, porphyrin metabolism as well as in glycolysis and glucogenesis, drug transmembrane transporter activity, target–ligand binding, and ion binding. The structural models of the two major protein interactions with chloroquine drug structure are shown in Fig. 20a–c. This result reveals the protein target sites that interact

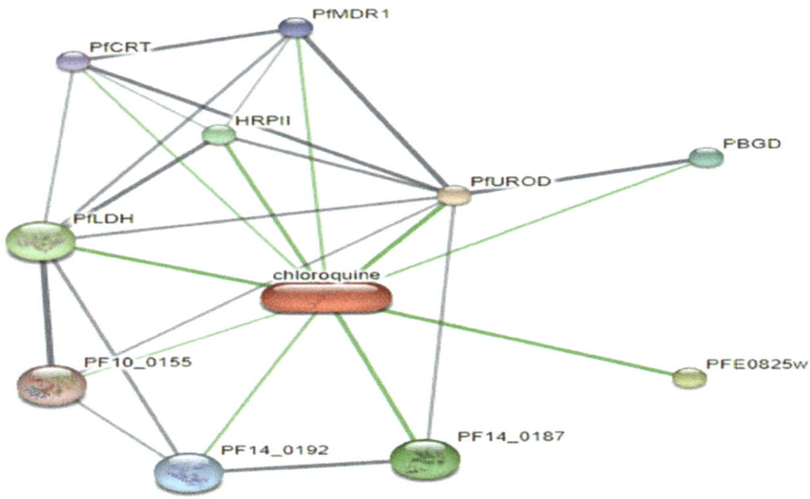

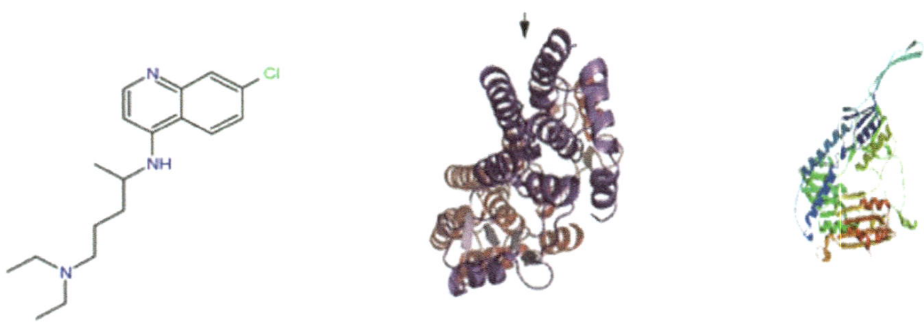

a b c

Fig. 20 **a** Chloroquine. **b** Glutathione S-transferase. **c** Glutathione reductase

with the chloroquine drug target compound for prevention, treatment, normalization, optimization, and complex formation.

*4.3 Results of Chemical–protein Interactions of Antimalarial Drug, Quinine Ligand, and Receptor Proteins of the Malaria Parasite (*Plasmodium Falciparum**)**

Quinine is a 2-aminoquinoline drug used in the treatment and prevention of malaria and related diseases. The results of quinine drug chemical and the malaria parasite receptor proteins show two major and direct interactions of quinine drug with L-Lactose dehydrogenase (PfLDH) and mesozite surface proteins (MSp1). The

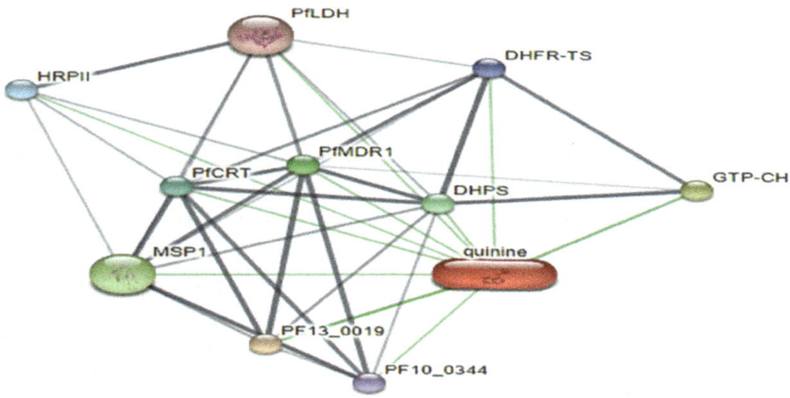

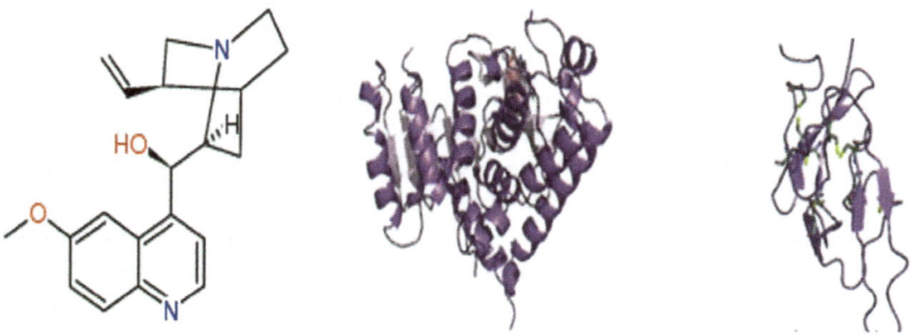

a b c

Fig. 21 **a** Quinine **b** L-Lactate dehydrogenase **c** Mesozite surface proteins

network of interactions also shows eight (8) indirect effects and minor interactions with histidine-rich proteins (HRP II), PfMDR1, P1CRT, PF13_0019, PF10_0344, DHRS, DHFR-TS, and the GTP cyclohydrolase (GTP). The quinine drug takes part in L-lactase dehydrogenase, biosynthesis of secondary metabolite, drug response signaling, folate synthesis, drug transmembrane transporter activity, target–ligand binding, and oxidation–reduction activity. The structural models of the two major protein interactions with quinine drug structure are shown above. This result reveals the protein target sites that interact with the quinine chemical drug target for prevention, treatment, bonding, normalization, and optimization.

Descriptors of networks: Big circles with thick lines indicate direct and major interactions while small circles with thin lines indicate indirect and minor interactions of drugs with the receptor proteins.

4.4 Discussion

This research focus is on translational science research methodology using artificial intelligence and bioinformatic applications on genomic data with positive outcome in small molecule drugs as plant-based malarial medicine and their repurposing for clinical trial efficacy for COVID-19. The results from this study research using molecular docking have shown that some of the antimalarial bioactive components (artemisinin, chloroquine, and quinine) found in local plants inhibited and fixed (repair) between 97.13 and 99.12% of the damaged protein and amino acid side chains, atoms, bad bonds, and bad angles in the tumor necrosis factor receptor protein target sites of the S-gene destroyed by the mutating coronavirus infections. The study approach has been able to ascertain the extent of inhibition of TNF rp2 of SARS-COV 2 by small molecule ligands and hence the need to repurpose these antimalarial compounds as herbal medicinal therapy for the much dreaded pandemic novel corona viruses as similar or higher results with in vitro cell cultures and in vivo clinical trials are highly anticipated and expected.

The data generated from this study showed that the COVID-19 strains are single-stranded pathogens whose genome size range from 26–32 kbp and contain about 15 functional protein types and each showing a motif size between 200 and 3000 bp. Most of the genomic contents are made up of introns (the non-coding regions of the S-gene). Molecular size of the corona virus strains plays important function in the protein structures of the viruses. They determine the site for chemical reactions with drug targets [24]. Presence of amino acids and molecular weight in the COVID-19 strains of corona viruses is an indication that they possess varying number of target sites for small molecules, candidate drugs, and ligands. This is very important in the design and discovery of drug as most drug targets are functional and effective at specific doses that correspond to the molecular weights and body mass of the recipient organism [25].

The strength and stability of SARS-COV 2 isolates are strongly determined and accounted for by the total number of atoms such isolate possess. This is because of the relationship between strength/stability and the binding energy affinities to the drug target. The free energy released is a measure of the acceptance or rejection of the drug by the protein target sites. The study revealed that the higher the number of atoms possess by an isolate, the higher the potential for binding affinities with small molecules toward forming the desired target ligand complexes. The connection table suitability is also a measure of possession of higher number of atoms [26].

This primary structure or sequence plays important role in chemical reactivity of drug target for computer-aided drug design for a variety of illnesses and related diseases including COVID-19 [27].

The modeling structures of the protein coat of the corona virus isolates can be used in the protein maneuvering of the lipid layer which occupies the outermost part of the corona viruses. This is important because it is a measure of substrate specificity and double bond positioning also known as regioselectivity, as the reaction products of the isolates and small molecules [28]. The output and conformation of the protein structure attributes more or less make way for the possible improvement in the efficacy of potential reaction processes and to develop proteins that can withstand any form of reactions at new positions in the aliphatic side chains of the drug molecules. An understanding of the protein structure features is essential as its tends to unveil the likelihood of synergizing newly formed fatty acid compounds with more ideal reactivity compounds thereby making it possible to design fatty compounds that could have been unavailable and non-reactive with most of the metabolic engineering strategies currently available, thereby creating a platform for drug discovery and delivery. This is in line with the position of [29–31].

The regular folding and organization of a polypeptide molecule into heavy and light chains of a protein is called its secondary structures. The secondary structures of a protein are the backbone or framework of the organism and determined to a large extent their thermal and genetic stabilities [32]. This attribute of the corona virus isolate is fundamental to the reactions that the virus protein coat undergoes and this is a measure of their thermal and genetic strengths. Alpha helices-dominated secondary protein structures provide a more stable complex and resilient pathway with drugs and small molecules when compared to the ones dominated by extended strand, random coil, and beta turns [33].

The 3-D tertiary protein folding structure can be modeled and optimized to provide the side chains and N-terminal amino acids as sites for targets by drugs and other small molecules (ligands) as revealed in this study. With the development of the structural 3-D model of the coding regions of the corona viruses, it is possible to develop therapy for the diseases using the computer-aided drug design which encompasses structure-based drug design and ligand-based drug design (CADD, SBDD, and LBDD) approaches [34] as adopted in this chapter.

The ability of proteins to withstand reactions with drug target depends largely on their genetic and thermal stabilities as may be determined by the aliphatic side chains. This also affects their reactivity with small molecules and ligands relative to reaction rates, complex formations, and decomposition in medium. Drug target should show comparative and complementary affinities with both protein and non-protein targets of the pathogen which, in this case, is the corona virus [35]. The half-life is the time it takes for half of the concentration of protein fractions in the cell to be denatured after translational synthesis of the cell cycle [36, 37]. This attribute plays a significant role on protein and drug target due to the fact that it determines the rate of complex formation between a potential drug target with the seemingly protein target of the host individual. The half-life reveals the mutation path and the rate of replication of the cells which give birth to new proteins. Its half-life depends markedly on the N-terminal amino acids also called N-end rule. This matches the half-life of a protein to the complement of its N- terminal amino acids molecule [38–40]. Hence, N-terminal amino acids were based on the assumptions that it plays a vital role in in vivo stability of a protein molecule.

This shows that a potential drug target must form complexes by binding together the target protein of the individual before the completion of the half-life. The genetic and thermal stabilities of isolates are further concerned with side chain terminal amino acids. It is the basis of the isolate stability and the fulcrum through which the decay constant and their decomposition in drug targets or small molecules are determined and extrapolated [41, 42]. This aspect of molecular epidemiology is tailored to utilize nucleotide or protein sequences of isolates or causative organisms in identifying means of expression, transmission, and distribution. The pathogens show variable level of transient and transverse mutations thus acquiring new attributes and characteristics as a mutant variant during the process of transmission from one infected individual to another [43–46] The novel COVID-19 corona viruses show that mutation occurs at an average rate of about two mutations per month [51]. As soon as one is infected, the virus is incubated for ~ $\pm$ 5 days before symptoms develop and transmission is set in. The study reports have shown that the occurrence interval of COVID-19 is ~ $\pm$ 7 days [47, 51]. This by implication suggests that it would take $\pm$ 7 days from one infection to the next. As the virus is transmitted from one host to another, it mutates at the rate of two mutations per month with each mutant strain remaining viable to infect the host protein and drug target sites. Exposure of the corona virus protein coats to high temperature denatures them due to their relative extinction coefficient. This affects the thermal and genetic stability indexes of the viruses. This further suggests that potential drug targets should be modeled to generate heat by increasing reaction temperature so that more heat energy will accompany the complex formation by target and ligands. The internal energy (entropy) of the drug target should be high to ensure the denaturing and thermal decomposition and dissociation of the target ligands complex formed with the corona virus [45, 49, 51].

5 Conclusion and Suggestions

The results of computer intelligence using bioinformatics and molecular docking of genomic sequences of the SARS-COV 2 sequence data revealed that the antimalarial compounds, artemisinin, chloroquine, and quinine were able to fix the 3301 damaged proteins of TNF rp2 of SARS-COV 2 to 99.12, 98.91, 95.91, and 99.72%, respectively. Hence, it is expected that similar results or more will be obtained from in vitro efficacy trials with the cell cultures of SARS-COV 2 in the laboratory and in vivo with experimental animal models like infected patients during clinical trials of these identified candidate plant-based antimalarial drugs, thus repurposing the antimalarial small molecule compounds as effective treatment for COVID-19, malaria, and related diseases. It is therefore recommended that local plants containing the antimalarial compounds as small molecules be repackaged, repositioned, repurposed, and used as potential therapeutic drug (single or in combinations) for the prevention and treatment of the much dreaded COVID-19 disease which for the moment remains the core aspirations and urgent need of the biopharmaceutical industries.

References

1. World Health Organization (2020) Global Update for COVID 19. https://www.worldhealthorganization.global.update.coivd-19
2. Cox E, Hooyberghs J, Pensaert MB (1990) Sites of replication of a porcine respiratory coronavirus related to transmissible gastroenteritis virus. Res Vet Sci 48(2):165–169
3. deGroot RJ, Horzinek MC (1995) Feline infectious peritonitis. In: Siddell SG (ed) The Coronaviridae. Plenum Press, New York, pp 293–315
4. Anonymous. (2003) Severe acute respiratory syndrome (SARS). Wkly Epidemiol Rec 78:81–83
5. Ballesteros ML, Sanchez CM, Enjuanes L (1997) Two amino acid changes at the N-terminus of transmissible gastroenteritis coronavirus spike protein result in the loss of enteric tropism. Virology 227(2):378–388
6. Baric RS, Fu K, Chen W, Yount B (1995) High recombination and mutation rates in mouse hepatitis virus suggest that Coronaviruses may be potentially important emerging viruses. Adv Exp Med Biol 380:571–576
7. Ikai A (1980) Thermostability and Aliphatic index globular ptoteins. J Biochem 88(6):1895–1988
8. Bonilla PJ, Gorbalenya AE, Weiss SR (1994) Mouse hepatitis virus strain A59 RNA polymerase gene ORF 1a: heterogeneity among MHV strains. Virology 198(2):736–740
9. Bost AG, Prentice E, Denison MR (2001) Mouse hepatitis virus replicase protein complexes are translocated to sites of M protein accumulation in the ERGIC at late times of infection. Virology 285(1):21–29
10. Brian DA, Hogue BG, Kienzle TE (1995) The coronavirus hemagluttinin esterase clycoprotein. In: Siddell SG (ed) TheCoronaviridae. Plenum Press, New York, pp 165–179
11. Brockway SM, Clay CT, Lu XT, Denison MR (2003) Characterization of the expression, intracellular localization, and replication complex association of the putative mouse hepatitis virus RNA dependent RNA polymerase. J Virol 77(19):10515–27.
12. Whittaker M (2004) The role of bioinformatics in target validation. Drug discovery to-clinical trial registration: a statement from the International Committee of Medical Journal Editors. Med J Australia 181:293–4
13. Lengauer M (2002) Bioinformatics. From genomes to drugs. Wiley-VCH, Weinheim, Germany
14. Lipinski D (2004) Lead and drug-like compounds: the rule-of-five revolution. Drug Discov Today Technol 1(4):337–341
15. Leeson PD, Davis AM, Steele J (2004) Drug-like properties: guiding principles for design—or chemical prejudice? Drug Discov Today Technol 1(3):189–195
16. Hou T, Xu X (2004) Recent development and application of virtual screening in drug discovery: an overview. Curr Pharm Des 10:1011–1033
17. Klebbe G (2004) Lead identification in post-genomics: computers as a complementary alternative. Drug Discov Today Technol 1(3):225–215
18. GisbertSchneider UF (2005) Computer-based de novo design of drug-like molecules. Nat Rev Drug Discov 4(8):649–663
19. Butte A (2011) The use and analysis of microarray data. Nat Rev Drug Discovery 1(12):951–960
20. Richards WG (1994) Computer-aided drug design. Pure Appl Chem 6(68):1589–1596
21. Kitchen DB, Furr DH, J R. Bajorath J, (2004) Docking and scoring in virtual screening for drug discovery: methods and applications. Nature reviews in drug discovery 3:935–949
22. Grabowski DJA, H G, (2007) The cost of biopharmaceutical R&D: is biotech different? Manag Decis Econ 28:469–479
23. Hajduk PJ, Huth JR, Tse C (2005) Predicting protein druggability. Drug Discov Today 10:1675–1682
24. Fauman EB, Rai BK, Huang ES (2011) Structure-based druggability assessment-identifying suitable targets for small molecule therapeutics. . Curr Opin Chem Biol 15:463–468
25. Laurie AT, Jackson RM (2006) Methods for the prediction of protein-ligand binding sites for structure-based drug design and virtual ligand screening. . Curr Protein Pept Sci 7:395–406

26. Becker OM, Dhanoa DS, Marantz Y, Chen D, Shacham S, Cheruku S, Heifetz A, Mohanty P, Fichman M, Sharadendu A (2006) An integrated in silico 3D model-driven discovery of a novel, potent, and selective amidosulfonamide 5-HT1A agonist (PRX-00023) for the treatment of anxiety and depression. . J Med Chem 49:3116–3135
27. Warner SL, Bashyam S, Vankayalapati H, Bearss DJ, Han H, Mahadevan D, Von Hoff DD, Hurley LH (2006) Identification of a lead small-molecule inhibitor of the Aurora kinases using a structure-assisted, fragment-based approach. . Mol Cancer Ther 5:1764–1773
28. Budzik B, Garzya V, Walker G, Woolley-Roberts M, Pardoe J, Lucas A, Tehan B, Rivero RA, Langmead CJ (2010) Novel N-substituted benzimidazolones as potent, selective, CNS-penetrant, and orally active M(1) mAChR agonists. Med Chem Lett 1:244–248
29. Buchan DW, Ward SM, Lobley AE, Nugent TC, Bryson K, Jones DT (2010) Protein annotation and modelling servers at University College London. Nucleic Acids Res 38:563–568
30. Martí-Renom MA, Stuart AC, Fiser A, Sánchez R, Melo F, Sali A (2000) Comparative protein structure modeling of genes and genomes. Annu Rev Biophys Biomol Struct 29:291–325
31. Schlitter J, Engels M, Krüger P (1994) Targeted molecular dynamics: a new approach for searching pathways of conformational transitions. . J Mol Graph 12:84–89
32. Grubmüller H (1995) Predicting slow structural transitions in macromolecular systems: conformational flooding. Phys Rev E Stat Phys Plasmas Fluids Relat Interdiscip Topics 52:2893–2906
33. Abrams CF, Vanden-Eijnden E (2010). Large-scale conformational sampling of proteins using temperature-accelerated molecular dynamics. Proc Natl Acad Sci USA 107:4961–4966
34. Sugita Y, Okamoto Y (1999) Replica-exchange molecular dynamics method for protein folding. Chem Phys Lett 1999(314):141–151
35. Liu M, Wang SMMCDOCK (1999) A Monte Carlo simulation approach to the molecular docking problem. . J Comput Aided Mol Des 13:435–451
36. Jones G, Willett P, Glen RC (1995) A genetic algorithm for flexible molecular overlay and pharmacophore elucidation. J Comput Aided Mol Des 9:532–549
37. Halgren TA (1996) Merck molecular force field. 1. Basis, form, scope, parameterization, and performance of MMFF94. . J Comput Chem 17:490–519
38. Böhm HJ (1992) The computer program LUDI: a new method for the de novo design of enzyme inhibitors. . J Comput Aided Mol Des 6:61–78
39. Rarey M, Kramer B, Lengauer T, Klebe G (1996) A fast flexible docking method using an incremental construction algorithm. . J Mol Biol 261:470–489
40. Jain AN (2003) Surflex: fully automatic flexible molecular docking using a molecular similarity-based search engine. J Med Chem 46:499–511
41. Shimada J, Ishchenko AV, Shakhnovich EI (2000) Analysis of knowledge-based protein-ligand potentials using a self-consistent method. . Protein Sci 9:765–775
42. Velec HFG, Gohlke H, Klebe G (2005) DrugScore (CSD)-knowledge-based scoring function derived from small molecule crystal data with superior recognition rate of near-native ligand poses and better affinity prediction. J Med Chem 48:6296–6303
43. DeWitte RS, Shakhnovich E (1997) SMoG: De novo design method based on simple, fast and accurate free energy estimates. J Am Chem Soc 119:4608–4617
44. Mitchell JBO, Laskowski RA, Alex A, Forster MJ, Thornton JM (1999) BLEEP-Potential of mean force describing protein-ligand interactions: II. Calculation of binding energies and comparison with experimental data. J Comput Chem 20:1177–1185
45. Feher M (2006) Consensus scoring for protein-ligand interactions. Drug. Discov Today 11:421–428
46. O'Boyle NM, Liebeschuetz JW, Cole JC (2009) Testing assumptions and hypotheses for rescoring success in protein-ligand docking. J Chem Inf Model 49:1871–1878
47. Becker OM, Dhanoa DS, Marantz Y, Chen D, Shacham S, Cheruku S, Heifetz A, Mohanty P, Fichman M, Sharadendu A (2006) An integrated in silico 3D model-driven discovery of a novel, potent, and selective amidosulfonamide 5-HT1A agonist (PRX-00023) for the treatment of anxiety and depression. J Med Chem 49:3116–3135

48. Johnson MA, Maggiora GM (1990) Concepts and applications of molecular similarity. Wiley, New York
49. Stumpfe D, Bill A, Novak N, Loch G, Blockus H, Geppert H, Becker T, Schmitz A, Hoch M, Kolanus W (2010) Targeting multifunctional proteins by virtual screening: structurally diverse cytohesin inhibitors with differentiated biological functions. Chem Biol 2010(5):839–849
50. Cramer RD, Patterson DE, Bunce JD (1988) Comparative molecular field analysis (CoMFA). Effect of shape on binding of steroids to carrier proteins. J Am Chem Soc 110:5959–5967
51. Kiaris H, Spandidos DA (1995) Mutations of Ras Genes in Human Tumors. Int J Oncol 7(3):413–421

Chapter 15
Explainable AI for Fighting COVID-19 Pandemic: Opportunities, Challenges, and Future Prospects

Kazeem Moses Abiodun, Joseph Bamidele Awotunde, Dayo Reuben Aremu, and Emmanuel Abidemi Adeniyi

Abstract Novel coronavirus illness 2019 (COVID-19) continues to spread exponentially and has incurred over 7,000,000 infections and 400,000 deaths worldwide. To build an efficient and easy diagnostic framework for COVID-19, artificial intelligence has revealed significant success in several activities, from image processing to natural language processing, particularly with the advent of deep learning (DL). The medical profession requires a high degree of precision and accuracy. Interpretations for system assumptions and forecasts are also required to explain their consistency. This needs better interpretability, which also implies the need to grasp the process behind the algorithms. Unfortunately, the black box existence of the DL remains unresolved, and certain system decisions are still badly known. The model's relevance to the black-box challenge comes under the eXplainable AI (XAI) area, there is explanation usage is generally accepted as a crucial aspect for the realistic application of AI models, especially during the COVID-19 outbreak. Therefore, this Chapter presents the applicability and prospect of AI during the current pandemic. Also, the chapter discusses the challenges facing using eXplainable AI in the healthcare system. A framework on which COVID-19 diagnosis can be based on eXplainable AI is also proposed in this chapter.

Keywords Artificial Intelligence · Deep Learning · eXplainable AI · COVID-19 pandemic · Diagnosis · Healthcare system

---

K. M. Abiodun · E. A. Adeniyi
Department of Computer Science, Landmark University, Omu-Aran, Nigeria
e-mail: moses.abiodun@lmu.edu.ng

E. A. Adeniyi
e-mail: adeyini.emmanuel@lmu.edu.ng

J. B. Awotunde (✉) · D. R. Aremu
Department of Computer Science, University of Ilorin, Ilorin, Nigeria
e-mail: awotunde.jb@unilorin.edu.ng

D. R. Aremu
e-mail: aremu.dr@unilorin.edu.ng

U. Kose et al. (eds.), *Computational Intelligence for COVID-19 and Future Pandemics*, Disruptive Technologies and Digital Transformations for Society 5.0,
https://doi.org/10.1007/978-981-16-3783-4_15

1 Introduction

The current coronavirus pandemic had a devastating effect on the healthy development of the world population [1, 2]. As of December 14, 2020, over seventy-one million people have been infected with COVID-19 and 1,608,648 death rate due to the illness were recorded. Latest research revealed that COVID-19, Coronavirus infection 2019, induced by Extreme Acute Respiratory Syndrome Coronavirus 2 (SARS-CoV-2), has transmitted steadily to more than 200 countries worldwide since the end of 2019.

Using reverse transcription, the reaction of the polymerase chain (RT-PCR) to detect viral ribonucleic acid is currently the widely available and effective method of diagnosing COVID-19 (RNA). However, this approach has been found not to be sufficiently sensitive for early detection of the disease and the quantity of RT-PCR testing kits is small in comparison to escalating the increase in contaminated instances. In addition, it takes hours for the whole model process before given accurate results that be used in reducing the spread of the outbreak. Using radiological imaging methods, like chest X-ray imaging and chest computed tomography (CT) imaging, is one of the most competitive candidates for solving this issue. These images can be read and interpreted by clinicians and radiologists to recognize specific distinctive sensory characteristics of the pulmonary related to COVID-19, like initial phase bilateral and peripheral ground glass (GGO) transparency and critical stage pulmonary consolidation visibility.

Moreover, in most clinics and hospitals, the analysis process is quick, well-pipelined, and reachable to many individuals. There are, however, still contradictions among the swift development of analysis individuals and the lack of clinicians. In addition, it takes time to read the radiological images, and the diagnosis of infection by radiologists is extremely subjective, which can be substantially influenced by specific views and professional practice.

Machine learning algorithms may be extended to radiological images for automatic diagnosis of COVID-19 due to the exponential development of artificial intelligence (AI). Various diagnostic technologies, like automated diagnosis [3], infection process detection [4], decision support tools in medical practice [5], etc., have been widely applied to deep learning techniques. Artificial Intelligence (AI) is at the heart of several sectors of activity which have incorporated modern information technology [6]. Though, the origin of AI date back to many years ago, there is a strong agreement on the overriding value of smart machines with understanding, reasoning, and adaptation capabilities nowadays. Because of these capacities, when working to unravel progressively multifaceted computational tasks, AI methods achieve unparalleled levels of success, which are essential for the upcoming growth of our society [7].

The complexity of systems driven by artificial intelligent has recently increased to such an extent that their design and implementation involve almost no human involvement. When conclusions resulting from such processes ultimately affect the lives of humans in fields such as medicine, law, or security for example. An urgent need is to understand how AI methods provide such conclusions [8].

The first AI systems were simple to interpret, and the emergence of Deep Neural Networks (DNNs) opaque decision systems has been observed in recent years. The empirical success of models such as DNNs for Deep Learning (DL) stems from a combination of successful learning algorithms and their enormous parametric space. There are hundreds of layers and millions of parameters in the latter space that makes DNNs model a difficult black box [9]. The quest of a straight understanding of the process by which a model operates is the opposite of black-box model [10]. Just as Machine Learning (ML) models of black-box are increasingly being used in critical contexts to make significant predictions, the call for transparency from the several patrons in AI is increasing [11]. The risk lies in the formation and use of decisions which are not justifiable, valid, or simply do not make it possible to obtain thorough explanations of their conduct [12].

In precision medicine, for example, where experts need much more model knowledge as oppose to simple binary forecast to support their conclusion [13], clarifications backing the performance of a model are essential. Self-driving cars in transportation, security, and finance are further examples. Human beings are usually hesitant to follow strategies.

In general, humans are reluctant to follow strategies that, considering the increasing request for ethical AI [3], are not explicitly interpretable, tractable, and trustworthy [14]. It is commonplace to assume that the systems would be increasingly opaque by concentrating solely on results. A trade-off always happens among a model's success and its transparency [15]. An enhancement in a system's understanding will however help in the modification of its short comings. The concern ability to view as an integrated concept catalyst can improve its execution when designing an ML model.

The remaining part of this chapter is organized thus: Sector 2 discusses the growth of eXplainable AI (XAI). Section 3 presents the artificial intelligence in fighting COVID-19 pandemic. Section 4 discusses related work of XAI in combating COVID-19 pandemic. The challenges of XAI in COVID-19 outbreak was presented in Sect. 6, while Sect. 6 discusses the prospects and a framework for XAI in fighting COVID-19 pandemic. Finally, Sect. 7 concludes the chapter.

2 eXplainable AI (XAI)

Several suggestions have been made indicating the general requirement, coherent notion that come near to an explanation's structure and purpose. However, when proposing such a general theory, nobody has stood up to criticism. Presently, the most agreed-upon thinking incorporates diverse approaches to clarity taken from various disciplines of knowledge. When approaching interpretability in AI, a similar issue is found. The literature indicates that there is still no general understanding of what interpretability or interpretation is.

Many contributions, however, say that interpretable models and techniques that empower explanability are achieved. In order to explain this lack of consensus, the reference point for the description of the word Explainable Artificial Intelligence (XAI) provided by Gunning in [12] may be interesting: "XAI will build a collection of machine learning methods that enable users to comprehend, belief, and handle the evolving era of artificially intelligent associates appropriately and effectively". This term puts together two definitions that need to be tackled in advance (understanding and trust). However, other purposes that drive the necessity for explainable AI models, such as trust, fairness, causality, easily transferable, information access, are not taken into account [5, 16–18]. A detailed full concept of explanability in AI sometimes slips from our fingertips, as exemplified by the definition above.

A wider re-formulation of this concept is "An Artificial Intelligence that describes its workings is one that generates explanations". This does not completely capture the definition, and it leaves vital aspects such as its purpose aside. A description of explanation is first needed in order to build on the completeness. Looking at the Cambridge Dictionary of English Language, "The details or reasons that someone offers to make things simple or simple to comprehend" [19] are an interpretation. This can be rephrased in the sense of an ML model as the particulars or reasons given by a model to make its operation flawless to comprehend easily."

It is at this stage where views begin to differ. Two ambiguities can be pointed out, inherently arising from the previous definitions. First of all, the motives used to justify them are largely dependent on the audience they are introduced to. Second, whether the idea has been made understandable or simple to understand by the interpretation often depends entirely on the user. The definition must therefore be rephrased to clearly represent the reliance on the users on the explanability of the model. To this end, a revised meaning could read as: Explainability refers to the facts and explanations provided by the models that make the mechanism clear or easily understandable view of a certain users. Since explaining, as arguing, can entail weighting, contrasting, or urging a user with arguments that is based on logic formalizations of counter [20], explanation can lead into the field of mental and explanatory psychology [7], the ability to measure objectively if information has been known or specifically set to be a challenging process.

However, it should be tackled critically to measure to what degree the internal dimensions of a model can be clarified. As a XAI approach, ways to reduce the difficulty of the model or to make its output easier should be considered. In terms of complexity or simplicity, how large this leap is will lead to way the explainable result model will be. An essential problem that is still unresolved is that it may not be easy to measure the interpretability benefit offered by such XAI approaches: for example, a simplification of the model may be measured based on declining the quantity of architectural characteristics or the range of parameter estimates itself (as often made, for instance, for DNNs). In other words, the use of software or natural language methods for the same purpose does not support a simplistic estimation of the enhancement of interpretability received. The derivation of general metrics for the evaluation of the value of XAI approaches remains a challenge that is open and should be the focus of researchers in the coming years.

Explainable AI is related to post-hoc explainability as it includes the approaches used to turn a prototype that is not interpretable into an explainable one. A model may be clarified, but something that comes from the nature of the model itself is the interpretability of the model. With these observations in mind, it is possible to describe explainable AI as follows: An Artificial Intelligence that is explainable for a user that brings information or explanations to make its process understandable. XAI according to [7] is to build a suite of machine learning systems that (a) generate extra explainable models and also retaining a great level of success in learning (e.g., accuracy of prediction) and (b) allow people to recognize, trust, and handle the new era of artificially intelligent associates properly and effectively. XAI also draws perspectives from the social sciences [12] and contemplates explanatory psychology.

So far the research operation around XAI has revealed numerous objectives to build on the success of an explainable model. Trustworthiness, causality, informativeness, trust, justice, accessibility, interactivity, and awareness of privacy are these.

3 The Artificial Intelligence

Artificial Intelligence (AI) enables computers make good decisions, react to new stimuli, and accomplish human-like activities. From chess-playing robots to self-driving cars, most of the AI descriptions you hear nowadays depend primarily on deep learning and natural language processing. Through analyzing large volumes of data and finding trends in data, computers may be equipped to perform complex tasks using these technologies. In both research and industrial applications, machine learning (ML), which is another branch of AI, has grown so large, particularly with the popularity of deep learning (DL) and neural networks (NNs), its influence and possible after-effects could no longer be considered for granted. Malfunction is never an alternative in some areas: even a brief dysfunctional machine vision application in an autonomous vehicle easily leads to destruction.

There are certainly human lives on the line in the medical profession. Detection at an early stage of the disease is also important for patients to recover or avoid the infection from progressing to more serious phases. While ML techniques, artificial NNs, and associated areas have lately shown encouraging results in the performance of health tasks, they are barely flawless [21–29].

Thus, ML algorithms' interpretability and explanability have become demanding topics: when things go wrong, who is responsible? Can we illuminate why things are going wrong? If things work well, we wonder how and why they can be additional leveraged? Several papers have proposed various interpretability steps and frameworks, and the subject of Explainable Artificial Intelligence (XAI) has been a hotspot in the scientific culture of ML. Common DL libraries, including Pytorch Captum and tensorflow tf-explain, have begun to provide their own XAI libraries. In addition, the abundance of parameters for interpretability evaluation like consistency, causality, and usability enables the ML group to preserve pathway of how algorithms are used and how their use can be strengthened, giving guidance for further developments [30–32].

In specific, visualization has shown to be proficient in assisting scholars identify inaccurate thinking in classification difficulties that might have been overlooked by many previous researchers [33]. There seems to be a lack of consistent acceptance within the research community of interpretability evaluation criteria, as mentioned above. Attempts were designed to define the principles of "usability," "explainability," "consistency," "authenticity," and other principles related concepts without specific descriptions as to how they should be implemented into the broad variety of ML model implementations; consider [30, 34–38].

3.1 Application of AI in COVID-19

Key Applications of AI to COVID-19 Disease outbreak is enormous and some of it are as follows:

(a) **Early diagnosis and identification of the infection**
AI can evaluate abnormal symptoms and other red flags rapidly and thus warning to patients and health departments [39, 40]. It allows, and is cost-effective, to make choices more easily. It helps to build, through useful algorithms, a novel diagnostic, and control system for COVID 19 cases. With the support of medical imaging technology including computed tomography (CT), magnetic resonance imaging (MRI) screening of human organs, AI is helpful in the diagnosis of sick patients.

(b) **Monitoring the treatment**
AI will build an intelligent framework for automatic tracking and forecasting of the propagation of this disease. In order to eliminate the spatial information of this condition, a neural network may also be created, which will assist in the appropriate controls and care of affected people [41, 43]. It has the capacity to provide patients with day-to-day updates and also to provide solutions to the pandemic of COVID-19.

(c) **Contact tracing of the individuals**
AI can help measure the degree of transmission by identifies the current and 'hot spots' of the viruses and can efficiently monitor and regulate the touch tracking of persons. The potential path and probable reemergence of this virus may be anticipated.

(d) **Projection of cases and mortality**
Accessible info, social networking and media channels, this technology can monitor and predict the existence of the virus, the risks of the infection and its possible spread. In addition, the number of positive cases and mortality in any area can be estimated. AI will help recognize and take steps accordingly for the most vulnerable territories, individuals, and countries.

(e) **Development of drugs and vaccines**
By evaluating the data available on COVID-19, AI is used for drug testing. It is valuable for the proposal and production of drug delivery. This technology is used to speed up real-time drug testing, where standard research requires a lot of time and hence seeks to significantly ask for an extension, which might not be possible for an individual [41, 42]. It may help classify effective medications for the treatment of COVID-19 victims. It has been an important medium for the development of therapeutic verification and validation and vaccinations [44–46]. AI helps to manufacture vaccines and medicines even faster than average and is also helpful for drug testing throughout vaccine improvement.

(f) **Lessening the workload of health professionals**
Health professionals had a very hectic schedule due to a sudden and dramatic growth in the number of people during the COVID-19 disease outbreak. AI is used here to reduce the burden of medical workers [47–51]. It aims to provide early diagnosis and treatment using digital tools and decision-making science at an early level and provides learners and physicians with the best preparation in this current disease [52, 53]. AI would have an effect on prospective health treatment and solve more emerging issues that relieve the workload of physicians.

(g) **Prevention of the disease**
With the aid of a real-time statistical analysis, AI can provide up-to-date useful information in preventing this outbreak. During this outbreak, it could be used to anticipate potential delivery points, the influx of viruses, the need for facilities and medical personnel. Previously mentored data on data was widespread at different periods, AI is helpful for the potential prevention of viruses and diseases. It recognizes features, causes, and reasons for infection to spread. This will become an important technology for battling other diseases and possible pandemics. It can provide a preventive measure for many other diseases and battle them. In the future, in offering more predictive and preventive healthcare, AI can play a crucial role.

4 Review of XAI for COVID-19 Pandemic

Several automated diagnostic networks were developed using AI technologies to enhance the efficiency of radiological image-based diagnoses. Various machine learning systems were used in the COVID-19 classification and recognition mission: [55] apply a basic 2D CNN architecture to differentiate COVID-19 from many other pneumonias generated by manually delineated radiological images; [56] introduce a more sophisticated CNN with leaky ReLU function to create a classification algorithm.

The various types of convolutional neural networks (CNN) were evaluated by [57] for the analysis of COVID-19 and it accomplished an accuracy of ninety-eight percent using a pre-trained ResNet50 model. In the study, they made the classification problem simple by just looking for healthy and non-healthy COVID-19 people, ignoring the trouble of differentiating normal pneumonia cases from COVID-19 conditions.

COVID-Net was proposed by [58] to detect distinct irregularities in images (CXR) among samples of people with non-COVID-19 virus and infectious agents and contrasts with stable patients. On a research set of ninety-two positive COVID-19 cases and approximately 300 others, COVID-Net achieves an average precision of 92.6% with 97.0, 90.0, and 87.1% sensitivity for standard, non-COVID-19, and COVID-19 cases, respectively. On the other side, COVID-Net produces PPV of 90.5, 91.3, and 98.9% in regular, non-COVID-19, and COVID-19 cases, respectively. However, the limited sample size does not make generalizable conclusions about the system reliability feasible. Outlined territories using 'GSInquire' are often not well placed in sensitive areas. Overall, this strategy has been hampered by practicing on mismatch images, lack of detailed image pre-processing, and bad decision modeling.

An Explained Attention-based Method (EXAM) for COVID-19 Automated Diagnostic with Graphical Interpretation was presented by [59]. The diagnostic method of radiological images is translated into an object detection concern that differs among instances of COVID-19, normal, and community-acquired pneumonia (CAP). The proposed approach can efficiently retrieve main traits and eliminate redundant data by combining channel-wise and spatial-wise focus processes. Experiment results and visualization demonstrate that the EXAM is outperforming and showing its interpretability for the new state-of-the-art versions. They made a real effort to calibrate the model components to make their results more reasonable and explainable, aside from the increased performance of both the chest X-ray image dataset and the chest CT image dataset relative to recent state-of-the-art simulations. They are also developing a Collaborative Diagnostic System for Human-AI, which somewhat provides an insightful result, but also offers a score of confusion by comparison to the subsequent decision made by the Commission (Fig. 1).

'DeepCOVIDExplainer' was proposed by [60] to exploit explainable prediction for COVID-19 based on images (CXR). Outcomes of the assessment indicate their strategy can classify COVID-19 with a 96.12% PPV and recall 94.3%, outperforming a recent method. In addition, as Curtis Langlotz12 pointed out, "AI will not substitute Radiologists, but radiologists who use AI will replace non-radiologists." They contend, though, that 'DeepCOVIDExplainer' is not intended to replace radiologists, but also to be examined in a therapeutic setting and not to replace human radiologists whatsoever. Human judgment, on the other hand, is important when patients' lives are at risk. In addition, they hope that their results will make a valuable impact to the battle against COVID-19 and to the growing approval and adoption in clinical practice of AI-assisted applications.

Karim et al. [61] developed a deep learning model that can better detect COVID-19 patients on both the CT scan and the image dataset of chest X-ray with greater precision. Eight different deep learning methods have been updated and tested in this work, such as VGG16, InceptionResNetV2, ResNet50, DenseNet201, VGG19, MobilenetV2, NasNetMobile, and ResNet15V2, on two datasets-one dataset contains

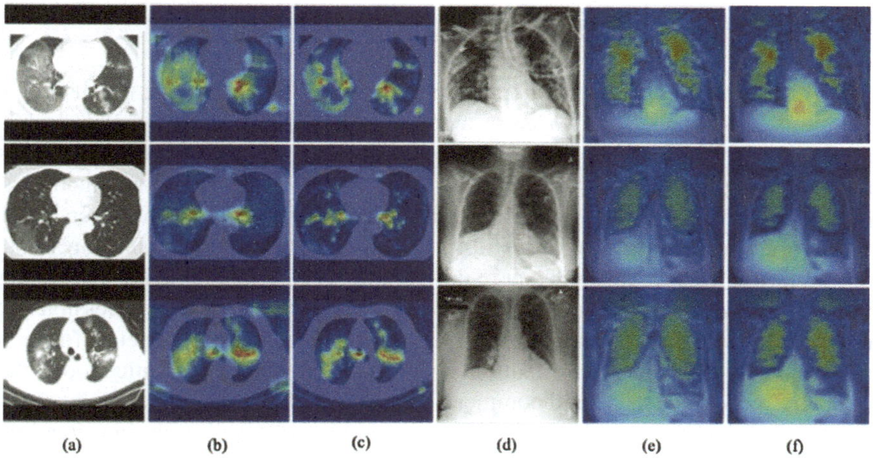

Fig. 1 Columns (**a**) and (**d**) are the main pictures and the X-ray pictures. Columns (**b**) and (**c**) are the concern graphs created after the first IAAM. Columns (**c**) and (**f**) are the awareness graphs provided after the double IAAM [59]

400 CT scan images, and another dataset contains 400 chest X-ray images. Results show that in terms of accuracy on CT scan (81.5–95.2%) and chest X-ray (95.4–100%) image data collection with a 95% confidence interval, NasNetMobile outperformed all other models. In addition, the interpretability of the model is explained using Local Interpretable Model-agnostic Explanations (LIME). Using LIME, test results show that it is conceivable to interpret top features that should have served to establish a trustworthy AI system to differentiate other patients from patients with COVID-19 symptoms (Fig. 2).

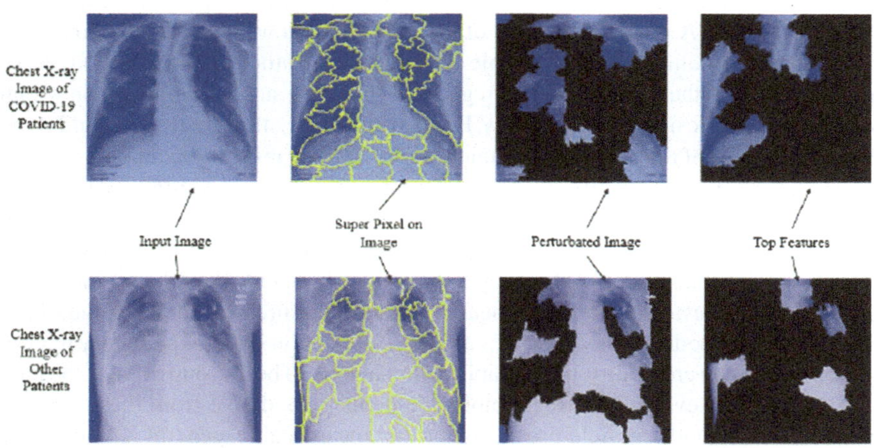

Fig. 2 Overall prediction analysis using LIME [61]

5 The Challenges of XAI

One of the problems facing AI explainability is that of performance versus interpretability. It is not necessarily true, as mentioned in [62], that more complex models are essentially more accurate. In cases where the data is well ordered and the functionality at our fingertips are of excellent quality and reliability, this assertion is incorrect. What can be considered real is that more difficult models have much more versatility than simpler models, making it possible to approximate more complex functions. Now, referring to the statement "more difficult models are more precise", assuming that the purpose to be estimated involves some difficulty, that the data accessible for analysis is widely dispersed throughout the sphere of acceptable parameters for each element, such that enough information is valid to manipulate a dynamic model, the argument is provided as a valid comment.

Trade-off can be found between efficiency and interpretability in this situation. Any attempt to solve issues not respecting the aforementioned principles would fall into the pit of trying to unravel an issue that has not a sufficient diversity of data (variance). Therefore, the model's additional difficulty can simply battle against the goal of solving the problem correctly. Once the performance comes with uncertainty, interpretability meets the situation on a downhill slope that seemed inevitable until now. The advent of more sophisticated explanatory approaches, however, could invert or at least cancel the slope.

The approximation problem is another factor worth noting because of its close relation to model interpretability and results: clarifications prepared for an ML model must be dramatic and close enough to fulfill the needs of its customers with whom they are being requested, to ensure that the explanations are representative of the prototype being analyzed and not to clarify its basics characteristics.

Another difficulty of AI's explainability is to have a common ground to fix solid metric problems to determine what is needed. In this respect, it is important that the XAI field enter an agreement joining the devastated work of a general field behind the same banner. A measure, or a set of them, should allow a significant correlation of how well the definition of explicable suits a model. Without such a method, every argument dilutes the literature in this regard, not having a stable foundation on which to stand. As classic metrics of accuracy, F1, sensitivity, etc., these metrics should show the performance of the model at a certain stage of explainability. Recent time, several efforts were made to quantify the XAI, as extensively examined in [63, 64].

In broad, the consistency, worth, and fulfillment of explanations, the enhancement of the audience's mental model caused by explanations of model, and the effect of elucidations on the model's efficiency and on the audience's confidence and confidence should be evaluated by XAI measurements. Measurement methods measured in [63, 64] like goodness checklist, explanation satisfaction scale, explanatory fidelity tests, explanation credibility, and reliability model look to be a strong drive in the path of XAI methods evaluation. Regrettably, the deductions derive from these surveys are inconsistent with our field prospects: further measureable, general XAI measure are actually required to help the current dimension processes and resources suggested by the group.

Another dilemma which needs urgent attention is the problem of meaning. However, plans are presently being made in the field of XAI. To obtain clarification in deep learning models meaning, there are still many challenges to be faced. Vocabulary agreement and the multiple meanings concerning XAI. For instance, the significance of words and the relevance of features relating to the same definition is seen here. In visualization techniques, without no continuity under what is called salience tables, salient caps, heat maps, neuron stimulation, attribution, and other methods, this is even more evident as XAI is a relatively young sector, there is no standardized terminology for the group yet. Clarity of thought does not have to be at the same level to give the viewer a understanding [10], even if the analyst is a professional in the field, a policy-maker or a consumer without machine learning expertise.

One of the challenge in XAI is how to set unbiased metrics on what establishes a good description, as stated above. An ability to minimize this subjectivity is to construct scientifically compelling theories inspired by research in human psychology, sociology, or cognitive sciences. Important conclusions to be taken into account when producing an explanatory AI model are emphasized in [12]: A precondition for a worthy clarification is that it not only indicates why a decision X was made by the model, it also explains why decision X was made rather than decision Y. Furthermore, they state that reasons are selective, implying that it is necessary to concentrate exclusively on key causes of a decision-making procedure. The use of counterfactual explanations has also been shown to help the consumer understand a model's decision [65–67].

It seems a favorable way of tackling this problem is to combine connectionist and symbolic paradigms [68–72]. Connectionist approaches, on the one hand, are more effective yet opaque. Symbolic methods, on the other hand, are usually considered to be less effective, although they give greater clarity, thus complying with the following conditions: (a) The ability to appeal to existing rules of reasoning makes it possible for symbolic procedures to be constrictive. (b) Use of a formalized KB, e.g., via ontology, can allow the processing of data directly in a qualitative manner. (c) For connectionist models, being selective is less transparent than for symbolic ones. Recalling that the mental model of the consumer, i.e., the representation of the external world using, among other things, symbols, needs to be influenced by a good explanation, it seems clear that it is reasonable to use the symbolic learning paradigm to provide an explanation.

Therefore, while maintaining or enhancing generic efficiency, neural-symbolic interpretability may provide compelling explanations [73]. As indicated in [16], a fully understandable model must not leave users with scores of explanations, which meaning depends on their context, whereby different explanations can be deduced. A model should be able to develop natural language meanings [68] by incorporating common sense logic and human-understandable characteristics by providing a semantic representation of information. In addition, it is important to make an effort to rigorously formalize measurement procedures before an objective metric has been adopted.

One approach should draw motivation from social sciences, by being dependable in the selection of the evaluation questions and the sample of the population used [74]. The problem that must be tackled by XAI procedures for deep learning is to provide explanations that are open to society, legislators, and as a whole the law. In particular, the transmission of explanations requiring non-technical knowledge will be vital both in coping with uncertainties and in establishing the social right to an explanatory right (not yet available) in the EU General Data Protection Regulation (GDPR) [75].

Another challenge facing explainable AI is security. Nothing was said about XAI-related confidentiality issues. Algorithm property and trade secrets [76] is of paramount importance to keep secure in XAI. Needed attention has not been paid to these principles, however. If the property that makes something secret is confidential, several elements involved in a model will keep this property in the AI sense. For instance, consider a model that has been developed by some organization over years of study in a particular field. The information synthesized in the built model may be considered secret, and only able to offer only input and output access [77], so that it security would not be compromised. The above demonstrates that the theft of data model features is feasible under limited assumptions. In [78], a method that assist in making deep learning models more resilient against exposure to academic property built on a set of non-accessible queries is discussed.

Recent research work shows the need to build XAI instruments that are proficient in clarifying machine learning prototypes while protecting the confidentiality of the prototype in mind. Ideally, inside a model AI, XAI should be able to clarify the details and think on what the model acts on. However, to produce more efficient adversarial contexts attacks aimed at confounding the model and to improve systems to better defend against private content disclosure by using such information, the information exposed by XAI techniques can be used. An attacks of adversarial [79] attempt to control an ML process after knowing what the specific data to be fed to the computer is to give it a certain output. For instance, aggressive actions aim to find the minimal modifications that can be made to input data in order to cause a certain category with regard to a supervised ML classification model.

Though XAI methods can be used to provide more efficient adversarial attacks or to expose the model itself confidential aspects, some latest researchers have exploited the likelihoods of Generative Adversarial Networks (GANs) [80], Variational Self-Encoders [81], and other generative structures to explain data-based decisions. Formerly trained, generative frameworks based on a noise input vector can produce instances of what they have learned, which can be construed as a latent illustration of the data at hand. The possibilities of getting insights and discovering unique patterns relevant to the class to be expected by influencing this latent illustration and analyzing its effect on the performance of the generative model is high.

6 Prospect of XAI in COVID-19 Pandemic

The prospect attached to explainable AI is so enormous, leaving all the challenges, XAI will help us to:

- Understand what occurs as applications of Machine Learning predict things. This, in essence, would continue to accelerate the mainstream acceptance of these schemes. New innovations often take time to be integrated, so it helps enormously if they are learned.
- Gets users more relaxed with technologies and eliminates the magic curtain that appears to envelop AI. Having consumers who value the technologies they utilize is of the extreme significance. Follow the criteria set out to accomplish the AI approach that we are implementing.
- In some crucial fields, including medicines, where AI can have such a great effect and significantly enhance our standard of living, it is important that the prototypes used will be confident without a shadow of doubt. Using a Netflix recommendation algorithm that often generates odd predictions does not have a very significant effect, but in the case of a medical condition, peculiar forecasts may be deadly. Supplying more detail to explain the forecast itself helps consumers to determine whether or not they believe the assumption.
- Explaining models can help their users make greater use of the results of those models, allowing them much more impactful in industry or decision-making. We should always bear in mind that, like every other invention, the aim of AI is to enhance our standard of living so that the more benefits we can gain from it, the better.

6.1 Framework/Steps to Follow in Building XAI

The data is an essential ingredient in the development of an XAI Model. Having the right data will help achieve the goals that we are so concern about. It gives credibility to the system and help to make correct predictions. The second tasks are actual training and building a model that is going to be able to meet the need of the user. It involves a series of training and model evaluation. There is various algorithm to select from using the best algorithm will enhance the model and makes it enjoyable to the user (Fig. 3).

Once we have a successful model, then we can improve the interpretability of the system in order to covey the result of the model to the intending users in a way that they would understand and relate with. At this stage, we introduce a loop to check if the explanation given by the system is good enough or not. If it is not good enough or explanatory enough, then the system goes back to the beginning to start the process again until a satisfactory model is developed.

Fig. 3 Steps to produce an explainable AI

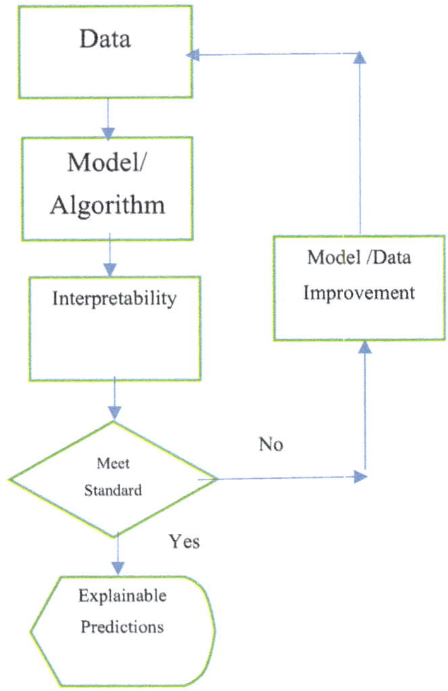

7 Conclusion

This chapter deals with eXplainable Artificial Intelligence (XAI), which has recently been established as the ultimate need for ML methods to be implemented in real-life uses. The study expanded on this subject by first clarifying various principles underlying the explicability of the model and also by illustrating the various reasons that drive the quest for more interpretable ML methods. The chapter also looks at XAI's difficulties and prospects for users, particularly in the case of Covid-19, where decisions need to be taken quickly because it could lead to death if not detected early. XAI's future is bright and once all the problems outlined have been tackled, all will enjoy the advantage.

References

1. Gozes O, Adar MF, Greenspan H, Siegel E (2020) Rapid AI development cycle for the coronavirus pandemic: initial results for automated detection and patient monitoring using deep learning CT image analysis. arXiv:2003.05037
2. Wang L, Wong A (2020) COVID-Net: a tailored deep convolutional neural network design for detection of COVID-19 cases from chest x-ray images. arXiv:2003.09871
3. Zhang Y, Wu H, Liu H, Tong L, DWang M (2019) Mitigating the effect of dataset bias on training deep models for chest X-rays. arXiv:1910.06745

4. Venugopalan J, Hassanzadeh H, Tong L, Wang MD (2020) Multimodal deep learning models for early detection of Alzheimer's disease stage. In: Nature scientific reports. In Press
5. Choudhary A, Tong Li, Zhu Y, Wang MD (2020) Advancing medical imaging informatics by deep learning-based domain adaptation. In Press, In Yearbook of Medical Informatics
6. Russell SJ, Norvig P (2016) Artificial intelligence: a modern approach. Pearson Education Limited, Malaysia
7. Awotunde, JB, Jimoh, RG, Oladipo, ID, Abdulraheem, M, Jimoh, TB, Ajamu, GJ (2021) Big data and data analytics for an enhanced COVID-19 epidemic management. In: Studies in systems, decision and control, vol 358, pp 11–29
8. Goodman B, Flaxman S (2017) European union regulations on algorithmic decision-making and a "right to explanation." AI Mag 38(3):50–57
9. Castelvecchi D (2016) can we open the black box of AI? Nat News 538(7623):20
10. Lipton ZC (2018) The mythos of model interpretability. Queue 16(3)30:31–30:57
11. Preece A, Harborne D, Braines D, Tomsett R, Chakraborty S (2018) Stakeholders in explainable AI
12. Gunning D (2017) Explainable artificial intelligence (xAI), Technical Report, Defense Advanced Research Projects Agency (DARPA)
13. Tjoa E, Guan C (2019) A survey on explainable artificial intelligence (XAI): towards medical XAI
14. Zhu J, Liapis A, Risi S, Bidarra R, Youngblood GM (2018) Explainable AI for designers: a human-centered perspective on mixed-initiative co-creation. In: 2018 IEEE conference on computational intelligence and games (CIG), pp 1–8
15. Dos˜ilovi ć FK, Brc˜i ć M, Hlupi ć N (2018) Explainable artificial intelligence: a survey. In: 41st international convention on information and communication technology, electronics and microelectronics (MIPRO), pp 210–215
16. Miller T (2019) Explanation in artificial intelligence: insights from the social sciences. Artif Intell 267:1–38
17. Doran D, Schulz S, Besold TR (2017) What does explainable AI really mean? a new conceptualization of perspectives
18. Doshi-Velez F, Kim B (2017) Towards a rigorous science of interpretable machine learning
19. Vellido A, Martín-Guerrero JD, Lisboa PJ (2012) Making machine learning models interpretable. In: European symposium on artificial neural networks, computational intelligence and machine learning (ESANN), 12, Citeseer, pp 163–172
20. Walter E (2008) Cambridge advanced learner's dictionary. Cambridge University Press
21. Besnard P, Hunter A (2008) Elements of argumentation. The MIT Press
22. Awotunde, JB, Folorunso, SO, Bhoi, AK, Adebayo, PO, Ijaz, MF (2021) Disease diagnosis system for IoT-Based wearable body sensors with machine learning algorithm. In: Intelligent systems reference library, vol 209, pp 201–222
23. Ronneberger O, Fischer P, Brox T (2015) U-Net: convolutional networks for biomedical image segmentation, CoRR, no 3, pp 234–241, Nov 2015. http://arxiv.org/abs/1505.04597
24. Dzindolet MT, Peterson SA, Pomranky RA, Pierce LG, Beck HP (2003) The role of trust in automation reliance. Int. J. Hum.-Comput. Stud. 58(6):697–718
25. Ayo FE, Awotunde JB, Ogundokun RO, Folorunso SO, Adekunle AO (2020) A decision support system for multi-target disease diagnosis: a bioinformatics approach. Heliyon 6(3):e03657
26. Çiçek O, Abdulkadir A, Lienkamp SS, Brox T, Ronneberger O (2016) 3D u-net: learning dense volumetric segmentation from sparse annotation, CoRR, pp 6–7, Aug 2016. http://arxiv.org/abs/1606.06650
27. Irvin J et al (2019) Chexpert: a large chest radiograph dataset with uncertainty labels and expert comparison, CoRR, pp 4–7, Jul 2019. http://arxiv.org/abs/1901.07031
28. Folorunso, SO, Awotunde, JB, Ayo, FE, Abdullah, KKA (2021) RADIoT: the unifying framework for IoT, radiomics and deep learning modeling. In: Intelligent systems reference library, vol 209, pp 109–128

29. Awotunde, JB, Folorunso, SO, Jimoh, RG, Adeniyi, EA, Abiodun, KM, Ajamu, GJ (2021) Application of artificial intelligence for COVID-19 epidemic: an exploratory study, opportunities, challenges, and future prospects. In: studies in systems, decision and control, vol 358, pp 47–61
30. Kelly CJ, Karthikesalingam A, Suleyman M, Corrado G, King D (2019) Key challenges for delivering clinical impact with artificial intelligence. BMC Med 17(1):195
31. Doshi-Velez F, Kim B (2017) Towards a rigorous science of interpretable machine learning. arXiv:1702.08608, http://arxiv.org/abs/1702.08608
32. Tonekaboni S, Joshi S, McCradden MD, Goldenberg A (2019) What clinicians want: contextualizing explainable machine learning for clinical end use, CoRR, pp 1–12, May 2019. http://arxiv.org/abs/1905.05134
33. Herlocker JL, Konstan JA, Riedl J (2000) Explaining collaborative filtering recommendations. In: Proceedings of the ACM conference on computer supported cooperative work (CSCW). Association Computing Machinery, New York, NY, USA, pp 241–250
34. Lapuschkin S, Wäldchen S, Binder A, Montavon G, Samek W, Müller K-R (2019) Unmasking clever hans predictors and assessing what machines really learn. Nat Commun 10(1):1096
35. Ribeiro MT, Singh S, Guestrin C (2016) Why should i trust you?: Explaining the predictions of any classifier. In: Proceedings of 22nd ACM SIGKDD international conference on knowledge discovery data mining. Association Computing Machinery, New York, NY, USA, Aug 2016, pp 1135–1144
36. Folorunso, SO, Awotunde, JB, Adeboye, NO, Matiluko, OE (2022) Data classification model for COVID-19 pandemic. In: Studies in systems, decision and control, vol 378, pp 93–118
37. Dosilovic FK, Brcic M, Hlupic N (2018) Explainable artificial intelligence: a survey. In: Proceedings of the 41st international convention on information and communication technology, electronics and microelectronics. (MIPRO), May 2018, pp 210–215
38. Gilpin LH, Bau D, Yuan BZ, Bajwa A, Specter M, Kagal L (2018) Explaining explanations: an overview of interpretability of machine learning. In: Proceedings of the IEEE 5th international conference on science and advanced analytics (DSAA), Oct 2018, pp 80–89
39. Arrieta AB et al (2020) Explainable artificial intelligence (XAI): concepts, taxonomies, opportunities and challenges toward responsible AI. Inf Fusion 58:82–115, Jun 2020. 3.1
40. Ai T, Yang Z, Hou H, Zhan C, Chen C, Lv W, Tao Q, Sun Z, Xia L (2020) Correlation of chest CT and RT-PCR testing in coronavirus disease 2019 (COVID-19) in China: a report of 1014 cases. Radiology. https://doi.org/10.1148/radiol.2020200642
41. Luo H, Tang QL, Shang YX, Liang SB, Yang M, Robinson N, Liu JP (2020) Can Chinese medicine be used for prevention of coronavirus disease 2019 (COVID-19)? A review of historical classics, research evidence and current prevention programs. Chin J Integr Med. https://doi.org/10.1007/s11655-020-3192-6
42. Haleem A, Vaishya R, Javaid M, Khan IH (2019) Artificial Intelligence (AI) applications in orthopaedics: an innovative technology to embrace. J Clin Orthop Trauma. https://doi.org/10.1016/j.jcot.2019.06.012
43. Biswas K, Sen P (2020) Space-time dependence of coronavirus (COVID-19) outbreak. arXiv: 2003.03149. Accessed 6 Mar 2020
44. Stebbing J, Phelan A, Griffin I, Tucker C, Oechsle O, Smith D, Richardson P (2020) COVID-19: combining antiviral and anti-inflammatory treatments. Lancet Infect Dis 27 Feb 2020
45. Sohrabi C, Alsafi Z, O'Neill N, Khan M, Kerwan A, Al-Jabir A, Iosifidis C, Agha R (2019) World health organization declares global emergency: a review of the 2019 novel coronavirus (COVID-19). Int J Surg 26 Feb 2020
46. Chen S, Yang J, Yang W, Wang C, Barnighausen T (2020) COVID-19 control in China during mass population movements at New Year. Lancet. https://doi.org/10.1016/S0140-6736(20)30421-9
47. Bobdey S, Ray S (2020) Going viraleCOVID-19 impact assessment: a perspective beyond clinical practice. J Mar Med Soc 22(1):9
48. Gozes O, Frid-Adar M, Greenspan H, Browning PD, Zhang H, Ji W, Bernheim A (2020) Siegel E Rapid AI development cycle for the Coronavirus (COVID-19) pandemic: initial results for

automated detection & patient monitoring using deep learning ct image analysis. arXiv:2003.05037. Accessed 10 Mar 2020
49. Pirouz B, ShaffieeHaghshenas S, ShaffieeHaghshenas S, Piro P (2020) Investigating a serious challenge in the sustainable development process: analysis of confirmed cases of COVID-19 (new type of coronavirus) through a binary classification using artificial intelligence and regression analysis. Sustainability 12(6):2427
50. Awotunde, JB, Jimoh, RG, AbdulRaheem, M, Oladipo, I D, Folorunso, SO, Ajamu, GJ (2022) IoT-based wearable body sensor network for COVID-19 pandemic. In: Studies in systems, decision and control, vol 378, pp 253–275
51. Wan KH, Huang SS, Young A, Lam DS (2020) Precautionary measures needed for ophthalmologists during pandemic of the coronavirus disease 2019 (COVID-19). Acta Ophthalmol 29 Mar 2020
52. Li L, Qin L, Xu Z, Yin Y, Wang X, Kong B, Bai J, Lu Y, Fang Z, Song Q, Cao K (2020) Artificial intelligence distinguishes COVID-19 from community-acquired pneumonia on chest CT. Radiology 19:200905
53. Smeulders AW, Van Ginneken AM (1989) An analysis of pathology knowledge and decision making for the development of artificial intelligence-based consulting systems. Anal Quant Cytol Histol 11(3):154e65
54. Gupta R, Misra A (2020) Contentious issues and evolving concepts in the clinical presentation and management of patients with COVID-19 infection with reference to use of therapeutic and other drugs used in Co-morbid diseases (Hypertension, diabetes etc). Diabet Metab Synd: Clin Res Rev 14(3):251e4
55. Gupta R, Ghosh A, Singh AK, Misra A (2020) Clinical considerations for patients with diabetes in times of COVID-19 epidemic. Diabet Metabc Synd Clin Res Rev 14(3):211e2. 4.0
56. Wang S, Kang B, Ma J, Zeng X, Xiao M, Guo J, Cai M, Yang J, Li Y, Meng X et al (2020) A deep learning algorithm using CT images to screen for corona virus disease (COVID-19). MedRxiv
57. Yildirim, Rajendra Acharya U (2020) Automated detection of COVID-19 cases using deep neural networks with X-ray images. Comput Biol Med 103792
58. Narin A, Kaya C, Pamuk Z (2020) Automatic detection of coronavirus disease using x-ray images and deep convolutional neural networks. arXiv:2003.10849
59. Ayo FE, Ogundokun RO, Awotunde JB, Adebiyi MO, Adeniyi AE (2020) Severe acne skin disease: a fuzzy-based method for diagnosis. Lecture notes in computer science (including subseries Lecture Notes in Artificial Intelligence and Lecture Notes in Bioinformatics), 12254. LNCS, pp 320–334
60. Shi W, Tong L., Zhuang Y, Zhu Y, Wang MD (2020) EXAM: an explainable attention-based model for COVID-19 automatic diagnosis. In: Proceedings of the 11th ACM international conference on bioinformatics, computational biology and health informatics, pp 1–6
61. Karim M, Döhmen T, Rebholz-Schuhmann D, Decker S, Cochez M, Beyan O (2020) Deep-covidexplainer: explainable covid-19 predictions based on chest x-ray images. arXiv:2004.04582
62. Ahsan MM, Gupta KD, Islam MM, Sen S, Rahman M, Hossain MS (2020) Study of different deep learning approach with explainable AI for screening patients with COVID-19 symptoms: Using ct scan and chest x-ray image dataset. arXiv:2007.12525
63. Rudin C (2018) Please stop explaining black box models for high stakes decisions
64. Hoffman RR, Mueller ST, Klein G, Litman J (2018) Metrics for explainable AI: Challenges and prospects
65. Mohseni S, Zarei N, Ragan ED (2018) A multidisciplinary survey and framework for design and evaluation of explainable AI systems
66. Goudet O, Kalainathan D, Caillou P, Guyon I, Lopez-Paz D, Sebag M (2018) Learning functional causal models with generative neural networks. In: Explainable and interpretable models in computer vision and machine learning. Springer, pp 39–80
67. Lopez-Paz D, Nishihara R, Chintala S, Scholkopf B, Bottou L (2017) Discovering causal signals in images. In: Proceedings of the IEEE conference on computer vision and pattern recognition, pp 6979–6987

68. Byrne RMJ (2019) Counterfactuals in explainable artificial intelligence (XAI): evidence from human reasoning, In: Proceedings of the twenty-eighth international joint conference on artificial intelligence, IJCAI-19, pp 6276–6282
69. Bennetot A, Laurent J-L, Chatila R, Díaz-Rodríguez N (2019) Towards explainable neural-symbolic visual reasoning. In: NeSy workshop IJCAI 2019, Macau, China
70. d'Avila Garcez AS, Gori M, Lamb LC, Serafini L, Spranger M, Tran SN (2019) Neural-symbolic computing: an effective methodology for principled integration of machine learning and reasoning
71. Garnelo M, Arulkumaran K, Shanahan M (2016) Towards deep symbolic reinforcement learning
72. Garnelo M, Shanahan M (2019) Reconciling deep learning with symbolic artificial intelligence: representing objects and relations. Curr Opin Behav Sci 29:17–23
73. Marra G, Giannini F, Diligenti M, Gori M (2019) Integrating learning and reasoning with deep logic models
74. Donadello I, Serafini L, Garcez AD (2017) Logic tensor networks for semantic image interpretation. Proc Twenty-Sixth Int Joint Conf Artif Intell IJCA I:1596–1602
75. Kelley K, Clark B, Brown V, Sitzia J (2003) Good practice in the conduct and reporting of survey research. Int J Qual Health Care 15(3):261–266
76. Wachter S, Mittelstadt B, Floridi L (2017) Why a right to explanation of automated decision-making does not exist in the general data protection regulation. Int Data Priv Law 7(2):76–99
77. Adadi A, Berrada M (2018) Peeking inside the black-box: a survey on explainable artificial intelligence (XAI). IEEE Access 6:52138–52160
78. Orekondy T, Schiele B, Fritz M (2018) Knockoffnets: stealing functionality of black-box models
79. Oh SJ, Schiele B, Fritz M (2019) Towards reverse-engineering black-box neural net- works. In: Explainable AI: interpreting, explaining and visualizing deep learning. Springer, pp 121–144
80. Goodfellow I.J, Shlens J, Szegedy C (2014) Explaining and harnessing adversarial ex- amples
81. Pan Z, Yu W, Yi X, Khan A, Yuan F, Zheng Y (2019) Recent progress on generative adversarial networks (gans): A survey. IEEE Access 7:36322–36333
82. Charte D, Charte F, García S, del Jesus MJ, Herrera F (2018) A practical tutorial on autoencoders for nonlinear feature fusion: Taxonomy, models, software and guide- lines. Inf Fus 44:78–96
83. Armstrong GL, MacCannell DR, Taylor J, Carleton HA, Neuhaus EB, Bradbury RS, ... Gwinn M (2019) Pathogen genomics in public health. New England J Med **381**(26):2569–2580
84. Schiavo R, May Leung M, Brown M (2014) Communicating risk and promoting disease mitigation measures in epidemics and emerging disease settings. Pathogens Global Health 108(2):76–94
85. De-Montjoye YA, Gambs S, Blondel V, Canright G, De Cordes N, Deletaille S, Krings G (2018) On the privacy-conscientious use of mobile phone data. Sci Data **5**(1):1–6
86. Ray EL, Reich NG (2018) Prediction of infectious disease epidemics via weighted density ensembles. PLoS Comput Biol **14**(2), e1005910
87. Cottle M, Hoover W, Kanwal S, Kohn M, Strome T, Treister N (2013) Transforming health care through big data strategies for leveraging big data in the health care industry. Institute for Health Technology Transformation. http://ihealthtran.com/big-data-in-healthcare

Chapter 16
Investigation on Some Aspects of Modeling, Forecasting, and Evaluating the Impact of Global Coronavirus Disease 2019

Abhijit Bora and Tulshi Bezboruah

Abstract In this study, a novel methodology is proposed to observe the characteristics and impact of coronavirus disease 2019 over 54 countries and territories in the world. The study of Daily Cumulative Rank for each observed country and territory is evaluated. It defines the daily number of average cases from the day of reporting till the day of the study carried out. The study also highlights the impact of the disease on six regions including Africa, America, Eastern Mediterranean, Europe, South-East Asia, and Western Pacific. The pattern of transmission cases of the disease for a set of 215 countries and territories is evaluated. It is observed that 75 countries and territories are affected by communal transmission. The applicability of the observation is carried out through statistical analysis. A detail discussion on some aspects of novel models for establishing the relationship among different managerial policies of the disease, supporting economically weaker section of the societies, deploying technological support for developing vaccines and treatment strategies, estimating the reliability of service for processing geographically separated data sets, and enhancing the behavioral impacts of international communities through collaborative model is also carried out.

Keywords COVID-19 · Daily cumulative rank · Mortality rate · Transmission case · Reliability of service

1 Introduction

The recent pandemic impact of coronavirus disease 2019 (COVID-19) has shown a critical vision and challenges for the societies worldwide. Based on different clinical and mortality scenario, the World Health Organization (WHO) has declared it as global pandemic and emergency health concern for the countries worldwide. From the first case of COVID-19 patient of Wuhan city, China that had shown the symptoms of severe acute respiratory syndrome of coronavirus 2 (SARS-CoV-2; earlier known

A. Bora (✉) · T. Bezboruah
Department of Electronic & Communication Technology, Gauhati University, Guwahati, India

U. Kose et al. (eds.), *Computational Intelligence for COVID-19 and Future Pandemics*, Disruptive Technologies and Digital Transformations for Society 5.0,
https://doi.org/10.1007/978-981-16-3783-4_16

as 2019-nCoV), the impact of COVID-19 with the number of cases is increasing exponentially and got the stamp of global pandemic [1–4]. As on dated May 11, 2020, total cases of 4,226,864 and deaths of 285,252 were reported [5]. Although the mortality cases of COVID-19 become stabilize in China, but in cases of United States, Italy, Iran, Europe, South Africa, India, etc., it is increasing gradually.

The pharmaceutical and hospital entities worldwide have become an important role for treatment of the COVID-19 infected patients. In the year 2020, such entities have experienced different cases based on patient's health conditions and their respective immunity system [2]. Each evaluation process of the treatment is generating different data sets worldwide that comprises important attributes, primarily required for pandemic disease management and developing clinical process. The characteristic of such data set includes clinical trials with safety and efficacy of chloroquine or related formulation. Cortegiani et al. [6] highlighted that a data set model of COVID-19 clinical cases primarily comprises patient population, intervention group, comparison group, primary outcome, study design, country, symptoms and contacts with suspected or confirmed cases [6]. However, based on isolated data sets in contrast to the clinical process of worldwide entities and results, accessing such data set through a collaborative measure has become an important concern from the perspective of clinical experts, patients, researchers, software practitioners as well as general societies.

1.1 Global Pandemic in Different Regions of the World

Globally as on May 22, 2020, the cumulative case of COVID-19 is observed to be 4,993,470 followed by 327,738 death cases [7]. The impact of COVID-19 on six regions has been observed including Africa, America, Eastern Mediterranean, Europe, South-East Asia, and Western Pacific. The highest impact of COVID-19 is observed in America with a cumulative case of 2,220,267. Out of that, a death record of 131,605 cases for COVID-19 is observed. For European region, the cumulative case of 1,966,244 and death case of 171,327 is observed. For Eastern Mediterranean region, the cumulative case of 389,588 and death case of 10,648 is observed. For South-East Asia region, the cumulative case of 173,159 and death case of 5347 is observed. For Western Pacific region, the cumulative case of 171,748 and death case of 6817 is observed. For Africa region, the cumulative case of 71,752 and death case of 1981 is observed. As such, the highest mortality rate (M_R) is observed to be 8.71% for European region, followed by 5.92% M_R for American region, followed by 3.96% M_R for Western Pacific region, followed by 3.08% M_R for South-East Asia region, followed by 2.76% M_R for Africa region, and 2.73% M_R for Eastern Mediterranean region. The characteristic study of 54 countries and territories is shown in Table 1.

The transmission of COVID-19 cases is evaluated based on the data set reported by WHO [7]. The data set can be classified into different categories including sporadic cases, clusters of cases, community transmission, pending, and no cases. The histogram of the recorded cases for transmission is shown in Fig. 1. It is observed

Table 1 Characteristics of 54 regions with COVID-19 cases recorded till May 22, 2020 [7]

Country/territory	Cumulative case no	Cumulative death no	M_R	Daily cumulative rank (DCR)
Asia				
Singapore	29,812.00	23.00	0.08	250.46
Republic of Korea	11,142.00	264.00	2.37	90.60
Japan	16,513.00	796.00	4.82	129.00
Australia	7081.00	100.00	1.41	60.12
Thailand	3037.00	56.00	1.84	23.33
Philippines	13,434.00	846.00	6.30	118.88
Malaysia	7059.00	114.00	1.61	60.44
Sri Lanka	1055.00	9.00	0.85	12.22
Cambodia	123.00	0.00	0.00	1.42
India	118,447.00	3583.00	3.02	1048.20
Vietnam	324.00	0.00	0.00	2.71
Nepal	487.00	3.00	0.62	5.64
China	84,520.00	4645.00	5.50	591.05
New Zealand	1154.00	21.00	1.82	13.74
America				
Brazil	291,579.00	18,859.00	6.47	3430.34
Canada	80,555.00	6062.00	7.53	692.45
USA	1,525,186.00	91,527.00	6.00	12,594.93
Europe				
Italy	228,006.00	32,486.00	14.25	2035.76
UK	250,912.00	36,042.00	14.36	2257.03
Germany	177,212.00	8174.00	4.61	1540.67
Switzerland	30,611.00	1637.00	5.35	355.93
France	141,590.00	28,164.00	19.89	1200.23
Spain	233,037.00	27,940.00	11.99	2098.11
Israel	16,690.00	279.00	1.67	185.36
Sweden	32,172.00	3871.00	12.03	290.08
Croatia	2237.00	97.00	4.34	26.01
Norway	8268.00	234.00	2.83	97.27
Romania	17,585.00	1151.00	6.55	206.88
Austria	16,332.00	633.00	3.88	189.92
Finland	6493.00	306.00	4.71	58.20
Greece	2853.00	168.00	5.89	33.56
Georgia	723.00	12.00	1.66	8.51

(continued)

Table 1 (continued)

Country/territory	Cumulative case no	Cumulative death no	M_R	Daily cumulative rank (DCR)
Russian Federation	326,448.00	3249.00	1.00	2925.91
Estonia	1800.00	64.00	3.56	21.18
Denmark	11,182.00	561.00	5.02	131.55
North Macedonia	1898.00	111.00	5.85	22.33
Netherland	44,700.00	5775.00	12.92	532.14
Lithuania	1594.00	61.00	3.83	18.98
Belarus	34,303.00	190.00	0.55	408.37
Belgium	56,235.00	9186.00	16.34	520.69
Eastern Mediterranean region				
United Arab Emirates	26,898.00	237.00	0.88	235.64
Kuwait	18,609.00	129.00	0.69	211.47
Oman	6370.00	32.00	0.50	73.22
Pakistan	50,694.00	10,367.00	20.45	596.40
Iran	129,341.00	7249.00	5.60	1405.88
Bahrain	8174.00	12.00	0.15	93.95
Iraq	3877.00	140.00	3.61	44.56
Egypt	15,003.00	696.00	4.64	154.22
Afghanistan	9216.00	205.00	2.22	105.93
Lebanon	1024.00	26.00	2.54	11.39
Africa				
South Africa	19,137.00	369.00	1.93	222.52
Algeria	7728.00	575.00	7.44	89.83
Nigeria	7016.00	211.00	3.01	83.52

DCR (cumulative cases/number of days between the first reported case and May 22, 2020)

that, for 75 countries and territories, the COVID-19 is due to community transmission. However, the 86 cases of clusters can be further categorized based on experiences, geographic locations, time frame, etc. As such, the communal transmission of the COVID-19 has marked as a serious issue for the world.

1.2 Related Work

Many authors have emphasized the aspects of different entities for COVID-19 global pandemic management [8–10. Cortegiani et al. [6] had discussed that the clinical test

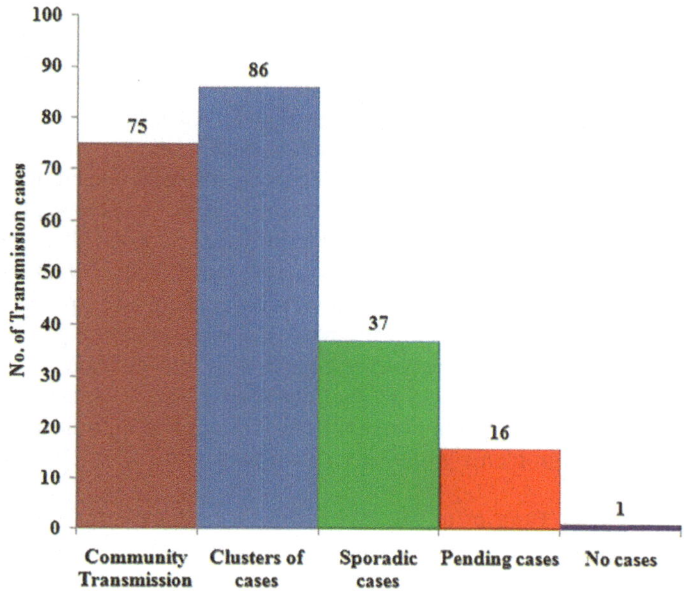

Fig. 1 Histogram of transmission cases for the observed records of 215 countries and territories

cases of COVID-19 are generating regional based different data sets based on blood samples, symptoms, population, histopathology, clinical images, and affected organs [6]. The clinical test reports of drug repositioning, antibody agents, anti malarial drug, chemical structure, and affect of hydroxyl chloroquine are observed to be important resources for development of vaccines and antibody injections [4, 5, 11]. Dhouha et al. [12] had discussed the importance of image data set for medical education and investigation [12]. Benjamin et al. [13] had discussed the importance of deploying 'ChatBot' software module for processing interactive data Mahmoud et al. [14] had elaborated the importance of clinical features for skin, oral cavity, eye and ear, upper and lower airway, cardiovascular, gut, urinary tract, nervous system, skeleton, and muscles. Jorge et al. [15] had proposed an empirical model to study the probability of spread of the pandemic disease COVID-19 in societies and native communities [15]. Espinosa-Serna [16] discussed the importance of bio safety measures, management strategies, mechanical ventilation management, patient route management, patient-wise ventilation mode to be used and the decision-making system based on available data reports [16]. Gilmiyarova et al. [17] had discussed a diagnostic tool to study the clinical remarks of oral fluid [17]. An empirical model was introduced to highlight the important characteristics of data set that can be categorized by blood groups. Lai et al. [4] had discussed the cumulative index parameters of COVID-19 over the societies [18]. They had highlighted the importance of infection control management technique to restrict the spread of the diseases. Medhi et al. [19] had introduced a novel method to study the response of data set processing system through service-oriented computing (SOC) [19]. Lee et al. [20] had emphasized the importance of the test case results of COVID-19 patients and the outcomes of clinical treatment applied [20]. The authors had discussed the pros and cons of the success and failure of the treatment for COVID-19 infected patients. Bora et al. [21] had discussed the importance of web service (WS) model for deploying it to generate clinical remarks. They had stated that the reliability and performance metric can play the key role for implementing WS-based system for medical industries. Lai et al. [4] had introduced a statistical model to evaluate the daily cumulative factors for COVID-19 infected patients in different countries [22]. The analysis was carried out by preparing a data set of cumulative cases, number of death, mortality rate, incidence per population, death per population, days since last reported and local transmission.

In the year 2020, Bezboruah et al. had elaborated the challenges of artificial intelligence for different problem domains [23]. The applicability of the algorithm and classification of problems while solving through artificial intelligence is also discussed.

In the same year, Bora et al. discussed the methodology for processing the COVID-19 data set through the paradigm of software service [24]. The applicability, reliability, and feasibility of such deployment is also discussed with experimental arrangement.

In the same year, Olorunshola et al. proposed a study based on software classification to evaluate dataset [25]. The study emphasized the possibility of reducing stress and saving time while solving problems. The novel defective model can predict the overall assessment of the software while processing the information of particular domain.

In the same year, Bora et al. proposed a novel model for processing pandemic disease through service oriented computing [26]. The performance issues of such deployment is also discussed to reflect the capability of the data processing during high of stress of usage over the network.

In the research community of medical practitioners, bio-medical engineering units and pharmaceutical industries, the accessing of geographically separated data sets is getting popularity, as it can enhance the clinical study for collaborative fight against COVID-19 [19–21, 27, 28]. As such, the deployment of software as a service (SaaS), Artificial Intelligence modules, machine learning algorithms, image processing, data science algorithms can play the vital role for providing collaborative report from different regions. However, the reliability of such deployment is also a matter of concern.

In this study, the author introduces a novel model for establishing the relationship among different policies and managing multiple entities for fighting against COVID-19. A novel model for supporting economically weaker section of the societies is also demonstrated and discussed. The study also discusses a model to involve technological support for developing vaccines and treatment strategies. A novel reliability evaluation framework is also introduced that can be deployed for evaluating the reliability metrics of such service for processing geographically separated data sets based on COVID-19 disease. The collaborative model is also elaborated that can grow the international relations for enhancing the behavioral impacts of international communities.

1.3 Computational Intelligence

Computational intelligence (CI) provides different methodology for processing information. The CI helps in sensing, learning, and thinking as like human processing information. Numerical approaches are applied for simulating like human brain. It also helps in optimizing problems while solving complexity of the information. Such deployment helps in increasing capability of the machine in learning and thinking while solving specific problem that mankind deal with.

In the domain of data mining, medical service, logistic areas, the deployment of Artificial Intelligence and its algorithm is observed to be working since 1990s [29–33]. Many researchers are giving their contribution while increasing the computational capacity of the machine while solving problem through scientific procedures and computational methodology [31–34]. Such methodology reviews the environment and evaluates the computational intelligence algorithms while solving the problem with increased success rate. The complexity analysis of such methodology depends upon the goal of the problem-solving algorithm. Different explicit and implicit goals can be set for the problem-solving. The model-based system can be developed that can replicate the real world event. The evolutionary method of model can help in predicting the outcome of the problem-solving. The training data set can play the vital role while processing information through such model [29–32].

Many researchers are working in classification of issues that a society faces. The benchmark for such work is set to achieve the goal [29–31]. The artificial intelligence contains a set instruction which can be simple or complex one while solving problems. It has the capability of learning and solving the problem from the existing datasets. Even, the algorithm can develop new algorithm for the specific goal during evaluation of the dataset. Mathematical function, decision tree, and classification of the logic can be deployed for such scenario. Hypothesis and mapping of dataset can also play an important role while solving problems. It also helps in finding the feasible solution from a set of possible solution for a specific problem [35–39].

2 Materials and Methods for Global Fight to Handle Pandemics, Poverty, Famine, Economic Impact, Vaccine Development, and Treatment

Many prominent authors have carried out the study over different models to establish the relation among different policies and managing multiple entities for fighting against COVID-19 [21, 40–45, 7]. A block diagram for the overall management of the COVID-19 cases is shown in Fig. 2. The block diagram shows the key entities that are playing important role for handling COVID-19 cases. The deployment of such model needs a management of database or registry. The records of such database will be prepared primarily through the processing of person entity. The object of person entity will go through primary clinical test for COVID-19. If positive results occur, the person will be assigned a unique identification number (ID). The person with that ID will go through clinical treatment as per symptoms. A parallel mechanism is also to be developed so that the database of the other person who was found to be in touch with that infected person ID can be observed. The database of such records needs to be processed and followed up properly through tele caller or field worker people. The quarantine mechanism policy is also to be developed for the in touch person of the infected cases. The clinical test for the quarantined person is to be carried out as per clinical policy. If positive COVID-19 infection is observed, then the ID is also given to them. On the contrary, the other person can be released after evaluation.

The advantage of the model is that the relationship among different entities can be established through relational database management system (RDBMS). One can easily track down the cases whenever required. However, the deployment of such methodology needs a technological and financial support. Such deployment can help different officials of hospital, police, municipality, social welfare, and other non-government organizations. Registrywise sanitization of the areas can be carried out through joint venture of different departments. However, the co-operation of

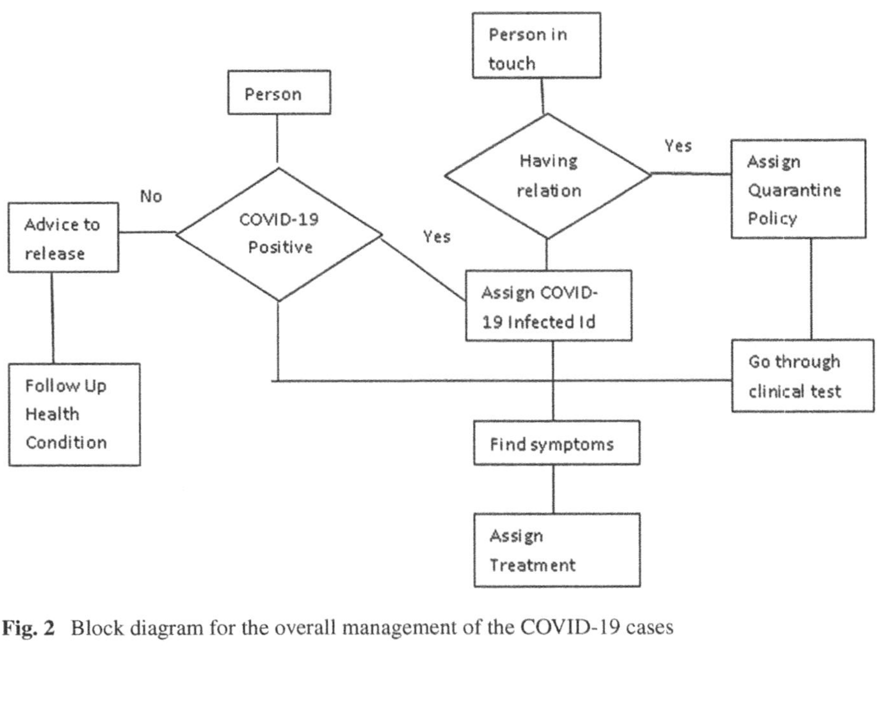

Fig. 2 Block diagram for the overall management of the COVID-19 cases

Government and localities is highly important in this case. The infection of COVID-19 can also impact the economic scenario of a country.

However, by considering the market risk of investment through mutual fund and share stock, the middle and lower class people can suffer a lot during this pandemic. As such, a novel model for supporting economically weaker section of the societies through polling fund from different sources is shown in Fig. 3. The co-operation of different societal entities can play a vital role for such situation. Figure 3 shows that the person of society A collects funds from his known resources in level 1. The resources may be a person or an organization from other society in level 2. Likewise, the person 2 from society B and person 3 from society C can pool fund from their respective sources. As such, chain of hierarchical tree structure can be formed in different areas of the world. Such pooling of fund can be used for managing economically weaker section during the epidemic situation. The advantage of the model is that, only reliable sources will be taken for managing reliable weaker section of the society. Deployment of such chain-based hierarchical tree can be made through societal level or through the help of community-based model of Internet technologies like Facebook, Whatsapp, etc. The support and involvement of technological aspects is highly demanded in such epidemic scenario. The development of vaccines and treatment strategy is based on experiences of COVID-19 cases over the world. Each case is generating different data sets for the medical units. Such dataset is the key source for developing the vaccines. However, compiling and processing such a huge data sets though network-based protocol is an important concern. In the research community, many prominent authors have discussed the importance and deployment of algorithms for development of vaccines and clinical strategy [13, 19, 21]. In Fig. 4, a model is discussed to involve technological support for developing vaccines and treatment strategies. The algorithm of data processing through the business logic of artificial intelligence, image processing, big data analysis, web service, network security policies, and data science can play the vital role as helping hand for the development of vaccines and treatment strategies in an efficient manner. The services

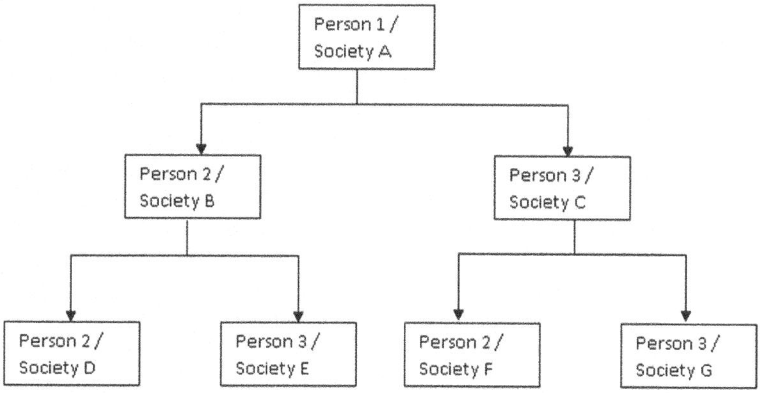

Fig. 3 A model for helping weaker section through pooling fund or contribution strategy

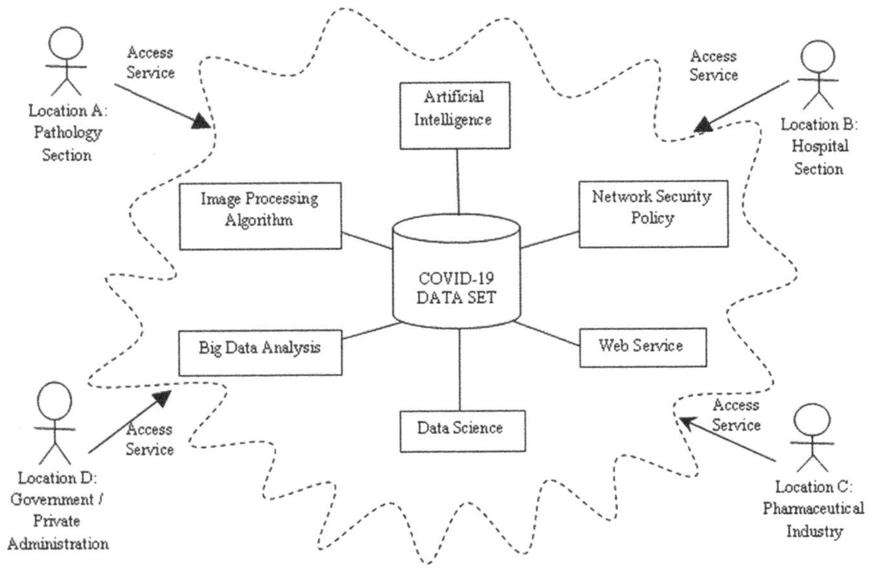

Fig. 4 A model for involving technological support while developing vaccines and treatment strategies

of such collaborative deployment can support the functional management of COVID-19 through the entity of pathology and hospital section, industrial units along with government and private administrations. With the help of technological deployment, the entities in different geographically separated locations can resolve the severe impact of COVID-19 in the societies.

However, the reliability of such technological processing is to be evaluated through a mechanism. The time between failure and fault count model can be followed while processing the failure rate of such deployment. Many authors have discussed the importance of evaluating technological support for processing data sets [21, 28, 46]. In Fig. 5, a novel cycle of reliability evaluation framework is shown. It can be deployed for evaluating the reliability metrics of such service for processing geographically separated data sets based on COVID-19 disease. The service is deployed for processing the disease data set. The test case environment and observing parameters can be set. The system is executed till failure record occurs. If failure record occurs, the data sample of size 'n' can be taken. The distribution of the data sample 'n' is observed to evaluate thenature of distribution of such processing. As such, the reliability and assessment can be carried out for the service. The advantage of the model is that, it shows the reliability through rigorous tests, as it is based on real case testing environment. However, the evaluation needs time period as the outcome is based on repetitive assessment of the clinical testing strategy. In this study, to demonstrate the reliability of such a system, a prototype experimental setup is followed. The setup is discussed elsewhere at [21]. The system is observed against

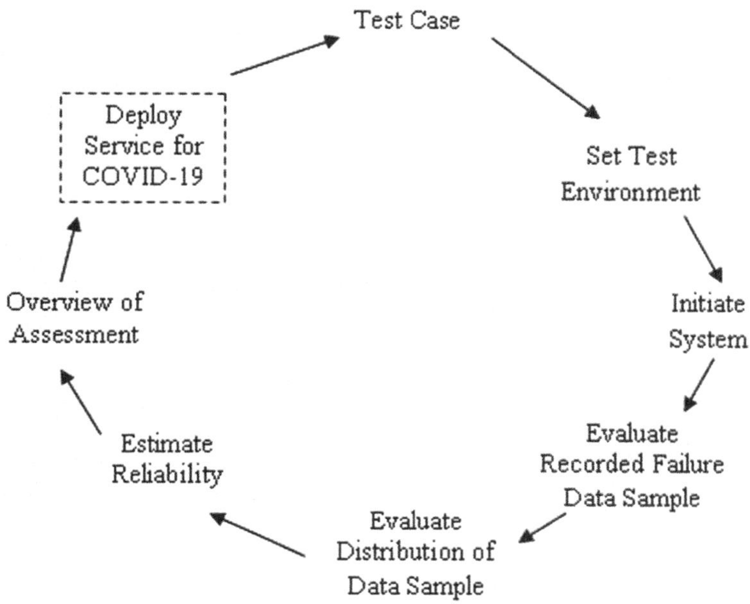

Fig. 5 Cycle of the reliability evaluation model for evaluating the metrics of service for processing geographically separated data sets based on COVID-19 disease

increasing load of consumers. The failure rate of the service execution is observed through Mercurry Loadrunner 8.1 [47]. The system is not showing any failure for the stress of 1700 simultaneous consumers. However, beyond that, it shows failure records. The reliability for each service execution is evaluated through Eq. (1) [28, 46, 48].

$$R = e^{ft} \tag{1}$$

The normal probability plot of the recorded failure rate and reliability for 1800 consumers is observed to evaluate the distribution of such execution. It is shown in Figs. 6 and 7. The normal probability plot reveals that the failure rate and reliability for the stress of 1800 consumers is normal in distribution.

3 Development of International Relation and Communal Behavior

The knowledge sharing can play the key role for establishing the international relation which in turn can enhance the behavioral aspects of international communities. It is observed that the COVID-19 pandemic cases are generating different data sets

Fig. 6 Normal probability plot of recorded failure rate against 1800 simultaneous consumers

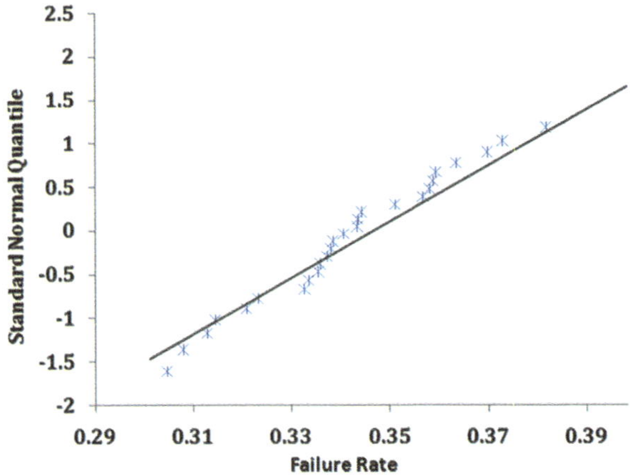

Fig. 7 Normal probability plot of evaluated reliability against 1800 simultaneous consumers

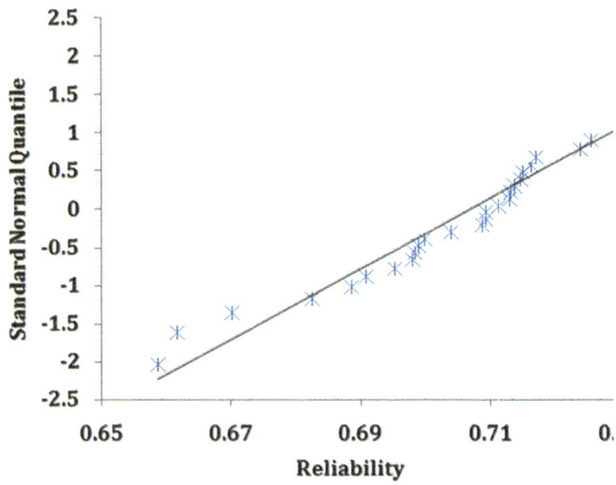

in different geographically separated areas. The data sets are the set of experience and observation of clinical and managerial reports of COVID-19 infected people. The development of vaccine needs a prolonged study over such data set. As such, the sharing of data set among clinical communities can enhance the development of societal infrastructure. A collaborative model for enhancing the behavioral impacts of international community and relations is shown in Fig. 8. Here, the medical practitioner accesses the data set of location A, which in turn can access the data set at location B and C. A hierarchical communication among different locations can be established through technological standards such as remote method invocation, service oriented architecture, extensible markup language (XML), and web service (WS).

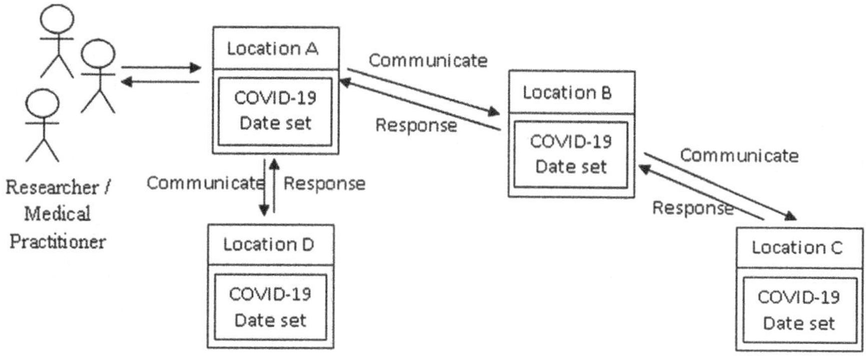

Fig. 8 The collaborative model for enhancing the behavioral impacts of international community and relations

The deployment of computational intelligence and artificial intelligence algorithm while processing the social networking places of mankind can predict a better communication. The mining of information based on dataset and establishing relationship through computational intelligence can deliver better outcome.

As such, the collaborative communication among the entities in different geographical areas can help to fight against COVID-19.

4 Overall Assessment

The study reveals that as on dated May 22, 2020, the M_R is recorded to be 6.56% for the global record of 4,993,470 cases of COVID-19. Out of six regions, the highest M_R is recorded to be 8.71% for European region followed by 5.92, 3.96, 3.08, 2.76, and 2.73% for America, Western Pacific, South-East Asia, Africa, and Eastern Mediterranean region, respectively. In this study, the DCR of 54 countries/territories is observed and evaluated for each. The highest DCR is recorded to be 12,594.93 for United States of America. However, from the histogram, it is observed that out of 215 countries/territories, the communal transmission of COVID-19 is suffered by 75 countries. The novel model for establishing the relationship among different policies and entities for fighting against COVID-19 is discussed. The proposed model for supporting economically weaker section of the societies is also demonstrated and discussed. The study had shown the possibility of involving technological support for developing vaccines and treatment strategies. The reliability evaluation framework for assessing such technological mean is observed to be viable. The strong reliability is observed for 1700 concurrent consumers. However, it degrades beyond that. The normality of the recorded failure rate of 1800 consumers is evaluated and observed to be normal in distribution. A methodology of knowledge sharing is shown, through which the possibility of enhancing the behavioral impacts of international communities can be achieved.

5 Conclusion

From December 31, 2019 to May 22, 2020, the COVID-19 has affected 4,993,470 patients in the globe and caused 327,738 deaths. The impact of COVID-19 is observed for the six regions including Africa, America, Eastern Mediterranean, Europe, South-East Asia, and Western Pacific. The highest impact of COVID-19 is observed in America with a cumulative case of 2,220,267 patients and cause of 131,605 death cases is recorded. However, the highest M_R is observed to be 8.71% for European region. In this study, some aspects of deploying modeling strategy is proposed for global pandemic management. The methodology for integrating collaborative hand for developing vaccines and treatment strategy is discussed. The viability and reliability of such model is observed through normality test. The study can help the medical practitioner and researcher for getting an in-depth idea about the global impact of COVID-19 with the collaborative managerial policies for fighting against COVID-19. However, an in-depth study is necessary to develop a monitoring mechanism that can be deployed to observe the COVID-19 cases. As such, it will be able to reveal the new possibilities of COVID-19 impact in the society.

References

1. Munster VJ, Koopmans M, van Doremalen N, van Riel D, de Wit E (2020) A novel corona virus emerging in China—key questions for impact assessment. N Engl J Med 382(8):692–694
2. Zhu N, Zhang D, Wang W, Li X, Yang B, Song J et al (2020) A novel corona virus from patients with pneumonia in China, 2019. N Engl J Med 382(8):727–733
3. Guan WJ, Ni ZY, Hu Y, Liang WH, Ou CQ, He JX et al (2020) Clinical characteristics of corona virus disease 2019 in China. N Engl J Med. https://doi.org/10.1056/NEJMoa2002032
4. Lai CC, Shih TP, Ko WC, Tang HJ, Hsueh PR (2020) Severe acute respiratory syndrome corona virus 2 (SARS-CoV -2) and corona virus disease-2019 (COVID-19): the epidemic and the challenges. Int J Antimicrob Agents 2020, Feb 17:105924. https://doi.org/10.1016/j.ijantimicag.2020.105924
5. Global COVID-19 Status. www.worldmeters.info/coronavirus/. Accessed 11 Mar 2020
6. Cortegiani A, Ingoglia G, Ippolito M et al (2020) A systematic review on the efficacy and safety of chloroquine for the treatment of COVID-19. J Crit Care. https://doi.org/10.1016/jjcrc.2020.03.005
7. World Health Organization (WHO) Corona virus disease 2019 (COVD-19). Situation report–123, Geneva, Switzerland: WHO; 2020.Data as reported by 10AM CEST 22 May 2020. http://www.who.int
8. Zhou P, Yang XL, WangXG HuB, ZhangL ZhangW et al (2020) Apneumonia outbreak associated with a new corona virus of probable bat origin. Nature 579:270–273
9. Mullard A (2012) Drug re purposing programmes get lift off. NatRev Drug Discov 11:505–506
10. Colson P, Rolain JM, Raoult D (2020) Chloroquine for the 2019 novel corona virus SARS-CoV-2. Int J Anti microb Agents 55:105923

11. Wang LH, Wang Y, Ye D, Liu Q (2020) A review of the 2019 Novel Corona virus (COVID-19) based on current evidence. Int J Antimicrob Agents. https://doi.org/10.1016/j.ijantimicag.2020.105948
12. Dhouha B, Ghofrane T, Wael F, Mejri A, Lasaad G (2020) Image in medical education: why, when and how to use it? A critical review of the literature. Biomed J Sci Tech Res 26(3). BJSTR. MS.ID.004343
13. Chaix B, Guillemassé A, Nectoux P et al (2020) Vik: a chatbot to support patients with chronic diseases. Biomed J Sci Tech Res 27(1). BJSTR. MS.ID.004453
14. Al Azab M, Safi M, Wang B (2019) ANCA-Associated vasculitis: advancement in pathogenesis, clinical features and management. Biomed J Sci Tech 21(2). BJSTR.MS.ID.003589
15. Jorge HV, Margarita Juarez N, Hector Daniel Molina R, Michiko Amemiya R, Cruz Garcia L et al (2020) Specification of a social intervention model against COVID-19. Biomed J Sci Tech Res 26(3). BJSTR. MS.ID.004366
16. Espinosa-Serna JS (2020) SARS-CoV-2, COVID 19 Useful information. Biomed J Sci & Tech Res 27(2). BJSTR. MS.ID.004484
17. Gilmiyarova F, Kuzmicheva V, Ryskina E, Kolotyeva N, Gusyakova O et al (2020) Saliva as a diagnostic tool for AB0 blood group determination. Biomed J Sci Tech Res 26(5). BJSTR. MS.ID.004424
18. Lai C-C, Wang C-Y, Wang Y-H, Hsueh P-R (2020) Global coronavirus disease 2019: what has daily cumulative index taught us? Int J Antimicrob Agents. https://doi.org/10.1016/j.ijantimicag.2020.106001
19. Medhi S, Bora A, Bezboruah T (2017) Investigations on some aspects of reliability of content based routing SOAP based windows communication foundation services. Int J Inf Retr Res 7(1):17–31. https://doi.org/10.4018/IJIRR.2017010102
20. Lee SH, Sonand H, Peck KR (2020) Can post-exposure prophylaxis for COVID-19 be considered as an outbreak response strategy in long-term care hospitals? Int J Antimicrob Agents. https://doi.org/10.1016/j.ijantimicag.2020.105988
21. Bora A, Bezboruah T (2020) Some aspects of implementation of web services in load balancing cluster-based web server. Int J Inf Retr Res 10(1):48–72
22. Lai C-C, Wang C-Y, Wang Y-H, Hsueh S-C, Ko W-C, Hsuehf P-R (2020) Global epidemiology of corona virus disease 2019(COVID-19): disease incidence, daily cumulative index, mortality, and their association with country health care resources and economic status. Int J Antimicrob Agents 55:105946
23. Bezboruah T, Bora A (2020) Artificial intelligence: the technology, challenges and applications. Trans Mach Learn Artif Intell 8(5):44–51. https://doi.org/10.14738/tmlai.85.8956
24. Olorunshola O, Irhebhude M, Evwiekpaefe A, Ogwueleka F (2020) Evaluation of machine learning classification techniques in predicting software defects. Trans Mach Learn Artif Intell 8(5):01–15. https://doi.org/10.14738/tmlai.85.8733
25. Bora A, Bezboruah T (2020) Evaluating the reliability of PwCOV: a loosely coupled software as a service for COVID-19 data processing system. Int J Eng 33(12):2496–2502. https://doi.org/10.5829/ije.2020.33.12c.09
26. Bora A, Bezboruah T (2021) PwCOV in cluster based web server: an assessment of service oriented computing for COVID-19 disease processing system, data science for COVID-19, Elsevier, Chapter 10, Edition 1, 2021 (In press)
27. Thompson R (2020) Pandemic potential of 2019-nCoV. Lancet Infect Dis 20(3):280

28. Rahmani M, Azadmanesh A, Siy H (2014) Architectural reliability analysis of framework-intensive applications: a web service case study. J Syst Softw 94:186–201
29. Bertini M, Del BA, Torniai C (2006) Automatic annotation and semantic retrieval of video sequences using multimedia ontologies. In: Proceedings of the 14th ACM international conference on Multimedia. 14th ACM international conference on Multimedia. ACM, Santa Barbara, pp 679–682
30. Mcgarry K (2005) A survey of interestingness measures for knowledge discovery. Knowl Eng Rev 20(1):39–61. https://doi.org/10.1017/S0269888905000408
31. Jordan MI, Mitchell TM (2015) Machine learning: trends, perspectives, and prospects. Science 349(6245):255–260. https://doi.org/10.1126/science.aaa8415
32. Cambria E, White B (2014) Jumping NLP curves: a review of natural language processing research. IEEE Comput Intell Mag 9(2):48–57. https://doi.org/10.1109/MCI.2014.2307227
33. Cadena C, Carlone L, Carrillo H, Latif Y, Scaramuzza D, Neira J, Reid I, Leonard JJ (2016) Past, present, and future of simultaneous localization and mapping: toward the robust-perception age. In: IEEE transactions on robotics, vol 32, Issue 6, pp 1309–1332. https://doi.org/10.1109/TRO.2016.2624754
34. Roberts J (2016) Thinking machines: the search for artificial intelligence. Distillations. 2(2):14–23
35. Schoenick C, Clark P, Tafjord O, Turney P, Etzioni O (2017) Moving beyond the turing test with the Allen AI science challenge. Commun ACM 60(9):60–64. https://doi.org/10.1145/3122814
36. Scassellati B (2002) Theory of mind for a humanoid robot. Auton Robot 12(1):13–24. https://doi.org/10.1023/A:1013298507114
37. Poria S, Cambria E, Bajpai R, Hussain A (2017) A review of affective computing: from unimodal analysis to multimodal fusion. Inf Fus 37:98–125. doi:https://doi.org/10.1016/j.inffus.2017.02.003
38. Chen M, Zalzala A (1997) A genetic approach to motion planning for redundant mobile manipulator systems considering safety and configuration. J Robot Syst 14(7):529–544
39. Najmaei N, Kermani MR (2011) Applications of artificial intelligence in safe human-robot interactions. IEEE Trans Syst Man Cybern-Part B 41(2):448–459
40. Leite H, Hodgkinson IR, Gruber T (2020) New development: healing at a distance-telemedicine and COVID-19. Public Money & Management. https://doi.org/10.1080/09540962.20201748855
41. Wang LH, Wang Y, Ye D, Liu Q (2020) A review of the 2019 Novel Coronavirus (COVID-19) based on current evidence. Int J Antimicrob Agents. https://doi.org/10.1016/j.ijantimicag.2020.105948
42. Wang CJ, Ng CY, Brook RH (2020) Response toCOVID-19 in Taiwan: big data analytics, New technology, and proactive testing. JAMA. https://doi.org/10.1001/jama.2020.3151
43. Ji YMZ, Peppelenbosch MP, Pan Q (2020) Potential association betweenCOVID-19 mortality and health-care resource availability. Lancet Glob Health. https://doi.org/10.1016/S2214-109X(20)30068-1
44. Wu Z, McGoogan JM (2019) Characteristics of and important lessons from the corona virus disease 2019(COVID-19) outbreak in China. Summary of a report of 72314 cases from the Chinese center for disease control and prevention. JAMA 24 Feb 2020. https://doi.org/10.1001/jama.2020.2648
45. Lozano R (2016) Measuring performance on the healthcare access and quality index for 195 countries and territories and selected sub national locations: a systematic analysis from the Global Burden of Disease Study 2016, GBD2016 Healthcare Access Qual Collab 391:2236–2271. https://doi.org/10.1016/S0140-6736(18)30994-2
46. Medhi S, Bora A, Bezbaruah T (2017) Investigations on some aspects of reliability of content based routing SOAP based windows communication foundation services. Int J Inf Retr Res 7(1):17–31
47. Mercury Load Runner. https://qageek.files.wordpress.com/2007/05/loadrunner_tutorial.pdf. Accessed 29 Mar 2020

48. Shooman ML (2002) Reliability of Computer Systems and Networks: Fault Tolerance, Analysis, and Design. John Wiley & Sons. https://doi.org/10.1002/047122460X

Chapter 17
"Horizontal World Treatment" and a Smart City Oriented Project for Total Recovery from COVID Outbreak: CORONAPOLISES

Mustafa Hulki Cevizoğlu

Abstract In this chapter, a general view on social and mental dimensions of the coronavirus epidemic was firstly provided by analyzing it in the context of Turkey. It is important that the appearance of this virus has caused many occurrences of both social and mental effects, and these should be discussed in detail, by looking at to the past and psychological roots of the humankind. After analyzing the social and mental dimensions, a solution of CORONAPOLIS was explained, and it was connected with Artificial Intelligence/Computational Intelligence aspects by considering Internet of Things (IoT) and smart city perspective. The chapter is a contribution for future pandemic city developments, which are considering societal and mental impacts of such massive health issues.

Keywords COVID-19 · Horizontal World Treatment · Smart City · CORONAPOLIS · Computational Intelligence · Internet of Things

1 Introduction

As can be seen in also other countries in the world, people in some regions of Turkey looted shops and a **looting culture** has revealed itself eventually. **Panic was infected before the virus.**

Behaviors are *"infectious"* when in a group (audience). Freud also introduces the concept of *mass psyche* (psyche: spirit, spirituality—Dr. Hulki Cevizoglu). In the *Heritage of Freud*, René Kaës quotes the following: If there was no assumption that accepts the *community psyche* that ensures the continuity of the psychic life of human, collective psychology and psychology of the peoples could not exist. *If the psychic phenomena of one generation were not transferred to the next*, did not continue with the next, every generation would have to start all over again to learn life. In his book: *The Group Mind*, McDougall describes the spiritual behavior of an unorganized group

M. H. Cevizoğlu (✉)
Nisantasi University, Istanbul, Turkey
e-mail: hulkicevizoglu@cevizkabugu.com.tr

U. Kose et al. (eds.), *Computational Intelligence for COVID-19 and Future Pandemics*, Disruptive Technologies and Digital Transformations for Society 5.0,
https://doi.org/10.1007/978-981-16-3783-4_17

Fig. 1 As Saramago expresses, the Blindness is "not like a sudden swell of sea, which takes everything with it, but like thousands of little streams, which suddenly invaded the whole after sneaking into the ground slowly…" [1] (The figure is an open-source material)

as extremely emotional, impulsive, violent, inconsistent, unstable, and undecided. Unlike Freud and McDougall, mass is not necessarily a "chaotic and pathological structure" in modern social psychology. Turner and Killian first introduced that as the *"Emerging Norm Theory."* According to the related theory, collective behavior (mass behavior) is limited to the norms. The idea that the community (mass) psyche, that is, the spiritual structure of society, is *transferred* to the next generation belongs to Freud. This spiritual (psychic) structure inherited from the previous generation is stimulated by some events of individual life in order to be impressive. In order to explain that situation, Freud developed the concept of *epigenesis* (not hereditary but not genetic through DNA) in 1914 [2] (Fig. 1).

There is the fact of irrational enemy creation in group behaviors. **In the days of the corona virus outbreak, that enemy has been the fear of hunger and starvation.** The fear of starvation was an enemy and the enemy was attacked accordingly. The looting of the shops (albeit limited time) meant the defeat of the enemy in the mind. In the words of Saramago [1], "fear without a good mentor" formed the source of distrust.

Many comments have been made on the spread of security concerns in areas where the fear of danger has not spread. In order to explain these states, stereotypical words that mean **"we live in an aggressive society"** are frequently used [3].

According to Furedi, **"fundamental existential concerns"** have led to the institutionalization of security in many areas of social life [3].

According to Jung, the spirit of primitive human is essentially collective, and therefore a large part is unconscious. Trotter, who is inspired from Freud, attributes the mental phenomena defined in the groups to an innate **"swarm instinct"** in humans as in other animal species. In the *Mass Psychology and Ego Analysis*, Freud argues that "The mass looks like the resurrection of the primitive human swarm." About half a century after Jung and Freud, McLuhan, an important communications theorist, also describes human as the **"tribal human of the global village"** [2].

The deterioration of social norms (in other words irregularity) constitutes an **"anomy"** in Durkheim's words. According to that the founding father of sociology, anomy is one of the collapses and suicide causes of the social psychology. The *"not touching each other,"* which one of the recommendations given to prevent from transmission of the corona virus, meant the destruction of one of the most basic existential needs of human.

According to the **phylogenetic theory**, the first human is a social creature, who has built its socialization by touching, since the existence. **Touching means to being exist** and keeps our existence alive!

1.1 "Fear of Norm" ® and "Dictator Societies" ®

In Turkey, on March 22, 2020, Sunday, **the curfew** has started for the people, who are over 65 years old and with chronic diseases (diabetes, heart, blood pressure), and the penalty of 3500 Turkish Liras (around 500 USD) was announced. The situation in the *Blindness* was happening.

The Evidence that the general mood was constantly deteriorating was that even the government changed its strategy twice within a week. Initially, it was believed that it would be possible to keep the evil that happens to us within certain limits by confining the blind and those who carry the blindness germ to special places, such as the mental hospital we are in. Then they began to think that the increasing incidence of blindness should not be tampered with, encouraging families to keep their blinds in their homes and not let them go out into the street, and that people who believe the white disaster is transmitted by eye contact should not be tampered with [4].

Everyone who had blindness, as well as those who had ***physical contact*** with them, would be gathered and isolated, so that the disease would not be transmitted to others later [4].

However, despite this, the curfew was not completely fulfilled. Because many people broke this ban. There were those who went out, ignoring the risk of major epidemic contamination in terms of their own health and ignoring the physical (safety) distance measures. We can (also) make the social and cultural anthropological explanation of this with the concept of *"maladaptation"* (incompatibility).

People tend to prefer diversity in all emotional approaches and often take dangerous risks to avoid boredom as they find it very unpleasant. As Ralph Linton believed, the need for diversity leads to creativity, but it can also lead to social disorders, deviant behavior as diversity violates what people experience as cultural constraints [5].

While some of our people over the age of 65, who were imposed with a curfew, violated the rule, some other people who took it "their duty" to obey them and to obey the rule appeared (Monday, March 23, 2020). He began to take the place of authority by shouting or throwing objects at them with naked voices or a microphone (with a children's play microphone) from their own balcony.

The Panopticon, which Bentham thought of as an "architectural model," turned into a security model (system) and began to be equated with dictatorship by these experts. The treatment of our people over the age of 65, seen in our country during the corona days, proved something else: **the dictatorship of society/individuals!** ® Here, the previous assessment is given:

There is no risk of panopticism turning into a dictatorship, because **the monitoring mechanism is now fulfilled by the society as a whole**, as people also observe someone/each other. While dictatorship is accepted as an individual institution and mechanism (…), we see that **society itself can become dictators**, even without realizing it. That is, we are no longer facing the reality of dictatorial rulers, but ***dictatorial societies***® [6].

The number of people over 65 years in Turkey, 7.5 million. The closure of millions of people at home has been suggested by scientists who have not yet known the enemy called the virus, thinking that they are "high-risk group." In the process, it has been observed that these and similar measures are successful in defeating the virus. Turkish Airlines has stopped accepting passengers over the age of 65 (although all flights have not yet been canceled for everyone). However, the issue that nobody discussed was why 65 people are considered old in this age when the average life expectancy is prolonged. When we look at the science of gerontology, we see that 65 years ago, which is considered as the beginning of old age, was taken by employers for economic reasons (and based on compensation and insurance calculations)!

Similar situations all over the world have brought us a new concept: It is a new concept that is formed by the individuals "*pushing out*" what is "in": "**Social/Communal Fascism**." ® It reproduces from a "*sense of alignment*" it creates. In other words, they are the result of a culturally implanted "**fear of the norm**" ® as a substance (Plato), a monad (Tarde). The concept of "*fear of norms*" that I have presented expresses the pressure of individuals to "adapt to social norms" to everyone familiar/unfamiliar.

1.2 The Power Defeating the Kings

Epidemic diseases have deeply affected societies throughout history. Organized social life has been disrupted, and also economic, psychological, and mass moral breakdowns have occurred. **Epidemic diseases that killed thousands, ten-thousands, and even hundred-thousands of people, displaced the emperors and overthrew kings**. Plague ("black death"), dysentery, smallpox, typhoid, tuberculosis, cholera, typhus, bovine plague, Justinian plague, Spanish flu, and swine flu in our age, SARS, mad cow, bird flu, and finally corona virus are all known epidemics. These outbreaks can occur within the natural flow of life, and have been especially seen within wars [7] (Fig. 2).

Fig. 2 The plague in Florence, 1348 (illustrating Boccaccio's Decameron) [8] (The figure is an open-source material)

In today's collective living institutions (schools, universities, dormitories, etc.), which we call "total institutions," measures can be taken, and meetings where people come together can be canceled to prevent an epidemic. However, in armies, which are the "**biggest total institution**," it cannot be said to "**go to your home**." And it has not been said generally.

1.3 From Antique to Present

Infectious diseases and global epidemics have caused people to be crushed in past historical periods. Throughout human history, more people have died as a result of infectious disease than famine and war.

Cholera, smallpox, measles, Ebola, AIDS, SARS, different types of influenza caused the death of millions of people in the region including Anatolia with different age and demographic characteristics. With the start of agricultural production in Anatolia about 12,500 years ago, the settlement was established, that is, the common life began. This brought feces and household waste in common areas, and people first encountered infectious diseases. Archaeological remains obtained from the Körtik Hill settlement revealed that people died at a young age during this period, and 65 percent of the population died before reaching the age of 30. About 5000 years ago, there was an epidemic of tuberculosis called thin disease. The lepers and leprosy epidemic, called the living dead or the sluggish, decimated people throughout the Middle Ages. The social impact of this was also great. Because the disease was feared so much that people were stigmatized with the letter L (Lepra), when these people approached someone, the rattles were worked and removed with sticks. The leprosy originating from Africa was transported to Asia with migrations. It spread to the west from there in 327 with the expedition of Alexander the Great to India. The famous "Silk Road" has become an effective transport route in the spread of infectious diseases [9].

In this chapter, a general view on social and mental dimensions of the coronavirus epidemic was firstly provided by analyzing it in the context of Turkey. It is important that the appearance of this virus has caused many occurrences of both social and mental effects, and these should be discussed in detail, by looking at to the past and psychological roots of the humankind. In the chapter, many new concepts were used as the first time and these concepts were indicated with the sign of ®. After analyzing the social and mental dimensions, a solution of CORONAPOLIS was explained, and it was connected with Artificial Intelligence/Computational Intelligence aspects by considering Internet of Things (IoT) and smart city perspective.

2 Looting Culture and Perspectives with Phenomenology

2.1 *Looting Culture and the "Extortion of the Need"* ®

In the face of the last corona virus threat, which was described as the "pandemic" with the official announcement by World Health Organization, people run to the shops in many places, and almost some of food products and colognies were looted. There are many reasons to act illegally and behave unethically. Those related to our subject are ego and power demonstration, greed, survival instinct, and individual gain. People bought 2–3 times more what they needed and did not think about others. This is called as an extortion event and I conceptualize it as the "**extortion of need**" ®.

The concept of "*norm conflict*" acknowledges that failure to comply with formal or informal norms and social values will drag the perpetrators into a conflict. "The individual has to comply with both formal and informal norms throughout his life. It is inevitable that they conflict with the law and legal order when they do not comply with the formal order. On the other hand, they conflict with the group (they live in) and the society (they belong to) when they do not comply with the informal order. In both cases, the individual becomes unhappy and restless" [10]. However, those who extorted the needs of others during the virus epidemic days were not bothered by this.

It is possible to explain that as a "*trauma effect*." The second part of Pavlov's "*conditional reflex*" experiments is generally not known. Pavlov was ringing the bell while giving food to the dogs at the same time, and when the bell rang without giving the food (meat), the dogs showed the same saliva secretion reflex. Pavlov, who wanted to repeat the experiment one day after a flood caused death of some dogs, he came across with a surprising result. The dogs did not show reflexes when the bell rang without giving them food. Thereupon Pavlov had made that not too much known conclusion: In cases of trauma (and intense stress), conditional reflexes disappeared as the animals did not show reflexes, and obey. Even neurosis was occurring. That scientific conclusion presented to us by Pavlov, who won the Nobel Prize for such studies on the psychology of animal and human behavior, also explained why people did not comply with the norms and values. The fear of not being able to find food as a result of being closed home to prevent from a deadly virus epidemic caused great trauma and the values were turned upside down.

That shows us the judgment and stereotype of "Values continue to exist even in times that are impossible to realize, (or in other words) to be adopted by people" [10] is not always true.

2.2 Understanding with the Phenomenology

Our thoughts and stereotypes are created by the society and the cultural universe, which we are in and the part of, respectively. Common assumptions and behaviors enable us to socialize and being cultured. However, phenomenology orders us to question the thoughts and behaviors that we accept without questioning and it works accordingly in this context.

Against the world, we all take part in a process of socialization with the patterns of thought that were adopted to us. Phenomenology asks us not to accept such patterns without exploring them. He asks us to question them (question directly our culture itself; in other words, question the way of seeing and welcoming the world we were adopted while growing up) [11].

In the *Phenomenology of Perception*, Merleau-Ponty emphasizes that phenomenology is a form of action before becoming a doctrine or a system and assumes the task of revealing the mystery of the world and the mind [12].

Phenomenology is the study of essences and, according to him, all problems relate to defining essences: for example, the essence of perception, the essence of consciousness. Phenomenology, however, is also a philosophy that puts essences into existence, arguing that we cannot grasp human and the world from their "realities" (*facticité*) [12].

According to Merleau-Ponty, "*Every consciousness is a consciousness of something*" and that is nothing new [12]. Regarding the "*perceptual belief*" that results from our natural perception, Merleau-Ponty says as:

"We see things themselves; the world is that what we see (VI, 17). Our perceptual belief is our belief in the existence of things and our formation in the world. Every moment we live we confirm it without speaking. However, this certainty provided by my perception is a certainty left in the dark. These two thoughts, which can be experienced simultaneously without contradicting each other in practical life, put us into the paradoxes of perceptual belief when they are turned into a thesis" [13] (Fig. 3).

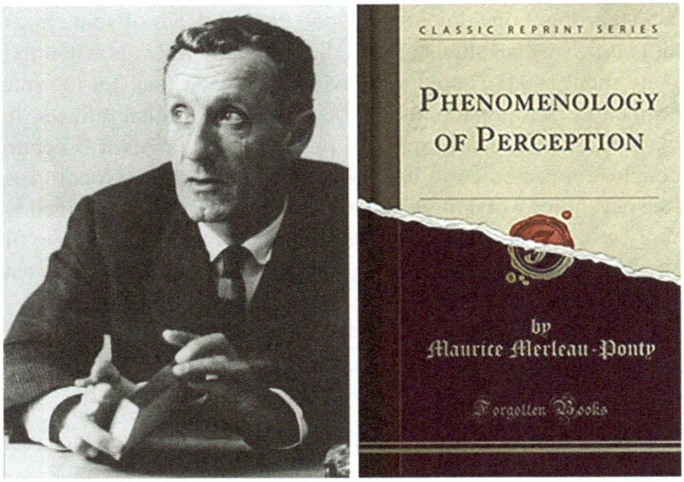

Fig. 3 According to Merleau-Ponty, "Every consciousness is a consciousness of something" and that is nothing new [12] (The figure is an open-source material)

3 Un-ethical Criminal and the "Produced Anxiety" ®

3.1 "Ethical Crime" and "Un-ethical Criminal" Against the Social Panic®

What is done with the logic of "I gave my money, I got it" is legal at the first stage. However, since there is a punishment for every extortion, we face here **an important sociological phenomenon and problem: "Social Ethical Crime."** ®

It is not an ethical behavior to "**buy someone else's right**" in situations of great social need. We need to consider that as a **"social ethical crime"** ® and to apply a sentence-penalty in return. This sentence should be decided together with the society, at least it should be a "**social condemnation**." This act of **"unethical stockpiling"** ® should be placed in the literature and laws as extortion of someone else's need, legal-looking looting, "**un-ethical behavior**" and be punished as an "**un-ethical crime**," in order not to harm social solidarity. So, that does not cause social "**moral panics**" at the end.

Even ignorant societies do not mind disasters, but rather reinforce their ignorance. Lesson learners are only a few, as seen throughout history.

3.2 Risk Society and the "Produced Anxiety" ®

The concept of "**risk society**" or "**world risk society**" belongs to the famous sociologist Ulrich Beck. According to Beck, people have shifted from industrialized society to risky/social structures. Modernity creates new risks from itself. And most importantly, these risks, which pose new threats for people, are again "**produced by the human hand**."

The panic atmosphere that emerged worldwide because of **corona virus epidemic** has justified once again Ulrich Beck. Panics throughout the history are increasingly experienced as a result of the human's own production. I think it is useful to summarize this situation as follows: **"Today, human has become his own Frankenstein!"** ®

Experts list the new risks of modernity as global warming, nuclear leakage hazard, wars produced by politicians in their own interests, food terrorism, and even "**marriages!**".

Modernity has brought the crises along with it, disrupted the psychologies, and transformed the society into an **"Anxiety Society"** ®, by making people anxious/worried as "**producing anxieties**." As it is seen during the corona virus outbreak, such societies have been often forced individuals to "**close in**," as not just during outbreaks.

4 Self-Ostracism® and the Coronavirus Hallucination

4.1 Self-Ostracism ®

The Corona epidemic led to mandatory quarantines, but also led to "voluntary" quarantines. Millions of people around the world have "closed" themselves to their homes for protection.

Human with a congenial behavior in nature also has the opposite behavior: exclusion behavior. Faced with a social exclusion, the individual's self-structure is disrupted and undergoes a cognitive destruction. The ejection of the ego out of society is called as *Social Exclusion* (*Ostracism*) [14].

One of the first examples of exclusion found in written texts can be seen in Athens democracy. In one of the series of reports on the Athens Agora excavations, Lang mentions an application called ostrakismos dating back to the fifth century BC. Every winter, Athens citizens write a name on pottery pieces called ostraka. Here, they were voting to decide whether the rulers, who had a tendency to dictatorship, became too strong or too rich, should be exiled or not, and individuals, who received 6,000 or more votes, were exiled for 5 to 10 years. A similar approach was observed in Sicily, and the voting here was done by writing on olive tree leaves [2].

I call the behavior of "voluntary exclusion" shown by millions of people as closing themselves to home because of the Corona epidemic, as **"self-ostracism."** ®

4.2 Hallucination of the Invisible

A human condition that applies not only in modern societies but also for all time is "the psychology of believing invisible." William James, who is the founder of American psychology and the representative of the pragmatist movement, examines the topic of "*Reality of the Invisible*" at a religious level in the *Varieties of Religious Experience* [15].

The fear of corona virus, which we could not see with our eyes and thought that it existed "everywhere," was a complete source of psychological panic in terms of public health.

James says that what we cannot see has its own psychological features [15]. According to the information in the *Journal of the Society for Psychical Research*, "**the sense of the existence of something turns into an evident hallucination within a few minutes**" [15]. The fear created by the corona virus, which we cannot see with our eyes, has also turned into such a hallucination in world societies. We started to see corona everywhere!

Eliminating a virus outbreak is not only about **biology** (the presence of the virus), **chemistry** (infection), and **infection medicine** (prevention of infection), but also related to **sociology** (imprisoning society at home, stopping life, closing places…etc.), **psychology** (the mood of those people), **anthropology** (above all human science, history), **political science** (political decisions to take), **international relations, economics** (economic losses arising from the stopped life), **food science** (what will be eaten and drunk), **education science** (closed schools)…etc.

5 Political Structure of the Coronavirus and the Cultural Racism

5.1 "Demo-Fascist" ®: Political Structure of a Virus

Corona epidemic has a social and spiritual dimension as well as a political dimension. I want to make a **sociopolitical definition** about that virus, which has always been given medical explanations and caused the masses to be trapped at homes.

The corona virus, which is coded as COVID-19, has a "**democratic**" structure in terms of not distinguishing between rich and poor, king and normal, unfamous and famous. Also, it is an imperative, non-verbal, "**fascist**" by imprisoning the masses, and the whole world.

When a structure carries both democratic and fascist qualities together, a new concept emerges. I call that concept as **"demo-fascist"** ® [16].

5.2 *"Age Discrimination" as Cultural Racism and the Mistake: "Social Distance" Instead of "Physical Distance"*

The global corona virus epidemic brought us new concepts. We started to misuse some of the concepts that already exist in the language of science, and even we do not know some of them at all. Initially, the related concept used incorrectly from 77-age to 7-age (I purposely reverse) was "**social distance**."

The most known meaning of that social distance concept, which is used to describe the "**physical distance**" that should be left between individuals to prevent from virus transmission, is related to racism. "**Social distance**" is the attitude followed towards people, who want to be kept away from their ideas/minds/social aspects, due to reasons on language, religion, race, color, etc. For example, **Hitler's attitude towards Jews** is "social distance" (Fig. 4).

Another approach that does not fit is to isolate **people over the age of 65** (briefly referred to as "**65 +**") from society. This is expressed in an unknown concept: "**Age discrimination!**" Other names are: "**cultural racism**," "**new racism**," or "**symbolic racism**." (Readers are referred to [18, 19], as two important sources about them).

The international name of "age discrimination" is "**ageism**." In Wikipedia, agism is defined as "the type of **discrimination** applied to any individual or group **due to their age**." Age discrimination can be **realized systematically or unconsciously**. These can be defined as "**corporate practices and policies** that support biased approaches, discriminatory attitudes and stereotyped perceptions." In other words, the closure of 65 + people to the home are within the scope of this concept, even if it is for health and protection reasons.

Fig. 4 "Social Distance" Instead of "Physical Distance" [17] (The figure is an open-source material)

5.3 *Disposal Communism Dystopia*

Due to the Corona epidemic, "tracking systems" and their applications in society have also improved. In Moscow, a system has been developed for people infected with the virus to give **a QR code** to their phones when they want to go out, tracking where they go outside, and how long they stay. This tracking–monitoring system has been criticized as a "**Virtual Gulag**" or "**digital concentration camp**" in Russia. After the epidemic, the **QR code tracking system** was transformed into the "**Ideological Control Device**" (ICD) of the state in accordance with Bauman's concept of "**fluid surveillance**" and there was a danger of spreading to all areas of life all over the world.

The multinational, global American internet search engine Google also proved—once again—that it has an important tracking technology. The work named "*COVID-19 Community Mobility Reports*" [20] published in April 2020 was a complete example of "fluid surveillance" (Bauman). In this report, Google has "followed" human movements in countries since the start of the coronavirus epidemic—as far as it leaked to the public; reported who went where and how far (going to work/supermarket/pharmacy/market/specialty food stores), how long he stayed at home based on "location information"! 131 countries were taken under control with this system, which is called "information technology," but actually used as a "tracking technology." The world was turned into a "digital concentration camp" for the reasons of the corona epidemic (we can call it a cover). Or, to put it in David Lyon's concept, "**Orwellian control**" has moved totalitarian fears from the borders of the government to "social channels" [7]. The next question that comes to mind is "**Who will be the global Hitler of these digital concentration camps**."

However, Slavoj Zizek's view is slightly different. Zizek's view that these control mechanisms have been the product of "disaster capitalism" for many years, that all states, whether socialist like China or capitalist like the USA, somehow control society, and Americans live under the illusion that they are only "free" is important. According to him, even the big states have been doing this for nearly 20 years. When talking about the global Hits of the digital concentration camps, Zizek talks about "**disaster communism**." This is because governments do and should do "by giving up market mechanisms" to organize the production of basic materials such as masks, test kits, and ventilators, to confiscate hotels/resorts and provide economic support to those who are unemployed [21].

In addition to the concepts such as Virtual Gulag, digital concentration camp, Orwellian control, global Hitler and disaster communism, the concepts of "**Global Socialism**," which includes a stateless or one world state, and "**Digitalism**," which includes digital tracking systems were added to the associated literature. All of these updates, in the end, gave rise to the dystopia in which the global fear of death and the search for remedies will turn the world into a totalitarian universe of life.

It is a scientific necessity that we evaluate this approach from different angles and within different contexts. The stigma produced by the COVID-19 pandemic, xenophobia against people and countries believed to be the source or spread of the virus, threat perceptions, social and psychological factors, social isolation, creative collective measures and technological developments in health systems for the global control of the epidemic. As can be seen in the opposition, there is a new type of civil activism. Bruno Latour's "Actor Network Theory," who has attacked sociology very harshly because he finds his basic classical dichotomies (dichotomies) such as agent/structure, macro/micro, subject/object, and old/new unsatisfactory, can help us in this regard. While classical sociology assumes fixed agents such as structures, apparatuses, unconscious, social pressure for the cause of action, in the actor-network theory, the source of the action is always uncertain, action is a surprise, an event, a new, a difference caused by many agents together. Bruno Latour replaces "Sociology" with "Associology" based on Gabriel Tarde.

Actor Network Theory, also known as "*the Sociology of Transformation,*" defends the idea that all human and non-human beings are shaped by the relationships they enter with other beings and acquire their own characteristics. Haraway joins Latour, arguing that all actors that cause change in society, that is, actors, are also dependent on non-human factors. "Of course, all these actors/actresses are not human beings; I agree that there is a sociology of machines" [22]. The "Actor-Network Theory" [23], developed by Bruno Latour with his friends and called ANT for short, is the term describing the techno-scientific problems and relations. COVID-19 virus can also be explained within this network structure. *The global epidemic created by the COVID-19 virus has created a global disease network.* Donna Haraway's statement of "non-machine, non-human beings" and explaining them as "non-human beings" [22] also supports this view.

In Latour's system, the social determinant (human and non-human) actors *exist in a network*, not alone. For example, the human being, an activist in the network, influences the network by building buildings. But the buildings that become part of the network also shape his life. According to the basic principle of this approach, there is a *generalized symmetry*, and all beings and entities, human and non-human (from technology, computer, laboratory equipment to human), are capable of acting equally. We will see this in our radical solution proposal below.

According to Jonas, the driver of technological progress may not depend on the capitalist system that is its proliferation area. It may even be better for *a socialist system as a totalitarian* [24]. It can be seen that the world has now become *"one body"* in many ways. The different opinions, support, and rejections that arise against the COVID-19 pandemic should also be evaluated in terms of "*perceptual beliefs.*" For example, according to Merleau-Ponty, "*the body in the world is what the heart is in the organism*" [25], *if there is no body, there is no world.* According to him, "perceptual belief" is formed as a result of our natural perception: "As a result of our perceptual belief, we assume that our perception ends in objects and at the same time is 'our perception'. These two conceptions, which can be lived simultaneously without contradicting each other in natural perception or in practical life, are put into thesis, it puts us into the paradoxes of perceptual belief" (cited in [26]). In other words, we should perceive the world, which is a single body, as a generalized symmetry or an asymmetric world that is not necessarily created by humans alone. We should say that non-human factors, such as all actors/actors that were initially non-living, such as the COVID-19 virus, which was alive or "revived when it penetrates into the human cell," together determine the changes in society. Just as today's global epidemic is reconstructing/rebuilding the society by upsetting it...

6 Foucault, "Big Locking Down" and "the World is a Great Madhouse"

As given by Michel Foucault, examples from history shed light on people's "being locked down at homes" today. Foucault expresses the examples of that as "locking down not only the mad, but also elderly, patients, unemployed, idlers, prostitutes, and everyone outside of the social order, in cities such as Hamburg, Lyon, Paris from 1650 to 1750" [27].

This is a "**filtering process**" [27]. There are **dangerous machines** in our society: **they sieve people**, and filter, pick up, lock down mental patients. Machines are considered to make these people normal. Question: Do they do their part? [27]. **The filtering machine** is blind to the person it is filtering [27]. **The conversion machine** does not know the limit of conversion. That is establishing **a power relationship that allows isolation, closure, suspending rights, and disrupting life**. It is also to hit an indelible **stamp** [27]. Foucault expresses that situation with the words **"The world (...) is a great madhouse"** [27]. "**Not to be included in the social body (...) To be expelled from the social space (...) Exclusion in its place!**" [27].

The Corona virus has turned the whole world into Foucault's "madhouse" and the "**panoptic prison**," which is watched by "**an invisible enemy/virus**."

6.1 Urban Empires

The world first evolved from the bipolar world of the "*cold war*" period to a unipolar world with the collapse of the Soviet Union. Then the electronics and internet revolution made the world a "*global village*." Geographical, physical, economic, and electronic boundaries have been fragmented. In Hardt and Negri's statements, "Empire" has been created: Hegemony has taken a new form, and this form consists of a series of national and supranational bodies united under a kind of ruling logic. This kind of new global hegemony is what we call empire [2].

The imperial age began after going through the twilight of modern domination. Unlike imperialism, empire does not create a land-based powerhouse; it also does not rest on fixed boundaries or obstacles. Empire is a centralized, decentralized governing mechanism that includes the whole world within its open and expanding borders. Empire manages hybrid identities, flexible hierarchies, and multiple exchanges through changing command networks [2].

These developments made metropolises that already existed in the liberal world more important. Alongside the developments at the global level, "*urban empires*" (first introduced by Dr. Hulki Cevizoğlu) are formed, which are called metropolis at the local level. While they include the stock market, financial centers, and international trade centers and corporate headquarters, they also remain political attractions. Being a metropolis doesn't just mean getting crowded; it also means being one of the capitalist centers.

If we use Foucauldian concepts, metropolises (urban empires) have become the application of "*control society*" and "*biopower.*" These urban empires contain not only "deadly weapons," not only those that occur in the global Empire of Hardt and Negri, but naturally styles such as "moral intervention" and "institutional intervention." There are crowds in "urban empires" called metropolises, but it is controversial whether there are "people." I believe that Deleuze generally says that "there is no more or no people ... the people are missing" (quoted by [2]) applies to metropolitan life as well.

This is where Bauman and Lyon's concept of "*liquid surveillance*" comes into play. Contemporary modern societies seem so volatile that it can be argued that these societies are in a "liquid" state. Contemporary citizens, workers, consumers, and travelers, who are always on the move but lack certainty and boundaries, are aware that their movements are being monitored, monitored and followed [4]. We can define the current situation as "*late*" modernity, "*postmodernity*" or, more colorfully, "*fluid*" modernity [4].

But the COVID-19 outbreak did not get caught in this "fluid surveillance" and caught people. Before moving on to the urban solution today, we need to address the metropolises we will take to get rid of COVID-19 and not to catch subsequent epidemics.

Cities that exceed a certain size are called metropolis/*main-city* (*mother-city*). The concept of metropolis in each country varies according to the population. While the neighborhoods and cities formed around the market in the feudal period became centers of agricultural products and trade, they also became industrial centers during the capitalism period. Under capitalism, which meant the separation of these from the feudal era based on a workplace-housing association, the population shifted from city centers to cheaper suburbs. As Kıray stated, "*The first, middle and upper-middle suburbs emerged of those who worked in the center of the city and returned to their homes at night on the ferry and railway route (Istanbul-Suadiye, Erenköy or Emirgan, Yeniköy, Yeşilköy). Later, residential areas and industrial suburbs were formed around the industry (Istanbul-Paşabahçe, Alibeyköy, Mahmutbey district)*" [11].

One of the features of metropolitanization is that it creates stratification and makes the lower- and upper-income groups more prominent. Over time, this separation reflects on the locations of the residences, and according to the stratification, upper, middle- and lower-income group areas are formed [7]. In less developed countries, the direction of urbanization towards metropolises takes place in two ways. The first is the overgrowth of one big city versus another. Cairo, Caracas, Tehran, Santiago,

Karachi, Dakar, and Istanbul ("*The single major city law*" is Mark Jefferson's conceptualization) [8]. Examples of such cities that are the "capital." The second is based on the existence of a relationship between the size of the city and the ranking of these cities in terms of size. There is no big difference between the largest city and the second and third largest cities. This is prominent in places governed by decentralization [9] ("*Primary City*" is the city that is disproportionately larger among cities in a country or region than others. It is the city that collects the largest part of the country. The place where political, intellectual, or economic life is concentrated. It is the dominant city in the settlement of the country; this is called as the king effect. Some countries may have more than one, while some countries do not).

The level of knowledge acquired by new technologies has become an economic commodity, thanks to the means of communication that each part of the economic commodity began to be produced in a different country, this caused the transformation of Fordist industry. Manufacturing and industry dependence is now minimal, spatial choice has become flexible and global centers are few in number but have increased since the 1990s and have spread around the world. Metropolises connected to the global system stand out. Spaces that cannot join the network also maintain their traditional structures [10].

Simmel is one of the sociologists who contributed greatly to the development of urban sociology by analyzing the effects of metropolises on individuals. Simmel points out that in modernity, which he perceives as the "*age of science and technology,*" the individual has lost his inner security and experienced tension anxiety and panic arising from the excitement and complexity of modern life. According to Simmel, who stated that this anxiety is mostly seen in urban life, the chaos, competition, infidelity of metropolises against personal relations and thoughts pushes individuals to despair and bring them to the brink of neurosis [5]. According to him, there is no other spiritual phenomenon other than the "*blasé*" attitude, perhaps directly related to the metropolis [5]. The city creates a deep contrast between small town and rural life in terms of emotional foundations of spiritual life with all its economic, professional, and social pace and diversity every time it passes through the street. The metropolis, as a being dependent on differences, *obliges the individual to more consciousness than rural life requires* [8].

6.2 "Urban Empire of Istanbul" in Turkey

Currently, many cities have become metropolitan in Turkey. Among these, Istanbul competes with other cities of the world. The phenomenon of metropolitanization brings to mind Lewis Mumford's thesis about the city. According to Mumford, there is a process that follows the "*growth*" and "*collapse*" period. It would transform from Polis to metropolis, later to megapolis and eventually to *necropolis, a cemetery town* [14]. The COVID-19 pandemic accelerated this at the beginning of the twenty-first century and at the same time undermined Simmel's view of the metropolis.

The metropolitan-type man develops an organ that will separate him from his roots and protect himself against threatening currents and incompatibility in the external environment. He provides his reactions with his brain, not his heart. Here an increased awareness of spiritual privilege assumes. That is to say, metropolitan life is at the basis of the increasing awareness and domination of the intelligence/intellectualism of the metropolitan people... What appears to be a direct separation in metropolitan life is only one of its most basic forms of socialization [5]. According to Simmel, *metropolitanization is also freedom!*

6.3 Marginalization and Depression Locations

On the other hand, there are many problems in metropolitan areas, especially overpopulation, and other problems with overpopulation. Among these problems, air pollution, safety, traffic, health problems, high cost of living, and dehumanization stand out.

Urban crisis areas emerge, separated from cities due to migration, poverty, and unemployment. Depression emerged in places where business areas were evacuated, with production facilities such as industrial facilities and workshops moved out of the city. In the city center, problems arising from poor residential areas where immigrants reside heavily are observed. These inner-city areas, attractive to immigrants and low-income groups, are places where poverty becomes permanent. Places such as Eminönü, Süleymaniye, Kasımpaşa, Gedikpaşa, Balat, and Tarlabaşı in Istanbul have become areas of depression [11].

6.4 Gentrification

The collapse was tried to be prevented by gentrification. The concept of gentrification took on a different meaning than it had in the 1970s. It refers to reclaiming urban areas by both protecting and re-functioning. The Business Development Districts in New York frees New York from waste, fragrance, and drug dealers in what might be called *"gentrification of the streets."* Meanwhile, the public spaces of the city are gradually closed to disadvantaged segments of society, such as the poor, the elderly, the homeless, the unemployed, and peddlers. According to Simmel, "*The metropolis has always been a place of economic exchange. and financial economy.*" Here, the multiplicity and intensification of economic exchange refers to the means of exchange that cannot be provided by the inadequacy of rural trade.

For a long time in Istanbul, big bourgeois groups were closing in on building a world for themselves in closed sites and high security places. They chose to stay in the center because they did not have a high income as the new middle-class bosses who replaced them or did not accept their newly preferred regular life, they created the demand for gentrification in Istanbul. This young class, inclined to cultural and artistic activities, who do not want to live in sheltered communities, started to live in these districts with their brothers settling in Cihangir, Asmalı Mescit, and Galata and thus started to form new communities. Gentrification started in the district of Kuzguncuk first, and following to that, the first big attempt was known as the Tarlabaşı demolitions of Mr. Dalan, the mayor of the Istanbul Metropolitan Municipality [18]. "Urban Transformation" projects and applications in recent years are the latest and intense examples of gentrification.

Locations of breakdown should now be perceived not only as physical but also mental structures. The "collapse" that begins in buildings and districts also occurs with various organized crime structures. Hence, in order to prevent any kind of collapse, the authorities (including all kinds of micro powers), which had the state's control devices, started to monitor the behavior. Here, people were subjected to "voluntary surveillance" as daily-life devices for self-utilization has been turned to tools for monitoring with the behavior of voluntary surveillance (or "*voluntary servitude*" as stated in 1550 by Etienne de La Boetie). (The concepts of "*voluntary surveillance*" "*self-utilization*" "*benefit those who monitor*" are all belonging to Dr. Hulki Cevizoglu). It is possible to indicate that the ban-opticons were replaced by panopticons. (Pan-opticon: *all* (pan) and *surveillance* (opticon)).

There is inspection and monitoring everywhere. Today, every contact we make on the internet is recorded, such as criminological forensics. Even though it is pretended to be asked to us under the name of "cookie," we have an out-of-control and "classification" in other words "filing." Buying all kinds of books, clothes, goods/objects we make; the topics we are interested in are categorized and our **"behavioral and mental profile"** ® is created and promotions suitable for these are presented to us. There is also one that is not removed.

"Do-it-yourself" surveillance tools, credit cards, mobile phones, e-mails, WhatsApp, Twitter, Facebook, Instagram, all kinds of internet usage, ATMs, televisions, Google, Yandex, parking cards, store cards, club or professional membership cards, license plates, national identification numbers, tax and real estate transactions, hospital and school records, medicines taken from pharmacies, cameras in public places, city surveillance cameras, that is every moment of "modern" (!) life ... If you don't want to, you have to give up the modern (!) life. The individuals that make up the society perceive/perceive the modernity offered to them not as Pandora's Box, but rather as the *Cornucopia of Zeus*. (Cornucopia: The carob bean absorbed by Zeus, the chief god of Greek mythology, turned into a self-filled request box—Dr. Hulki Cevizoğlu). Looking to the future with fear and arguing that the only solution is to stop the masses by controlling it, Charles Darwin revealed a causal relationship that the capitalist world uses most to justify itself. Darwin's assumption of *"the survival of the fittest"* is one of the cornerstones of today's ideology of domination, as described by Thomas Hobbes. (...) One of the areas where this manifested itself was the field of culture [28].

After the attack on the Twin Towers in New York on September 11, the metropolises of the USA and Europe increased their security measures to the extent that life was overwhelming. The crowd of metropolises in the West that are under threat from international (or interstate) terrorism require major security measures.

The technologies for "*liquid surveillance*" (as explained by Bauman and Lyon), become widespread in European and US metropolises as they also replaced the whole social life. That state pushes Western societies and the associated groups to the ghettoization, by forming also the danger of becoming further introverted. This creates a danger of *being introverted* by dragging both the Western societies and, for example, Muslims in this society into "ghettoization."

Agier... was right in placing urban ghettos in the category of "corridors of exile." The legal and illegal residents of such places all have the same characteristic definitions: They are all unnecessary. They are all denied and leftovers of society. In short, they are somehow waste [4].

6.5 Power as a Body Technology ®

We can add the "power technologies" dimension to this approach. The powers have now begun to shape, correct, and reshape the body, just like the interior and prostheses articulated in the body. The sociologist Foucault says that "old" powers are replaced by such "new" powers and makes them the subject of study. In this case, we can now say that *powers are also body technologies* ®.

With his analysis in the 70s, Foucault explains how to move from a concept of power in which the body is first of all examined as a "recording surface of tortures and punishments" to another concept of power that "tries to shape, rectify and reshape the body" on the contrary, tries to understand [29].

According to Foucault, the human body has been turned into a "machine" as a result of the domination of powers, thus "the subject is dead" and only "utopian bodies" remain. For this reason, Foucault takes the subject of "self-technologies" and the self under the control and domination of power, instead of the individual (self) as the "subject." According to him, since "*the subject* -the individual being-*dies*" and is now governed by the power, it is no longer possible to speak of its consistency, consciousness, and rationality. According to him, "subject" is a social construct produced through social discourses shaped within a power domain and a specific set of practices. The "individual" is not the creative source of meaning. For this reason, it cannot be mentioned that the non-independent self is a conscious, self-controlling, internally integrated, consistent, and rational being. It expresses this with "the death of the subject" [30].

According to Foucault, the body becomes a "useful power" when it is both a "productive body" and a "subordinated body." Power imposed on a body should be understood as a part of the body's political technology [31], just as body technologies transform the body into a useful power. The power, which Foucault sees as the "political technology of the body," here I call it differently and see "powers as a body technology" ®.

The body is defined by Merleau-Ponty as "*the power of a certain world*," "the means of being-in-the-world": For a living being, having a body means articulating a certain environment, mixing with some designs and engaging itself in them continuously [32]. It is possible to draw *a new conclusion* from this: Just as power is a "body technology," the body is also a "world power" in the world in which it is located/articulated.

The body is the first and most natural tool of the human being, more precisely, without using the word tool, we can say that the human body is the first and *most natural object* and the *technical tool* of the human being [33]. This is *the most natural object/technical tool* of the body, its "techniques in itself" activities such as swimming, walking, and running. Mauss classifies the bodily techniques of the human as gender, age, productivity, biographical, and "transference" of techniques [33]. The body is also "the primary means of being present and of all communication. The body transfers and receives the basic messages (visual, auditory and kinesthetic) that give meaning to communication" [34].

6.6 Technical and Technological Body

Body technologies are not the same as "body's own techniques." I call the "own" techniques of the body as the "*technique in the body itself®*", inspired by the concepts of "*class for itself*" and "*class in itself*" that Marx used in describing the state of consciousness.

Marcel Mauss calls the body's own/natural techniques "bodily techniques": "I call bodily techniques, since a theory of bodily technique can be created from a study of bodily *technique*, a presentation, or a pure and simple description. What I understand from this statement is the way people traditionally benefit from their own bodies from one society to another" [33].

The body, losing its personality and integrity with diseases and wars, is evaluated in a mechanical structure in the 1950s with the effect of modernization and industrial production. A person who can use a saw with a mechanical hand and a forearm prosthesis that can be equipped with work equipment when necessary has a more functional arm than the arm he lost. A disease has not been cured, a missing body has not been remedied, and a higher model has been adopted as a body [35]. Portable Person is a vision of the future of a mobile, life-supporting earthling, equipped with a technology-inspired body extension. The *Portable Person* was prophetic with many devices and inventions on it. Some devices are still under development, such as digital displays built into "ordinary glasses." *Portable Person* is performance-/transaction-based and all elements of the system are integrated into a "functioning techno-person" [36].

Human biology and mentality began to change radically with the invention of the mobile phone in 1983. The mobile phone is the main tool of personal, social, and professional life and is increasingly destroying the distinctions between these areas. People are tilting their heads towards this device as if they were in a religious ritual. Eye contact is made with the cell phone instead of other people. One does not only carry, look, and touch a cell phone; because man has become unthinkable without this device. Mobile phone has become an indispensable part of self-perception [37]. Inventions in the nineteenth century were the evidence that humanity is moving towards a better society, even a perfect society [38]. However, two centuries later today, in the twenty-first century, the situation seems to be reversing.

It can be said (including many other examples) that we are dealing with a new *"techno-culture"* in the relationship between material culture and technology. As Robins said, techno-culture unconditionally embraces *new image technologies.* Techno-culture suggests that we need a new cyberspace and *cyber-culture anthropology* today [39].

The subject of "body technologies" also includes optimistic and pessimistic approaches to technology, and this has been included in the study of philosophical and social anthropology. Technology contains two opposite views (duality–dilemma) in itself because of the problems it creates rather than the problems it solves. One of them is "technological optimism" (*technophilia*), the other is "technological pessimism" (*technophobia*). At the core of his optimistic and pessimistic opposition; there is an opinion that technologies are not neutral, that they are linked to facts and value judgments, and that they are evaluated together with them. In other words, there is also a conflict between objectivity (certainty, clarity, clarity) and subjectivity (uncertainty, interpretativeness) [40].

Ellul, the foremost proponent of the pessimistic technology view, argues that "technique" enslaves everything. "To protect man from the domination of machines and technocrats is a commendable undertaking, but if machines are full of people who find their salvation there, such a protection is only absurd" (cited in [41]). Sennett asks: "Is it a friendly tool that replaces the human hand or an enemy tool?" [42]. For him, in the economic history of skilled handicraft, mechanization often began as a friend and ended as an enemy.

Although we cannot agree with Ellul's thought that technique enslaves "everything," it can be said that he attempted to enslave many things. We even believe that medical technology that disrupts (cuts, adds) our body integrity or tends to "repair" the body that has already been degraded makes us "slaves of medicine" (makes us "dependent"), or that we are moving towards becoming "medical robots" ® (robots of medicine). It is possible to say easily. Mounting inorganic metals in our organic body makes us "cyborg slaves," although it does not make us "mechanical slaves" [43].

Whether the COVID-19 virus is produced in a laboratory in the city of Wuhan, the variety of protection methods, the ways of transmission, the drugs used in the treatment, which type of vaccine is more useful, the enemy's invisible before catching you and making you sick, threatening your mental health, etc. The fact that the dimensions turn all humanity into "slaves of medicine," "addicts," and almost "medical robots" ® supports the above-mentioned views. After the vaccine production was carried out at super speed compared to the classical vaccine studies, it brought other discussions in societies, including mental worries. The main agenda of the society was whether the vaccine should be hit, class, status, age group, or those with torpedoes should be given priority.

Technological vision of the future seems to be split into two poles between affirmation and negation. On the one hand, there is the technological utopia idea that celebrates the arrival of the ideal/perfect life by preventing diseases, aging, and mortality with the development of technology, and on the other hand, there is a dystopian approach that looks at the technology of the future with despair, based on the idea that machine–human hybridization will corrupt the human essence [44]. Technology is changing the traditional context of man and life: Could life be synthetic rather than biological? And this change can bring along a post-humanist perspective of life where life, human beings, existence, and the world can be reconsidered from a non-anthropocentric perspective [44].

It is possible to explain the COVID-19 pandemic period with Heidegger. According to Heidegger, whether we affirm the technique with passion or oppose (negate it), *"everywhere we lack freedom and we are committed to technique."* When we view technique as something neutral, we surrender to technique in the worst possible way. Or, as Lovitt interprets Heidegger, *"We are all captive and condemned to technology.* Heidegger attributes this to our ignorance of our true relationship to technology. This requires to know exactly that 'technology and technology are not the same essence' [45]. The question of Jonas [46] years ago whether he/she performs an engineering art based on biological knowledge, in which *the human being is his/her object*, *"taking our evolution into our own hands"* and *"transforming human beings"* regarding the future of man in the fight against COVID-19, shows that his thinking is no longer fantasy and taboo.

7 A Total Solution: "Horizontal World Treatment" and a Smart City Oriented Project: Coronapolises

Attaching certain logical meanings to technologies requires some work. However, some features of technologies are absolutely autonomous and not open to interpretation [47]. According to the *"Social Construction of Technology"* analysis, the success of a product depends on the power and size of the group that adopts and develops it [47].

According to Winner, technology integrates with the social and economic system and this is called *social determinism* [48]. Contrary to this view, *the political theory of technology* focuses on the properties of technical objects and the meanings of these properties. For him, the point is that machines, structures, and modern material culture systems are precisely not only their contribution to efficiency and productivity and their positive/negative side effects on the environment, but also to contain *certain forms of power and authority*. Today, some of the most interesting research on *technology and politics* are those that try to show in detail and concretely that the seemingly harmless design features of technologies actually mask social preferences that have enormous meanings. All kinds of authority and power implicitly "impose" the norms that it has built in accordance with its own purposes, and can present this as the desire and benefit of society. Objects and technological projects are politicized by total and local authorities. For this, it is sufficient to create a "perception of reality and need" in the society. Modern man (here the authority) can use "the perceptions he produces" as a mask. The emergence of the COVID-19 virus and the efforts to eradicate it have shown these forms of power and authority and the phenomenon of social determinism to the world public opinion and caused great discussions that are still ongoing. The fact that some of these are evaluated in the category of "conspiracy theories" is a part of the "social reality" phenomenon.

In his article titled "*Democratic Rationalization / Rationalism: Technology, Power and Freedom,*" Feenberg states that technology is one of the most important sources of state power in modern societies, and that political democracy that determines the decisions in our daily lives is largely dependent on the masters of technical systems. These great technical system masters can be big company executives or military leaders, or professional groups such as doctors and engineers. Such leaders and groups have more impact on urbanization, housing and transportation projects, innovation, employees, patients, and consumers than all government agencies have together. According to Feenberg, *technology is neither determinative nor neutral,* modern methods of hegemony rely on various technical means in various fields of social activity such as production, medicine, education, or the military, and democratization of society requires not only political changes but also fundamental technical changes. Is the COVID-19 global epidemic caused by Wuhan, which is being discussed today, is the global work of the political technical system masters in the field of great medicine and biology?

According to Heidegger's "*technology problem*" and Ellul's "*technical phenomenon*" theories, *human beings are not much different from technical objects and people have become part of the mechanisms they have created.* Yes, technology and any power source that uses it "enslave us" in order to control us, but technological progress is also progressing steadily for the progress of humanity and "new modernities," "higher modernities."

7.1 Sheltered Smart Cities Against the Epidemic (a Model)

To date, solutions against the global virus epidemic have been based on mask-physical distance and cleaning methods to "*not to catch – to escape,*" and all of the discussions and studies have focused here. Now, as the whole world, we have come to the stage of "*attacking*" with vaccine production against the COVID-19 virus enemy. However, due to the panic generated by the emergency, we could not think of a fundamental (fundamental) solution by raising our heads and adopting a "holistic approach."

COVID-19 virus is transmitted by close contact and distance. Man is a social being. It is not possible for him to live without contact and doing "gathering behavior" (I call this the *mediocre behavior®* because he exhibits the behavior of constantly and systematically gathering and dispersing). Our living spaces are also arranged according to this need. In other words, the COVID-19 outbreak caught the human being in their "*adjacent world* "®. Almost adjoining, adjacent, vertical buildings, vertical workspaces, schools, military, trade unions, etc. Caught in "total institutions."

We have seen above that in Latour's *Actor Network System*, the actors that determine social life are not only people, but social activists are non-human factors and entities. In other words, viruses, bacteria, etc., that affect human and life. Non-human material assets such as biological beings and genius buildings *exist in a network*. Here, all entities/entities that constitute a wide range of non-human beings such as biotechnology, computers, laboratory materials, and their production formed a *generalized symmetry*. In my search for solutions, I produce the concept of "*a generalized symmetrical habitat*" ®. We will see this once again in our analysis below.

During the global epidemic, sometimes the lightning-fast closure to homes (the "big closure") has also changed the settlement structure in the cities—temporarily. Large numbers of people "migrated" from big cities to small settlements, seaside resorts, or rural areas. Bodrum in Turkey or flock are two examples of them emigrated to San Francisco in the United States. There has been a "reverse immigration" in the USA. "During the corona virus epidemic process, many people's definition of quality of life, expectations from life and lifestyles have changed. These changes have caused some to make radical decisions that change their homes and even the cities they live in. This forced migration occurred especially in regions where the cost of living was much lower" [49].

By arranging our living spaces (*Umwelt* in German), I propose a solution that will both prevent the contamination and subsequent contamination and revive the economies that have come to the point of collapse due to the epidemic: To build "*coronapolises*" (corona cities) with "*Horizontal World Therapy*"! Before moving on to the "horizontal world treatment" ® let's touch the vertical world.

The current fashion of an era was the idea of building high-rise buildings to "save" (!) Cities. Great progress has been made on the road set out with this "faith." However, the COVID-19 outbreak has falsified this belief. Professor Edward Glaeser from Harvard, who is regarded as the most important urban economist in the world, explains the basis of the idea of "saving" cities with tall buildings differently. By quoting the Bible, Glaeser quotes the following words of those who built the Tower of Babel: "*Let's build a city and a tower with its head in the sky. Let our reputation be that we will not be scattered all over the world*" [50]. Unlike the epidemics in previous ages, the COVID-19 global epidemic caused the "*luxurious dwellings,*" people, and the people to scatter and scatter. Because living together has been one of the most dangerous living areas for the virus to be transmitted.

Graham says that high-rise buildings are a "*top view policy,*" and that ethical considerations that relate height and depth to the value/status in society, have accumulated in this policy pool. Such buildings have a "*vertical pathology*" and "*blind windows*" [50]. Tower-style high-rise buildings have a modernist esthetic obsession, and psychological and social alienation. "Vertical collective housing is understood as structures that represent 'urban problems' rooted like a chronic disease, namely crime, poverty, gang violence, ghettoization and drug use" [50]. Geographer Maria Kaika defines and criticizes today's "iconic" high towers as "*autistic icons*" or "*serial objects*" (cited in [50]).

Defined as buildings of vanity and violence, skyscrapers are perceived as symbolic representatives of the "vertical hierarchies" of giant companies and their financial power, wealth, and corporate identities. Feminist criticisms have also been made of skyscrapers. According to this approach, skyscrapers and high-rise buildings are masculinized and heroized "*clearly thought as penis-oriented*" and "*inflated masculine egos*" [50]. The French philosopher Paul Virilio describes the twin towers with the World Trade Center, which was destroyed on September 11, as an example of "Babylonizing the city" [50].

We need to solve both the COVID-19 global epidemic and the transmission of other types of outbreaks that may arise from now on, not only with medical measures, but with a "*vital network*" ® approach. For this, we need to save our living spaces (*umwelt*) from being dead cities (necropoleis) from "vertical pathologies" and establish "coronapoles" (*corona cities*) with "*Horizontal World Therapy*" ®. We should plan a "ring-shaped city" as similar to ancient Rome style ("*All roads lead to Rome*") or the "*Republic Villages*" model by the great Turkish leader Atatürk. Right in the center (in the middle of the ring) there should be "official zone" (state and public institution buildings, hospitals, universities, etc., total institutions), and the rings expanding outwards should be "*special/formal zones*" (residences, residential areas, living areas). In order to prevent contamination in case of any contamination danger—just like closing the castle doors—the masses should stay in their houses in the outer rings (which should be built with a horizontal architecture and small-scale gardens that can grow their own vegetables), isolating from the official area in the center, so that in case of an epidemic, it should be "isolated."

During the Second World War, 20 million homes in the United States grew their own vegetables and produced up to 44 percent of the total annual American agricultural harvest. As another example, in the face of the US embargo, a significant part of the food need in Cuba's cities can be provided by urban agriculture [51].

With such a planning, the personnel required for the maintenance of the state services will carry out their work in that region, and contact with the government departments, which are scattered among private regions today, will be prevented. If proximity to total institutions such as schools and universities is preferred, instead of a single ring, self-sufficient "public cities" consisting of many "*ring shaped cities*" that are self-sufficient with vital and social units will be built.

7.2 Politicalized Life Preferences

It is obvious that objects and technological projects that organize, determine, and even determine the lives of the subjects are preferred by central and local political/private authorities. Even wrong life choices are possible by creating a "perception of reality and need" in the society. In terms of our subject, the whole world has experienced this reality behind urbanization that is not protected against epidemics.

As a solution outside of medicine, the "*generalized symmetrical habitat*"® is the "coronapolis" which is a circular, ringed city model. The Halkalı city is also the

"Horizontal World"; a horizontal and distant world... Thus, our neighborhoods and homes will not be "*our own prison,*" but will be safe and maintain "physical distance" with neighbors. Mental (mental) and physical health will be protected, crowded city centers that are vulnerable to epidemics will be avoided, "nodal points" will not be formed due to compulsory contacts, and sick buildings and districts will become healthy. In other words; there will be no "*subject and object pathologies*" ®.

This solution project is a non-human (construction-based) entity that provides human benefit and a "*generalized symmetrical habitat*" ® (English generalized symmetrical habitat) project. Thus (by creating a "horizontal world"), the inequalities in the fields of politics and identities caused by the vertical world will be able to be eliminated to a certain extent. The construction sector, which is still the locomotive of the economies in the countries that are not fully industrialized in the world, will gain a brand-new momentum and revive the subordinate economies. This also means that the masses that the epidemic left unemployed are getting back to work.

As the architecture of the coronapolis model, this chapter proposes to consider Atatürk's "*Ideal Republic Villages*" model on an urban scale (Fig. 5 provides view for the Atatürk's model).

7.3 Computational Intelligence and Internet of Things (IoT)-Based Background

By considering the explanations so far, the technical framework of the *CORONAPOLIS* model can be shaped with Computational Intelligence and IoT components. It is important that the model should be carefully managed by humans to provide necessary ethical and logical feedback, in order to prevent the system from working maliciously and turning into a *system with its own rules as* ***digital dominance*** over humans. Of course, that system is controlled by policy makers and the power side of the nation(s), so the algorithmic structure of the model should be blended with societal and political background (and in a flexible architecture to be used accordingly after policy updates/upgrades).

A CORONAPOLIS is a mixture of IoT sections managed from a central unit dealing with communication of the sections and flow of the massive data (in terms of ***big data***). IoT-based structure of the CORONAPOLISES may include the related components and data support as briefly explained in Table 1.

Focusing on Table 1, it is possible to extend the used components, data types according to changing locations-structures of coronapolises, and the IoT sections may be differing according to the latest policies. For example, for future pandemics, IoT sections may be including ***Health Units*** for better analysis of massive health data, *and managing the health state in each Home Unit*. Here, some essential Computational Intelligence/Artificial Intelligence models can be used for certain purposes:

- **Fuzzy Logic**: As a form of applying fuzzy numbers for transforming digital systems into human-like decision makers [53, 54], Fuzzy Logic will be a key

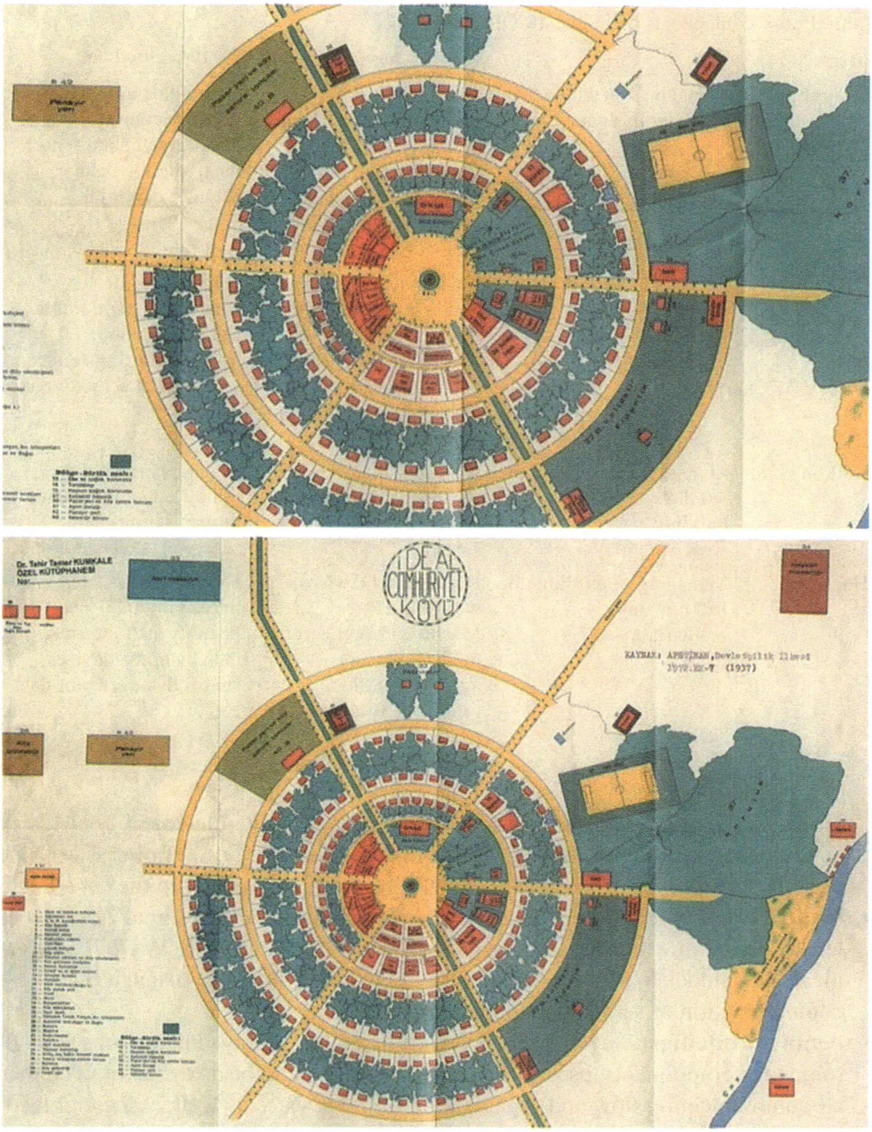

Fig. 5 The model of the Atatürk's Ideal Republic Villages [52] (The figure includes open-source materials)

Table 1 IoT components of a typical CORONAPOLIS

IoT section	Role	Components	Used/required data
Central unit	Managing the whole coronapolis area	Fuzzy Logic, Machine/Deep Learning models, Swarm Intelligence models, sensors, actuators, M2M protocols, authenticated protocols	Communication data of the district units, policy data, lawful data, sensor data, satellite data, actuator data
District unit	As associated with the near districts, controlling the human-system interactions	Environmental devices, Ambient Intelligence, Fuzzy Logic, Machine/Deep Learning models, Sensors, actuators, wide area protocols	Human-outside component data, traffic lights, cameras, infrastructure system data, sensor data, actuator data
Building unit	Managing-controlling the IoT flow in a smart building	Building smart devices, sensors, actuators, in-door protocols	Business oriented data, policy data, lawful data, human actions data, sensor data, actuator data
Home unit	Managing-controlling the IoT flow in a home/rooms	Home smart devices, wearables, Fuzzy Logic, Machine/Deep Learning models, sensors, actuators, in-door protocols, short area protocols	Human health data, human digital phenotypes/footprints, data by home devices, sensor data, actuator data

model for short-term effective system runs. For direct rule-based controls of components and the IoT units. That will be a better way to apply power side policies and societal arrangements–lawful interactions directly to the *societies*.

- **Swarm Intelligence**: Swarm Intelligence algorithms are used for real world problems with optimization perspective [55]. In this context, the *CORONAPOLIS* organization includes IoT-based interactions, so it will be better to manage especially communication of sections with mathematical models.
- **Ambient Intelligence**: As the model of coronapolis employs interactions of environmental components, use of *Ambient Intelligence* will be a key factor to control all communication flow and organization of the devices as well as societal feedback. Ambient Intelligence is known as the future of IoT-based systems, by covering a common use of Computational Intelligence for creating the atmosphere of smart cities [56].
- **Machine/Deep Learning**: As autonomously learning models are important components of Computational Intelligence, massive (*big*) data will be requiring use of essential Machine/Deep Learning models such as Artificial Neural Networks, Decision Trees, or Bayesian Networks [57–59].

7.4 Control of Coronapolises

Control of the proposed *CORONAPOLISES* may not be too simpler in terms of applying it. Especially, societal impacts-issues explained before the seventh section will be key points while controlling the role of the model in different *society types, cultures, and regions* of the world. So, control of the model may include the following methods generally:

- Technical evaluations such as ***accuracy, error-based metrics*** for better examination of the models for target problems.
- *Computational running and resource usage* evaluations.
- Data oriented ***cyber-security controls*** and manage by policy makers, software companies...etc.
- Societal evaluations with *voting* methods to run some **consensus**-based mechanisms to track good or malicious use of the model.
- Individual and massive political participant based solutions in terms of determining the present and future state of the coronapolises (*considering nations, regions, different groups of humans*...etc.).

8 Conclusions

Artificial Intelligence/Computational Intelligence applications that provide "big data" analysis against the complexity and difficult problems (to solve in human life) provide a great improvement [60]. Considering that dozens of philosophers such as Plato, Aristotle, John Locke, Karl Marx, Jean-Jacques Rousseau, Immanuel Kant, Friedrich Engels continue to have an effect on the governments of the country, we can experience the excitement that intellectual wealth awaits us in the age of artificial intelligence and robots [61]. Technology offers new opportunities, conditions, advantages and disadvantages for human life in the era of artificial intelligence. While developing the plans and programs needed for the country, the link between legislation and artificial intelligence should be dominant [61]. Artificial Intelligence and the related computational applications will also be one of our biggest assistants in the construction of smart city formations (which we call as Coronapolis in this chapter) and recovery from global epidemics. Against the non-human virus, non-human Artificial Intelligence could be the hope of humans (us) all over the world. As it is observed nowadays, Internet of Things (IoT) oriented advancements are somehow big steps of all these innovative movements towards the future.

References

1. Collins R, Makowsky M (2014) The discovery of society, 8th (edn), Translation, Translation Editor: Prof Dr Nurgün Oktik, Ankara: Nobel Publications.

2. Hardt M, Negri A (2015) Empire, (Empire) Translated from English by: Abdullah Yılmaz, 8th edn. Detay Publications, Istanbul
3. Tanyeli U (2017) Making by demolition (Base for An Anarchist Architectural Theory), (Building by Demolition, A Base for an Anarchist Architectural Theory), 2nd edn. Metis Publications, Istanbul
4. Saramago J (2020) Blindness, 17th edn, Translated by: Işık Ergüden. Red Cat Publishing House, Istanbul
5. Edgerton RB (2016) Sick societies (A Challenge to the Legend of the Primitive Order), 4th edn, Translated by: Harun Turgut. Buzdağı Publications, Ankara
6. Cevizoğlu H (2019) Body and technology (Philosophical and Anthropological Investigation). Science and Art Publications, Ankara
7. Bauman Z, Lyon D (2013) Liquid surveillance, Translated from English by: Elçin Yılmaz. Details Publications, Istanbul
8. Simmel G (2015) Individualism and culture, (Individualism and Culture) Translated by: Tuncay Birkan, 2nd edn. Metis Publications, Istanbul
9. Erdal YS (2020) Brief summary of the biocultural history of some infections in Anatolia. J Sci Utopia, May 2020, Issue 311. Istanbul: Bilkoop Publication
10. Gürel A, Balta Y (2011) On migration event of Istanbul and ethnic life, (On Migration Event of Istanbul and Ethnic Life). Marmara Sosyal Araştırmalar Dergisi, Istanbul, No 1
11. Kıray M (2007) Urbanization writings, (Essays on Urbanization). Bağlam Yayınları, İstanbul
12. Kıray M (1998) In the least developed countries metropolitanization processes (Metropolitanizatio-in Underdeveloped Countries) 75 year changing urban and architecture (Changing City and Architecture in the 75th Year). Turkey Economic and Social History Foundation Publications, Istanbul
13. Keleş R (2002) Urbanization policy. Image Publishing House, Ankara
14. Demir K, Çabuk S (2019) Population development of metropolitan cities in Turkey (Demographic Development of Metropolitan Cities). http://www.acarindex.com/doslick/article/acarindex-1423879631.pdf. Accessed 26 May 2019
15. Şen B (2012) Depression tendencies in the metropolitan urban centers, (Case of Istanbul, Eminönü, Süleymaniye District) SDÜ. FEAS J (2)
16. Coşkun M (2008) Village associations as migrants, (Graduate Thesis) master thesis, Ankara Bilkent University, Cited in: Ayça Kurtoğlu, Immigration as a Spatial case and associations as a place of citizenship, (Fellow Countrymen as a Spatial Phenomenon and Associations as a Space for Fellow Countrymen). Eur J Turkish Stud 2, AKÜ, J Acad Stud C.3(1)
17. Özkiraz A, Acungil Y (2012) The role of fellow countrymen associations in urbanization process, the tokat case, HU. J Fac Lett 29(1)
18. Özerkmen N, Kahya Y (2008) The role of fellow countrymen's relations in the formation of criminal organizations. J Police Sci, C10(4). http://www.acarindex.com/polis-bilimleri-dergisi/Theroleoforganizing-suc-organizations-hemsehrylic-relations-22007#.V0ehGpGLTIU. http://www.acarindex.com/dosyalar/makale/acarindex-1423911085.pdf. Accessed 26 May 2019
19. Sağlam S (2006) Internal migration and urbanization in Turkey (The Phenomenon of Internal Migration and Urbanization in Turkey), TRC (HU Institute of Turkish Studies), Ankara, Issue 5
20. OdaTV (2020) Google reports human activities. https://odatv4.com/google-insanlarin-tum-hareketliligi-raporlayip-acikladi-06042013.html. Accessed 16 Apr 2020
21. Zizek S (2020) We are drawn into a disgusting barbarism. Interview with Tomasz Kurianowicz, Trans Nurcan Dikme Yaşar, Birgün Newspaper. https://www.birgun.net/haber/marksist-sosyolog-filozof-ve-kultur-elestirmeni-slavoj-zizek-igrenc-bir-barbarliga-dogru-suruklenenuz-325363. Accessed 05 Dec 2020
22. Haraway DJ (2010) Elsewhere, Translated by: Güçsal Pusar. Metis Publications, Istanbul
23. Latour B (2014) In: Ed. Robert CS, Val D (eds) Actor-Network theory (ANT), Philosophy of technology (The Technological Condition An Anthology), 2nd edn. Wiley Blackwell, UK, pp 278–288

24. Wolin R (2012) Heidegger's children (Hannah Arendt, Karl Löwith, Hans Jonas and Herbert Marcuse), Translated by: Hüsamettin Arslan. Paradigma Publishing, Istanbul
25. Merleau-Ponty M (2017) Phenomenology of perception, Translators: Emine Sarıkartal and Eylem Hacımuratoğlu. Ithaki Publications, Istanbul
26. Savaşçın Z (2017) Perceptual Belief. The skin of the world (Studies on Merleau-Ponty Philosophy), 3rd edn, by: Zeynep Direk. Metis Publications, Istanbul, pp 127–153
27. Enlil ZM (2000) Refunctioning and gentrification: Retrieval of Old (Urban Centers and Historical Residance Fabric as a Class Project) domus m, No 8
28. Erdoğan İ, Alemdar K (2010) Other theory, (Other Theory) renewed, 3rd edn. Pozitif Matbaacılık, Ankara
29. Cevizoğlu H (2017) Compositio (On Sociology, Psychology, Philosophy of Religion and Astronomy), 4th edn. Ceviz Kabuğu Yayınları, Ankara
30. Baran AG, Suğur S (eds) (2014) Contemporary sociological theories, 3rd edn. Anadolu Üniversitesi Yayınları, Eskişehir
31. Butler J (2012) Bodies and power, repetition, Translated by Şeyda Öztürk, Cogito, Issue: 70–71. Yapı Kredi Publications, Istanbul
32. Barbaras R (2017) From the phenomenology of the body to the ontology of the skin. Maurice Merleau-Ponty, Translated by: Gül Tonak, Cogito. Yapı Kredi Publications, Istanbul, pp 154–192
33. Mauss M (2013) Sociology and anthropology, 4th edn, Translated by: Özcan Doğan. East West Publications, Ankara
34. Melucci A (2013) Actor self, Translated by: Başak Kıcır. Sel Publishing, Istanbul
35. Curatorial Team (2016) In "Homo Cellular", are we human, 3rd Istanbul Design Biennial (22 Oct–20 Nov 2016). İKSV Publication, Istanbul, pp 270–275
36. Works S (2016) In "Portable Person", we are human, 3rd Istanbul Design Biennial (22 Oct–20 Nov 2016). İKSV Publication, Istanbul, pp 422–425
37. Curatorial Team (2016) Are we human, 3rd Istanbul Design Biennial (22 Oct–20 Nov 2016). İKSV Publication, Istanbul
38. Basalla G (2013) Evolution of technology, 14th edn, Translated by: Cem Soydemir. East West Publications, Ankara
39. Robins K (2013) Image (Culture and Policy of Seeing), 2nd Edition, Translated by: Nurçay Türkoğlu. Details Publications, Istanbul
40. Tiles and Oberdiek (2014) In: Robert CS, Val D (eds) Conflicting visions of technology, philosophy of technology (The Technological Condition an Anthology), 2nd edn. Wiley Blackwell, UK, pp 249–259
41. Latour B (2008) We have never been modern (Symmetrical Anthropology Essay), Translated by: İnci Uysal. Norgunk Publishing, Istanbul
42. Sennett R (2013) Craftsman, 2nd edn, Translated by: Melih Pekdemir. Details Publications, Istanbul
43. Kakoudaki D (2017) Robot Anatomy (Artificial Human in Literature, Cinema and Cultural Studies), Translated by: Deniz Aras. Collective Book, Istanbul
44. Tanyeri E (2018) Crisis of coexistence in stiegs thought: dystopia, robots and pharmacological technology. In: Cogito: today's dystopia (Focus: Citizenship Regime, Sexualities and Body Politics). YKY, Istanbul
45. Lovitt W (2017) A talk on technology with heidegger, heidegger: technology and the future of humanity, compiled and Translated by: Ahmet Aydoğan. Say Publishing, Istanbul, pp 78–111
46. Jonas H (2014) In: Robert CS, Val D (eds) Toward a philosophy of technology, philosophy of technology (The Technological Condition an Anthology), 2nd edn. Wiley Blackwell , UK, pp 210–223
47. Sismondo S (2016) Science and technology research approach: foundations, Translated by: Serkan Sayğan- Ümit Tatlıcan. Epos Publications, Ankara
48. Winner L (2014) In: Robert CS, Val D (eds) Do artifacts have politics, philosophy of technology (The Technological Condition an Anthology), 2nd edn. Wiley Blackwell, UK, pp 668–680

49. Tınaz Ö, ve Cevgen C (2020) The epidemic triggered reverse migration in the USA. https://www.amerikaninsesi.com/a/salgin-abd-de-tersine-gocu-tetiklenen/5687801.html. Accessed 17 Dec 2020
50. Graham S (2020) The vertical world (From Satellites to Shelters), Translated by: Ali Karatay. Koç University Press, Istanbul
51. Constable H (2020) Coronavirus: how cities can be built against epidemics? https://www.bbc.com/turkce/haberler-dunya-52531299. Accessed 17 Dec 2020
52. Onedio (2020) Atatürk's ideal republic villages. https://onedio.com/haber/enginlere-sigmayan-dehasiyla-ataturk-un-tasarladigi-super-proje-ideal-cumhuriyet-koyu-760648. Accessed 10 Dec 2020
53. Hájek P (2013) Metamathematics of fuzzy logic, vol 4. Springer Science & Business Media
54. Kose U (2012) Fundamentals of fuzzy logic with an easy-to-use, interactive fuzzy control application. Int J Mod Eng Res 2:1198–1203
55. Köse U (2017) Development of artificial intelligence based optimization algorithms (Doctoral dissertation, Selçuk University Institute of Science, Konya, Turkey)
56. Weber W, Rabaey J (2005) Ambient intelligence. Springer Science & Business Media
57. Mohri M, Rostamizadeh A, Talwalkar A (2018) Foundations of machine learning. MIT Press
58. Sra S, Nowozin S, Wright SJ (eds) (2012) Optimization for machine learning. MIT Press
59. LeCun Y, Bengio Y, Hinton G (2015) Deep learning. Nature 521(7553):436–444
60. Karaca B, Telli G (2019) The role of artificial intelligence in various processes and prediction function. In: Telli G (ed) Artificial intelligence and the future. Doğu Kitabevi Yayınları, Istanbul, ss 172–186
61. Demir O (2019) Artificial intelligence for sustainable development. In: Telli G (ed) Artificial intelligence and the future. Doğu Kitabevi Yayınları, Istanbul ss 44–65

Chapter 18
Selection of the Best Neonatal Ventilator in Patients with COVID-19 Using Multicriteria Analysis

Alberto Ochoa-Ortiz, Oliverio Cruz-Mejía, and José Mejia

Abstract The multicriteria analysis is a useful tool that helps us to solve problems in which it is necessary to analyze the various variables present. In this work, the model of AHP and Grand Prix will be used to select the best neonatal ventilator among seven options of neonatal ventilators; to achieve this, different variables will be analyzed that will help us determine which is the best ventilator according to the criterion taken. When using both multicriteria methods, it was obtained that by AHP the ventilator that presents a better performance is the SynoVent E5 widely used in the current pandemic named COVID-19. With the Grand Prize model, the best option corresponds to the Humming Vue ventilator. It is of utmost importance to make the correct decision associated with the best ventilator, largely because the infants affected with COVID-19 require a constant flow of oxygen and with an adequate temperature for the timely breathing of their fragile bodies, including the wear and tear from the effort to breathe adequately. Neonates with serious problems may be in an incubator with periods from 7 weeks as a maximum limit to values of 27 weeks in the case of infants weighing <0.500 kg.

Keywords Neonatal ventilator · Mechanical ventilation · High-frequency ventilation · Inspiratory time · Expiratory time · Neonates with COVID-2019

A. Ochoa-Ortiz · J. Mejia
Universidad Autónoma de Ciudad Juarez, Ciudad Juárez, México
e-mail: alberto.ochoa@uacj.mx

J. Mejia
e-mail: jose.mejia@uacj.mx

O. Cruz-Mejía (✉)
Universidad Autónoma del Estado de México, Nezahualcóyotl, México
e-mail: ocruzm@uaemex.mx

U. Kose et al. (eds.), *Computational Intelligence for COVID-19 and Future Pandemics*,
Disruptive Technologies and Digital Transformations for Society 5.0,
https://doi.org/10.1007/978-981-16-3783-4_18

1 Introduction

The 2019–2020 coronavirus disease pandemic is a 2019-initiated coronavirus disease pandemic (COVID-19), caused by the severe acute respiratory syndrome coronavirus (SARS-CoV-2). It was first identified on 1 December 2019 in the city of Wuhan, the capital of Hubei Province in central China, when a group of people were reported to have pneumonia of unknown cause, linked mainly to workers in the southern Chinese wholesale seafood market of Wuhan. The WHO recognized it as a global pandemic on 11 March 2020. As of 8 April 2020, over 1.43 million cases of IDOC-19 have been reported in over 227 countries and territories, resulting in over 82,100 deaths and over 301,000 cases of recovery. The United States is at the top of the list of countries with the highest number of infected, followed by Spain, Italy, Germany, France and China. The virus is usually transmitted from one person to another via the tiny droplets, known as Flügge droplets that are emitted when talking, sneezing, coughing or breathing out. It is mainly not only spread when people are in close contact but can also be spread by touching a contaminated surface and then touching one's face. The incubation period is usually 5 days but can range from 2 to 14 days. The most common symptoms are fever, dry cough and breathing difficulties. Complications can include pneumonia, acute respiratory syndrome or sepsis. There is no vaccine or specific antiviral treatment yet. The main treatment is symptomatic and supportive therapy. Recommended prevention measures include hand washing, covering the mouth when coughing, the physical distance between people and the use of masks, as well as self-isolation and follow-up for those suspected of being infected. Currently in Mexico, 7 of every 1,000 children die in the country in the first 4 weeks of life, this corresponds to 0.7% of births per year. In recent years, the neonatal mortality rate has been decreasing thanks to investments in the health sector and the development of new medical technologies. In the 1990s and the early 2000s, the neonatal mortality rate was 1.6%, according to information provided by OMS, although this number has been decreasing, the neonatal mortality rate worldwide corresponds to 41% of the deaths of children under 5 years of age is considered an increase because, in the early 90s, the mortality rate corresponded to 37%.

The three main causes of death in newborns according to the World Health Organization [1] are the following: preterm births, asphyxia, serious infections such asepsis and atypical pneumonia.

With an increase in the care of mothers and babies during the critical moment, at least two thirds of the neonatal deaths present in the country could be prevented. Based on the values of the mortality rate and the causes of neonatal death described above, good neonatal care can increase the life chances of the newborn. Prevention is directed toward information, monitoring and preventive treatment of the population to avoid health problems. The actors involved can be the public, social groups with a propensity or tendency toward certain health problems, doctors, specialists and government agencies, the pharmaceutical industry, etc. The area of health applications is moving toward the integration and personalization of functions, where prediction is also an important part of prevention. The use of TIC'S and environmental intelligence systems for monitoring health and behavior related to health can provide information to make the decision. Through the development of "lab on a chip" technologies for intelligent environments, many devices for self-diagnosis will be available. In the neonatal intensive care unit, we can find several medical teams that help the patient during their treatment, in this work, seven different neonatal ventilators will be analyzed in order to choose, which has a more optimal functioning and offers better parameters for the early recovery of the newborn. To start analyzing the different proposed ventilators, AHP matrices will be created first, which allows us to compare the different variables to be analyzed between pairs, these will be punted according to the determined criteria, and the values of scores obtained will be analyzed in order to take the best possible decision. The Grand Prix model will allow us to evaluate and determine the impact of the human factor on our product, this model will be used since the neonatal ventilators are operated by doctors and nurses, also the Grand Prix model will allow us to analyze the associated economic variables. The lung function of premature babies is compromised by several factors, among which are central neurological immaturity and respiratory muscle weakness, associated with a lung with poor alveolar development, lack of surfactant synthesis and increased thickness of the alveolar-capillary membrane [2]. At the respiratory level, alveolar collapse and atelectasis develop due to the low production of surfactant; characteristic data of immaturity of the respiratory center, responsible for apnea crisis, leading to require in most cases ventilator support [3]. Failure or respiratory failure can be of two types:

(a) In oxygenation, characterized by a decrease in the partial pressure of oxygen in the arterial blood.
(b) In ventilation, characterized by an increase in the partial pressure of CO_2 in the arterial blood.

For the treatment of the first one, it is possible to resort to the increase in the partial pressure of the oxygen in the inspired air. In contrast, since the partial pressure of CO_2 in ambient air is close to zero, the only way to remedy the second is to improve alveolar ventilation. In addition, an increase in alveolar ventilation will also improve oxygenation [4]. Ventilators are controlled electromechanical pneumatic devices for the treatment of respiratory insufficiency by means of external support to ventilation coupled with an increase in the partial pressure of oxygen in the inspired air. The objectives of artificial ventilation are to achieve an adequate exchange of O_2 and CO_2, reduce the patient's respiratory work and minimize damage to the lungs and airways. A ventilator is a medical device that generates a flow of gas in the patient's airways. The ventilator controls both the direction and magnitude of the flow, as well as its pressure, humidity, temperature and gas mixture. You can control the duration of the different phases of inspiration and expiration and monitor various parameters during them. The respiratory rate in the neonatal patient can range between 30 and 120 breaths per minute, so the neonatal ventilator requires a rapid response in the inspiratory and expiratory phase change and a very precise control in the volume and pressure, for which reason it is necessary to maintain a constant flow and to activate a valve so that the flow is directed intermittently toward the patient and the atmosphere. In the ventilators for neonates, there is not the same variety of modalities of artificial ventilation as in the ventilators for older children or adults, instead, the so-called high-frequency ventilation, greater than 120 breaths per minute, is used, in which the current volume may be less than anatomical dead space, but gas exchange is achieved by complicated dispersion phenomena and turbulent flow. High-frequency ventilation has the advantage of producing lower alveolar pressure than low frequency ventilation, but it is not equally effective in any pathological condition. It can be used as a single mode or in combination with low-frequency ventilation. The decision about the use of one or another modality depends on the conditions of the neonate and on the experience and clinical judgment of the operator [5]. Since there is a wide variety of ventilators, it is important to classify them according to their function and the ventilation mode they provide.

1.1 Conventional Mechanical Ventilation

The conventional term is used to distinguish current ventilation from high-frequency ventilation. Traditional ventilation modes describe how mechanical respirations are delivered to patients as well as the relationship between mechanical and spontaneous breaths. These modes include Intermittent Mandatory Ventilation (IMV), Intermittent Mandatory Synchronized Ventilation (SIMV), Assist-Controlled Ventilation (A/C), Pressure Support Ventilation (PSV), Ventilation triggered by the patient's effort.

1.2 High-Frequency Ventilation

This ventilatory modality achieves adequate alveolar ventilation using very low current volumes, equal to or lower than the dead space (Vd), i.e. 2 ml/kg, at frequencies well above the physiological (more than 3 Hz). Theoretically, it presents a series of advantages derived from the low volume used and the high frequency of ventilation. Thanks to the vibratory energy that it produces, the mobilization of pulmonary secretions is favored and also an effective exchange of CO_2 and O_2 is achieved, with lower peak pressures at the alveolar level, minimal variations in pressures and ventilation volumes, maintaining lungs with a relatively constant volume, above their residual functional capacity thanks to the application of a stable mean airway pressure (MAP), with a lower impact on cardiopulmonary function [6]. The first high-frequency ventilator was patented by John Emmerson in 1959, this was an airway vibrator. Subsequently, Luckehmeiker in 1972, studying thoracic impedance in apneic dogs, fortuitously discovered that he could maintain normocapnia with a small volume of air in the airway in animals with frequency of 23–40 Hz (1 Hz = 60 cycles per minute). Subsequently, several investigators demonstrated that adequate alveolar ventilation was possible with current volumes smaller than the anatomical dead space, with supraphysiological frequencies that characterize the high-frequency ventilator [7]. In the present research, seven different models of neonatal ventilators are analyzed, the ventilators to be analyzed are the following:

1.2.1 SLE5000

Developed in collaboration with nurses and doctors, the SLE5000 ventilator uses a unique valve system that allows both a manageable range of parameters and the possibility of having active exhalation. Of course, the SLE5000 has all conventional ventilation modes, which means that it can be used in practically any baby in the intensive care unit [8]. Some of the technical features of this model are: the ability to preset parameters in all modes of operation, powerful HFO with active expiration to cover a wide range of patients, integral flow monitoring, measuring lung mechanics and displaying of loops and waveforms, trending of measured parameters.

1.2.2 VIASYS Bear Cube 750PSV

The VIASYS Bear Cube 750PSV infant ventilators are designed to provide ventilator support for critical care of patients from small neonates, 500 g or more, to the small pediatric patient (up to 30 kg) with compromised lung function. This model has integrated synchronized ventilation and tidal volume monitoring. It also includes independent base and inspiratory flows and internal battery backup [9]. Some additional include integrated synchronized ventilation, tidal volume monitoring, independent base and inspiratory flow, exclusive volume limit, assist control, flow cycled AC, SIM/IMV, flow cycled SIMV, CPAP and SIMV with pressure support.

1.2.3 Dräger Babylog 8000 Plus

The Dräger Babylog 8000 plus is a modern device obtained in neonatal ventilation, due to internal qualities that will benefit the neonate patient. The optional features of pressure support and guaranteed volume, for example, provide a new dimension in ventilation therapy and facilitate the process of removing the baby from ventilator assistance. Both characteristics contribute to offer this ventilator the most valuable resource: its ability to enhance harmonious breathing [10].

1.2.4 Hamilton-C1

The Hamilton-C1 mechanical respirator needs minimal space and combines invasive and noninvasive modes, as well as high oxygen flow therapy with maximum mobility. The integrated high-performance turbine gives the HAMILTON-C1 total independence from compressed air, making it the ideal companion for all patient groups.

Requiring minimal space, the Hamilton C-1 ventilator combines invasive and non-invasive modes, as well as high flow oxygen therapy with maximum mobility. The integrated high-performance turbine gives to this ventilator complete independence from compressed air. In addition, it provides a maximum flow up to 260 l/min, ensuring optimal performance even with non-invasive ventilation [11].

1.2.5 Mindray SynoVent E5

The SynoVent E5 ventilator is developed with features that can meet all your needs. It not only provides advanced ventilation modes and functionality for patients ranging from infant to adult but also features a modern, easy-to-use interface to enhance the efficiency of your clinical workflow. The intuitive interface, it allows quick and easy control over ventilator parameters and settings. The 12.1 inch screen can display up to four simultaneous waveforms or two loops. With a smartly designed user interface that combines a touch screen with direct access knobs, users can quickly learn to navigate the SynoVent E5 and put all of its features into clinical use. E5 ventilator supports invasive, non-invasive ventilation and high flow O_2 therapy, which can fulfill different clinical needs. It supports multiple advanced ventilation modes such as Duolevel, APRV, PRVC and PRVC-SIMV. Mindray's leading gas analysis technology can be integrated to enable $EtCO_2$ monitoring [12].

1.2.6 Neumovent Graphnet Advance

Respirator provides invasive and non-invasive ventilation adaptable to all patient categories: Wide variety of ventilatory modes and monitoring capacity. The GraphNet advance respirators present a clear and complete visualization of vital variables through its integrated 12-inch LED technology screen, user-friendly and intuitive programming, touch-sensitive screen, hot keys, upper alarm indicator light. In this ventilator, it is possible to monitor the most important parameters, from the beginning of the therapy, with P/V Flex, as well as the intermediate phases toward weaning, with Primax among other test variables. It also reports exhaled CO_2 and display of volumetric curves over time, as well as measurements of physiological dead space [13].

1.2.7 Nihon Kohden Humming Vue

Nihon Kohden Humming Vue additionally counts on HFO, with SIMV and CPAP modes that allow to cover all the ventilatory needs of patients with a single respirator. For patients from 250 g to 30 kilos (neonatal and pediatric), HFO ventilation provides a sufficient and sustained exchange of gas when the stroke volume is smaller than the anatomical dead space. HFO pressure waves are sinusoidal and symmetrical with respect to the mean airway pressure (MAP) axis. After passing through the endotracheal tube, the amplitude of the HFO pressure wave falls drastically upon entering the trachea so there is minimal pressure variation at the airway periphery near the pulmonary alveoli. Compared with continuous mandatory ventilation (CMV), HFO ventilation has much less pressure swing at any MAP and this greatly reduces physical injury to the pulmonary alveoli [14].

2 Introduction

For the presentation of results, the name of each neonatal ventilator was changed by the corresponding option. Each option corresponds to the following neonatal ventilator (Table 1).

For the selection of the best neonatal ventilator through AHP and Grand Prix, the following criteria and sub-criteria were used:

(A) Ventilation modes
(B) Inspiratory time
(C) Inspiratory pressure
(D) PEEP pressure.

Electric specifications

(E) Battery
(F) Battery operation time.

Graphical data

(G) Waveform
(H) Dimensions
(I) Price.

We can observe the classification of each sub-criteria in Fig. 1.

Table 1 Brand of the neonatal ventilators used

Option number	Ventilator
Option 1	SLE5000
Option 2	VIASYS Bear Cube 750PSV
Option 3	Dräger Babylog 8000 plus
Option 4	Hamilton-C1
Option 5	Mindray SynoVent E5
Option 6	Neumovent Graphnet Advance
Option 7	Nihon Kohden Humming Vue

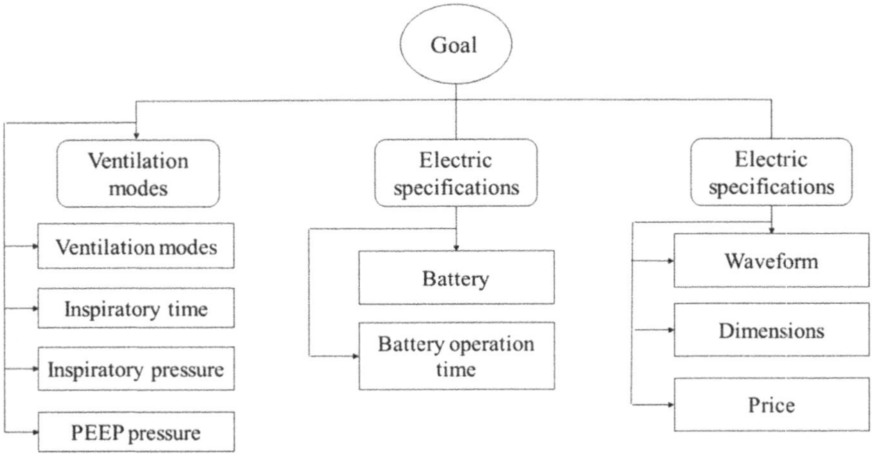

Fig. 1 Criteria and sub-criteria diagram

3 Results

3.1 Analytic Hierarchy Process

For the multicriteria analysis using AHP, matrices were made, and in each of them, a score was assigned to each of the neonatal ventilators.

As can be seen in the matrix corresponding to the price sub-criterion, option 2 is the one with the lowest price, this price is 4.7 times better than that of option 5 and 2 times better for option 4, 6 and 7. The ventilator with the worst price corresponds to option 3, which is 0.53 times better than option 2 (Table 2).

Making the matrix corresponding to the price sub-criterion, the Bear Cube 750 PSV ventilator was obtained as the best option with a score of 0.269209 (Fig. 2).

Table 2 Weighting of the price criterion

Criteria price							
	Option 1	Option 2	Option 3	Option 4	Option 5	Option 6	Option 7
Option 1	1.00	0.29	0.50	0.56	1.37	0.56	0.48
Option 2	3.50	1.00	1.90	2.00	4.70	2.00	2.00
Option 3	2.00	0.53	1.00	1.10	2.50	1.10	0.91
Option 4	1.80	0.50	0.91	1.00	2.40	1.00	0.83
Option 5	0.73	0.21	0.40	0.42	1.00	0.42	0.36
Option 6	1.80	0.50	0.91	1.00	2.40	1.00	0.83
Option 7	2.10	2.10	1.10	1.20	2.80	1.20	1.00

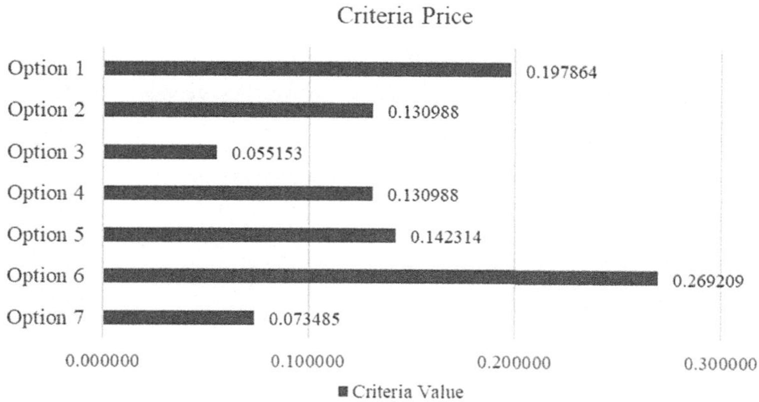

Fig. 2 Price

For the sub-criterion of ventilation modes, the option that obtained the highest score was option 6, this option is 4.7 times better than option 2 and 3 times better than option 1. sub-criterion, option 2 is the one with the lowest price, this price is 4.7 times better than that of option 5 and 2 times better for option 4, 6 and 7. The ventilator with the worst price corresponds to option 3, which is 0.53 times better than option 2 (Table 3).

For the ventilation mode sub-criterion, the best option corresponds to the Neumovent Graphnet Advance ventilator with 0.2566, while the worst ventilator for this criterion is the Bear Cube 750 PSV (Fig. 3).

When making the matrix corresponding to the inspiration time sub-criterion, it was found that options 3 and 5 are better compared with the others, both being 5 times better than options 1, 2, 4, 6 and 7 (Table 4).

In the inspiratory time sub-criterion, there was a tie between the Dräger Babylog 8000 plus and Mindray SynoVent E5 ventilators with 0.3333 points (Fig. 4).

Table 3 Weighting of the ventilation modes criterion

Criteria ventilation modes							
	Option 1	Option 2	Option 3	Option 4	Option 5	Option 6	Option 7
Option 1	1.00	1.70	0.80	0.42	0.60	0.40	0.70
Option 2	0.70	1.00	0.50	0.25	0.40	0.20	0.20
Option 3	1.20	2.00	1.00	0.50	0.80	0.40	0.90
Option 4	2.40	4.00	2.00	1.00	1.50	0.90	1.70
Option 5	1.60	2.70	1.40	0.70	1.00	0.60	1.20
Option 6	3.00	4.70	2.30	1.20	1.80	1.00	2.00
Option 7	1.40	2.30	1.17	0.60	0.90	0.50	1.00

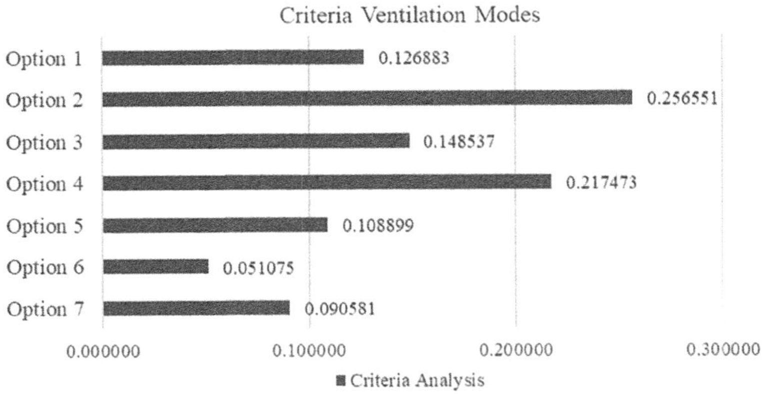

Fig. 3 Ventilation modes

Table 4 Weighting of the inspiratory time criterion

Criteria inspiratory time							
	Option 1	Option 2	Option 3	Option 4	Option 5	Option 6	Option 7
Option 1	1.00	1.00	0.20	1.00	0.20	1.00	1.00
Option 2	1.00	1.00	0.20	1.00	0.20	1.00	1.00
Option 3	5.00	5.00	1.00	5.00	1.00	5.00	5.00
Option 4	1.00	1.00	0.20	1.00	0.20	1.00	1.00
Option 5	5.00	5.00	1.00	5.00	1.00	5.00	5.00
Option 6	1.00	1.00	0.20	1.00	0.20	1.00	1.00
Option 7	1.00	1.00	0.20	1.00	0.20	1.00	1.00

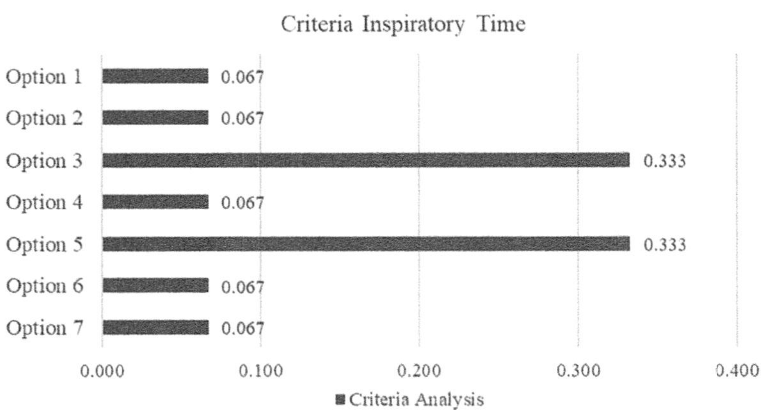

Fig. 4 Inspiratory time

As can be seen in the following table, options 5 and 6 are the best for this sub-criterion, both are 1.7 times better than option 4 and 1.6 than option 1. The worst option is option 4, this is 0.6 times better than the options 5 and 6 (Table 5).

The Mindray SynoVent E5 and the Neumovent Graphnet Advance ventilators present the best inspiratory pressure according to the established sub-criterion (Fig. 5).

Options 5 and 6 are the best for the PEEP pressure sub-criterion, option 6 is 3.2 times better than option 3 while option 5 is 2.8 times better than option 3. The worst option for this sub-criterion is option 3, which is 0.3 times better than option 6 and 0.4 times better than option 6 and 0.4 times better than option 5 (Table 6).

The neonatal Mindray SynoVent E5 ventilator presented a better PEEP pressure than the other ventilators with which it was compared obtaining a score of 0.2168 (Fig. 6).

In the sub-criterion corresponding to the battery, we can find that option 5 is 4.2 times better than option 4, option 7 is 0.44 times better than option 3 and option 3

Table 5 Weighting of the inspiratory pressure criterion

Criteria inspiratory pressure							
	Option 1	Option 2	Option 3	Option 4	Option 5	Option 6	Option 7
Option 1	1.00	0.90	1.10	1.00	0.70	0.80	0.80
Option 2	1.20	1.00	1.10	1.20	0.80	0.70	0.90
Option 3	1.10	1.10	1.00	1.30	0.70	0.70	0.90
Option 4	0.90	0.90	0.90	1.00	0.60	0.60	0.75
Option 5	1.60	1.40	1.40	1.70	1.00	1.00	1.25
Option 6	1.60	1.40	1.40	1.70	1.00	1.00	1.25
Option 7	1.20	1.20	1.20	1.40	0.80	0.80	1.00

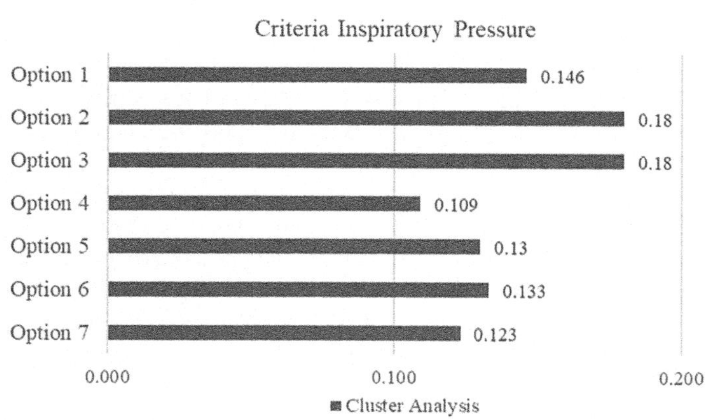

Fig. 5 Inspiratory pressure

Table 6 Weighting of the PEEP pressure criterion

Criteria PEEP pressure							
	Option 1	Option 2	Option 3	Option 4	Option 5	Option 6	Option 7
Option 1	1.00	0.70	1.30	0.80	0.50	0.40	0.70
Option 2	1.23	1.00	1.90	1.20	0.70	0.60	1.00
Option 3	0.80	0.50	1.00	0.60	0.40	0.30	0.60
Option 4	1.20	0.80	1.60	1.00	0.60	0.50	0.90
Option 5	2.20	1.50	2.80	2.30	1.00	0.90	1.50
Option 6	2.50	0.70	3.20	2.00	1.20	1.00	1.70
Option 7	1.50	1.00	1.90	1.20	0.70	0.60	1.00

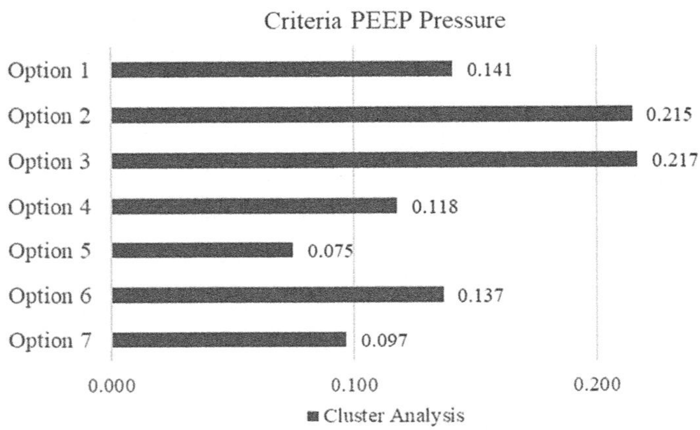

Fig. 6 PEEP pressure

is 4.5 times better than option 4. The worst option for this sub-criterion is option 4 (Table 7).

Another sub-criterion considered was the voltage of the batteries, for this sub-criterion, it was found that the Dräger Babylog 8000 plus had a better voltage obtaining a score of 0.253929 (Fig. 7).

The sub-criterion of battery operation time helps us to identify how long the ventilator can last in operation after being disconnected from the power supply. The ventilator that has a longer operating time with battery is the option 4, this option is 5.33 times better than option 6 and 8 times better than option 2 (Table 8).

Battery operating time was one of the most important sub-criteria when performing this analysis. The ventilator that offers a longer operating time with batteries is the HAMILTON-C1/T1/MR1 (Fig. 8).

For this sub-criterion, the number of waveforms that each ventilator can display on its screen is evaluated, while the largest number of waveforms that can be seen is

Table 7 Weighting of the battery criterion

Criteria battery							
	Option 1	Option 2	Option 3	Option 4	Option 5	Option 6	Option 7
Option 1	1.00	1.30	0.50	2.50	0.60	1.40	1.30
Option 2	0.80	1.00	0.40	2.00	0.48	1.20	1.00
Option 3	2.00	2.30	1.00	4.50	1.08	2.50	2.30
Option 4	0.40	0.50	0.20	1.00	0.24	0.60	0.50
Option 5	1.70	2.00	0.93	4.20	1.00	2.30	2.00
Option 6	0.72	1.10	0.40	1.80	0.40	1.00	1.10
Option 7	0.80	1.00	0.44	2.00	0.48	1.11	1.00

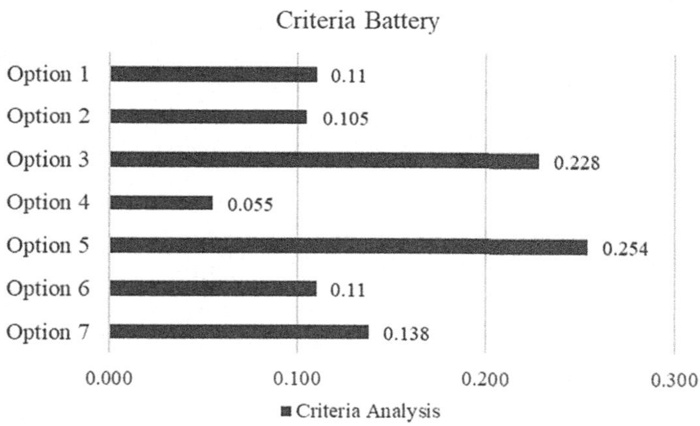

Fig. 7 Battery

Table 8 Weighting of the battery operation time criterion

Criteria battery operation time							
	Option 1	Option 2	Option 3	Option 4	Option 5	Option 6	Option 7
Option 1	1.00	2.00	0.67	0.25	0.33	0.50	1.33
Option 2	0.50	1.00	0.33	0.13	0.17	0.25	0.67
Option 3	1.50	3.00	1.00	0.38	0.50	0.75	2.00
Option 4	4.00	8.00	2.67	1.00	1.33	2.00	5.33
Option 5	3.00	6.00	2.00	0.75	1.00	1.50	4.00
Option 6	2.00	4.00	1.33	0.50	0.67	1.00	2.67
Option 7	0.75	0.15	0.50	0.19	0.25	0.38	1.00

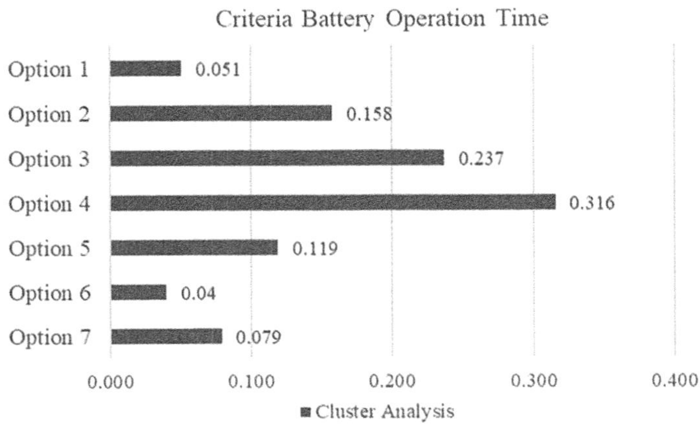

Fig. 8 Battery operation time

Table 9 Weighting of the waveforms criteria

Criteria waveforms							
	Option 1	Option 2	Option 3	Option 4	Option 5	Option 6	Option 7
Option 1	1.00	3.00	1.50	0.60	1.00	0.50	0.60
Option 2	0.10	1.00	0.10	0.10	0.10	0.10	0.10
Option 3	0.67	2.00	1.00	0.40	0.67	0.33	0.40
Option 4	1.67	5.00	2.50	1.00	1.67	0.83	1.00
Option 5	1.00	3.00	1.50	0.60	1.00	0.50	0.60
Option 6	2.00	6.00	3.00	1.20	2.00	1.00	1.20
Option 7	1.67	5.00	2.50	1.00	1.67	0.83	1.00

better. The best option for this criterion is the one corresponding to the Bear CUBE 750 PSV ventilator (Table 9).

The waveform criterion refers to how many graphs the ventilator can show; the ventilator that has more waveforms is the GraphNet advance (Fig. 9).

Option 7 is the one that has a better performance for this sub-criterion, this option is 3.2 better than option 3 and 2.08 better than option 1. The worst option is number 3, this option is 0.5 times better than option 5 and 0.84 better than option 2 (Table 10).

The dimensions of the ventilator are an important sub-criterion since a bigger screen allows to visualize a better way the graphs that the ventilator offers, the ventilator with better dimensions is the Nihon Kohden Humming Vue (Fig. 10).

When analyzing each of the sub-criterion, a comparison matrix was made in pairs where the best neonatal ventilator was obtained, the Mindray SynoVent E5 ventilator with a score of 0.1973. The purpose of the matrix of comparison by pairs is to score each of the criteria to determine the order of importance for making this decision. In

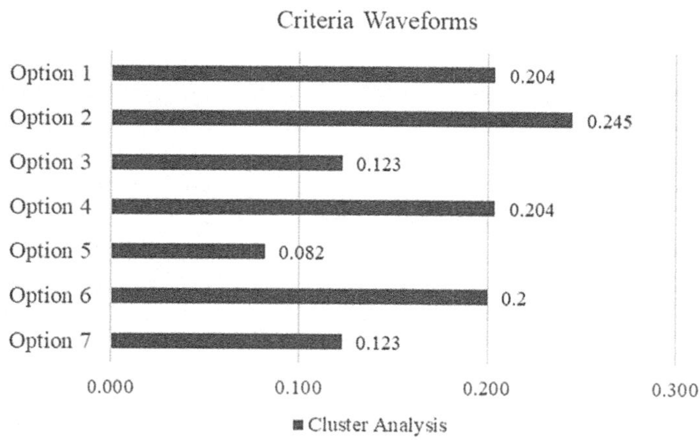

Fig. 9 Waveforms

Table 10 Weighting of the dimensions criteria

Criteria dimensions							
	Option 1	Option 2	Option 3	Option 4	Option 5	Option 6	Option 7
Option 1	1.00	1.30	1.55	1.07	0.76	1.10	0.50
Option 2	0.80	1.00	1.20	0.82	0.60	0.83	0.37
Option 3	0.65	0.84	1.00	0.69	0.50	0.70	0.31
Option 4	0.94	1.22	1.45	1.00	0.70	1.02	0.45
Option 5	1.30	1.70	2.00	1.39	1.00	1.42	0.63
Option 6	0.92	1.20	1.43	0.98	0.70	1.00	0.44
Option 7	2.08	2.70	3.20	2.20	1.60	2.25	1.00

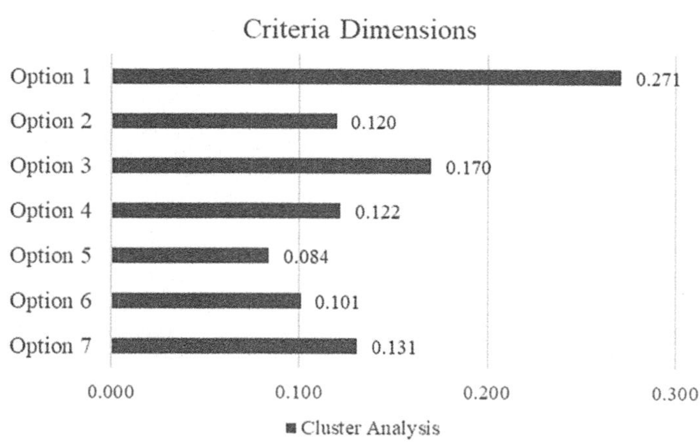

Fig. 10 Dimensions

the tables presented below, the weight of each criterion can be observed with respect to the other (Table 11).

In the matrix of comparison by pairs for the criterion of electrical specifications, it is observed that the most important sub-criterion is the operating time with battery (Tables 12 and 13).

The score obtained by each ventilator for each group of criteria and sub-criteria can be seen in the following tables (Table 14).

The score obtained by each ventilator in the criterion of electrical specifications is shown (Table 15):

Table 11 Matrix of comparison by pairs graphical data criteria

Matrix of comparison by pairs graphical data			
	Price	Waveforms	Dimensions
Price	1.00	0.25	1.00
Ventilation Modes	6.00	2.00	0.50
Inspiratory time	4.00	2.00	0.33
Inspiratory Pressure	5.00	4.00	0.50
PEEP pressure	2.00	6.00	0.20
Battery	0.25	0.20	0.50
Battery operation time	0.33	2.00	6.00
Waveforms	4.00	1.00	0.25
Dimensions	1.00	4.00	1.00

Table 12 Matrix of comparison by pairs electric specification criteria

Matrix of comparison by pairs electric specifications		
	Battery	Battery operation time
Price	4.00	3.00
Ventilation modes	5.00	1.50
Inspiratory time	3.00	3.00
Inspiratory Pressure	4.00	2.00
PEEP pressure	6.00	4.00
Battery	1.00	0.17
Battery operation time	6.00	1.00
Waveforms	5.00	0.50
Dimensions	2.00	0.17

Table 13 Matrix of comparison by pairs ventilation modes criteria

Matrix of comparison by pairs electric specifications		
	Battery	Battery operation time
Price	4.00	3.00
Ventilation modes	5.00	1.50
Inspiratory time	3.00	3.00
Inspiratory Pressure	4.00	2.00
PEEP pressure	6.00	4.00
Battery	1.00	0.17
Battery operation time	6.00	1.00
Waveforms	5.00	0.50
Dimensions	2.00	0.17

Table 14 Score for each ventilator in the graphical data criterion

Graphical data			
Model alternative	Price	Waveforms	Dimensions
SLE5000	0.0735	0.0953	0.1314
Bear Cub 750 PSV	0.2692	0.1153	0.1013
Dräger babylog 8000 plus	0.1423	0.1661	0.0844
HAMILTON- C1/T1/MR1	0.1310	0.1447	0.1219
SynoVent E5	0.0552	0.1999	0.1701
GraphNet advance	0.1310	0.1588	0.1202
Humming Vue	0.1979	0.1200	0.2707
Weighing	0.0643	0.0620	0.1853

Table 15 Electric specifications

Electric specifications		
Model alternative	Battery	Battery operation time
SLE5000	0.1377	0.0791
Bear Cub 750 PSV	0.1101	0.0395
Dräger babylog 8000 plus	0.2539	0.1186
HAMILTON- C1/T1/MR1	0.0550	0.3162
SynoVent E5	0.2282	0.2372
GraphNet advance	0.1046	0.1581
Humming Vue	0.1105	0.0513
Weighing	0.0237	0.1253

Table 16 Ventilation modes

Ventilation modes				
Model alternative	Ventilator and modes	Inspiratory and time	Inspiratory pressure	PEEP pressure
SLE5000	0.0906	0.0667	0.1226	0.0967
Bear Cub 750 PSV	0.0511	0.0667	0.1329	0.1374
Dräger babylog 8000 plus	0.1089	0.3333	0.1303	0.0751
HAMILTON -C1/T1/MR1	0.2175	0.0667	0.1087	0.1180
SynoVent E5	0.1485	0.3333	0.1798	0.2169
GraphNet advance	0.2566	0.0667	0.1798	0.2148
Humming Vue	0.1269	0.0667	0.1460	0.1411
Weighing	0.0988	0.0999	0.1899	0.1509

Table 17 Final score

Final score	
Model alternative	Total
SLE5000	0.1016
Bear Cub 750 PSV	0.1085
Dräger babylog 8000 plus	0.1361
HAMILTON- C1/T1/MR1	0.1475
SynoVent E5	0.1974
GraphNet advance	0.1614
Humming Vue	0.1476

For the criterion of ventilation modes, the score obtained by each ventilator for each sub-criterion can be observed (Table 16).

After evaluating each of the criteria and sub-criteria, it was observed that the best choice of neonatal ventilator is the SynoVent E5 ventilator with a score of 0.1974 (Table 17).

In the following figure, we can see the score obtained by each ventilator according to the AHP method (Fig. 11).

3.2 *Grand Prix Model*

As explained above, the Grand Prix model allows us to select the best neonatal ventilator considering the human and economic factors. The best choice of neonatal

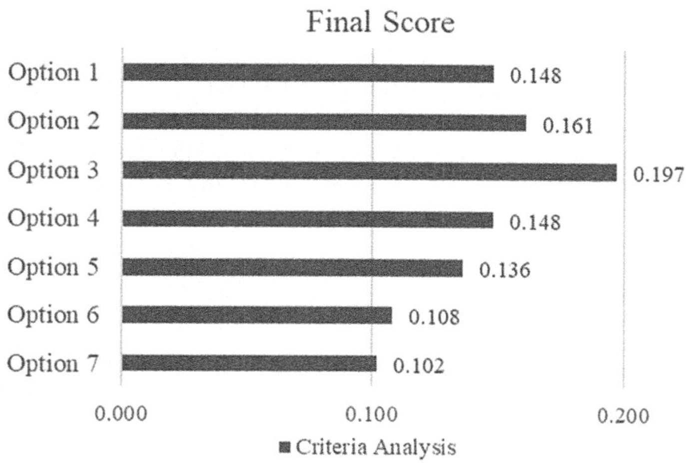

Fig. 11 Final score

ventilator using AHP was the option corresponding to the SynoVent E5 ventilator with a score of 0.1974; but the next two best options do not find with a score so far away from this ventilator. These options are GraphNet advance with 0.161 points and the next option is a tie with 0.148 points between the Humming Vue and HAMILTON-C1/T1/MR1 ventilators (Fig. 12).

Analyzing these three ventilators and comparing them at the price, it is observed that the ventilator with the lowest price is the Humming Vue with a price of $70,000 mx, followed by the GraphNet advance ventilator with a price of $ 80,000 mx. Considering the economic factor, the best option obtained by AHP is not the best option with the Grand Prix model since the cost of that air is around $ 1,000. When analyzing the human factor, we look for the selected option be the one with the most graphical interface that allows the user to learn to operate the equipment in a shorter

Fig. 12 Best options with AHP

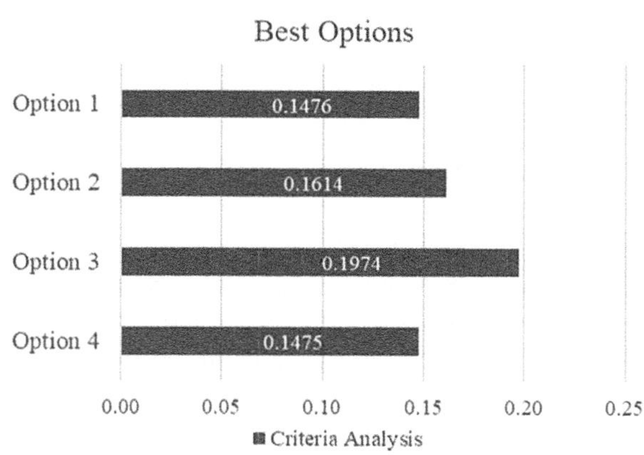

time, this, with the purpose of using the equipment as soon as the purchase, is made and not wait a long time for users to learn how to handle it. The ventilator that offers all these characteristics is the ventilator HAMILTON-C1/T1/MR1, since not being a ventilator so complex allows the learning to happen in a faster way compared with others. With the Grand Prix model, we get two better options of ventilators; these options are the HAMILTON-C1/T1/MR1 and the ventilator Humming Vue.

4 Further Results

An area of opportunity to improve this work would be to implement new criteria that allow us to make a more exact decision, as well as add more options for ventilators to be evaluated, with this new improvement we could analyze more aspects of each ventilator and therefore the criteria and sub-criteria would change.

5 Discussion

As can be seen for both methods of multicriteria analysis, the results obtained are different, this is because when making a decision using AHP, we try to find the perfect option that satisfies all our needs. For this work, that option was the one corresponding to the SynoVent E5, which was the one that has more attached to an ideal score. The Grand Prix model was made after finding the best options of fans with AHP, this in order to compare between those options that are better for the user according to the price and the human factor. After analyzing these two determining factors, it was found that the fans that meet them best are HAMILTON-C1/T1/MR1 and the Humming Vue fan. The Grand Prix model was made after finding the best options of fans with AHP, this in order to compare between those options that are better for the user according to the price and the human factor. After analyzing these two determining factors, it was found that the fans that meet them best are HAMILTON-C1/T1/MR1 and the Humming Vue fan.

6 Conclusions

The multicriteria analysis allows us to make decisions in a better way, that is because all the important criteria for making the best decision are analyzed. With multicriteria analysis, we can adapt our choice to the conditions that need to be met so that the choice is considered and satisfying the needs of each problem.

In the hospital environment, the decisions to buy equipment are mostly in the administration of the hospital, this is a great disadvantage since an administrator doesn't have enough training to know the medical team and know what parameters are important to make the purchase decision Based on this problem, the creation of an IGU that calculates the best option for the purchase of medical equipment is very useful since all the possible criteria that affect this selection are considered in the process to make a correct decision [15]. Considering that the pandemic, if it continues its upward spiral, will have as patients more than two million people with different features [16] in terms of age, sex, health and other attributes, it is necessary to specify a type of ventilator for each one of them from the paradigm of Ambient Intelligence.

References

1. Salud OMDI (2013). Estadisticas Sanitarias Mundiales, OMS
2. JNH (2002) Neonatologia. Editorial Universitaria, Santiago de Chile
3. Social IMDS (2013) Intervencion de Enfermeria en la Atencion del Recien Nacido Prematuro, Mexico: IMSS
4. Hall JE (2011) Tratado de fisiologia medica. Elsevier, Mexico
5. ABM (2003) Ventilación de alta frecuencia en el recién nacido: un soporte rexpiratorio necesario. Rev Chil Pediatr 74(5):475–486
6. EJH (1959) Apparatus for vibrating portions of a patient-s airway. E U Am Pat 1959(2):918–927
7. Salud CNDETES (2004) CENETEC, Agosto. http://www.cenetec.salud.gob.mx/descargas/biomedica/guias_tecnologicas/5gt_ventilador.pdf
8. SLE, "SLE," (2019) https://www.sle.co.uk/files/library/files/products/brochures/SLE5000_-_Spanish.pdf
9. VIASYS, «VIASYS » (2019) https://www.medonegroup.com/pdf/manuals/techManuals/VIASYS-Bear-Cub-750-VentilatorService-Manual.pdf
10. Medical D, "Ingenieria Bahia" (2019) http://www.ingenieriabahia.com.ar/manuales/BabyLog8000Plus.pdf
11. Hamilton, "Hamilton Medical" https://www.hamiltonmedical.com/es/Services/Downloads.html#tags=5db4d6d0–6fed-4b59–806eedd19510d2cf&DataTables_Table_0=od3
12. Mindray, "Procesos Biomedicos," (2014) http://procesosbiomedicos.com/wpcontent/uploads/2014/08/Ventilador-MINDRAYSYNOVENT-E5.pdf
13. Neumovent, "Mecamedcorp," (2013) http://www.mecamedcorp.com/shop/hospitalario/13ventiladores-respiradores/graphnet-advance/
14. Kohden N (2019) "Indomedik," https://www.indomedik.com/file/brochure_hummingvue.pdf
15. Liao YH, Wang ZC, Zhang FG, Abbod MF, Shih CH, Shieh JS (2019) Machine learning methods applied to predict ventilator-associated pneumonia with *Pseudomonas aeruginosa* infection via sensor array of electronic nose in intensive care unit. Sensors 19(8):1866
16. Rehm GB, Kuhn BT, Nguyen J, Anderson NR, Chuah CN, Adams JY (2019) Improving mechanical ventilator clinical decision support systems with a machine learning classifier for determining ventilator mode. CoRR abs/1904.1296

Chapter 19
Blockchain-Based Secure Biomedical Data-as-a-Service for Effective Internet of Health Things Enabled Epidemic Management

Amirhossein Peyvandi, Babak Majidi, and Soodeh Peyvandi

Abstract The Medical Infrastructure-as-a-Service (MIaaS) allows the healthcare providers to share the often expensive and rare smart medical infrastructures to deal with a significant number of patients requiring medical expertise and achieve accurate and rapid diagnosis and treatment. However, in epidemics, many of these computational intelligence-based diagnosis systems require up-to-date trained machine learning models, which are dependent on receiving current data. In this chapter, the concepts of Biomedical Data as a Service (BDaaS) for providing real-time data for training the medical machine learning models, Model Training as a Service (MTaaS) for providing off-site training of machine learning models on Internet of Health Things (IoHT) and Computational-Intelligence-as-a-Service (CIaaS) for using the trained models are proposed and investigated, and a blockchain-based framework for secure BDaaS+MTaaS+CIaaS is proposed. The advantages of decentralized BDaaS+MTaaS+CIaaS are security, agility, cost-effectiveness, data quality, computational intelligence quality, data equality and computational intelligence equality. The proposed framework uses the blockchain network for secure decentralized transfer and sharing of biomedical data and machine learning models on IoHT. Different scenarios exploring the complex dynamics of COVID-19 pandemic and the applications of the proposed framework are investigated.

Keywords Data-as-a-service · Computational intelligence · Blockchain · Machine learning · Pandemic management · Internet of health things

---

A. Peyvandi · B. Majidi
Department of Computer Engineering, Khatam University, Tehran, Iran

B. Majidi (✉)
Emergency and Rapid Response Simulation (ADERSIM) Artificial Intelligence Group, Faculty of Liberal Arts & Professional Studies, York University, Toronto, Canada
e-mail: b.majidi@khatam.ac.ir

S. Peyvandi
Business Intelligence, University of Applied Sciences Upper Austria, Steyr, Austria

U. Kose et al. (eds.), *Computational Intelligence for COVID-19 and Future Pandemics*, Disruptive Technologies and Digital Transformations for Society 5.0,
https://doi.org/10.1007/978-981-16-3783-4_19

1 Introduction

On March 11, 2020, the World Health Organization (WHO) assessed that COVID-19 could be characterized as a pandemic. The fast spread of COVID-19 infected more than 66 million people and killed more than 1.5 million patients as of early December 2020. It is crucial to notice that these numbers may not reveal the real situation, as there are undoubtedly many unknown cases of infected individuals and many deaths, which, especially in early months, attributed to other complications. The lack of preparedness is the leading cause of the inadequacies of many decisions and actions taken at the local, national and international levels to contain the virus. In order to manage the pandemic, biomedical data play a vital role in smart health solutions. However, medical information is the most sensitive and private data about a person and its privacy is paramount. On the other hand, currently, many hospitals are not working under a universal system by which different medical records can be shared across the globe, and if a patient travels to a foreign country or even another hospital, the doctor examining that person does not have access to the previous records. Furthermore, smart healthcare systems are needed in pandemic conditions in order to deal with a significant number of patients requiring medical expertise and to achieve an accurate and rapid diagnosis and treatment. In newly developed epidemics such as COVID-19 pandemic, these smart medical systems require up-to-date trained machine learning models and current data. Biomedical Data as a Service (BDaaS) can provide real-time data for training these medical machine learning models. Furthermore, machine learning Model Training as a Service (MTaaS) can provide off-site model training on Internet of Health Things (IoHT) and Computational-Intelligence-as-a-Service (CIaaS) can be used for transferring these trained models across the IoHT.

In this chapter, a blockchain-based framework for secure BDaaS+MTaaS+CIaaS is proposed, which can provide security, agility, cost-effectiveness, data quality, computational intelligence quality, data equality and computational intelligence equality for smart medical infrastructure across the IoHT. The proposed framework uses the Stellar blockchain network for secure decentralized transfer and sharing of biomedical data and machine learning models on IoHT. First, a framework based on a blockchain network in which data owners can share their data without violating their privacy is introduced. In the proposed decentralized framework, the models will be trained on the data owner's side and off-chain. Only the learned model parameters will be shared on the blockchain. The applications of machine learning-based models for predicting the spread and infection dynamics of the COVID-19 pandemic are investigated. As there is a large probability that the next waves of the pandemic will be more prominent due to coinciding with the colder seasons and influenza epidemic, the modeling and forecasting of the spread and infection dynamics of COVID-19 is an important topic. Therefore, in the results section of the chapter, first, a data-driven analysis of COVID-19 pandemic is performed. The infection dynamics and spread and the responses of various countries and continents to the pandemic are visualized. Then, using the proposed framework, a deep learning-based model for predicting the spread of the virus is developed. As the specific case study, the spread of the virus in Iran as one of the initial hotspots of the pandemic is modeled and compared to the top 10 affected countries. Then, the total deaths and confirmed cases are modeled by a new deep neural network architecture. Various loss functions and optimization methods are compared for achieving the best predictions. The experiments show that the model trained by MAE and RMSProp results in better forecasts for total confirmed cases. The MAE and Adam trained model have better results for the total reported death prediction. The models trained by MAE as loss function and Adam as optimizer showed more promising results than the other models. The results of these models demonstrated that the complexity of infection dynamics and the spread of the virus is much more than anticipated, and many parameters should be considered for accurate modeling of COVID-19 dynamics. In order to train more accurate models and tackle more complex issues encountered in the pandemic, larger datasets are needed. The collection of these datasets can be crowdsourced and the contact tracing applications can be used for the collection of various forms of data. However, this raises serious issues regarding privacy.

The rest of this chapter is organized as follows. The related works are discussed in Sect. 2. The proposed framework is detailed in Sect. 3. The experimental design, simulation scenarios and acquired results are discussed in Sect. 4. Finally, Sect. 5 concludes the chapter.

2 Related Works

Since the beginning of the pandemic, researchers have dedicated their time to understand the dynamic spread of COVID-19 and its characteristics. Several researchers utilized Machine Learning (ML) and Deep Learning (DL) in their studies, while others investigated other areas, such as blockchain applications in the current pandemic. However, the main issue of having access to an extensive and accurate dataset while holding the privacy of data owners remains. Many of the works that will be discussed later on, which are dedicated to understanding the COVID-19 spread, and providing governments with equipment and tools for the current task, can still be improved given more data, or perhaps other researchers can achieve better results if they have access to the data. However, the aforementioned statement is not likely achievable because of privacy issues.

2.1 COVID-19 and Computational Intelligence

Jamshidi et al. [1] discuss how artificial intelligence can enable the medical community to combat the current pandemic. Therefore, different deep learning methods such as Generative Adversarial Networks (GANs) and LSTM are explored regarding the aforementioned goal. They discuss that AI is able to present an integrated bioinformatics method in which it can systemize different features of information from a variety of data sources in order to form user-friendly platforms for the medical community, and discuss that the primary benefit of these platforms is that they are able to speed up the process of diagnosis and treatment of COVID-19 patients. Regarding the application of AI in COVID-19, a modified stacked auto-encoder was developed by Hu et al. [2], which then was utilized by authors in order to model the transmission dynamics of the current pandemic. In order to investigate the transmission structure in China, they utilized the latent variables and clustering algorithms in their proposed auto-encoder to group different provinces or cities. Next, their proposed model was applied to real-time forecasting of confirmed cases of COVID-19 across China, and they predicted that the pandemic would be over by the middle of April. Their predicted date is somewhat arguable, however, the fact that China has managed to successfully control the COVID-19 while the world still struggles [3].

Arora et al. [4] utilized deep learning-based models for predicting the number of novel COVID-19 confirmed cases for 32 states and union territories of India. They utilized different DL methods, such as recurrent neural network (RNN)-based LSTM variants like Deep LSTM, Convolutional LSTM and Bidirectional LSTM, and applied the aforementioned method on their dataset in order to predict the number of confirmed cases. Based on the results and prediction errors, the bi-directional LSTM model gives the best results, and the convolutional LSTM trained model offers the worst. Zheng et al. [5] proposed a hybrid AI model for COVID-19 prediction. In this approach, since classical pandemic models treat all individuals with coronavirus as having the same infection rate, they proposed an improved susceptible-infected model in order to estimate the infection rates for analyzing the trend. The authors embed the natural language processing (NLP) module and LSTM network into their susceptible-infected model for building their proposed hybrid model. The results of this approach show that individuals with COVID-19 have a higher infection rate within the third to eighth days after they were infected, which is in agreement with the actual transmission laws of the pandemic. Moreover, the results of their study suggest that compared with classical pandemic models, their proposed model can notably reduce errors of the prediction. Kang et al. [6] focused on chest computed tomography (CT) of patients infected with COVID-19. They proposed to identify COVID-19 autonomously using a collection of extracted features from CT images and AI techniques. In this study, they used a total of 2522 CT images in which 1495 samples were acquired from COVID-19 infected patients and 1027 samples from community-acquired pneumonia (CAP).

2.2 *Blockchain for COVID-19 and Privacy*

In Ref. [7], the authors discuss that the sudden growth of the COVID-19 pandemic has exposed the limitations in modern healthcare systems that handle emergencies, and they believe adopting innovative technologies such as blockchain can enable effective planning and resource deployments. They state that the spread of disinformation has been multiplied in the course of the pandemic, and current platforms are insufficient to verify the credibility of data, which may lead to public panic. Therefore, developing a blockchain-based tracking system is crucial for ensuring that the received information is reliable. Therefore, they developed a tracking system for the COVID-19 data collected from various sources, which is based on blockchain and uses smart contracts and oracles in order to track reported data related to the number of new cases, deaths and recovered cases obtained from trusted sources. Fusco et al. [8] focused on integrating blockchain technology in healthcare, and the authors suggested a trace-route for a COVID19. They believe that the application of blockchain technology with AI systems allows the creation of an applicable predictive system, which could contribute to the containment of pandemic risk. Moreover, the authors discuss that blockchain can play a major part in future digital healthcare, and specifically, it may work to improve COVID19-safe clinical practice. They discuss a concept in which medical data are gathered from different sources, such as hospitals, clinical labs and primary care physicians. Next, the data will be placed on a blockchain network.

Torky et al. [9] proposed a blockchain-based framework, which is used for verifying and detecting the unknown infected cases of COVID-19 virus. The main goal of their study was to investigate the possibility of developing a novel, innovative approach for verifying and detecting the unknown infected cases of COVID-19. Their proposed framework enables citizens to predict the infection risk of COVID-19 virus within people or within public places through a novel design of peer-to-peer Mobile Application. Zwitter and Gstrein [10] also discussed that during a pandemic, the usage of geo-location data presented by telecommunication and technology companies becomes an applicable option, and legal regulations rarely safeguard the privacy of people against governmental and corporate exploitation. The authors debate that this matter leaves little safeguards to guarantee individual and collective autonomy. Therefore, the authors tackled the legal gaps in data protection and privacy in the current pandemic. It is evident that by filling these gaps and offering solutions for them, or by utilizing different technologies (such as blockchain) for safe data sharing, we can take a step further toward offering systems that provide practical solutions for managing the current or future pandemics while holding the privacy of citizens and data owners.

Gerke et al. [11] also explore how the development of contact-tracing applications raises privacy concerns. Therefore, the authors recommend interventions to ensure safety and review regulatory requirements for privacy and security. They also discuss that even though home monitoring technologies have considerable potential to decrease personal contacts between people—which is crucial during the current pandemic—yet, this kind of monitoring poses concerns regarding people's privacy.

By combining such systems with privacy-preserving frameworks, a medical data-sharing platform that utilizes AI methods such as ML and DL and provides physicians and medical experts with valuable models using a combination of BDaaS, MTaaS and CIaaS is proposed in this paper. The proposed framework can present solutions for the management of current and future pandemics.

3 Blockchain-Based Internet of Health Things

The blockchain and smart contract-based technologies can provide a solution for the decentralized transfer of data. In the decentralized approach, data remain distributed on individual devices, and therefore, data will remain anonymized. Using the proposed framework, data holders who previously were worried about privacy and abuse of their data can participate in training models without their data being stored anywhere other than their own storage. Data holders, e.g., government bodies, companies, hospitals, and citizens, can provide datasets and training machine learning models without worrying about what will happen to their data.

3.1 Preliminaries

Before describing the proposed framework, its utilized aspects and technologies, namely blockchain, smart contract and IPFS, will be discussed.

3.1.1 Blockchain

A blockchain consists of a number of blocks, in which except for the genesis block (the first block), every block is connected to its previous block through a backward reference, which is the hash of its previous block. Moreover, a block also holds other values, such as a timestamp and a piece of cryptographic information [12]. The blockchain offers valuable features, such as immutability, which means that the manipulation of data stored on a blockchain is difficult since even the slightest modification can lead to invalid hash values, which lead to the block being useless. The next feature is non-repudiation, which utilizes digital signatures and cryptographic algorithms in order to assure that transactions will not be repudiated. The third feature offered by a blockchain is traceability, which means the origin of data can be traced by monitoring the publicly available blockchain data given its respective timestamp. Finally, the fourth feature is the fact that the decentralization of blockchain allows transactions to be validated by the majority of peers distributed in the entire system without the need for any central agency [13].

3.1.2 Smart Contract

A smart contract (SC), which can be traced back to the 1990s [14], is a piece of code that runs on top of blockchain and automates the execution of predefined terms when certain conditions are met by running different coded functions. After SC's codes are compiled into machine code and stored on a blockchain, it will take advantage of the immutability of the blockchain. SCs can simplify the predefined process and improve the efficiency of its activities [15]. On the other hand, nowadays, SC technology has begun to modernize traditional industrial and technical processes. An SC is self-executable, it verifies itself and is embedded into the blockchain. SC removes the need for any third-party system and thus reduces the costs of different services. Moreover, SC improves the productivity of systems and decreases security risks [16]. By utilizing this technology in the proposed framework of this chapter, the integrity and security of the process will be ensured, and possible abuses and frauds will be eliminated.

3.1.3 InterPlanetary File System

Since storing large content and data on a blockchain is costly and inefficient, the InterPlanetary File System (IPFS) [17] is utilized in the proposed framework in order to store data. It is a decentralized peer-to-peer protocol, which will disseminate files between a list of trusted nodes known as bootstrap nodes and make them available to other users by Content Identifiers (CIs) and content-based addressing. This protocol does not have the deficiencies of the current centralized (client–server) HTTP architectures. Using IPFS, each file will be given a unique fingerprint, which is a 32-byte cryptographic hash and is stored on a node of the network. With IPFS, large amounts of data can be addressed, and become immutable, permanent links in transactions. Moreover, time-stamping and securing content can be achieved while there is no need to set the data on-chain.

3.2 The BDaaS+MTaaS+CIaaS Framework

In the proposed framework, a data owner can be anyone (real or legal) that has useful data, such as a patient, hospitals, companies, citizens, doctors, other researchers or companies who have valuable and sensitive data and are not willing to share their data because of privacy concerns under current systems. By utilizing the proposed framework, various sources of IoHT data, including various multimedia data, e.g., video, audio, images, or textual and sensor data, can be shared. In the proposed framework, the data owners do not actually share their data, and instead, they will keep their data locally and off-chain. However, if the need arises, the data owner can share and send a part of its data to another user. Model owners are the researchers or anyone who trains a machine learning model using the data offered by the data owner. If a previously trained model exists, the model is shared and retrained, and enhanced by various machine learning and deep learning algorithms on the data owner's side. While the model is being trained, the model's checkpoints will be saved periodically. The final part of the proposed framework is the validating node, which is the smart contract of the framework. It is used to control the contract signed by the Data Owner and Model Owner and transfer the new parameters of the retrained model. Moreover, it is used to observe and monitor the process and checkpoints periodically. This entity

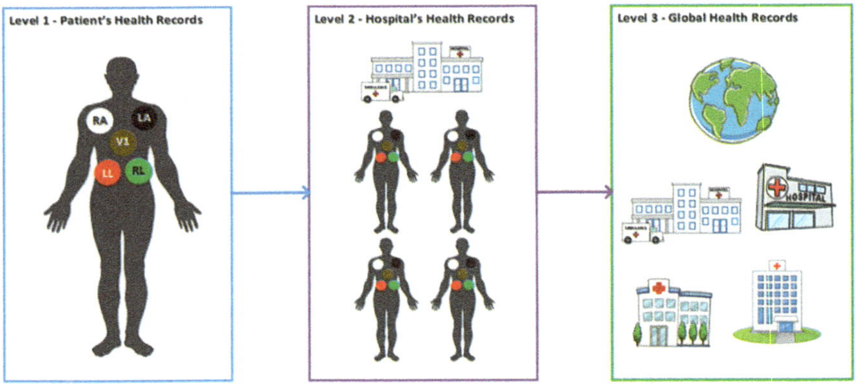

Fig. 1 Decision fusion in the proposed framework

works as a proxy between the other two entities. Using the proposed framework, the biomedical data fusion can also be achieved at the patient level, hospital, emergency and government level as even as strategic decision level as shown in Fig. 1.

In order to demonstrate the applications of the proposed framework, several scenarios are investigated. The scenario takes place in the city of London and is shown in Fig. 2. In this scenario, the patient who is located in Newington needs an emergency medical assistant. An ambulance was dispatched and now is en route, taking the patient to London Bridge Hospital located in Southwark, near the London Bridge. According to Google Maps, it is a 3 km drive, which takes 11 min under the current road traffic.

The patient's condition is critical, and the ambulance needs more info on him regarding his past medical records in order to comprehend the situation more accurately and eliminate any possible mistakes. If there is any information available on the patient such as name, (National Insurance Number) NINO, phone number, or even if a relative is with him on the ambulance and can offer the patient's NINO, everything is set for the next step. However, if the identity of the patient is unknown (he has no wallet or ID on him) and he cannot offer any info, or when the ambulance went to pick him up, no one knew him, or there are no relatives and friends with him in the ambulance, the solution is a comprehensive system that can be used to identify the patient using facial detection, and the patient will be identified, and the NINO of the patient is transferred to the emergency system (e.g., AB 12 34 56 C). The same process can be used for sharing the trained machine learning models, as described in Sect. 4.

Department of Health and Social Care is the department that, in its database, holds which patients have visited which medical centers or hospitals. Note that this is not a centralized scenario, and the department only holds the aforementioned records, not medical records themselves, as shown in the table in Fig. 2. Therefore, the nurse on the ambulance will send the NINO to the department via blockchain and SC. This SC, written in Solidity programming language, is utilized to handle data sharing between

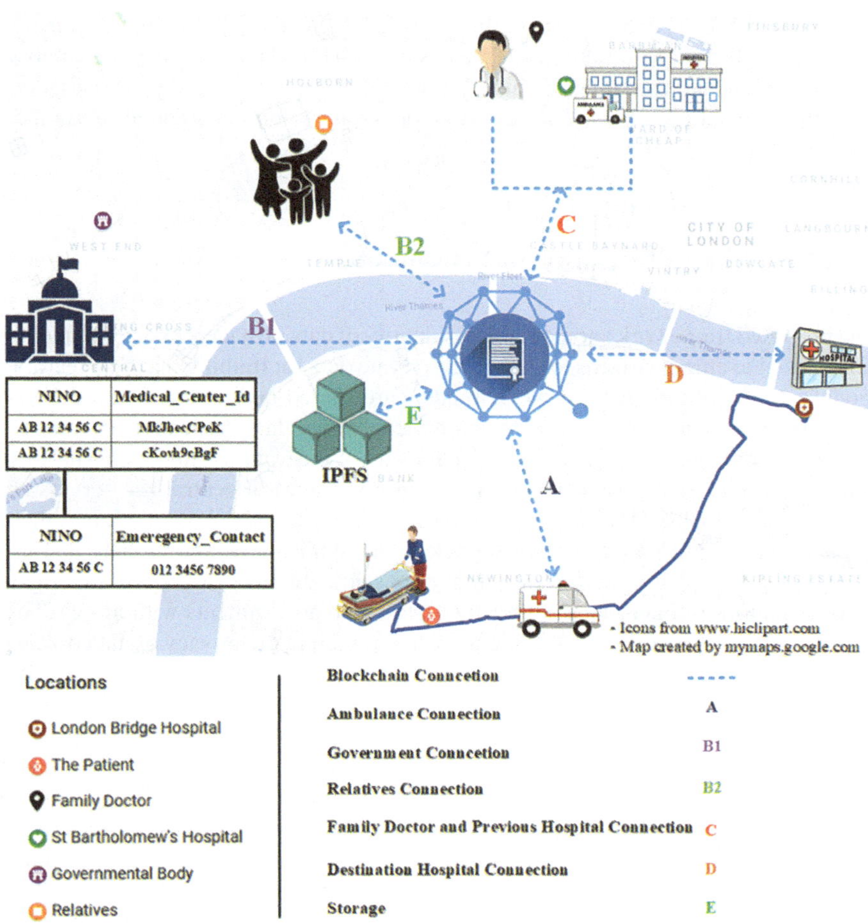

Fig. 2 The decentralized framework in the emergency scenario

different nodes, such as hospitals, health department and the ambulance. First, the nurse will send (*A*) the patient's NINO using a block to the health department (*B1*) and also inform them if there is an escort with the patient or not. If there is no one, the smart contract will be used to notify the relatives via the emergency contact number. Meanwhile, after receiving the NINO, the ID of each medical center visited by the patient, plus the unique ID of the family doctor, will be given to the smart contract. Now, the SC will request the patient's medical records from previously visited medical centers (*C* in Fig. 2). In this example, the patient usually goes to St Bartholomew's Hospital in Farringdon, and his family doctor is also located there. These medical centers accept the request and will share the requested data. However, since there are many records, it is impossible to share such data via blocks in the blockchain. Therefore, St Bartholomew's Hospital will upload the file containing the records on IPFS (E) and only share the hash via the SC (Block 2 in Fig. 2). Finally, the London Bridge Hospital will use the hash to download patient's records, analyze them to understand the situation, make things ready on their end (e.g., if operations and other crucial tests are needed), and guide the nurse on the ambulance via their communication channel.

4 Experimental Results and Discussion

The proposed framework anonymizes the identity of data providers and consequently can be used to attract citizens to share their private data for training machine learning models. In the proposed framework, models are shared between different entities, and data owners are not required to share their private data. Instead, they train the shared model on their data locally, and at the end of the training, they only share the learned parameters. By utilizing IPFS, the disadvantages of centralized servers are eliminated, and IPFS is used to store the learned models' checkpoints. Later, these trained models will be aggregated together. The applications and models that use this kind of data need to be as transparent as possible. In this chapter, we investigate the applications of the proposed framework to equip governments with more robust tools to combat the COVID-19 pandemic while retaining the privacy of data owners.

In Sect. 2, researches to achieve a more nuanced understanding of the COVID-19 pandemic using different AI techniques are presented. However, most of the data used in these researches regarding the pandemic are not publicly available, and those willing to share their data are few. There are many valuable datasets at the disposal of hospitals, labs and companies, which can help the community to gain more understanding of the COVID-19 dynamics. However, these entities' legal obligations about the privacy of their patients do not allow them to share their data. Furthermore, various multimedia and sensory data are owned by individuals, and if the number of individuals willing to share their data is large enough, these data may help prevent the further spread of the virus. The proposed framework can help with this issue. As the first case study of the proposed framework, a deep learning-based system for modeling and forecasting of epidemic spread and infection dynamics of COVID-19 is presented.

4.1 Deep Learning-Based Modeling and Forecasting of Epidemic Spread and Infection Dynamics of COVID-19

In this section, the deep learning-based modeling and forecasting of epidemic spread and infection dynamics of COVID-19, and the dataset utilized for that task, are investigated.

Table 1 Examples of CSSE dataset (Total confirmed cases)

State	Country	Lat	Long	5/20/20
–	USA	37.0902	−95.7129	1,551,853
Ontario	Canada	51.2538	−85.3232	25,595
–	Iran	32.0	53.0	129,341
Victoria	Australia	−37.8136	144.9631	1593

4.1.1 Dataset

In order to investigate the applications of machine learning-based models for the prediction of spread and infection dynamics of COVID-19, the reported confirmed and death cases, the dataset provided by the Centre for Systems Science and Engineering (CSSE) at Johns Hopkins University (JHU) [18, 19] is utilized in this case study, and an example of the dataset is shown in Table 1. The provided datasets are updated daily around 23:59 (UTC), and they consist of reported confirmed and death cases, each in a separate dataset.

These data are publicly available, however, in order to discuss the implementation of the proposed framework, we assume that the JHU did not share their data publicly, and only accepts to train the user model on their side. The first network, shown in Fig. 3, is a simple deep neural network (DNN) architecture. For the network layers, only dense (fully connected) layers [20] were used. The input data had only one dimension. In the first dense layer, it was connected to 128 neurons. Furthermore, in the second hidden layer, each of the neurons in the previous layer was connected to 256 neurons. To compact the architecture, the third layer has 128 neurons again, and finally, the output layer is a dense layer with only one neuron since the model only works with one task, which is either confirmed cases or death cases. It should be noted that we trained different models for each of these tasks.

The second architecture is based on LSTM [21] and is shown in Fig. 4. The used LSTM architecture in this study is composed of two LSTM layers. The first layer has 128 units, whereas the second and final LSTM layer has 64 units, and finally, the output is a dense layer.

For training, the deep neural network models, Keras [22] was used with TensorFlow [23]. The models were trained for a range of epochs, from 100 to 5,000, and the default learning rate of 0.001 was used.

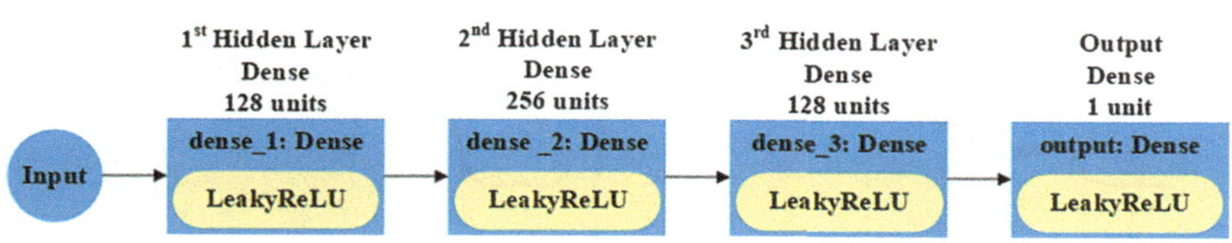

Fig. 3 The dense DNN architecture

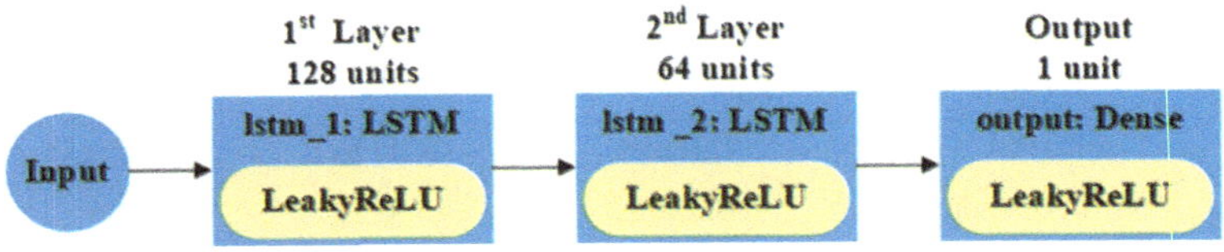

Fig. 4 The LSTM architecture

4.2 Results

In this paper, first, a descriptive data analysis of the COVID-19 pandemic in different countries, regions and environments is performed. The analysis is done by the latest data, which was updated on December 3, 2020. As mentioned before, COVID-19 has spread to 218 countries across the globe. So, first, we will investigate the spread of confirmed cases and reported deaths. Figure 5 shows the spread of global confirmed and death cases.

From March 14, the world suffered high numbers of confirmed and death cases. Furthermore, this ascendant trend continued its course and peaked in November, with more than 500 thousand confirmed cases reported daily. Furthermore, on November 14, 667,917 cases were reported. Figure 6 shows the global daily new confirmed cases and global daily death, both in thousand.

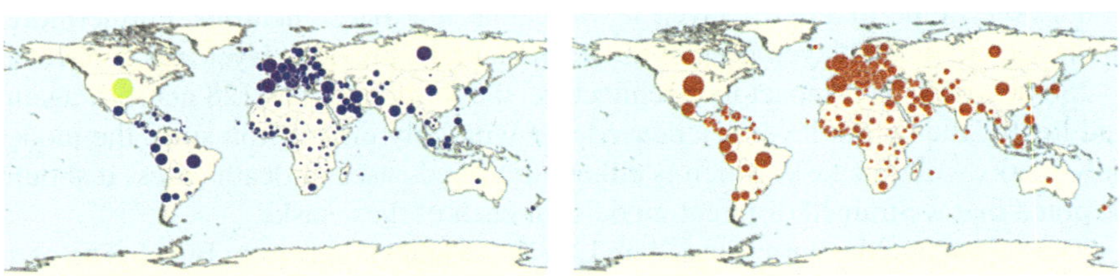

Fig. 5 Confirmed cases and deaths of COVID-19 across the globe as of December 2020

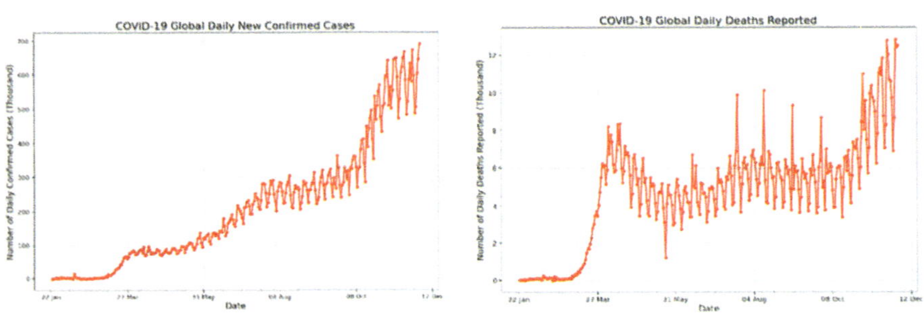

Fig. 6 COVID-19 Global Daily New Confirmed Cases and Deaths

Table 2 Summary comparison of different continents

Continent	Confirmed	Death	Recovered	Active	Mortality rate
Africa	2,217,878	52,855	1,893,342	271,681	23.8
Asia	17,038,755	295,324	15,100,038	1,643,393	17.3
Australia	30,755	942	28,253	1,560	30.6
Europe	18,043,608	413,590	7,527,950	9,057,293	22.9
North America	16,484,490	413,774	7,163,081	8,907,635	25.1
Others	230,600	3,898	171,524	55,178	16.9
South America	11,349,748	329,161	10,200,370	820,217	29

Until December 3, 2020, there have been more than 65 million confirmed, 1.5 million deaths and 20 million recovered cases reported. Table 2 shows a summary comparison of different continents.

As shown, the most populated continent and where the virus started to spread managed to get a hold of its spread and compared with other continents with less population, such as Europe, managed much better in case of mortality rate. Moreover, the virus had the most definite impact on Europe and North America, respectively. The most reported death cases belong to Europe, with a population of approximately 747 million [24], and also North America, with a population of 368 million [25]. The total confirmed and death cases reported in these two continents are somewhat similar and very close. The results suggest that Australia has the highest mortality rate, with 30 deaths reported per 1,000 confirmed cases. To better understand the above table, different comparisons of confirmed cases and reported deaths for four continents are plotted in Fig. 7. For Europe, North America and Asia, the number of cases is scaled down to a million.

A trend comparison of confirmed cases in different continents can be seen in Fig. 8. Day zero means the first day that the first case in the respective continent was discovered. This approach was taken to see a trend of how each continent coped after *n* days passed of COVID-19's arrival.

According to Table 3, the USA has the most confirmed, death, recovered and active cases. The highest number of recovered cases belongs to India.

As the table suggests, with 94 deaths reported per 1,000 confirmed cases, Mexico has the highest mortality rate among the most affected countries. After Mexico, Iran has suffered the most, with a mortality rate of 48, among these countries. Moreover, the United Kingdom has the least recovered number of cases. Figure 9 shows a summary visualization of a few countries as an example.

After the first case of COVID-19 was discovered in Iran on February 19 (Day Zero), Iran followed the same trend as other countries, a steep ascending slope for the first 4 weeks (until Day 28), taking a somewhat linear trend. Figure 10 shows daily reported deaths and confirmed cases of Iran. Iran is placed 14th in confirmed cases and 8th in total deaths reported.

Trend comparison of the top 10 countries and Iran for confirmed cases and deaths is presented in Fig. 11.

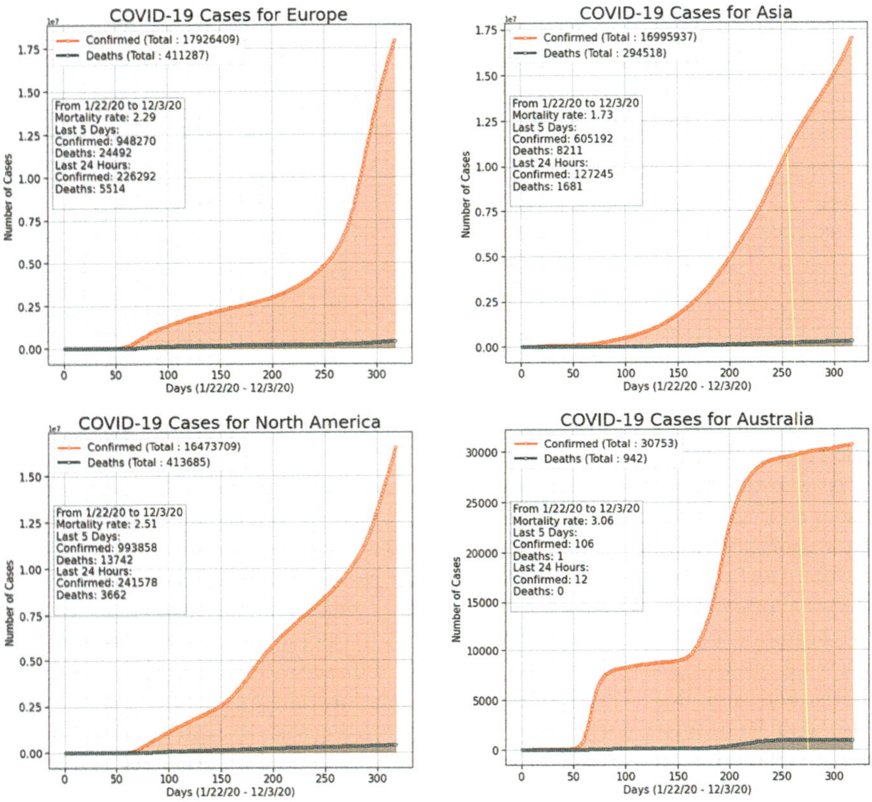

Fig. 7 Confirmed versus death cases of Europe, North America, Asia and Australia

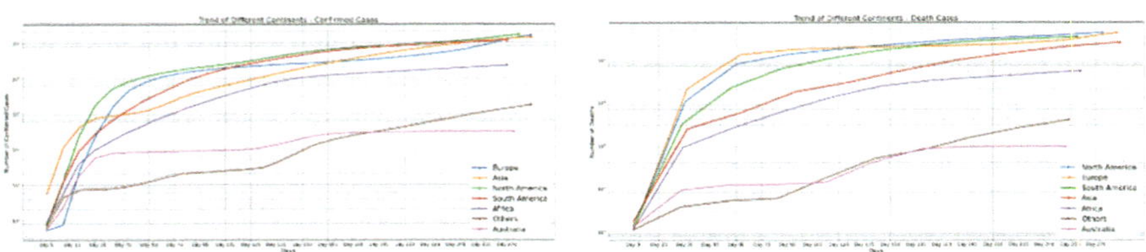

Fig. 8 Trend comparison of different continents—confirmed cases and deaths

Figure 12 shows the COVID-19 impact on Iran since the outbreak compared with the rest of the world.

In order to demonstrate the application of the proposed BDaaS+MTaaS+CIaaS framework for training the previously discussed models, we focus on training LSTM models, however, the process for DNN models will be the same. Here, for the sake of argument, we assume that the dataset discussed in Sect. 4.1.1 is not publicly

Table 3 Country level analysis of COVID-19 dynamics

Country	Confirmed	Deaths	Recovered	Active	Mortality Rate
USA	14,149,770	276,406	5,404,018	8,469,346	19.5
India	9,571,559	139,188	9,016,289	416,082	14.5
Brazil	6,487,084	175,270	5,792,539	519,275	27.0
Russia	2,382,012	41,730	1,872,214	468,068	17.5
France	2,310,271	54,231	172,574	2,083,466	23.5
United Kingdom	1,678,423	60,210	3,632	1,614,581	35.9
Spain	1,675,902	46,038	150,376	1,479,488	27.5
Italy	1,664,829	58,038	846,809	759,982	34.9
Argentina	1,447,732	39,305	1,274,675	133,752	27.1
Colombia	1,343,322	37,305	1,233,115	72,902	27.8
Germany	1,145,645	18,135	819,735	307,775	15.8
Mexico	1,144,643	108,173	843,231	193,239	94.5
Poland	1,041,846	19,359	666,413	356,074	18.6
Iran	1,016,835	49,695	708,106	259,034	48.9

available. However, the JHU agrees to train our model on their dataset, using the proposed framework in this chapter.

In this scenario, first, we (the researcher or Model Owners) upload the python file that holds the code for training the LSTM model on IPFS using it's Desktop App [26], and we are given a hash, which points to the file. Next, after signing a contract using the SC (or Validating Node), we will share this hash. Here, a transaction will be created in which both sides' IDs (Wallet IDs, by which we are connected to the blockchain). Then it will be recorded on the blockchain, and one of these two entities will pay the transaction fee—if there is any, depending on the contract and the blockchain platform. Then, SC will check if the hash is available on the IPFS. If so, then the hash will be sent to the data owner (i.e., JHU) and they can download the file, execute it, and train the model on their own side, off the blockchain. Meanwhile, if, in our algorithm, we added a few lines of code that saves a checkpoint of the model, say after each epoch, the SC can check if the checkpoint is available, periodically. Finally, when the training is over, the data owner will upload the final model file (or files) on IPFS, and then shares the hash via the SC. Once again, SC will check if such a file (hash, to be precise) exists. If so, then we (Model Owners) will receive the hash to the model file and download it. It should be noted that only IPFS and SC are located on the blockchain, and the rest of the process is done off-blockchain.

Considering the fact that for predicting a complex matter such as the spread of a virus, a vast amount of accurate data is needed, the proposed model has acceptable predictions. Perhaps, other data owners with more data will be willing to share it through the framework later on. It must be noted that the previous analysis of data shows that the numbers reported by many governments may not reveal the real situation, as there are undoubtedly many unknown cases of infected individuals and

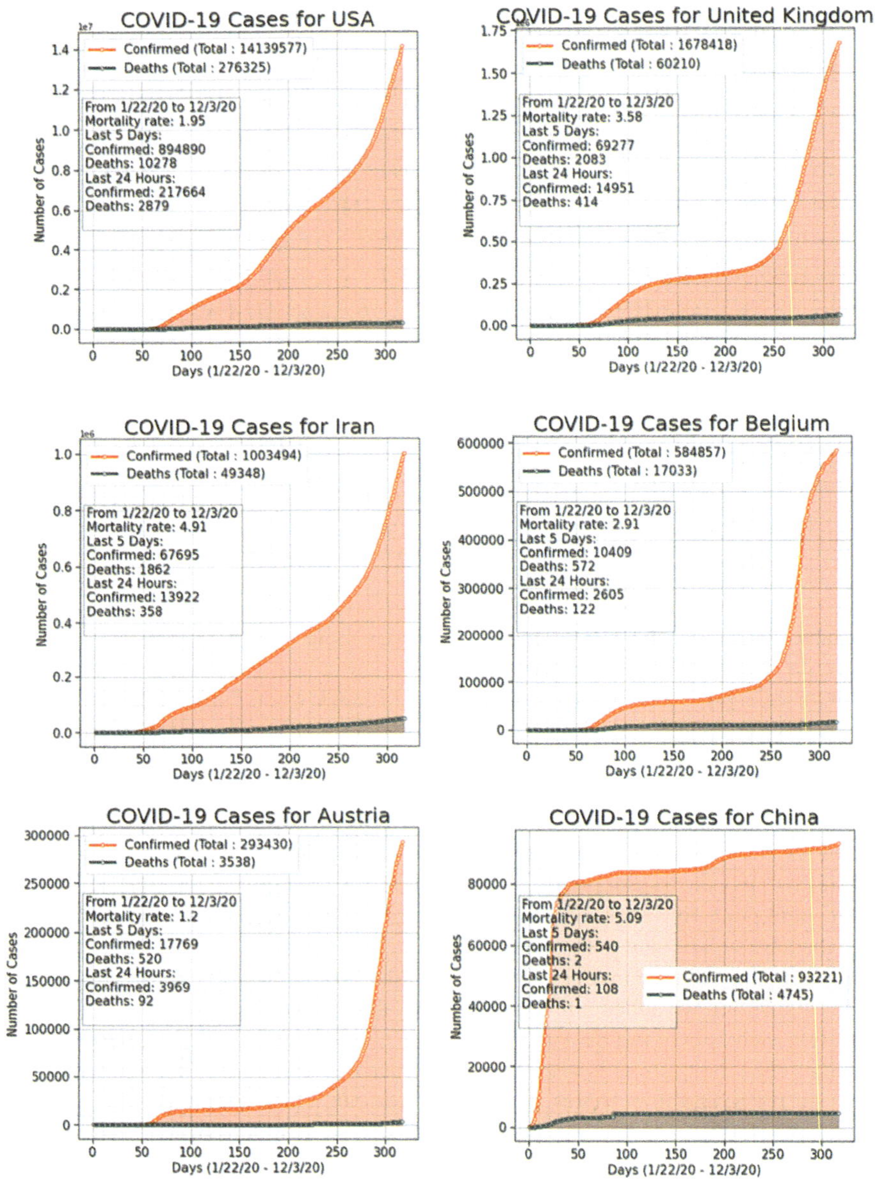

Fig. 9 A summary visualization of a few countries and Iran

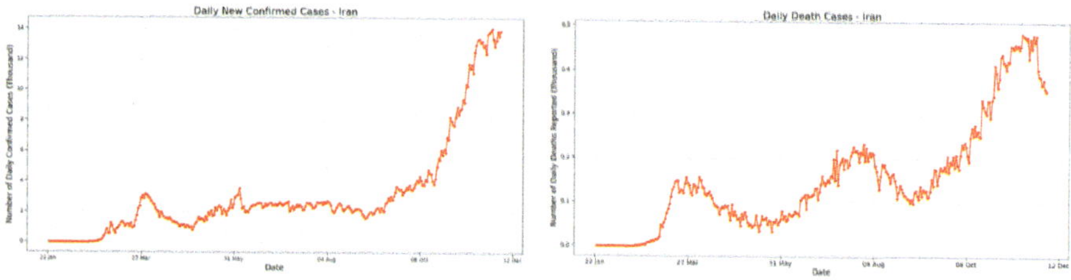

Fig. 10 Daily reported deaths and confirmed cases of Iran

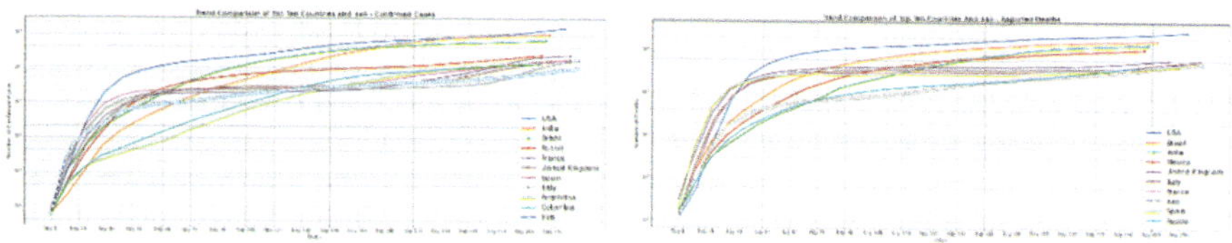

Fig. 11 Trend comparison of top 10 countries and Iran—confirmed cases and deaths

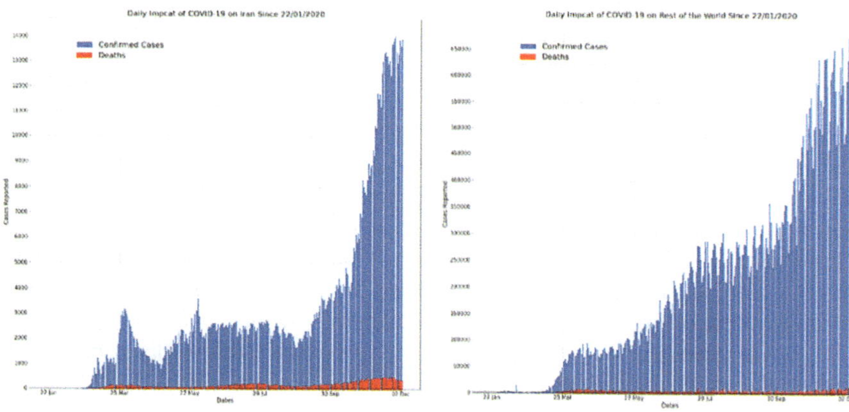

Fig. 12 The COVID-19 impact on Iran since the outbreak compared with the rest of the world

many deaths attributed to other complications. In total, 16 models with a combination of different architecture+optimizer+loss function were trained. Overall, the best performance was shown by DNN architecture trained models for predicting total confirmed cases and total death cases. Moreover, detailed results suggest that among LSTM trained models, for the given task of predicting total confirmed and death cases, the Adam optimizer can be set as the preferred optimizer when compared with RMSProp. As for the loss function, MSE and MAE both show similar results. On the other hand, among DNN trained models, for the loss function, MSE offers the best results.

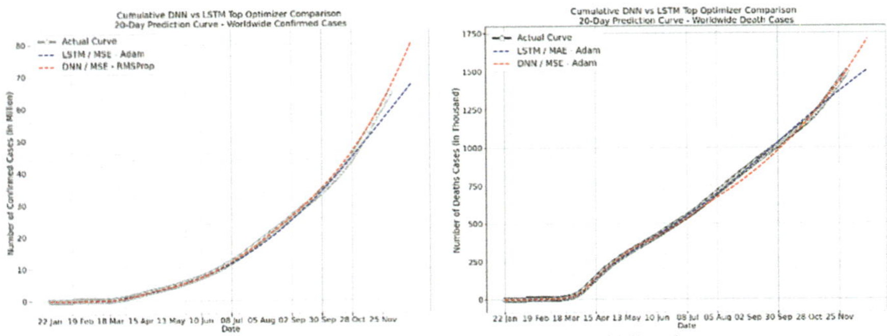

(a) Confirmed cases (b) Death cases

Fig. 13 A comparison of best models trained to predict **a** confirmed and **b** death cases

In LSTM architecture, the MSE+Adam trained model offered better performance when predicting total confirmed cases, and the MAE+Adam trained model performed better compared with other LSTM models when predicting total reported deaths. In DNN architecture, results suggest that the MSE+RMSProp model is the best for predicting total confirmed cases among other trained models in this architecture, and as for predicting total reported death, the MSE+Adam model offered the best performance. The comparison of the aforementioned best-performed models is shown in Fig. 13. Here, the models are tasked with predicting for the next 20 days, from December 2, 2020, until December 22, 2020. Moreover, observing the decreasing trend of loss value while the models were trained suggests that MSE and MAE offer similar and close values. However, the MSE managed to offer values less than 0.001, whereas it was not the case for MAE since it could not offer values less than 0.003.

Finally, Table 4 shows a 10-day comparison of the aforementioned best-performed models, which both belong to DNN architecture. The table compares real values with predicted numbers, from November 28, 2020, up to December 12, 2020.

Besides the management of the pandemic, the proposed framework can be utilized in many other applications such as Internet of Robotic Things (IoRT) [27, 28], natural environment management [29, 30], e-marketing [31] and surveillance [32].

5 Conclusion

This chapter proposes a framework for decentralized BDaaS+MTaaS+CIaaS, which would encourage data owners to participate in different researches and offer their valuable data without actually sharing it. Using the proposed framework, data privacy and security would be guaranteed, which would also encourage different patients to participate in different scientific activities. Applications of the proposed framework were investigated in different scenarios. Moreover, applications of machine learning

Table 4 A 10-day comparison of best trained models

Date	Real		MSE+RMSProp	MSE+Adam
	Confirmed	Death	Confirmed	Death
28 November	62,286,269	850,535	63,549,000	1,456,880
29 November	62,772,731	857,015	64,187,000	1,466,740
30 November	63,278,795	863,028	64,829,999	1,476,680
1 December	63,875,627	868,733	65,479,000	1,486,680
2 December	64,508,175	874,369	66,135,000	1,496,750
3 December	–	–	66,798,000	1,506,890
4 December	–	–	67,465,000	1,517,090
5 December	–	–	68,138,000	1,527,360
6 December	–	–	68,817,000	1,537,710
7 December	–	–	69,502,000	1,548,120
8 December	–	–	70,193,000	1,558,610
9 December	–	–	70,892,000	1,569,060
10 December	–	–	71,598,000	1,579,580

algorithms for modeling the epidemic spread and infection dynamics of COVID-19 are explored.

References

1. Jamshidi M et al (2020) Artificial intelligence and COVID-19: deep learning approaches for diagnosis and treatment. IEEE Access 8:109581–109595
2. Hu Z, et al (2020) Artificial intelligence forecasting of covid-19 in china. arXiv:2002.07112
3. Burki T (2020) China's successful control of COVID-19. Lancet Infect Dis 20(11):1240–1241
4. Arora P, Kumar H, Panigrahi BK (2020) Prediction and analysis of COVID-19 positive cases using deep learning models: a descriptive case study of India. Chaos, Solitons Fractals 139:110017
5. Zheng N, et al (2020) Predicting covid-19 in china using hybrid AI model. IEEE Trans Cybern
6. Kang H, et al (2020) Diagnosis of coronavirus disease 2019 (covid-19) with structured latent multi-view representation learning. IEEE Trans Med Imaging
7. Marbouh D, et al (2020) Blockchain for COVID-19: review, opportunities, and a trusted tracking system. Arab J Sci Eng 1–17
8. Fusco A et al (2020) Blockchain in healthcare: insights on COVID-19. Int J Environ Res Public Health 17(19):7167
9. Torky M, Hassanien AE (2020) COVID-19 blockchain framework: innovative approach. arXiv: 2004.06081
10. Zwitter A, Gstrein OJ (2020) Big data, privacy and COVID-19–learning from humanitarian expertise in data protection. Springer
11. Gerke S et al (2020) Regulatory, safety, and privacy concerns of home monitoring technologies during COVID-19. Nat Med 26(8):1176–1182
12. Dai H-N, Zheng Z, Zhang Y (2019) Blockchain for internet of things: a survey. IEEE Internet Things J 6(5):8076–8094

13. Dai H-N, Imran M, Haider N (2020) Blockchain-enabled internet of medical things to combat COVID-19. IEEE Internet Things Mag 3(3):52–57
14. Chang SE, Chen Y-C, Lu M-F (2019) Supply chain re-engineering using blockchain technology: a case of smart contract based tracking process. Technol Forecast Soc Chang 144:1–11
15. Zheng Z et al (2020) An overview on smart contracts: challenges, advances and platforms. Futur Gener Comput Syst 105:475–491
16. Gupta R et al (2020) Smart contract privacy protection using AI in cyber-physical systems: tools, techniques and challenges. IEEE Access 8:24746–24772
17. Benet J (2015) Inter planetary file system. https://ipfs.io/
18. JHU (2020) COVID-19 data repository by the center for systems science and engineering (CSSE) at Johns Hopkins University
19. JHU (2020) Github repository for the data
20. Huang G, et al (2017) Densely connected convolutional networks. In: 2017 IEEE conference on computer vision and pattern recognition (CVPR)
21. Hochreiter S, Schmidhuber J (1997) Long short-term memory. Neural Comput 9(8):1735–1780
22. Chollet F, Keras (2015)
23. Team GB (2015) Tensor flow. https://www.tensorflow.org/
24. WorldoMeter (2020) Europe population. https://www.worldometers.info/world-population/europe-population/
25. WorldoMeter (2020) Northern America population. https://www.worldometers.info/world-population/northern-america-population/
26. IPFS Desktop App (2020). https://github.com/ipfs-shipyard/ipfs-desktop
27. Abbasi MH, et al (2019) Deep visual privacy preserving for internet of robotic things. In: 2019 5th conference on knowledge based engineering and innovation (KBEI). IEEE
28. Norouzi A, Majidi B, Movaghar A (2018) Reliable and energy-efficient routing for green software defined networking. In: 2018 9th international symposium on telecommunications (IST). IEEE
29. Nazerdeylami A, Majidi B, Movaghar A (2019) Smart coastline environment management using deep detection of manmade pollution and hazards. In: 2019 5th conference on knowledge based engineering and innovation (KBEI). IEEE
30. Majidi B, Patra JC, Zheng J (2014) Modular interpretation of low altitude aerial images of non-urban environment. Digit Signal Process 26:127–141
31. Nedaei D, et al (2018) Inbound e-marketing using neural network based visual and phonetic user experience analytics. In: 2018 4th international conference on web research (ICWR). IEEE
32. Qezavati H, Majidi B, Manzuri MT (2019) Partially covered face detection in presence of headscarf for surveillance applications. In: 2019 4th international conference on pattern recognition and image analysis (IPRIA). IEEE

The manufacturer's authorised representative in the EU is Springer Nature Customer Service Centre GmbH, Europaplatz 3, 69115 Heidelberg, Germany. If you have any concerns regarding our products, please contact ProductSafety@springernature.com

Printed and bound by CPI Group (UK) Ltd, Croydon, CR0 4YY
30/09/2026
02225291-0015